BATON ROUGE, NOV 2019

TO AN HONEST, SMART, STRONG AND VERY ATTRACTIVE YOUNG LADY!

THANK YOU FOR YOUR FRIENDSHIP, STACEY!

Michael

Crime Modeling and Mapping Using Geospatial Technologies

Geotechnologies and the Environment

VOLUME 8

The "Geotechnologies and the Environment" series is intended to provide specialists in the geotechnologies and academics who utilize these technologies, with an opportunity to share novel approaches, present interesting (sometimes counter-intuitive) case studies, and most importantly to situate GIS, remote sensing, GPS, the internet, new technologies, and methodological advances in a real world context. In doing so, the books in the series will be inherently applied and reflect the rich variety of research performed by geographers and allied professionals.

Beyond the applied nature of many of the papers and individual contributions, the series interrogates the dynamic relationship between nature and society. For this reason, many contributors focus on human-environment interactions. The series are not limited to an interpretation of the environment as nature per se. Rather, the series "places" people and social forces in context and thus explore the many sociospatial environments humans construct for themselves as they settle the landscape. Consequently, contributions will use geotechnologies to examine both urban and rural landscapes.

For further volumes:
http://www.springer.com/series/8088

Michael Leitner
Editor

Crime Modeling and Mapping Using Geospatial Technologies

Springer

Editor
Michael Leitner
Department of Geography and Anthropology
Louisiana State University
Baton Rouge, LA, USA

ISBN 978-94-007-4996-2 ISBN 978-94-007-4997-9 (eBook)
DOI 10.1007/978-94-007-4997-9
Springer Dordrecht Heidelberg New York London

Library of Congress Control Number: 2012956191

Printed on acid-free paper

Springer is part of Springer Science+Business Media (www.springer.com)

Editorial

The idea for editing a book on *Crime Modeling and Mapping Using Geospatial Technologies* goes back to fall 2009, when the editor of this book organized paper sessions for the Association of American Geographers (AAG) conference to be held in Washington DC in April 2010. The sessions were titled "Crime Modeling and Mapping" and the call for papers included the following note: "It is planned that authors can revise their presentations for publication in a special content issue of *Cartography and Geographic Information Science* (*CaGIS*) or an edited book." The call for papers for the AAG conference attracted 25 presentations, filling five sessions. Ten of those 25 presentations were later revised and are now included as chapters in this book.

On December 29 2010, a first solicitation of a chapter proposal submission for a book on *Crime Modeling and Mapping Using Geospatial Technologies* to be published by Springer in its popular *Geotechnologies and the Environment* series (http://www.springer.com/series/8088) was sent out. This solicitation was distributed to members of different list servers, including those from the University Consortium of Geographic Information Science (UCGIS), Cartography and Geographic Information Society (CaGIS), three different Specialty Groups of the AAG (Cartography, Geographic Information Science and Systems, and Spatial Analysis and Modeling), and geography-and-crime@googlegroups.com. Members from the geography-and-crime@googlegroups.com list server generated the most interest and submitted the highest number of chapter proposals of any of the above mentioned list servers.

By March 15, 2011, the deadline for chapter proposal submission, a total of 40 proposals was submitted. The authors of 39 of those 40 proposals were invited to submit a chapter manuscript. This decision was made solely by the editor. By the end of September 2011, a total of 26 manuscripts were submitted. Each chapter manuscript was subsequently reviewed (double-blind) by at least two expert reviewers, with one of the two reviewers being selected from the list of authors that submitted a manuscript for this book publication. Reviews were based on standard Springer review forms. The innovative aspect and the scientific quality of the research weighted heavily on the decision whether or not a manuscript was accepted or rejected. In case major revisions were requested, a second review of the revised manuscript by the same original reviewers was conducted. Eleven of the 18 chapters included in this book underwent a second review thus guaranteeing innovative research and high scientific quality of all contributions.

The 18 chapters included in this book are written by 48 different authors, with two authors contributing to two different book chapters. This is an average of 2.7 authors per book chapter, in part reflecting the interdisciplinary and collaborative nature of current crime modeling and mapping research. The majority of contributors are from the USA (36) with others coming from Mexico (seven), Belgium (four), England (two), and Canada (one). Surprisingly, only 3 of the 18 chapters involve direct collaboration with the police or forensic sciences organizations and have personnel from those organizations listed as co-authors.

Eighty-eight percent (44 out of 50) of authors come from academia, of which 75% (33 out of 44) can be attributed to departments with a geographic and 16% (7 out of 44) to departments with a criminology or criminal justice focus. Fifty-two percent (23 out of 44) of all academic authors are university professors, 27% (12 out of 44) are students, and the rest staff researchers. Most of the crime data are from urban areas in the USA. Vancouver in Canada, Leeds and Manchester in the UK, Charleroi in Belgium, and Mexico City are the only study areas outside of the USA. Philadelphia, PA, is the most studied city, with three book chapters applying crime data from this urban study site.

To summarize, *Crime Modeling and Mapping Using Geospatial Technologies*, is a topic of much interest mostly to academia, but also to the private sector and the government. Research in this area is foremost carried out by geographers, but also by criminologists and criminal justice experts. While research is carried out in different regions world-wide, the focus of study areas seems to be urban areas in the USA. In the following, the content of each chapter and its contribution to crime modeling and mapping will be briefly discussed part by part.

The 18 chapters included in this book are classified into the following five main parts:

- Fundamental Spatial Problems (two chapters)
- Crime Analysis (five chapters)
- Crime Modeling (six chapters)
- Crime Mapping (three chapters)
- Applications and Implementations (two chapters)

The allocation of the chapters into the five parts is reflective of the logical structure when writing a research paper: Starting with a (research) *problem*, proceeding with the *analysis* and/or *modeling* of the data, and finally *mapping* and *applying/implementing* the results. Additionally, two of the five part headings also appear in the book title.

Part I on *Fundamental Spatial Problems* includes two chapters. Starting off is the chapter by **Andresen and Malleson**, who investigate the phenomenon of spatial heterogeneity or multiple spatial scales of analysis in the context of spatial point patterns changing over time. This research contributes to a better understanding of the modifiable areal unit problem (MAUP) in spatial crime analysis. In the second chapter, **Johnson and Ratcliffe** propose the nearest neighbor hierarchical clustering technique as an alternative method to create distinct boundaries of drug market activities. As such, their research addresses the so-called boundary problem, a long-standing problem in geography/criminology, of how to identify valid boundaries of social phenomena.

Part II on *Crime Analysis* includes five chapters. To begin with, **Murray et al.** analyze sex offender residential movement patterns over a two and half year period in Hamilton County, Ohio. Their study evaluates sex offender policies and their collateral consequences, such as implications of spatial restrictions on housing availability and residential mobility. The next chapter by **Herrmann** explores both spatial and temporal variations in violent crime at the street level. Utilizing street segments as units of analysis allows the identification of hot streets and the detection of unique temporal patterns, which both can assist police departments in crime prevention and control strategies. Chapter 5 by **Murray and Grubesic** reviews non-hierarchical spatial cluster analysis for crime hot spot detection that can incorporate both event attributes and neighborhood characteristics (i.e., spatial lag) as a modeling parameter. Their research thus contributes to a better understanding of the distributions and morphologies of crime that is needed for proactive policing, predictive hot spotting, and crime forecasting strategies. The final two chapters in Part II contribute to a better understanding of criminals' journey-to-crime travel behavior. In Chap. 6, **Kasprzyk et al.** apply multi-criteria analysis, among other methods, to reconstruct the most probable journey-to-crime travel and the vehicle's withdrawal site of a series of crimes committed by the same group of offenders. In Chap. 7, **Levine and Lee** examine the influence that gender and age have on offenders' journey-to-crime trips. The results indicate that simple generalizations about criminal travel are suspect. Instead, crime travel must be understood as reflecting the interaction of the type of crime, the characteristics of the metropolitan structure, the presence of accomplices, and offender characteristics. Their findings might aid police not only in identifying the specific area to be searched for a criminal, but might also increase the accuracy of the search.

Part III on *Crime Modeling* includes six chapters. In Chap. 8, **Kim et al.** propose a conceptual model named *Spatial Configurations of Homicide Crime* (SCHC), which categorizes sequences of homicides from a geographical perspective that includes the offender's and the victim's residences and the murder and disposal locations of the victim. Connecting these locations with social relationships among victims and offenders, the context of the crimes and the ethnic composition of people within a community provide law enforcement agencies with a better understanding of the dynamics of crime in a geographical perspective. Chapter 9 by **Curtis et al.** utilizes the spatial video method, a low cost mobile data collection strategy, to capture information on abandoned/blighted buildings and the returnee rate in the Lower 9th Ward of New Orleans for the period 2010 and 2011. When linking this information with individual crime, this research contributes to a better understanding of the relationship that dynamic fine scale environments have with criminal activities. The research by **Mennis and Harris** presented in Chap. 10 investigates the influence that spatial contagion and neighborhood socioeconomic character have on drug offense juvenile delinquency and recidivism among urban male youth. As expected, results indicate that neighborhoods with concentrated disadvantage enhance both drug delinquency and repeat drug delinquency. As such, this research not only lends substantial support to theories of neighborhood and social contextual mechanisms of juvenile drug offending, but also provides valuable information to the juvenile justice system.

In Chap. 11, building upon the neighborhood life cycle model and environmental criminology theories, **Lee and Wilson** apply the newly developed Urban Crime Simulator (UCS) software toolbox to simulate changes in property crime as a response to increased foreclosure rates. It is considered a more flexible and easier to use alternative to existing simulation approaches, such as cellular automata or agent-based modeling. The UCS is especially geared toward city administrators, planners, and law enforcement agencies to examine the possible impacts from urban changes on current crime patterns and to be more proactive toward preventing the emergence of new patterns. In Chap. 12, **Groff** builds a geoprocessing model to quantify the exposure of street blocks to drinking places in Seattle, WA. In addition to operationalizing new, geographically-based measures of exposure, **Groff** demonstrates that geoprocessing models can also automate and document complex processes, such as the examination of measure performance across multiple thresholds, making them more transparent, and increasing the ease of replication. The final chapter (Chap. 13) in Part III by **Gilbreath** uses a spatial regression model to assess the relationship between methamphetamine lab seizures and county characteristics in the states of the Midwest High-Intensity Drug Trafficking Area (HIDTA) for the years 2000–2010. A key research question is whether the 2005 Combat Methamphetamine Epidemic Act exhibits any influence on the covariates of methamphetamine lab seizures. Results show that while the Act caused a dramatic dip in the number of labs seized, both their spatial distribution as well as their covariates are not much affected by the Act. **Gilbreath** concludes by arguing that the strategy for policing methamphetamine does not need to be significantly different from that for other drugs and drug markets.

Part IV on *Crime Mapping* includes three chapters. In Chap. 14, **Fuhrman et al.** introduce the bivariate mapping technique to compare campus crime data with participants' self-reported cognitive fear of crime maps. Their research shows that student fear of crime is aligned with data for burglary and theft but is over predicted, compared to the reported cases, for harassment and sexual assault. The authors conclude that the bivariate crime maps are effective representations for university administrators and university police in decision and policy making tasks, for crime mitigation, and public awareness purposes. The overall goal of the research presented in Chap. 15 by **Morgan and Steinberg** is to assess the practicality of using time geography within a crime mapping context. More specifically, this chapter evaluates the usability of 3D space-time cube maps for representing crime patterns and considers the utility of the time-geographic framework for exploring crime events that occur at unknown points in space and time. The authors conclude that with training, even the most complex time-geographic maps could achieve large-scale adoption by the crime-mapping community. In Chap. 16, **Guo and Wu** use a data mining and visual analytics approach called VIS-STAMP to analyze crime data of Philadelphia, PA, from January 2007 to June 2011. This exploratory approach can aid police to discover and understand complex crime patterns that involve multiple perspectives such as spatiotemporal trends across different crime types.

The final Part V on *Applications and Implementations* includes two chapters. In Chap. 17, **Elmes and Roedl** apply geospatial information technology in a collaboration between university researchers and practitioners from two adjacent law enforcement jurisdictions. The purpose is to examine crime incidents for spatiotemporal trends stretching across jurisdictional borders. One major implication for this partnership suggests that researchers and multiple law enforcement jurisdictions can work together to identify and solve community problems. In the final Chap. 18, **Martínez-Viveros et al.** report on an academia-government collaboration aimed to implement the geospatial dimension into the information systems and decision making processes of Mexico City's Public Safety Ministry. More specifically, this collaboration includes the implementation of a geospatial data infrastructure (GDI) to enable the seamless integration of data from different sources, platforms, and systems. Additionally, an open-source interactive solution that retrieves and displays geospatial patterns and trends is developed as well as models for space and space-time analysis of crime incidents.

Contents

Part I
Fundamental Spatial Problems

Chapter 1
Spatial Heterogeneity in Crime Analysis

Martin A. Andresen and Nicolas Malleson

Abstract Issues related to the modifiable areal unit problem are well-understood within geography. Though these issues are acknowledged in the spatial crime analysis literature, there is little research that assesses their impact. In fact, much of the cited spatial crime analysis literature that investigates the impact of modified areal units suggests that there is no problem—there is, however, an alternative literature. In this paper, we employ a new area-based spatial point pattern test to investigate the impact of modified areal units on crime patterns. We are able to show that despite the appearance of similarity in a (spatial) regression context, smaller units of analysis do show a high degree of variation within the larger units they are nested. Though this result in and of itself is not new, we also quantify how much spatial heterogeneity is present. This quantification is undertaken using multiple crime classifications and in a cross-national comparison.

Keywords Spatial crime analysis • Spatial heterogeneity • Modifiable areal unit problem (MAUP) • Point pattern analysis

M.A. Andresen (✉)
School of Criminology, Institute for Canadian Urban Research Studies, Simon Fraser University, Burnaby, BC, Canada
e-mail: andresen@sfu.ca

N. Malleson
School of Geography, Centre for Spatial Analysis and Policy, Centre for Computational Geography, University of Leeds, Leeds, UK

M. Leitner (ed.), *Crime Modeling and Mapping Using Geospatial Technologies*, Geotechnologies and the Environment 8, DOI 10.1007/978-94-007-4997-9_1,

1.1 Introduction

Over the past 180 years, the geography of crime literature has moved to ever finer spatial scales of resolution. Beginning with the work of Quetelet (1831, 1842) and Guerry (1833), this literature has moved from French Departments, to counties, towns, neighborhoods and now the street segment (Glyde 1856; Burgess 1916; Shaw and McKay 1931, 1942; Sherman et al. 1989; Weisburd et al. 2004, 2009). The drive for analyses to be undertaken at these ever finer spatial scales is the discovery of significant heterogeneity within smaller spatial units of analysis: there are safe places within bad neighborhoods and dangerous places within good neighborhoods (Sherman et al. 1989).

An obvious question to emerge within this geography of crime literature because of this finding is: what is the appropriate spatial scale of analysis? Indeed, those that advocate for smaller spatial units of analysis state that micro-places are now deemed appropriate whereas larger spatial units of analysis are not (Andresen and Malleson 2011). But how much does this issue really matter? Yes, there may be significant spatial heterogeneity, but does this impact the analysis?.[1]

A small branch of literature has investigated this question. The results most frequently cited show that the choice of the spatial unit of analysis is irrelevant (Land et al. 1990; Wooldredge 2002). Because of this finding, much of the literature that follows has used this as a justification for only analyzing one type of spatial unit (Schulenberg 2003; Bernasco and Block 2009; Matthews et al. 2010; Osgood and Anderson 2004). But is this a reasonable assumption to be made in all contexts? We argue that it is not.

In this paper, we use calls for service and recorded crime data from police forces in two municipalities (one in Canada and another in England) and a similarity-based spatial point pattern test. We are able to show that despite similarities in the results of global analyses, the results are significantly different at alternative spatial scales of analysis. Because of the nature of this spatial point pattern test we are able to show how results change when the spatial unit of analysis is changed. Previous research has investigated this phenomenon, but we explicitly show the results using two different spatial units of analysis: census tracts and dissemination areas in Canada and middle layer super output areas and output areas in England. Specifically, we are able to quantify the spatial heterogeneity within larger units of analysis for multiple crime classifications and in a cross-national comparison: Vancouver, Canada and Leeds, England.

[1] We define spatial heterogeneity being present when a large spatial unit of analysis has smaller spatial units of analysis within it that do not all exhibit the same properties.

1.2 Scale and Spatial Crime Analysis

In geography, scale matters: changing the size or shape of the spatial unit under analysis may lead to unexpected and substantial changes in results (Blalock 1964; Clark and Avery 1976; Gehlke and Biehl 1934; Fotheringham and Wong 1991; Openshaw 1984a, b). This is referred to as the modifiable areal unit problem (MAUP). Faced with the MAUP, there are three possible scenarios that may emerge when modifying the spatial units of analysis under study. First, there may be no impact. In other words, the results are identical (or differences are statistically insignificant) at all spatial scales of analysis. This is clearly the ideal situation. Second, there may be a quantitative impact on the results, but the qualitative results are the same. In this situation, the estimated parameters for the variables in an analysis may change (with statistical significance, so bias is present) but those estimated parameters do not change signs (positive to negative, become statistically insignificant, or negative to positive); as such, variables may be thought to have a stronger or weaker relationship with the dependent variable than is actually the case, but the qualitative interpretations are the same. Third, there may be a qualitative impact in the results. If this occurs, the results may lead the researcher to make substantively incorrect statements: rejecting or accepting a theory when they should not, and/or making incorrect statements regarding a policy initiative in an evaluation. This is the worst-case scenario and is the possibility outlined by Fotheringham and Wong (1991).

Another, but related issue emerges when one makes inference based on an a analysis at one spatial scale and applies it to another spatial scale; when the inference is based on a larger spatial unit and applied to a smaller spatial unit it is referred to as the ecological fallacy (what it true of the whole is not necessarily true of its parts) and when the inference is based on a smaller spatial unit of analysis and applied to a larger spatial unit it is referred to as the atomistic fallacy (what is true of the parts is not necessarily true of the whole). Such problems in inference have been known for a long time and are most often in the context of assigning neighborhood characteristics/relationships to individuals, the ecological fallacy (Robinson 1950). Because of the ecological fallacy, change that occurs at a larger spatial scale may be driven by a small number of the smaller spatial scale units within the larger spatial scale unit. Consequently, there may be variations in the spatial patterns at different scales.

Of course, there may be limitations in the geography of crime when it comes to the choice of spatial scale that are beyond the control of the researcher. For example, when using census data, issues of confidentiality may arise that lead to missing data values and preclude the analysis at a particular geography—Andresen (2006) was unable to undertake an analysis at a smaller spatial scale because almost 25% of the census boundary units were missing data because of confidentiality issues.[2]

[2] There is also the issue of missing data because of underreporting of crime and/or systemic biases in reporting crime. This may or may not have spatial implications, but we are unaware of any research that addresses this issue.

Additionally, there may be a number of factors in a decision-making process for research that leads to the use of only one spatial scale. First, data availability may prevent the use of multiple scales of analysis; because of confidentiality concerns, a police department may only provide counts of crime based on one spatial unit. Second, there may be specific spatial scales of interest for those performing the analysis; in such a research context, other spatial scales are simply not of interest or relevant. Third, the researchers may be interested in replicating (being consistent with) previous research that is only concerned with one spatial scale. Though not an exhaustive list, these examples do show that spatial scale is not necessarily being ignored by researchers. Barring situations such as this, we did expect to find the use of multiple scales of analysis in the geography of crime literature in order to investigate the role of the MAUP. However, we found that this is not the case.

For example, Wooldredge (2002), comparing census tracts to administrative neighborhoods, found that the substantive results for different spatial units of analysis are the same. This led Wooldredge (2002, 681) to refer to the "(ir)relevance of aggregation bias" in the context of the MAUP, for the geography of crime.[3] Despite the increasing availability of crime data as points (addresses, street intersections, and x-y coordinates), aggregation will still be a concern as long as those using these data aggregate points in order to analyze crime relative to other data that are only available as area polygons, such as census data. More importantly for this issue, it is not that a small number of studies have found such a relationship, but that these studies, particularly Wooldredge (2002), are used as justification for only using one spatial unit of analysis. The use of Wooldredge (2002) for this purpose was picked up almost immediately (Schulenberg 2003), and continues in a variety of contexts (see, for example, Bernasco and Block 2009; Matthews et al. 2010; Osgood and Anderson 2004).

Despite this rapid adoption of Wooldredge's (2002) conclusion, there is another side to this literature. Ouimet (2000) showed that using census tracts versus neighborhoods does impact the results; specifically, the choice of spatial aggregation impacts the theory that is supported by the data. More recently, and similar to Ouimet (2000), Hipp (2007) showed that explanatory variables exhibit different effects on crime and disorder based on the level of aggregation. Consequently, it is curious that Wooldredge (2002) is almost always cited to support the use of only one level of spatial aggregation.

We are in no way being critical of the work done by Wooldredge (2002) and others. In fact, for the case of Vancouver using the crime data described below, we find that the choice of spatial unit of analysis matters little for the substantive results of a spatial regression. Rather, we are asking if we can simply dispense with multiple spatial units of analysis when studying the geography of crime? In other words, is there any spatial heterogeneity and does it matter? We are unable to find any research that quantifies the degree of spatial heterogeneity, so this is our task in this paper.

[3] Wooldredge (2002) is not the first to make this type of claim (see Land et al. 1990, for example), but is the most often cited research on this topic.

1.3 Data and Methods

Data for Vancouver, Canada and Leeds, England are used in the analysis below. We use data from these two cities for three reasons. First, these are the police data available to us for analysis. Second, we know these cities and are, therefore, able to make interpretation using local knowledge. Third, and most significantly, the inclusion of data from two different countries aids in our ability to make generalizations rather than relying on one set of data that may produce spurious results.

1.3.1 Vancouver and Its Data

The Vancouver data used in the analysis below are for the years 1991 and 2001. The Vancouver Census Metropolitan Area (CMA) is the third largest metropolitan area in Canada, based on population (currently approximately two million people), and the largest metropolitan area in western Canada. In 2001, the City of Vancouver had a population of 546,000. In recent years, Vancouver has experienced substantial growth in its resident population: 431,000 in 1986, 472,000 in 1991, and 514,000 in 1996. This high rate of growth is often attributed to the 1986 World Exposition on Transportation and Communication that led to Vancouver receiving worldwide attention and is expected to continue because of the most recent 2010 Winter Olympics held in the Vancouver CMA. With an area of approximately 115 km^2, the City of Vancouver has 110 census tracts (CTs) and 990 dissemination areas (DAs), defined by Statistics Canada. Census tracts are relatively small and stable geographic areas that tend to have a population ranging from 2500 to 8000—the average is 4,000 persons. Dissemination areas are smaller than census tracts, equivalent in size to a census block group in the U.S. census—approximately 400–700 persons, composed of one or more blocks.[4]

Though Vancouver has had a decreasing crime rate from 1991 to 2001, its crime rate remains substantially higher than the national average. In fact, the Vancouver CMA had the highest crime rates among the three largest metropolitan areas in Canada at 11,367 criminal code offences per 100,000 persons in 2001, more than doubling the rate found in Toronto (5381 per 100,000 persons) and almost doubling that in Montreal (6979 per 100, 000 persons). The same relative standing held for the 2001 violent crime rate in the Vancouver CMA (1058 per 100,000 persons) in comparison to the Toronto CMA (882 per 100 000 persons) and the Montreal CMA (886 per 100,000), but to a lesser degree. These differences in crime rates between these three cities have been decreasing in recent years (Kong 1997; Savoie 2002; Wallace 2003).

[4] Prior to the 2001 census, these census boundaries were called enumeration areas.

Table 1.1 Counts and percentages for crime types

Vancouver				
	1991		2001	
	Count	Percent	Count	Percent
Assault	16,556	20.1	7,643	13.4
Burglary	18,068	22.0	13,025	22.9
Robbery	1,421	1.7	1,251	2.2
Sexual Assault	672	0.8	440	0.8
Theft	16,862	20.5	11,255	19.8
Theft of Vehicle	5,957	7.2	6,273	11.0
Theft from Vehicle	22,728	27.6	16,991	29.9
Total	82,264	100.0	56,878	100.0
Leeds				
	2001		2004	
	Count	Percent	Count	Percent
Assault	5,830	9.3	15,323	23.4
Burglary	13,610	21.6	13,838	21.1
Robbery	2,289	3.6	1985	3.0
Sexual assault	493	0.8	807	1.2
Theft	22,043	34.9	27,458	41.9
Theft of Vehicle	9,098	14.4	7354	11.2
Theft from Vehicle	15,493	24.6	14,167	21.6
Total	63,026	100.0	65,609	100.0

Note: The substantial increase for assaults in Leeds is due to a change in recording practices

All crime data used below come from the Vancouver Police Department's Calls for Service Database (VPD-CFS Database) generated by its Computer Aided Dispatch system. The VPD-CFS Database is the set of requests for police service made directly to the VPD, through the 911 Emergency Service and allocated to the VPD, and calls for service made by the VPD members while on patrol. The VPD-CFS Database contains information on both the location and the complaint code/description for each call. For each call, there are two codes: the initial complaint code and a complaint code filed by the officer on the scene. The code provided by the officer is always taken to be correct. Though the VPD-CFS Database is actually a proxy for actual crime data because not all calls for service represent actual crimes, the primary advantage of the VPD-CFS Database is this raw form—these data are not dependent on a criminal charge. It should be noted, however, that few calls for service are subsequently unfounded by the VPD. The crime classifications of assault, burglary, robbery, sexual assault, theft, theft of vehicle, and theft from vehicle are all analyzed below.

The counts and percentages of these crime classifications are presented in Table 1.1. In Vancouver, there has been a notable decrease in the counts of crime, consistent with the international crime drop phenomenon (Tseloni et al. 2010; Farrell et al. 2011). Despite this significant decrease in crime (31% drop), the distribution of the different crime classifications has remained rather constant; assault has experienced a decrease, with corresponding increases in theft of vehicle and theft from vehicle. Leeds has experienced an increase in crime from 2001 to 2004, just under 4%. However, it should be noted that there was a change in recording practices in Leeds between these time periods. Also, the two data sets for Leeds are only 3 years apart. As such, this increase in crime counts could simply be a result of year-to-year fluctuations. Regardless, aside from the assault classifications, the distribution of the different crime classifications has remained relatively constant, aside from an increase in theft.

1.3.2 Leeds and Its Data

The Leeds data used in the analysis are for the years 2001 and 2004. Ideally the same time period would be used for both countries, but data constraints mean that the most reliable crime data in Leeds are only available for these years—reliability refers to the standardization of crime data, discussed below. Leeds is the third largest city in the UK, after London and Birmingham, with a population estimated to be approximately 812 000 in 2011 (Office for National Statistics 2010). Spatially, Leeds is the second largest city in the UK, covering an area of approximately 550 km^2. As a consequence of hosting two universities, Leeds has a very large student population that has had a strong influence on the development of the city in order to cater for a large number of student migrants. The student population is also highly concentrated into a relatively small area to the north of the universities that has a substantial effect on crime patterns. Spatially, the Leeds area can be subdivided into 108 medium-level super output areas (MSOAs) and further into 2440 output areas (OAs). An output area is the smallest 2001 census geography available and contains a minimum of 40 households or 100 people, but the recommended size is approximately 125 households. MSOAs are a larger geography which contain 7,200 people on average and have been designed to fit to the borders of OAs to allow for data aggregation.

In terms of crime, Leeds has generally followed the UK national trend and has seen consistent yearly reductions in most types of crime since 1997. Leeds has higher than average crime rates compared to the average for England and Wales, although this is not unexpected given its demographic and socioeconomic characteristics. The most unusual observation is that rates of residential burglary are particularly high: almost double the national average (18.4 crimes per 1,000 people compared to 9.6) and it has not exhibited the decline that the other types of crime have shown. The explanation for this is largely tied in with the effects of the student population who generally suffer a disproportionate number of burglaries.

The crime data used in the analysis below consist of all crimes recorded by the police in the Leeds area. The data cover the time periods 1st^t April 2000–31st^t March 2001 (hereby abbreviated to '2001') and 1st^t April 2003–31st^t March 2004 ('2004'). The data are coded by crime type and are stored with a location address that can be geocoded. There are numerous implications for using this type of data in research;

namely not all crime is reported to the police in the first place and, even if it is reported, the crime might not necessarily be *recorded* by the police. In fact, recording practices varied substantially across police forces so to standardize them the National Crime Recording Standard was phased in through 2001–2004. The new standard followed a more "victim centred" approach so that a crime should be recorded even if there is no evidence that it has taken place. This led to an apparent increase in some types of crime, particularly violent crimes: assault and, to a lesser extent, sexual assault. The counts and percentages of these crime classifications are presented in Table 1.1.

1.3.3 Geocoding

Geocoding has the potential to introduce error into any analysis. Previous research has noted that geocoding algorithms are not only inaccurate at times, but are also at risk of not locating all street addresses or street intersections for (criminal) incidents (Ratcliffe 2001; Cayo and Talbot 2003; Zandbergen 2008). Consequently, potential for spatial bias is present. Ratcliffe (2004) addresses this issue through the identification of a minimum acceptable hit/success rate of 85%. The geocoding procedure used for the current data generated 93 and 94% success rates for 1991 and 2001 in Vancouver. In Leeds, the data were already geocoded by the police and although actual hit rates are not known the data were put through an extensive cleaning process and can be assumed to be sufficiently high. It should be noted that error may still be present: geocoding to the wrong address, being placed to a centroid, or a correct match may be aggregated to the incorrect spatial unit. With our success rates exceeding the minimum acceptable success rate generated by Ratcliffe (2004) and the indication that improper address records are random, the analysis is undertaken with little concern for spatial bias.

In addition to acceptable hit/success rates in geocoding, there are a number of potential problems that are country specific. In Vancouver – where the 'block' street system means that building location can be estimated from its number on a street – long streets may be arbitrarily broken into segments that are not based on intersections; events are placed on the street segment using an interpolation process that may place the event in the wrong place on the street segment; a geocoding match may be made on an areal unit and subsequently misplaced on the wrong street segment; and there is variation in street segment length that may skew the analysis.

In the UK, the street system is not regular so it is not possible to estimate a location based on a building number. Instead, a lookup table is used to match an address directly to some spatial coordinates. The Leeds data were then matched directly to the coordinates of the building at which the crime occurred, or they were assigned manually in places where no building was available to link to. The data were cleaned considerably (both manually and using computer software) before use so we are confident that geocoding issues will not influence the analysis.

Therefore the Vancouver data used in the current analysis are geocoded to the street network and the Leeds data are geocoded directly to points. Both data are then subsequently aggregated to their respective census boundary units using a spatial join function.[5] We use the same (most recent) street network or address lookup table in each city for geocoding different years of data; this avoids the problem of not being able to find new streets or buildings but has the potential of old roads being closed and old buildings being torn down. No such street closures occurred in Vancouver and, as mentioned above, if no building was present (from a possible tear-down) points were manually assigned to the spatial units of analysis. Though shorter street segments may have a lower probability of having a criminal event, *ceteris paribus*, the randomization process in the spatial point pattern test minimizes the potential for having less scope for change than longer street segments. Lastly, the interpolation issue of geocoding algorithms, whereby a point's position on a street segment might be inaccurate, is not a concern here because no inference is made at a finer scale than the dissemination area (Vancouver) or the output area (Leeds).

1.3.4 The Spatial Point Pattern Test

In order to investigate spatial heterogeneity within larger spatial units of analysis (census tracts in Vancouver and MSOAs in Leeds), a testing methodology that identifies changes in spatial crime patterns at multiple scales is necessary. The spatial point pattern test developed by Andresen (2009) serves this purpose well because it can be used to independently identify changes in the spatial patterns of crime at different spatial scales and the output may then be used to quantify spatial heterogeneity. The change for each smaller spatial unit of analysis (DAs and OAs) can be assigned to its respective larger spatial unit of analysis (CTs and MSOAs) and then spatial heterogeneity (or homogeneity) can be assessed by counting the number (percentage) of smaller spatial units within their larger units of analysis that have the same classification of change. This spatial point pattern test has been applied to investigate pattern changes in international trade (Andresen 2010) and for testing the stability in crime patterns (Andresen and Malleson 2011).

The Andresen (2009) spatial point pattern test is area-based[6] and is concerned with the similarity between two different spatial point patterns at the local level. This particular spatial point pattern test is not concerned with null hypotheses of random, uniform, or clustered distributions, but may be used to compare a particular

[5] The street network in Vancouver recognizes to which side of the street a point is geocoded. If that particular street is a boundary for a spatial unit, the point is assigned to the census unit on the appropriate side of the street in the spatial join.

[6] There are two general forms of spatial point pattern tests: area-based and distance based. See Andresen (2009) for a discussion of their respective benefits and limitations.

point pattern with these distributions. An advantage of the test, as we demonstrate here, is that it can be calculated for different area boundaries using the same original point datasets. In order to simplify the process of calculating the test we developed a computer program that is freely available from the authors. The test is computed as follows:

1. Nominate a base dataset (1991 assaults, for example) and count, for each area, the number of points that fall within it.
2. From the test dataset (1996 assaults, for example), randomly sample 85% of the points, with replacement.[7] As with the previous step, count the number of points within each area using the sample. This is effectively a bootstrap created by sampling from the test dataset.
3. Repeat (2) a number of times (in our analysis below we used 200 iterations).
4. For each area in the test data set, calculate the percentage of crime that has occurred in the area. Use these percentages to generate a 95% nonparametric confidence interval by removing the top and bottom 2.5% of all counts (5 from the top and 5 from the bottom in this case). The minimum and maximum of the remaining percentages represent the confidence interval. It should be noted that the effect of the sampling procedure will be to reduce the number of observations in the test dataset but, by using *percentages* rather than the *absolute counts*, comparisons between data sets can be made even if the total number of observations are different.
5. Calculate the percentage of points within each area for the base dataset and compare this to the confidence interval generated from the test dataset. If the base percentage falls within the confidence interval then the two datasets exhibit a similar proportion of points in the given area. Otherwise they are significantly different.[8]

The purpose of this spatial point pattern test is to create variability in one dataset so that it can be compared statistically to another dataset. The 85% samples generated, each maintain the spatial pattern of the test dataset and allows for a "confidence interval" to be created for each spatial unit that may be compared to the base dataset. Therefore, statistically significant changes/differences are identified at the local level.

[7] An 85% sample is based on the minimum acceptable hit rate to maintain spatial patterns, determined by Ratcliffe (2004). Maintaining the spatial pattern of the complete data set is important so we used this as a benchmark for sampling. An 85% sample was for the purposes of generating as much variability as possible while maintaining the original spatial pattern. Also note that "replacement" in this context refers to subsequent samples; any one point may only be sampled once per iteration in this procedure to mimic Ratcliffe (2004).

[8] The program written to perform the test uses double precision that has at least 14 decimal points when dealing with numbers less than unity. The smallest number that we have to deal with in the current analysis (regardless of scale) is 0.000034553. This is well within the limits of double precision.

The output of the test consists of two parts. First, there is a global parameter that ranges from 0 (no similarity) to 1 (perfect similarity): the index of similarity, S, is calculated as:

$$S = \frac{\sum_{i=1}^{n} s_i}{n},$$

where s_i is equal to one if two crimes are similar in spatial unit i and zero otherwise, and n is the total number of spatial units. As such, the S-Index represents the proportion of spatial units that have a similar spatial pattern within both data sets. Second, the test generates mappable output to show where statistically significant change occurs; i.e. which census tracts, dissemination areas, middle layer super output areas ,and output areas have undergone a statistically significant change. Though this spatial point pattern test is not a local indicator of spatial association (LISA, see Anselin 1995) and there is much more to LISA than being able to produce maps of results, it is in the spirit of LISA because the output may be mapped.[9]

A number of tests for similarity are performed. For each crime classification and each spatial unit of analysis, indices of similarity are calculated for 1991–2001 (Vancouver) and 2001–2004 (Leeds). These indices are then used to quantify the degree of spatial heterogeneity present with the changes of the spatial point patterns at the different scales of analysis.

1.4 Results

Before we turn to the results for the examination of spatial heterogeneity, it is important to examine the Indices of Similarity within each of the different spatial units of analysis for each crime classification. These results are presented in Table 1.2. In the case of Vancouver, census tracts do not exhibit much similarity over time, with most values of S being less than 0.300; the results for the dissemination areas are most often more similar over time, close to twice that of census tracts in half of the crime classifications. Noteworthy here is the high degree of similarity for sexual assault, especially within census tracts; sexual assault also has the same proportion of criminal events in both years. In the case of Leeds, the S indices are most often much greater magnitude than in Vancouver. However, this is expected as the time frame for the Leeds crime data is much shorter: 3 years instead of 10 years. And aside from the crime classifications for robbery and sexual assault the S values for the output areas are similar to those for the middle layer super output areas.

[9] It should be noted that the role of local spatial analysis has been growing in interest in recent years (Lloyd 2011).

Table 1.2 Indices of similarity

	Vancouver, 1991–2001	
	Census tracts	Dissemination areas
Assault	0.300	0.335
Burglary	0.155	0.299
Robbery	0.327	0.662
Sexual assault	0.509	0.691
Theft	0.136	0.237
Theft of vehicle	0.300	0.332
Theft from vehicle	0.146	0.261
	Leeds, 2001–2004	
	Middle layer super output areas	Output areas
Assault	0.769	0.639
Burglary	0.667	0.726
Robbery	0.667	0.283
Sexual assault	0.667	0.148
Theft	0.722	0.701
Theft of vehicle	0.898	0.677
Theft from vehicle	0.796	0.718

With regard to spatial heterogeneity, the results are remarkably similar across not only crime classifications but also across municipalities. In the case of assault (Table 1.3), the number of larger areas with zero smaller areas in Vancouver and Leeds is always zero. This is the expected result. In fact, if (when) this occurs, it is highly problematic; such a situation is further discussed below. However, the number of larger areas with all smaller areas having the same classification is also zero in most cases—all cases in Leeds. When this does occur (2001 < 1991 and insignificant change, in Vancouver), it occurs in very few cases. Overall, the average percentage of smaller areas with the same larger area classification is surprisingly low. The best case scenario, for both Vancouver and Leeds, is that a little more than half of the smaller areas have the same larger area classification. Though this may be viewed positively, it also means that a little less than half do not have the same classification. This is a substantial degree of spatial heterogeneity that must be considered when inference is being made at only one level of analysis. The results for burglary (Table 1.4) are similar to those for assault and require little further discussion. The primary result to note here is that assault and burglary have similar results despite these two crime classifications exhibiting different patterns over time in Table 1.1: relatively speaking, assault is decreasing in Vancouver and increasing in Leeds, but burglary in both cities is constant. As such, the degree of spatial heterogeneity does not necessarily depend on other changes in a crime's distribution.

The results for robbery (Table 1.5) and sexual assault (Table 1.6) have similar results for the average percentage of smaller areas with the same larger area classification, but some of the other results are worthy of note. In both Leeds and Vancouver, robbery and sexual assault have some larger areas with zero smaller

Table 1.3 Spatial heterogeneity, assault

Vancouver			
	Census tracts	Census tracts	Census tracts
	2001 > 1991	2001 < 1991	Insignificant change
Number (percentage) of CTs with ***zero*** DAs having the same classification	0	0	0
Number (percentage) of CTs with ***all*** DAs having the same classification	0	1 (2.2)	4 (12.1)
Average percentage of DAs with same CT classification	0.35	0.61	0.43
Total number of CTs with this classification	31	46	33
Leeds			
	MSOAs	MSOAs	MSOAs
	2004 > 2001	2004 < 2001	Insignificant change
Number (percentage) of MSOAs with ***zero*** OAs having the same classification	0	0	0
Number (percentage) of MSOAs with ***all*** OAs having the same classification	0	0	0
Average percentage of OAs with same MSOA classification	0.58	0.43	0.27
Total number of MSOAs with this classification	48	35	25

Notes: *CTs* census tracts, *DAs* dissemination areas, *MSOAs* middle layer super output areas, *OAs* output areas, total CTs = 110; total MSOAs = 108

areas having the same classification: 2001 > 1991, for both cases in Vancouver, and insignificant change for both cases in Leeds. Such a result is particularly problematic because the nature of the spatial heterogeneity is such that the smaller spatial units of analysis have nothing in common with the larger spatial units of analysis. A problem emerges here specifically in the context of policy. If policy is being implemented based on global results and the larger area is used as a reference point for policy implementation, the policy may be applied in error. This will lead to a misallocation of resources, at best, or aggravate the original situation that policy-makers are trying to correct, at worst.

Turning to the three classifications of theft—theft (Table 1.7), theft of vehicle (Table 1.8), and theft from vehicle (Table 1.9)—the results are more promising in terms of the magnitude of within larger spatial unit spatial heterogeneity. The average percentages of smaller areas with the same larger area classification are of the same magnitude as the other crime classifications. Though theft from vehicle (Leeds) and theft of vehicle (Vancouver) do have a small number of larger areas with zero small areas with the same classification, Vancouver has promising results

Table 1.4 Spatial heterogeneity, burglary

Vancouver			
	Census tracts	Census tracts	Census tracts
	2001 > 1991	2001 < 1991	Insignificant change
Number (percentage) of CTs with ***zero*** DAs having the same classification	0	0	1 (5.9)
Number (percentage) of CTs with ***all*** DAs having the same classification	1 (2.4)	0	3 (17.6)
Average percentage of DAs with same CT classification	0.50	0.57	0.45
Total number of CTs with this classification	41	52	17
Leeds			
	MSOAs	MSOAs	MSOAs
	2004 > 2001	2004 < 2001	Insignificant change
Number (percentage) of MSOAs with ***zero*** OAs having the same classification	0	0	0
Number (percentage) of MSOAs with ***all*** OAs having the same classification	0	0	0
Average percentage of OAs with same MSOA classification	0.66	0.49	0.14
Total number of MSOAs with this classification	55	37	16

Notes: *CTs* census tracts, *DAs* dissemination areas, *MSOAs* middle layer super output areas, *OAs* output areas; total CTs = 110; total MSOAs = 108

for the number of larger areas with all corresponding small areas having the same classification. The magnitudes of the percentages are not that great, ranging from 2.8 (theft of vehicle) to 18.8 (theft from vehicle) percent, but this is a definite improvement over the results for the other crime classifications.

1.5 Discussion

In this chapter we have investigated the phenomenon of spatial heterogeneity in the context of spatial point patterns changing over time. Though this is only one dimension of change that may be investigated, the results are strong enough to cause some concern over the lack of sensitivity analyses in the geography of crime literature—the lack of using multiple spatial scales of analysis. The general result is that, on average, approximately one-half of smaller spatial units of analysis have the same classification as their larger counterparts. Though this may translate into an irrelevant effect when using a global statistical technique, as it does using the data in the current

Table 1.5 Spatial heterogeneity, robbery

Vancouver			
	Census tracts	Census tracts	Census tracts
	2001 > 1991	2001 < 1991	Insignificant change
Number (percentage) of CTs with *zero* DAs having the same classification	3 (11.1)	0	0
Number (percentage) of CTs with *all* DAs having the same classification	0	0	8 (22.2)
Average percentage of DAs with same CT classification	0.21	0.36	0.74
Total number of CTs with this classification	27	47	36
Leeds			
	MSOAs	MSOAs	MSOAs
	2004 > 2001	2004 < 2001	Insignificant change
Number (percentage) of MSOAs with *zero* OAs having the same classification	0	0	2 (5.6)
Number (percentage) of MSOAs with *all* OAs having the same classification	1 (2.4)	0	0
Average percentage of OAs with same MSOA classification	0.61	0.39	0.35
Total number of MSOAs with this classification	42	30	36

Notes: *CTs* census tracts, *DAs* dissemination areas, *MSOAs* middle layer super output areas, *OAs* output areas; total CTs = 110; total MSOAs = 108

analysis, the magnitude of the spatial heterogeneity cannot be ignored. Therefore, spatial heterogeneity in the presence of an irrelevant effect in a particular context does not mean there are no aggregation biases present, generally speaking. As such, we as researchers cannot simply assume that aggregation bias is not present and only perform analyses at one spatial scale because of a small number of research projects have not found evidence for aggregation bias; aggregation is present, it just does not manifest itself in particular contexts using particular techniques.

The case of sexual assault in Vancouver is of particular interest here. Figure 1.1 shows the results from the spatial point pattern test. All four census tracts shown in Fig. 1.1 have statistically significant increases, 2001 > 1991. The two middle census tracts are likely representative of the presence of spatial heterogeneity: some DAs exhibit increasing trends, some DAs exhibit decreasing trends, and some DAs exhibit insignificant change. In these cases, there are a small number of DAs (one in the case of the CT on top of the map) that are driving the results for the larger CTs. However, for the CTs on either side of Fig. 1.1, there is clearly something else going on. In each case, there are no DAs that exhibit increasing trends; rather, most have statistically insignificant change with a small number of decreasing trends. How can this be the case?

Table 1.6 Spatial heterogeneity, sexual assault

Vancouver			
	Census tracts	Census tracts	Census tracts
	2001 > 1991	2001 < 1991	Insignificant change
Number (percentage) of CTs with ***zero*** DAs having the same classification	7 (43.8)	0	0
Number (percentage) of CTs with ***all*** DAs having the same classification	0	0	8 (14.3)
Average percentage of DAs with same CT classification	0.11	0.31	0.69
Total number of CTs with this classification	16	38	56
Leeds			
	MSOAs	MSOAs	MSOAs
	2004 > 2001	2004 < 2001	Insignificant change
Number (percentage) of MSOAs with ***zero*** OAs having the same classification	0	0	5 (13.9)
Number (percentage) of MSOAs with ***all*** OAs having the same classification	11 (37.9)	0	0
Average percentage of OAs with same MSOA classification	0.53	0.51	0.35
Total number of MSOAs with this classification	29	43	36

Notes: *CTs* census tracts, *DAs* dissemination areas, *MSOAs* middle layer super output areas, *OAs* output areas, total CTs = 110, total MSOAs = 108

As it turns out for the CTs on the sides of Fig. 1.1, there are DAs with statistically insignificant changes that have increasing trends. And these increasing trends are close to being statistically significant; if a 90% confidence interval had been chosen, for example, the results of those DAs would have been statistically significant and increasing. But the point of this discussion is not in regard to the choice of statistical significance. Rather, the point is that insignificant changes at the level of a smaller spatial unit of analysis may become statistically significant with a larger spatial unit of analysis. In other words, there is an aggregation effect.

Comparing the results from Tables 1.3, 1.4, 1.5, 1.6, 1.7, 1.8, and 1.9 to Table 1.1, an interesting relationship emerges. The crime classifications that had the most problematic results (robbery and sexual assault) had the lowest counts and percentages for both Vancouver and Leeds, and the crime classifications that had the most promising results (burglary, theft, and theft from vehicle) had the greatest counts and percentages for both Vancouver and Leeds—theft of vehicle also had promising results. Therefore, it would appear that if the event is more common, the results are less problematic. This does not mean that spatial heterogeneity is not an issue when there are more crimes, just that the issue does not appear to be as great. This result

Table 1.7 Spatial heterogeneity, theft

Vancouver			
	Census tracts	Census tracts	Census tracts
	2001 > 1991	2001 < 1991	Insignificant change
Number (percentage) of CTs with ***zero*** DAs having the same classification	0	0	0
Number (percentage) of CTs with ***all*** DAs having the same classification	2 (8.7)	5 (6.9)	2 (13.3)
Average percentage of DAs with same CT classification	0.46	0.70	0.45
Total number of CTs with this classification	23	72	15
Leeds			
	MSOAs	MSOAs	MSOAs
	2004 > 2001	2004 < 2001	Insignificant change
Number (percentage) of MSOAs with ***zero*** OAs having the same classification	0	0	0
Number (percentage) of MSOAs with ***all*** OAs having the same classification	0	0	0
Average percentage of OAs with same MSOA classification	0.53	0.36	0.29
Total number of MSOAs with this classification	45	33	30

Notes: *CTs* census tracts, *DAs* dissemination areas, *MSOAs* middle layer super output areas, *OAs* output areas; total CTs = 110; total MSOAs = 108

relates to the discussion above regarding the ecological fallacy. Variations in spatial patterns may be more evident when the count of points in the spatial pattern is less. Such a situation is understood intuitively: a spatial pattern with fewer points is more likely to have zero values in spatial units, leading to more spatial heterogeneity withing larger spatial units. This is confirmed in Table 1.2 for Vancouver that has the highest *S*-Index values for the low-count crime classification of robbery and sexual assault. Therefore, the degree of concern for spatial heterogeneity should be inversely related to the number of points in the spatial pattern. Consequently, if an analysis (for the purposes of pure academic interests, policy, or a combination of both) is restricted to one spatial unit of analysis, results may have to be tempered depending on the number of points under analysis.

There are a number of obvious directions for future research. Though we have performed this analysis in two municipalities that are quite distant from one another, more replication is always preferable. We claim that too often research relies on a small number of other studies that claim aggregation bias is minimal or non-existent as justification for only performing analyses at one spatial scale. Consequently, we wish to be careful with our generalizations. The form of replication needs to be

Table 1.8 Spatial heterogeneity, theft of vehicle

Vancouver			
	Census tracts	Census tracts	Census tracts
	2001 > 1991	2001 < 1991	Insignificant change
Number (percentage) of CTs with ***zero*** DAs having the same classification	0	0	1 (3.0)
Number (percentage) of CTs with ***all*** DAs having the same classification	1 (2.8)	2 (4.9)	3 (9.1)
Average percentage of DAs with same CT classification	0.42	0.59	0.42
Total number of CTs with this classification	36	41	33
Leeds			
	MSOAs	MSOAs	MSOAs
	2004 > 2001	2004 < 2001	Insignificant change
Number (percentage) of MSOAs with ***zero*** OAs having the same classification	0	0	0
Number (percentage) of MSOAs with ***all*** OAs having the same classification	0	0	0
Average percentage of OAs with same MSOA classification	0.71	0.52	0.11
Total number of MSOAs with this classification	56	41	11

Notes: *CTs* census tracts, *DAs* dissemination areas, *MSOAs* middle layer super output areas, *OAs* output areas, total CTs = 110, total MSOAs = 108

varied as well. Not only should further investigations into spatial heterogeneity be in other urban areas, but suburban and rural areas as well. Because rural areas tend to have less crime than urban areas, the spatial heterogeneity may be more of a problem in rural areas than suburban and urban areas. Similarly, the crime mix likely varies across urban, suburban, and rural areas, so may the issue of spatial heterogeneity. The more context we have regarding spatial heterogeneity, the better choices we can make regarding spatial scale.

Though the current analysis is instructive, the format of quantifying spatial heterogeneity should be performed in different ways. For example, it would be most useful to investigate spatial heterogeneity in the context of standard spatial theories of crime. It may be the case that a small number of small area (DAs/OAs) are driving the results for their aggregate areas (CTs/MSOAs). Specifically, once we have more information regarding the role of spatial scale and spatial heterogeneity we may be able to further develop/refine/test spatial theories of crime. Not only may a small number of small areas be driving aggregate results, but the way we think about particular theoretical frameworks may change.

Table 1.9 Spatial heterogeneity, theft from vehicle

Vancouver			
	Census tracts	Census tracts	Census tracts
	2001 > 1991	2001 < 1991	Insignificant change
Number (percentage) of CTs with ***zero*** DAs having the same classification	0	0	0
Number (percentage) of CTs with ***all*** DAs having the same classification	1 (5.0)	3 (4.1)	3 (18.8)
Average percentage of DAs with same CT classification	0.46	0.67	0.43
Total number of CTs with this classification	20	74	16
Leeds			
	MSOAs	MSOAs	MSOAs
	2004 > 2001	2004 < 2001	Insignificant change
Number (percentage) of MSOAs with ***zero*** OAs having the same classification	0	1 (2.6)	0
Number (percentage) of MSOAs with ***all*** OAs having the same classification	0	0	0
Average percentage of OAs with same MSOA classification	0.68	0.53	0.24
Total number of MSOAs with this classification	48	38	22

Notes: *CTs* census tracts, *DAs* dissemination areas, *MSOAs* middle layer super output areas, *OAs* output areas; total CTs = 110; total MSOAs = 108

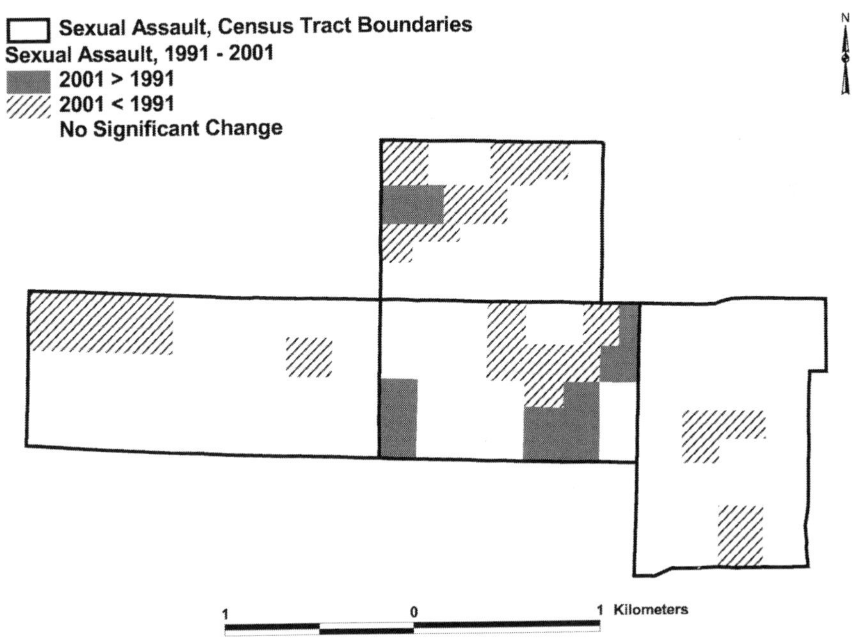

Fig. 1.1 Sexual assault, census tract to dissemination area

References

Andresen MA (2006) A spatial analysis of crime in Vancouver, British Columbia: a synthesis of social disorganization and routine activity theory. Can Geogr 50(4):487–502
Andresen MA (2009) Testing for similarity in area-based spatial patterns: a nonparametric Monte Carlo approach. Appl Geogr 29(3):333–345
Andresen MA (2010) Canada – United States interregional trade: quasi-points and spatial change. Can Geogr 54(2):139–157
Andresen MA, Malleson N (2011) Testing the stability of crime patterns: implications for theory and policy. J Res Crime Delinq 48(1):58–82
Anselin L (1995) Local indicators of spatial association – LISA. Geogr Anal 27(2):93–115
Bernasco W, Block R (2009) Where offenders choose to attack: a discrete choice model of robberies in Chicago. Criminology 47(1):93–130
Blalock HM (1964) Causal inferences in nonexperimental research. University of North Carolina Press, Chapel Hill
Burgess EW (1916) Juvenile delinquency in a small city. J Am Inst Crim Law Criminol 6(5):724–728
Cayo NR, Talbot TO (2003) Positional error in automated geocoding of residential addresses. Int J Heal Geogr 2(1):1–12
Clark WAV, Avery KL (1976) The effects of data aggregation in statistical analysis. Geogr Anal 8(4):428–438
Farrell G, Tseloni A, Mailley J, Tilley N (2011) The crime drop and the security hypothesis. J Res Crime Delinq 48(2):147–175
Fotheringham AS, Wong DW (1991) The modifiable areal unit problem in multivariate statistical analysis. Environ Plan A 23(7):1025–1044
Gehlke CE, Biehl K (1934) Certain effects of grouping upon the size of the correlation coefficient in census tract material. J Am Stat Assoc Suppl 29(185):169–170
Glyde J (1856) Localities of crime in Suffolk. J Stat Soc Lond 19(2):102–106
Guerry A-M (1833) Essai sur la statistique morale de la France. Crochard, Paris
Hipp JR (2007) Block, tract, and levels of aggregation: neighborhood structure and crime and disorder as a case in point. Am Sociol Rev 72(5):659–680
Kong R (1997) Canadian crime statistics, 1996. Statistics Canada, Canadian Centre for Justice Statistics, Ottawa
Land KC, McCall PL, Cohen LE (1990) Structural covariates of homicide rates: are there any invariances across time and social space? Am J Sociol 95(4):922–963
Lloyd CD (2011) Local models for spatial analysis, 2nd edn. CRC Press, Taylor & Francis Group, Boca Raton
Matthews SA, Yang T-C, Hayslett KL, Ruback RB (2010) Built environment and property crime in Seattle, 1998–2000: a Bayesian analysis. Environ Plan A 42(6):1403–1420
Office for National Statistics (2010). 2008-based subnational population projections for England. Newport, Office for National Statistics. Report available on-line http://www.ons.gov.uk/ons/. Accessed July 2011
Openshaw S (1984a) The modifiable areal unit problem. CATMOG (Concepts and Techniques in Modern Geography) 38. Geo Books, Norwich
Openshaw S (1984b) Ecological fallacies and the analysis of areal census data. Environ Plan A 16(1):17–31
Osgood DW, Anderson AL (2004) Unstructured socializing and rates of delinquency. Criminology 42(3):519–549
Ouimet M (2000) Aggregation bias in ecological research: how social disorganization and criminal opportunities shape the spatial distribution of juvenile delinquency in Montreal. Can J Criminol 42(2):135–156
Quetelet LAJ (1831) [1984] Research on the propensity for crime at different ages. Anderson Publishing, Cincinnati

Quetelet LAJ (1842) A treatise on man and the development of his faculties. W. and R. Chambers, Edinburgh

Ratcliffe JH (2001) On the accuracy of TIGER type geocoded address data in relation to cadastral and census areal units. Int J Geogr Inf Sci 15(5):473–485

Ratcliffe JH (2004) Geocoding crime and a first estimate of a minimum acceptable hit rate. Int J Geogr Inf Sci 18(1):61–72

Robinson WS (1950) Ecological correlations and the behavior of individuals. Am Sociol Rev 15(3):351–357

Savoie J (2002) Crime statistics in Canada, 2001. Statistics Canada, Canadian Centre for Justice Statistics, Ottawa

Schulenberg JL (2003) The social context of police discretion with young offenders: an ecological analysis. Can J Criminol Crim Justice 45(2):127–157

Shaw CR, McKay HD (1931) Social factors in juvenile delinquency. U.S. Government Printing Office, Washington, DC

Shaw CR, MacKay HD (1942) Juvenile delinquency and urban areas: a study of rates of delinquency in relation to differential characteristics of local communities in American cities. University of Chicago Press, Chicago

Sherman LW, Gartin PR, Buerger ME (1989) Hot spots of predatory crime: routine activities and the criminology of place. Criminology 27(1):27–56

Tseloni A, Mailley J, Farrell G, Tilley N (2010) Exploring the international decline in crime rates. Eur J Criminol 7(5):375–394

Wallace M (2003) Crime statistics in Canada, 2002. Statistics Canada, Canadian Centre for Justice Statistics, Ottawa

Weisburd D, Bernasco W, Bruinsma GJN (2009) Putting crime in its place: units of analysis in geographic criminology. Springer, New York

Weisburd D, Bushway S, Lum C, Yang S-M (2004) Trajectories of crime at places: a longitudinal study of street segments in the City of Seattle. Criminology 42(2):283–321

Wooldredge J (2002) Examining the (ir)relevance of aggregation bias for multilevel studies of neighborhoods and crime with an example of comparing census tracts to official neighborhoods in Cincinnati. Criminology 40(3):681–709

Zandbergen PA (2008) A comparison of address point, parcel and street geocoding techniques. Comput Environ Urban Syst 32(3):214–232

Chapter 2
When Does a Drug Market Become a Drug Market? Finding the Boundaries of Illicit Event Concentrations

Lallen Johnson and Jerry H. Ratcliffe

Abstract The difficulties of forming valid measurements of social phenomena have been well documented in social science research (Blalock 1971; Denton and O'Malley 1999; Murphy and Arroyo 2000). As the concept under study becomes more abstract, so too does its measurement. The spatial world is no exception to this problem as we frequently rely on convenient spatial boundaries such as census areas to compartmentalize a phenomenon in a meaningful way. In this chapter we illustrate this problem through the conceptualization and operationalization of drug markets. After we have explained some of the nuances of drug market construction and 'creation' in detail, we argue that many of the current measurements used to spatially define them are subject to validity issues. We therefore propose a hierarchical clustering methodology that provides a more refined indicator of market activity. We conclude with a summary of implications for crime analysts, police resource allocation, and theory testing.

Keywords Hierarchical clustering • Validity • Drug market • Operationalization • Conceptualization

L. Johnson (✉)
Department of Criminal Justice, University of Central Florida,
Orlando, FL, USA
e-mail: lallen.johnson@ucf.edu

J.H. Ratcliffe
Department of Criminal Justice, Temple University,
Philadelphia, PA, USA

M. Leitner (ed.), *Crime Modeling and Mapping Using Geospatial Technologies*, Geotechnologies and the Environment 8, DOI 10.1007/978-94-007-4997-9_2,

2.1 Introduction

The illegality involved in drug markets places buyers and sellers in complicated situations (Eck 1995). Because both seller and buyer are breaking the law and neither wants to be apprehended by law enforcement, they must make a personal connection in such a way that they are reassured the other is not a police officer, and in such a way that the drug buyer is reassured the dealer is unlikely to rip them off and steal their money. Furthermore, they must converge in time and space in order for the seller to make a profit and the buyer to satisfy a narcotic dependency. While police enforcement of street drug markets can promote an indoor drug market economy (Rengert et al. 2005), conducting activity on the street still provides for some buyer security (St. Jean 2007) and sufficient dealer access to potential buyers (Eck 1994, 1995). Illegal street drug markets therefore provide opportunities for all of these needs to be addressed. Considering the variation across places as shown by environmental criminology, it comes as no surprise that certain locations are more amenable to drug markets than others. Therefore, it is important that researchers are able not only to define what a drug market is but also be able to effectively outline their locations. These are research methodology issues of conceptualization and operationalization, respectively.

The current chapter begins by discussing how drug markets are conceptualized and operationalized. After we have explained some of the nuances of drug market construction and 'creation' in detail, we argue that many of the current measurements used to spatially define them are subject to validity issues. Namely, much of past research has merely aggregated totals of drug sale incidents to block group areas—an approach that ignores the spatial concentration of incidents and may exaggerate the nature of the problem. Upon covering issues of validity we briefly explore additional hotspot techniques, noting the strengths and weaknesses of each as well as suitability for the current work. We finally suggest an alternative method to create distinct boundaries of drug markets by identifying areal concentrations of criminal incidents using a nearest neighbor hierarchical clustering technique. This approach is demonstrated with a case study of drug sale incidents from the Philadelphia (PA) Police Department.

2.2 Literature Review

This chapter exists because, in part, researchers have failed to agree on how to conceptualize drug markets. In the most general sense, markets are merely ways in which buyers and sellers relate to one another through the sale/purchase of desired items such as antiques, foods, or illicit drugs. Markets also imply places where exchanges occur such as antique shops, grocery stores, or street corners (Reuter 2000). Murji (2007) argues that "at the most banal level, all drug transactions are market relations in that they entail an exchange between sellers, buyers, traders, and so on" (p. 782). Reuter (2000) proposed a similar idea, defining a drug market as "... a place to which a drug user can go with fair confidence of finding a willing seller, perhaps even one whom he does not know" (p. 7).

The conceptualization of drug markets is, however, highly dependent upon the disciplinary lens used (Ritter 2006). Economic studies define markets as exchanges for commodities based upon principles of supply and demand. Criminological research focuses on the illegality of drug exchanges, behavioral and sociological theories explaining those exchanges, and the responses of law enforcement. Ethnographic work, however, describes the experiences of those involved in and/or exposed to illicit drug exchanges within the broader social context. These related yet distinguishable definitions bring confusion to drug market research because "the absence of a unifying definition is important—one cannot presume to know what a particular discipline or researcher is referring to when the term 'drug market' is used" (Ritter 2006, p. 460).

Understanding how researchers conceptualize drug markets is important because it has distinct implications as to how drug markets are operationalized as a variable and thus how results are to be interpreted. Perhaps the primary reason that researchers have operationalized drug markets in so many different ways is because—as illustrated above—not only does the disciplinary paradigm influence one's perception of drug markets, but within disciplines the drug market concept is continually modified to address innovative research inquiries.

Within criminology, drug markets have been conceptualized in a number of ways. Eck (1995) alludes to the systemic nature of drug markets, arguing that they are adaptations to the unique situations buyers and sellers encounter while making transactions. These include: avoiding police attention, the inability of relying on the police to settle disputes, and the need to conduct transactions in secure places. He goes on to provide a classification of drug markets. What he calls 'social network markets' are those in which buyers only purchase from screened sellers or those recommended by a mutual acquaintance. This technique provides security because buyers and sellers know one another and can be confident that the other party is not a representative of law enforcement. From a policy perspective, social network markets are hard to detect because they usually serve few buyers and are less likely to be located on street corners, but rather on private premises away from the surveillance of police. Conversely Eck's 'routine activity markets' are open to a larger number of customers and generally are found in public places. They are usually found along major arterial routes and demonstrate high place attachment. Such markets usually afford less protection from enforcement interdiction due to their more open nature (Eck 1995).

May and Hough's (2004) 'open markets' are similar to Eck's (1995) routine activity markets in that they are 'open' (i.e. available) to all buyers willing to purchase a product. Sellers are able to maximize their customers' access by being in the same location on a regular basis, making them sensitive to the spatial pattern of demand. However, open markets make buyers and sellers vulnerable to policing because they are more likely to be located outdoors. Because of this, certain areas (such as the Kings Cross commercial district in London) afford buyers and sellers protection due to the high population concentration and the difficulty in determining who is in the area for legitimate or illegitimate reasons (May and Hough 2004). In this way drug markets often sprout up near transportation hubs.

'Closed markets' are similar to social network markets because both parties know one another (May and Hough 2004). Sellers only sell to people they trust or to those who are vouched for through third parties. The risk of police apprehension is less than that of open markets, and buyers like closed markets due to the stability of supply, quality of drugs purchased, and trust between themselves and sellers. Furthermore, technology has changed the face of drug markets making them more flexible and convenient. Buyers and sellers can make appointments to meet at specified places using 'Pay as you go' phones that don't retain a record of the user's home address (Beckett et al. 2006).

Drug markets can also be conceptualized within an economic framework in that they involve exchanges where sellers position themselves closer to communities with more potential customers (geographic perspective) or closer to more disorganized communities (social disorganization perspective) (Robinson and Rengert 2006). It is therefore clear that how drug markets are conceptualized from a theoretical or disciplinary framework has the potential to influence the choice of data used to identify a market, and the analytical regime employed. For example, closed or social network markets are more likely to utilize qualitative data (Warner and Coomer 2003) or social network links to identify central nodes, whereas open markets may be better identified through official data, such as arrest records or relevant calls for service.

Because of the variation in conceptualization, the operationalization of drug markets within criminal justice is just as varied. By 'operationalization' we mean the manner by which social phenomena are measured for inclusion in empirical analysis (Blalock 1971). Unfortunately while features such as neighborhood socioeconomic status have relatively standard constructions (commonly measured by creating an index of educational attainment, household income, and poverty status), no such mechanism exists for drug markets. The myriad variants of official data, while adding richness and texture to the picture of criminality, do not immediately suggest a uniform approach. Oakland's Beat Health Program identified drug market locations using emergency call data and contacts from community organizations at the spatial scale of the street block level (Mazerolle et al. 1998, 2004). Evaluations of the program took place at the street block level using surveys of place managers and on-site observations of social and physical disorder (Mazerolle et al. 2004). In the Jersey City Experiment, drug markets were defined as hot spots of crime by mapping narcotic sale arrests and drug-related emergency calls for service to street intersection areas (Weisburd and Green 1995). Researchers defined the hot spot areas by seeking input from narcotics detectives and then created market boundaries based on those data.

Sophisticated continuous surface mapping techniques such as kernel density estimation (KDE) have been used to define drug markets using arrest data (Lum 2008). Mapping techniques effectively view a drug market as a hotspot of individual crime events and seek to identify these crime hotspots. Hotspots can be individual locations among groups of victims, streets, or areas (Eck et al. 2005). Chainey and colleagues argued that an approach using KDE provided the best opportunity to create statistically robust crime hotspot maps (Chainey et al. 2003, p. 29).

Using KDE, the locations of geocoded crime incidents are overlaid with a fine grid of digital cells. After that, a three-dimensional function of a specified radius is superimposed over the center of each cell and an inverse distance weight is calculated for each point within the kernel's radius. Points closer to the outer edge of the radius have a lower scaling factor. In most cases the value for each crime event is one; therefore the weighting becomes the *de facto* value for the crime event for that particular cell calculation.[1] The final cell value comprises the summed value of all points within the radius of the superimposed function scaled against their individual inverse distance weighting (Chainey and Ratcliffe 2005; Eck et al. 2005; Ratcliffe and McCullagh 1999a). The end result is a continuous surface of fine grid cells that cover the study area, each cell containing a value indicative of the intensity of crime around the location. In this case, the term 'around' is important because the cell size is often selected to be smaller than the radius of the kernel function to ensure a continuous surface where every point is included in the study. A cell size of 0.8r (radius) was proposed over a decade ago as a reasonable scaling to ensure overlap (Ratcliffe and McCullagh 1999b) but this is largely a decision of the analyst. Cartographically, extreme clustering is usually indicated by bright eye-catching colors such as red or yellow while dispersion is indicated by less vibrant colors, such that the eye is drawn to the crime hotspots.

Drug markets pose a particular challenge to law enforcement, but they are not everywhere across our urban areas. A considerable body of research has focused on trying to explain why some areas have more drug-related incidents than others. Taniguchi et al. (2009) set out to determine whether drug markets demonstrate agglomeration effects similar to legitimate business firms. In other words, drug markets were viewed as businesses that cluster in space to appreciate benefits (intentionally or unintentionally) that would not be realized if they were more spatially dispersed. In Taniguchi and colleagues' research drug markets were operationalized as a count of drug sale arrests in each of Philadelphia's 1,816 block groups. Agglomeration was identified using a spatial lag variable to determine whether block groups with high drug sale arrests cluster near others with high drug sale arrests. They found that agglomeration was predictive of higher drug sale arrests within block groups, controlling for local demand, social disorganization, concentrated disadvantage, and land use correlates.

Other research has aggregated counts of drug sale arrests to block groups to explain changes over time (Robinson 2008; Robinson and Rengert 2006) and has examined the influence of social disorganization and criminal opportunity on the locations of drug markets by aggregating drug arrest counts to block groups (McCord and Ratcliffe 2007). It appears that single female-headed households, low educational attainment, and the percentage of minorities are positively related to drug arrest counts, while (surprisingly) the percentage of renter-occupied households and male unemployment are negatively related to drug arrest counts (McCord and Ratcliffe 2007).

[1] However while it is usual to count a crime incident as 1.0 for the purposes of scaling with the weighting factor, it is possible to create crime hotspot maps based on other characteristics of the crime, such as the value of property stolen. The inverse distance weight in this case would be scaled against the value of the good stolen to create a map showing hotspots of property value lost.

Alterative spatial units such as census tracts have also been used to outline drug markets. Rengert et al. (2000) argued that drug markets are sites where exchanges of drugs take place but their location depends on whether customers are local or regional. Retail marketing would suggest that dealers would want to locate near suspected drug-using populations. Research shows that such populations tend to be young, without a high school diploma, and unemployed. Using census data and operationalizing drug markets as the number of drug sale arrests per square mile of each census tract, they found that local drug markets tend to locate near tracts with the greatest proportions of young members, unemployed, with less than a high school education while regional markets tend to locate near highway on/off ramps.

Recent research by Ratcliffe and Taniguchi (2009) and Taniguchi et al. (2011) took an innovative spatial approach to drug market identification. In their research, a drug markets were conceptualized as street corners where known gang members were witnessed selling drugs, the street corner being an original unit of spatial analysis for this context. To investigate whether drug corners controlled by gangs are more violent than those not controlled by gangs, the researchers constructed Theissen polygons around each corner in the city of Camden, New Jersey. The counts of drug incidents within polygons became the operationalization of drug markets.

Some have questioned how well arrest data are indicators of real drug activity. Warner and Coomer (2003) used a regression model that incorporated resident self-report surveys on the witnessing of drug activity as a predictor of official drug arrest data across 66 neighborhoods (block groups). They found that the survey measure was significant and positively associated with drug trafficking arrests, but demonstrated no real relationship with drug possession arrests. This suggests not only that respondents were able to distinguish trafficking from possession, but possibly that drug trafficking is a more visible and readily identifiable drug crime than possession. Other research at the city level has also supported the use of official drug arrest data, with construct validity tests indicating strong relationships with public health data (Rosenfeld and Decker 1999). Finally, Rengert et al. (2005) found that drug-related calls for service data co-varied with illegal drug arrests.

In line with past research, we also use drug sale incident data to identify drug market areas. Arrests for drug selling are more likely than those for drug possession to take place within or near a drug market, given that ethnographic evidence suggests that outdoor drug sellers are territorial in nature (St. Jean 2007). A drug possession incident could simply indicate that an arrestee possessed—but not necessarily purchased—drugs. It is possible that an officer didn't witness the arrestee buy drugs, but that after a search due to some other suspicious activity the officer found drugs on the arrestee's person. Such an arrest *could* happen at or near a drug market area, but it could also happen anywhere else, making it more theoretically difficult to spatially link a drug possession arrest with that of a drug market. A review of the literature indicates that research has relied heavily on aggregation of official police data to census features to operationalize drug markets. The following sections illustrate the problems with this approach and provide an alternative measure that deals with some of the threat to construct validity evidenced by past research.

2.2.1 What Methodologies are Available?

In this work, we seek to identify drug market areas and non-drug market areas, attempting to draw a boundary around the former in order to distinguish the market area. For the purpose of this illustration drug markets are required to be clearly delineated from other areas by evidence of statistically significant spatial concentrations of drug sale incidents. Furthermore we are interested in areas, rather than necessarily single drug sale locations. Therefore this research considers drug *markets*, areas where a drug user could expect to find more than one drug sale operation and could hope to encounter a drug seller within a general area. We also seek a definitive spatial ordering (drug market area or not), rather than a fuzzy classification. This conceptualization of drug markets therefore requires a binary spatial classification technique.

Often, kernel density estimation would be considered under these circumstances. While the technique is aesthetically appealing and commonly used among the crime analysis community, it does have some limitations. The use of a smoothing algorithm generates a smoothed display that may not indicate sudden changes in crime distribution. Furthermore, the technique often counts a single crime event into the calculation for multiple grid cells, limiting the range of statistical tests that can be applied due to the problem of independence. Finally, the technique does not suggest a definitive cut-off point whereby an analyst can determine whether a location is inside or outside a drug market. Instead, the surface map indicates an intensity value on a continuous range, and any cut-off point would be an arbitrary selection of the crime analyst.[2] This arbitrary limit can be highly subjective given that, due to the smoothing algorithm, cells just outside the drug market area would have KDE values close to their neighboring cells just inside the area. The approach therefore has limited applicability when trying to determine the fixed boundaries of drug markets in a robust and replicable manner. KDE does, however, have the advantage of indicating gradual declines in value as distance from a hotspot increases, reflecting a fuzzier boundary that is likely to mimic the more mutable nature of many urban drug environments.

Point location techniques such as the mode and fuzzy mode of the CrimeStat software package also fail to meet the needs of this research because they merely total the number of events taking place at or within a certain distance of user-determined features such as transit nodes or other point locations (Levine 2004). These techniques would be appropriate for assessing the count of robberies in and around a specified set of alcohol-serving establishments but not for determining statistically significant concentrations of drug selling incidents. The inability to cover the whole study area is a considerable limitation.

Partitioning cluster tools typically assign all events to a user-defined number of clusters, with each point belonging to one cluster exclusively (Levine 2004).

---

[2] Some analysts determine the dimensions of a crime hotspot simply on the basis of a change in color in the choropleth map classification system. Given the ease with which these can be manipulated with modern software, this approach is most definitely not recommended if a more robust statistical analysis is desired.

However, such techniques are likely to identify multiple clusters that are not necessarily hot spots simply because of the constraint to assign *every* point to a cluster. Also problematic is the requirement for a user-defined *a priori* estimate of the number of likely clusters. These conditions make the results arbitrary and overly subject to user specifications. Additionally, partition cluster approaches have the ability to mask variation within hot spots potentially leading to significant data reduction (Grubesic 2006). This is inadequate for the current research which seeks to discriminate significant clusters of drug arrests from those that are dispersed across space.

Similar to partitioning cluster tools is the aggregation of crime incidents to census features. Like partition clustering, every crime incident is allocated to a particular enumeration feature (e.g. block, block group, or tract). The strength of this technique is that it allows the researcher or analyst to easily blend census data with crime counts, and to measure the extent of spatial autocorrelation (Taniguchi et al. 2009). On the other hand, aggregation draws concerns of the modifiable areal unit problem (MAUP), in that the results of aggregation techniques will vary according to the number of areal units and the nesting of smaller areal units into larger ones (Openshaw and Taylor 1979). While the use of census features is convenient, they are designed for census collection and not drug market delineation. When block groups and tracts are used as units of analysis, individual areas in their entirety have the potential to take on the drug market characterization. This is unlikely to be an accurate representation of drug markets. Environmental criminology has revealed that hot spots of crime are not hot all the time, and that cold spots may be temporally intertwined within hot spots (Eck et al. 2005). High drug crime block groups or tracts may not have uniformly high drug activity within their units of analysis.

Grubesic (2006) notes that partition clustering essentially ignores the potential presence of spatial outliers; however, this is a problem that applies to aggregation techniques as well. The inclusion of *every* point (including outliers) compromises the validity of partition clustering and renders the use of areal aggregation as nothing more than a total count of crime incidents within areal features. According to Grubesic and Murray (2001) "Broadly defined, cluster analysis is a method of classification that places objects in groups based on the characteristics they possess" (p. 5).

In the section that follows, we adopt a spatial variant of a hierarchical clustering approach. Hierarchical clustering techniques have existed for many decades (Johnson 1967; Sneath 1957; Ward 1963) in both parametric and non-parametric forms (D'Andrade 1978). Hierarchical clustering routines begin by calculating some measure of dissimilarity (commonly distance) between each point and all other points in a population (Bailey and Gatrell 1995). An average of the nearest distance among all points is typically computed and used as a threshold for grouping points across space (random nearest neighbor distance). Two or more points that have a distance less than the nearest neighbor distance are then grouped into a series of first-order clusters.[3] Second-order clusters are created by repeating the same

[3] Other measurements can be used to define the threshold such as the minimum distance and maximum distance. See D'Andrade (1978).

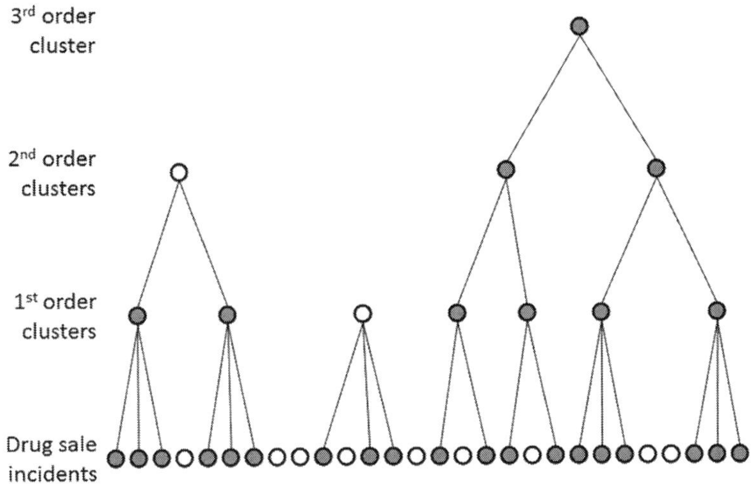

Fig. 2.1 Hierarchical clustering technique (Adapted from Levine 2004)

process on the primary clusters that were generated. According to Levine (2004) second-order clusters are grouped into larger ones, and this process is repeated until no additional clusters can be identified. Visually, hierarchical clustering resembles an inverted tree diagram (Fig. 2.1), yet it is important to note that not all points are grouped into a cluster (Levine 2004). In the diagram, drug incidents that are not part of a higher cluster are shown as white, while incidents shown in grey are part of a higher order cluster. This useful feature of hierarchical clustering will be relevant later, preventing crime events outside of drug markets to be artificially drawn into a drug market area by virtue of the analytical technique.

2.3 A Two-Stage Clustering Approach

Prior research has therefore shown that not only are there varied definitions of drug markets but that operational definitions can suffer from validity concerns. A review of the literature indicates that research has relied heavily on aggregation of official police data to census features to operationalize drug markets. The purpose of the current section is to suggest an alternative measure that deals with some of the threats to construct validity evidenced by past research. We demonstrate this with an example case study using 5 years of drug sale incident data from the Philadelphia (PA) Police Department.

Recent research has suggested that using more refined, spatially sensitive measures of drug markets may better our understanding of the extent to which they coexist in violent areas (Lum 2011); therefore, we identify drug market areas and non-drug market areas, attempting to draw a boundary around the former in order to distinguish the market area using a two-stage nearest neighbor hierarchical clustering approach. The aim of this is to clearly delineate drug markets from other areas by evidence of statistically significant spatial concentrations of drug sale incidents.

Nearest neighbor hierarchical clustering (Nnh)[4] is a technique available within the CrimeStat spatial statistics software package used to outline the clustering of point data (Levine 2004). Similar to the general description of hierarchical clustering above, CrimeStat's Nnh technique uses a threshold distance to group points into a cluster. Two main criteria guide the clustering process. The first criterion is the selection of a threshold distance. Nnh allows users to select their own fixed distance threshold, or to use the random nearest neighbor distance for first-order clusters. The random nearest neighbor distance is defined as:

$$d(ran) = 0.5\sqrt{\left(\frac{A}{N}\right)}$$

where A is the size of the study area (defined by the user) and N the number of spatial events (Levine 2004).

The threshold distance is determined by selecting the appropriate one tailed confidence interval around the random nearest neighbor distance. Therefore, the confidence interval for the random nearest neighbor distance equals the random nearest neighbor distance plus or minus the standard error of the mean random nearest neighbor distance:

$$0.5\sqrt{\left(\frac{A}{N}\right)} \pm t\left(\frac{0.26136}{\sqrt{\frac{N^2}{A}}}\right)$$

where A is the size of the study area, N is the number of spatial events, t is the Student t-value for a given probability level, and 0.26136 is a constant. Here, the confidence interval is used to determine the probability that the distance between any pair of events would be less than the random nearest neighbor distance (assuming the data are randomly distributed across space). In other words, if the data are *randomly distributed* and a user-selected significance is at $p < .05$ then about 5% of the pairs of events could be expected to be closer than the random nearest neighbor distance (for more details, see Levine 2004).

The second criterion is the selection of a minimum number of points necessary to create a cluster. This criterion is necessary to reduce the number of very small clusters that would otherwise be created by chance. A large dataset can result in

[4] We use the abbreviation Nnh throughout the chapter for continuity with the CrimeStat manual.

many clusters if the only requirement is to have points within a specified distance of one another. As a result of this criterion, points will only be clustered if the distance between them is less than the set threshold *and* if the number of points in the cluster is greater than or equal to the minimum set by the user (Levine 2004). At present, the choice of a threshold is an arbitrary researcher-driven decision.

For the purposes of this study Nnh was used to identify significant clusters of drug activity using the CrimeStat 3.0 software package. A Nnh analysis was run on the drug sale incident data using the following parameters. The land area value used to calculate the random nearest neighbor distance was 135 square miles. A 95% confidence interval was used to determine the probability that the distance between any pair of events would be less than the random nearest neighbor distance (assuming the data are randomly distributed across space). Ten (10) was chosen as the minimum number of points necessary to create a cluster (the default option).

Before running the Nnh analysis the user can choose to map the resultant spatial clusters as convex hulls or ellipses. The problems with standard deviational ellipses that rarely mimic the geography of the underlying crime events have been known for some time (Ratcliffe and McCullagh 1999b); therefore, our results are mapped as convex hull polygons. Grubesic (2006) makes three observations with regard to convex hull polygons; convex hulls are constructs of the smallest polygons necessary to bound a set of clustered points, the use of convex hulls minimizes the area necessary to group a set of clustered points, and convex hull polygons lend themselves to crime density calculations. As a result, the conservative construction of clusters using convex hulls more accurately represents real crime clusters and demonstrates better construct validity when compared to hot spots constructed using areal features. A technique to blend multiple spatial hierarchies of convex hull polygons is a further advantage, as will be demonstrated below.

2.4 Data and Results

Drug sale incident data were sourced from the Philadelphia Police Department's Incident Transmittal System in March of 2011, covering 5 years (2006–2010). These incident data detail the date of the incident, UCR code, X/Y coordinates of the incident location, and a unique incident number. UCR codes (a code that indicates whether the crime involved buying or selling drugs, and the type of drug involved) were used to extract incidents for the sale of any of the following: opium, marijuana, synthetic/manufactured narcotics, dangerous non-narcotic drugs, powder cocaine, and crack cocaine. A total of 18,299 drug sale incidents occurred from 2006 to 2010 in the city of Philadelphia.

Nnh analysis revealed 329 first-order, 34 second-order, and 2 third-order clusters, shown in Fig. 2.2. For purposes of clarity, the map on the left side of the figure shows the first- and third-order clusters. First-order clusters appear as small specks or dots on the map (Fig. 2.2a). Third-order clusters appear to resemble large drug market regions (Fig. 2.2a). They are located close to one another in North

Fig. 2.2 1st and 3rd order clusters of drug sales incidents (*left*), 2nd order clusters (*right*)

Philadelphia, in an area historically known as 'The Badlands'. The map at (b) on Fig. 2.2 displays the locations of the 34 second-order clusters (revealed by the Nnh analysis) which are concentrated throughout North and West Philadelphia.

There were a number of operational issues with both first- and second-order spatial clusters. First-order clusters, when examined for a city such as Philadelphia, identify a large number of very small areas often concentrated around individual street intersections. Even though they were within the random nearest neighbor distance, and constituted sufficient offenses to overcome the minimum number of points threshold, these first-order areas were more representative of drug selling corners (which can also be hot spots of drug activity). Table 2.1 provides descriptive statistics for each of the cluster-orders produced from the hierarchical clustering process. First-order clusters possess a median area of about 5,000 square feet, indicating that these clusters are generally very small and cover an area resembling the size of a street intersection. The minimum area value of zero is the result of some first-order clusters resembling "problem" addresses that have had multiple calls for service for drug activity. For example, over the 5-year study period, 61 separate incidents for the sale of illegal drugs occurred near the corner of N 61st Street and W Thompson Street in West Philadelphia.

A total of 34 second-order clusters were produced by the Nnh analysis, and were substantially larger than those within the first-order. Second-order clusters had a median area of slightly over 730,000 square feet, or about 5–6 city blocks. Second-order clusters collated various first-order clusters into more cohesive units, but did so based on the centroid of the first-order clusters. Depending on the orientation of the first- and second-order spatial units, a considerable number of crime events

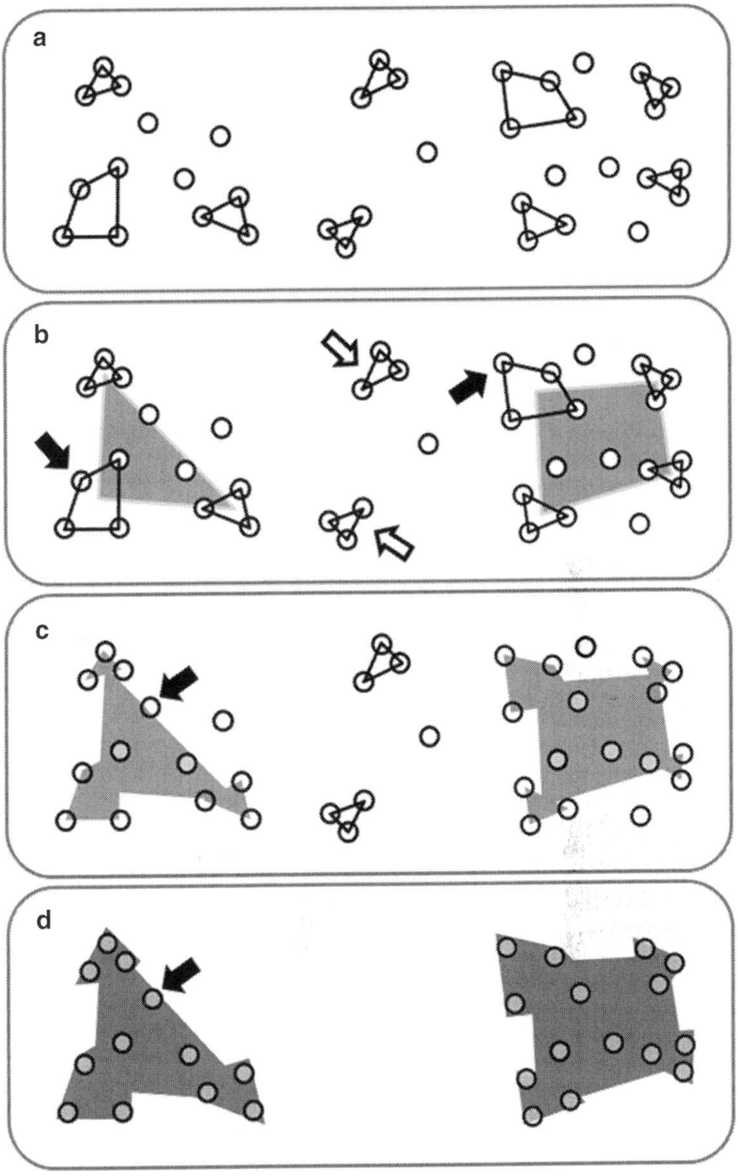

Fig. 2.3 Modification of nearest neighbor hierarchical clustering process

within first-order clusters were excluded from second-order clusters, *even* when they were members of a first-order cluster that was part of the second ordering. This problem is demonstrated in Fig. 2.3, where (a) shows a pattern of crime events as circles, among which are a number of first-order cluster areas, indicated by black angular shapes linking certain points. The grey area in (b) shows the second-order

Table 2.1 Descriptive statistics of hierarchical clusters

Clusters	n	Mdn. area	Min.	Max.	Mean area	St. Dev. area
1st order	329	4,775	0	63,612	10,239	12,614
2nd order	34	730,452	33,289	3,400,609	1,015,249	801,344
Merged	34	757,923	39,673	3,447,044	1,045,814	807,534
Buffer	34	1,896,528	450,016	4,486,090	2,153,066	972,246
3rd order	2	13,491,178	9,514,580	17,467,776	13,491,178	5,623,759
Phila. BG	1,816	1,022,856	149,391	95,735,799	2,191,157	5,313,473

Note: Values are in square feet

hierarchical spatial relationship between a number of the first-order clusters, based on the centroid of the first-order areas. Two hollow arrows indicate a couple of areas that were first-order spatial clusters but were too remote from other areas to be included in a second-order cluster. These isolated first-order clusters were not considered further.

The black arrows in (b) indicate two problematic areas, from the perspective of drug market identification. These are crime event clusters that were included in the first-order clusters. Centroids of first-order clusters are used by the Nnh analysis to construct the shape of the second-order clusters. Because of this problem, some events can be inadvertently excluded. For example, if it was determined to select all of the points within a second order cluster, the crime events identified with the black arrows would not be included in the identified areas, even though they were part of a first order cluster that contributed to the second order spatial pattern.

The solution we employed was to spatially merge the first- and second-order convex hull areas so that points that were contained within an entire first order cluster were not lost when the second-order hierarchy was applied. In other words, a convex hull was created for each level 1 cluster, and then these were spatially joined to the relevant level 2 cluster convex hull. This is demonstrated in part (c) of Fig. 2.3.

One issue to consider at this stage is shown by the arrow in Fig. 2.3c. The arrow indicates a crime event that is very close to the drug market area but which is just outside the formal spatial polygon. Therefore for both purposes of visual clarity, and to avoid any potential ambiguities, the final combined first- and second-order polygon areas have had a small buffer applied. This process allows the drug market area to retain all of the original first-order events that clustered to create the crime hotspot area in the first place, as well as any points right on the edge of the drug market area. This final stage is shown in Fig. 2.3d, where the arrow now shows the drug incident included in the drug market area.

Descriptive statistics of merged first- and second-order clusters (Table 2.1) indicate that the median size of the merged clusters is not substantially different from the second-order clusters. This is because the first-order clusters generally covered very small areas and contributed very little additional area to the second-order clusters, once they were merged. Conversely, the two third-order clusters appear to denote large drug regions of the city. The smallest of such clusters has an area of over 9.5 million square feet or .34 mi^2, and the largest is .63 mi^2.

2.4.1 Analytical Considerations

2.4.1.1 Rigidity of Convex Hulls

According to Levine (2004) the advantage of using convex hulls is that they reflect an outline of clustered points, but the disadvantage is that they may exclude areas that should be included in the hot spot. Furthermore, there could be incidents related to the drug market but outside of its rigid boundaries. Arguably, this is not a concern from a strictly spatial statistics perspective; however, a little latitude in the boundary can alleviate concerns of geocoding accuracy as well as other anxieties of accuracy in relation to micro-geography. We addressed this problem by creating a small buffer around each cluster to ensure that we captured drug incidents associated with the drug market, as well as provided for the chance to include points very close to the drug market cluster but originally excluded by the convex hull. If the analyst chooses to take this route, the selection of an appropriate buffer size requires the user to balance the need to include points within a theoretically relevant distance of the clusters while not making buffer areas so large that they overlap one another or become so large as to be unrealistic indications of the drug market extent. This factor was noticeable in the North Philadelphia – Badlands area where some of the clusters are within one city block of another (Fig. 2.2).

Similar to the approach taken in Fig. 2.3d, a 200 ft buffer[5] was applied to the merged Philadelphia clusters[6] to provide a chance to capture additional incidents that may be just outside the boundaries of the drug market, but nonetheless most likely related. Figure 2.4 shows a map of Philadelphia level 2 cluster drug markets including a 200 ft buffer. The dashed rectangle indicates the inset region displayed on the right of Fig. 2.4. The shaded polygons indicate areas where two markets overlap one another. Statistically speaking, overlapping areas violate the assumption of independent observations. This may not be a problem for crime analysts, but it does cause problems for researchers looking to take an analysis to the next level. A solution to this issue, in line with the research by Haberman et al. (in press), involves measuring the distance from the centroid of each overlapping area to that of its respective drug markets. Each overlapping area is then merged with its closest drug market. This is done in Fig. 2.4.

---

[5] The median length of a street segment in Philadelphia is 277 ft. Using a buffer a little under this size would theoretically mean that any incident within 200 ft of a drug market's boundary is attributable to the dynamics of that drug market's immediate block.

[6] Buffers were applied to the merged clusters for visual clarity. The decision to apply a buffer as well as considerations of which order clusters they should be applied should be guided by practicality and the theoretical interests of the research at hand.

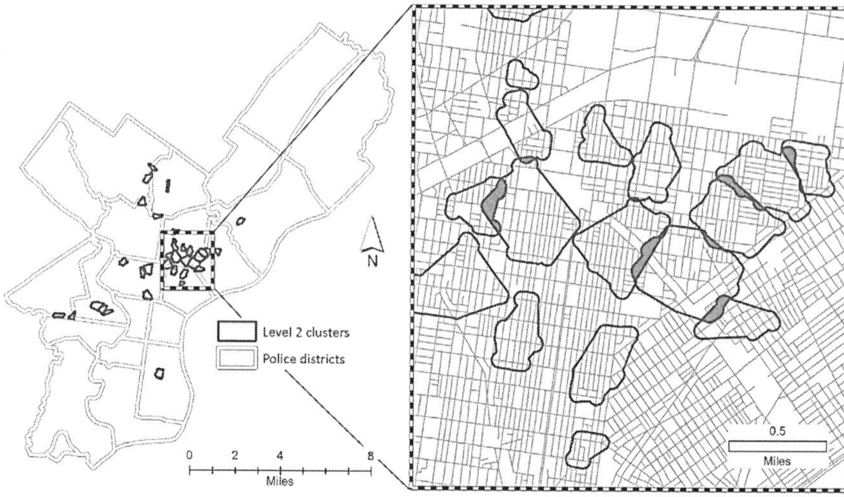

Fig. 2.4 Drug markets with 200 ft buffer

2.4.1.2 Statistical Significance

The second consideration with the Nnh analysis concerns how to determine statistical significance of the clusters. The test to determine whether two points are clustered by chance is traditionally the confidence interval around the random nearest neighbor distance. If the probability level is set to .05 then one could expect that 5% of all point pairs are grouped by chance. However, Nnh groups more than just two points within a cluster, depending on the minimum number of points set by the user. A Monte Carlo simulation can address this problem by constructing confidence intervals based on the first-order clusters. It is essentially a test of the null hypothesis that the data are *not* spatially clustered. This occurs by randomly assigning the same number of cases to a rectangle containing the same area as the study region. For each simulation (the total number of simulations is determined by the user) the number of expected clusters created is calculated and reviewed against the observed number.

The process is demonstrated here for Philadelphia. A Monte Carlo simulation was run on the data, using 1,000 iterations. An equivalent number (to the 18,299 drug sale incidents) of randomly assigned points were thrown on a plane with an area matching the square footage of Philadelphia. The same nearest-neighbor distance and cluster minimum frequency (10) was applied to the simulation data and tested for the presence of clusters. This random allocation and test process was repeated 1,000 times. Across all iterations of spatial randomization, zero clusters were produced, suggesting that it is extremely unlikely that any of the first- and second-order clusters actually observed were produced by chance ($p < 0.001$).

Table 2.2 Sensitivity analysis

Min. # of points	1st-order clusters	% change	Clusters obtained by chance (95th percentile)
8	420	–	1
9	375	−10.71	1
10	329	−12.27	0
11	307	−6.69	0
12	266	−13.36	0

2.4.1.3 Arbitrariness of Parameter Selections

The third consideration is that the selection of a minimum number of points for cluster determination does render Nnh slightly arbitrary (Grubesic and Murray 2001) and research has yet to empirically guide the selection of a suitable minimum number. The purpose of a choice such as 10 (the default CrimeStat option) is to minimize the number of extremely small clusters (Levine 2004). In other words, the aim is for the cluster detection to identify small neighborhood clusters rather than collections of a couple of points at a street corner; however, it is recognized that the requirement to pre-determine a minimum cluster density is a limitation of the process.

To determine the sensitivity of this option, the analysis was re-run using 8, 9, 11, and 12 points as the minimum. There appears to be an inverse relationship between the minimum number of points selection and the number of first-order clusters produced (see Table 2.2), which substantiates (personal) communication with Levine (2010). Increasing the minimum number of points necessary to form a cluster from 9 to 10, decreases the number of first-order clusters produced by 12%. Using 11 points instead of 10 points decreases the number of clusters by a further 7%. Monte Carlo simulations were run for each additional analysis using 1,000 iterations to determine the number of clusters that would be obtained by chance, assuming a random spatial distribution. Results indicate that when setting the threshold to eight or nine points, using 18,299 incidents it is possible that one cluster (or in this case drug market) is a statistical artifact. Selecting any number of points greater than nine reduces the chance of statistically artificial clusters to essentially zero. This suggests that clusters revealed by setting the minimum number of points to ten are robust clusters of drug sales activity. Therefore, the appropriate minimum number of points (for the Philadelphia data at least) is any option of at least ten points, noting that there is an inverse relationship between the minimum number of points and the number of first-order clusters that will be identified.

Returning to the issue of validity, we investigated whether census features or clusters from the Nnh analysis produce a more valid measure of drug markets. Specifically for the purpose of this illustration we compared the drug sale crime densities of block groups to the buffered clusters. Considering that the Nnh clusters delineate the hottest clusters of drug sale incidents they were only compared to

Table 2.3 Density measures of cluster analyses

	Median	Min.	Max.	Mean	St. Dev.
Merged	2,416.88	1,379.56	8,796.87	3,113.07	1,987.26
Phila. BG	831.77	0	12,455.47	1,394.13	1,833.06

Note: Values reflect incidents per square mile

census block groups they encompass or intersect (n=221). Drug sale crime density was calculated by dividing the total count of incidents within each spatial unit feature by the respective feature area. Higher density values indicate a larger number of incidents per square mile than lower values.

Table 2.3 displays descriptive statistics of drug sale density calculations for block groups and the merged 1st and 2nd order clusters (including the 200 ft buffer). The median density value for the merged Nnh drug market cluster is 2,417 drug sale incidents per square mile, while the median for comparable block groups is 832 drug incidents per square mile. The density of drug incidents within Nnh clusters is about three times greater than that of comparable block groups indicating that they are indicators of more highly concentrated areas of drug crime than areal aggregation to census boundaries. Only three block groups had density values greater than the maximum Nnh density value of 8,797 per square mile. A visual inspection of those block groups indicated that two are entirely encompassed by Nnh clusters, and about 95% of the third block group coincided with a Nnh cluster

2.5 Discussion

We have proposed a new method to spatially operationalize drug markets. Much of the past research on drug markets has accomplished this by aggregating counts of drug sale data to census features. Although this allows researchers to conveniently append census data, such an operationalization has the potential to be over-inclusive and runs the risk of labeling large tracts as drug markets when the reality is quite different. This becomes apparent when we consider research indicating that neighborhood drug problems can stem from places as geographically small as one address (Mazerolle et al. 2004). Using official drug crime data from the Philadelphia Police Department, we have shown that the nearest neighbor hierarchical clustering technique addresses this limitation by identifying significant clusters of drug incidents at multiple geographic levels ranging from street intersections to areas of over half of a square mile.

Although our example of the two-stage nearest neighbor hierarchical clustering technique used drug sale incident data, the process can be applied to other crime categories, such as violence. Perhaps the most detrimental aspect of urban drug markets is the violence by which they are often characterized. Concentrated disadvantage appears to be strongly related to drug market activity, with drug market activity in turn having a strong causal connection with robbery rates (Berg and Rengifo 2009). Such communities tend to be socially disorganized and unable to regulate drug crime and the related violence that it engenders (Berg and Rengifo 2009). However, even controlling for sociodemographic factors such as instability, heterogeneity, and deprivation, drug activity still has a significant positive effect on assault and robbery rates (Martínez et al. 2008). Ousey and Lee (2002) found that increases in drug arrest rates were positively related to homicide rates; however, that relationship is contingent on the preexisting level of resource deprivation.

While studies such as these contribute to our understanding of violence, they are subject to how drug markets are operationalized and conceptualized. A positive relationship between neighborhood drug arrest rates and violence within census tracts is operationally different from empirically derived statistically significant clusters of drug sale arrests and the violence that occurs within. The former examines community correlation between drug arrest rates and violence while the latter focuses on the significance of the most problematic drug crime areas (hot spots) and how they engender violence.

The use of two-stage nearest neighbor hierarchical clustering presents a new way for studying drug markets, while addressing issues of the MAUP, over-inclusion, and construct validity. The crime density comparison of Nnh clusters and the aggregation technique indicated that Nnh was about three times better at aggregating incidents and identifying hot spots (Table 2.3). Nnh may be preferable for researchers and analysts when the desire is to visually outline areas of highest drug sales activity.

Although clustering techniques such as the one described here are fundamentally descriptive in nature and often the starting point for an analysis rather than an end point, hierarchical clustering could provide crime analysts with a more accurate method to inform the targeting of police resources across multiple levels of police organization. For example, Fig. 2.1 suggests that 1st order clusters which tend to outline street intersections or problem addresses may be of interest to beat officers who routinely patrol assigned areas and need to be aware of potential threats. District commanders may see more value in larger, 2nd order clusters that outline larger areas of drug activity for the direction of additional officers, or for planning neighborhood level community oriented policing or situational crime prevention efforts. Such information could aid in the planning of police crackdowns and the targeting of specific drug sellers or selling organizations that may be the cause of violence in the community. These areas are also more conducive to community organization.

Finally, 3rd order clusters may be of interest to police executives and federal agencies that plan for the assignment of personnel at scales larger than local police districts. Although targeted police efforts to hotspots may cause some displacement, interdiction of the most spatially advantageous sites may displace sellers to less advantageous locations in turn making the market less profitable (Robinson and Rengert 2006).

The technique shown here therefore has the capacity to indicate at least three levels of drug market organization that roughly conform to commonly used drug market terminology;

- Level one Nnh spatial clusters equate to *drug places* (such as around an intersection)
- Level two Nnh spatial clusters equate to *drug markets* (groups of blocks with drug problems)
- Level three Nnh spatial clusters equate to *drug regions* (such as the Philadelphia Badlands)

These spatial units have operational benefits to crime science and crime prevention, given that the first two spatial clusters operate at scales that are amenable to local policing and problem-solving approaches, while the third scale indicates a region or area than may need a broader investment to combat the drug problem.

It is recognized that while we have endeavored to bring an objective approach to the spatial identification of urban illicit drug markets, there still remains the issue of the arbitrary minimum number of points required to form a level one cluster. We hope that additional research will identify some guidelines that can inform future decisions. In the meantime, we recommend that analysts using this approach clearly publish the selected number in any maps and publications.

A final theoretical note deserves comment. It is possible that the drug markets constructed from the nearest neighbor hierarchical clustering technique are not real, but abstractions of reality. Recent work by Taylor (2010) argues that there are several inconsistencies that hot spot policing researchers must address before hot spot policing can advance to a national policy. Although it is not in the theoretical interest of this work to address the nuances of hot spot policing at the national level, Taylor's point does bring into question the construct validity of hot spots techniques. Analytically and methodologically this work has taken steps to address some of these concerns, and communication the Philadelphia Police Department during the summer of 2009 has confirmed that many of the drug markets outlined in this work are areas of high drug activity. The reader is referred to Taylor's (2010) interesting polemic for more detail.

2.6 Limitations

First, it is worth noting that the use of a 200 ft buffer around second-order polygons is arbitrary. The mere selection of a specified buffer has implications for the size and intensity of the cluster. As we argued earlier, for strictly spatial statistical applications, it may not be necessary, though this would assume that (1) the drug market does not extend beyond the combined convex hull polygons, and (2) the points that are the basis for the convex hull polygons have been accurately geocoded. These are constraints that suggest caution with all but the most carefully generated data sets. With regard to buffers around crime points (and not drug markets *per se*), Guerette (2009) argues that large buffers may encompass too much data and exaggerate statistical relationships, while small buffers may fail to capture incidents theoretically associated with the location in question. These observations are equally applicable to our modest buffer at the edge of the drug markets.

A further concern with an arbitrary buffer refers to the fact that the current approach ignores the physical geography of the areas surrounding the hot spots. Admittedly buffers that consider the physical geography are ideal; however, this would involve researchers venturing to each drug market to identify physical and perceived barriers. Even physical barriers are hard to estimate from remotely accessed data. For example, while a railway line can often be a barrier, in some parts of Philadelphia residents routinely cross the lines, rendering this physical barrier inconsequential. Perceived barriers are even harder to estimate. The fieldwork to more accurately use the urban mosaic to reflect 'real' boundaries was beyond the scope of the project, and we have had to settle for an arbitrary distance.

A second limitation is that our analytical approach assumes that the drug markets are stable over time. It other words it is believed that the drug markets are not contracting or expanding spatially. The validity of such an assumption is subject to empirical investigation. This concern was mitigated by using 5 years of data to identify ongoing drug problem areas rather than using 1 or 2 years of data that may be as reflective of temporary policing initiatives as long-term chronic drug markets. The advantage of the technique used here is that it allows for the identification of historically stable drug market areas. Nonetheless, drug markets—like other social phenomena—may change over time. Therefore aggregate use of multiple years of data may amount to data reduction and thus mask the extent to which markets contract or expand over time.

One method by which to address the spatial rigidity of drug markets would be to systematically perturb the incident locations (Murray 2003). Perturbing works by selecting a random distance (limited up to a predetermined threshold) *and* direction to move each point within a dataset. Iterations of perturbed locations are produced in separate spatial datasets that can be compared to the locations included in the original dataset by way of the nearest neighbor index. Values less than 1 suggest significant clustering, while values greater than 1 suggest dispersion (Levine 2004). In turn, nearest neighbor hierarchical clustering routines can be run on perturbed datasets to remove the rigidity of market boundaries and account for greater flexibility.

A final limitation with regard to the hierarchical clustering routine is the grouping of incidents with other incidents in order to form a first cluster level. As noted by Grubesic (2006) the weakness of this form of clustering is that this process happens on a pair-by-pair basis. In other words, although two grouped points may be within the threshold distance of one another, each respective point may not satisfy the threshold distance requirement when being compared to other points within the group. This not only draws concerns for the validity of second and third order clusters, but also suggests that hierarchical clustering routines may be more accurate at indicating local versus global clustering. This limitation is acknowledged, though it should be noted that our research is inherently interested in more local scales within a larger study area. Our research here is a first step to a more methodologically robust method, however further research in this area is warranted.

2.7 Conclusion

Perhaps one of the most significant advancements in recent criminology (and crime science specifically) is its geographical focus on unique places through the science of GIS. Although an accomplishment in its own right, spatial criminology is also subject to the problems of construct validity that have been well documented in other social sciences. As the 'cone of resolution' (Brantingham et al. 1976) continues to shrink, aggregations of crime to broad geographic areas simply for analytical convenience will increasingly be drawn into question. Clustering tools that influence the targeting of scarce police resources must be as geographically specific as possible. The use of hierarchical clustering may represent a methodological step in the right direction.

References

Bailey TC, Gatrell AC (1995) Interactive spatial data analysis. Longman Scientific and Technical, London

Beckett K, Nyrop K, Pfingst L (2006) Race, drugs, and policing: understanding disparities in drug delivery arrests. Criminology 44(1):105–137

Berg MT, Rengifo AF (2009) Rethinking community organization and robbery: considering illicit market dynamics. Justice Q 26(2):211–237

Blalock HM (1971) The measurement problem: a gap between the languages of theory and research. In: Blalock HM, Blalock AB (eds) Methodology in social research. McGraw-Hill Book Company, New York, pp 5–27

Brantingham PJ, Dyreson DA, Brantingham PL (1976) Crime seen through a cone of resolution. Am Behav Sci 20(2):261–273

Chainey S, Ratcliffe JH (2005) GIS and crime mapping. Wiley, Chichester

Chainey S, Reid S, Stuart N (2003) When is a hotspot a hotspot? A procedure for creating statistically robust hotspot maps of crime. In: Kidner DB, Higgs G, White SD (eds) Socio-economic applications of geographic information science. Taylor and Francis, London, pp 21–36

D'Andrade RG (1978) U-statistic hierarchical clustering. Psychometrika 43(1):59–67

Denton B, O'Malley P (1999) Gender, trust and business: women drug dealers in the illicit economy. Br J Criminol 39(4):513–530

Eck J (1994) Drug markets and drug places: a case–control study of the spatial structure of illicit drug dealing. University of Maryland, College Park

Eck J (1995) A general model of the geography of illicit retail marketplaces. In: Wiesburd DE, Eck JE (eds) Crime and place, vol 4. Criminal Justice Press, Monsey, pp 67–93

Eck J, Chainey S, Cameron JG, Leitner M, Wilson RE (2005) Mapping crime: understanding hot spots. National Institute of Justice, Washington, DC

Grubesic T (2006) On the application of fuzzy clustering for crime hot spot detection. J Quant Criminol 22(1):77–105

Grubesic T, Murray AT (2001) Detecting hot spots using cluster analysis and GIS. Paper presented at the Crime Mapping Research Conference, Dallas, TX. http://www.tonygrubesic.net/hot_spot.pdf

Guerette RT (2009) Analyzing crime displacement and diffusion. Center for Problem Oriented Policing, Washington, DC

Haberman,C, Groff ER, Taylor RB (in press) The variable impacts of public housing community proximity on nearby street robberies. J Res Crime Delinq

Jean PKB (2007) Pockets of crime: broken windows, collective efficacy, and the criminal point of view. University of Chicago Press, Chicago
Johnson SC (1967) Hierarchical clustering schemes. Psychometrika 32(3):241–254
Levine N (2004) CrimeStat III: a spatial statistics program for the analysis of crime incident locations. National Institute of Justice, Houston
Lum C (2008) The geography of drug activity and violence: analyzing spatial relationships of non-homogenous crime event types. Subst Use Misuse 43(2):179–201
Lum C (2011) Violence, drug markets and racial composition: challenging stereotypes through spatial analysis. Urban Stud 48(13):2715–2732
Martínez R, Rosenfeld R, Mares D (2008) Social disorganization, drug market activity, and neighborhood violent crime. Urban Aff Rev 43(6):846–874
May T, Hough M (2004) Drug markets and distribution systems. [Article]. Addict Res Theory 12(6):549–563
Mazerolle L, Kadleck C, Roehl J (1998) Controlling drug and disorder problems: the role of place managers. Criminology 36(2):371–403
Mazerolle L, Kadleck C, Roehl J (2004) Differential police control at drug-dealing places. Secur J 17(1):61–69
McCord ES, Ratcliffe JH (2007) A micro-spatial analysis of the demographic and criminogenic environment of drug markets in Philadelphia. Aust N Z J Criminol 40(1):43–63
Murji K (2007) Hierarchies, markets and networks: ethnicity/race and drug distribution. [Article]. J Drug Issue 37(4):781–804
Murphy S, Arroyo K (2000) Women as judicious consumers of drug markets. In: Clarke R (ed) Crime prevention studies, vol 11. Criminal Justice Press, Monsey, pp 101–120
Murray AT (2003) Site placement uncertainty in locational analysis. Comput Environ Urban Syst 27:205–221
Openshaw S, Taylor PJ (1979) A million of so correlation coefficients: three experiments on the modifiable areal unit problem. In: Wrigley N (ed) Statistical applications for the spatial sciences. Pion, London, pp 127–144
Ousey GC, Lee MR (2002) Examining the conditional nature of the illicit drug market-homicide relationship: a partial test of the theory of contingent causation. Criminology 40(1):73–102
Ratcliffe JH, McCullagh M (1999a) Burglary, victimisation, and social deprivation. Crime Prev Communit Saf Int J 1(2):37–46
Ratcliffe JH, McCullagh M (1999b) Hotbeds of crime and the search for spatial accuracy. Geogr Syst 1(4):385–398
Ratcliffe JH, Taniguchi TA (2009) Is crime higher around drug-gang street corners? Two spatial approaches to the relationship between gang set spaces and local crime levels. Crime Pattern Anal 1(1):23–46
Rengert GF, Chakravorty S, Bole T, Henderson K (2000) A geographic analysis of illegal drug markets. J Crime Prev Stud 11:219–239
Rengert GF, Ratcliffe JH, Chakravorty S (2005) Policing illegal drug markets: geographic approaches to crime reduction. Criminal Justice Press, Monsey
Reuter P (2000) The measurement of local drug markets. Paper presented at the National Institute of Justice Conference on Drug Markets, Washington, DC, 14 February
Ritter A (2006) Studying illicit drug markets: disciplinary contributions. Int J Drug Policy 17(6):453–463
Robinson J (2008) Measuring the impact of a targeted law enforcement initiative on drug sales. J Crim Justice 36(1):90–101
Robinson J, Rengert GF (2006) Illegal drug markets: the geographic perspective and crime propensity. West Criminol Rev 7(1):20–32
Rosenfeld R, Decker SH (1999) Are arrest statistics a valid measure of illicit drug use? The relationship between criminal justice and public health indicators of cocaine, heroin, and marijuana use. Justice Q 16(3):685–699
Sneath JP (1957) The application of computers to taxonomy. J Gen Microbiol 17(1):201–226

Taniguchi TA, Rengert GF, McCord ES (2009) Where size matters: agglomeration economies of illegal drug markets in Philadelphia. Justice Q 26(4):670–694

Taniguchi TA, Ratcliffe JH, Taylor RB (2011) Gang set space, drug markets, and crime: violent and property crimes around drug corners in Camden. J Res Crime Delinq 48(3):327–363

Taylor RB (2010) Hot spots do not exist, and four other fundamental concerns about hot spot policing. In: Frost NA, Freilich JD, Clear TR (eds) Contemporary issues in criminal justice policy: policy proposals from the American Society of Criminology Conference. Wadsworth, Belmont, pp 271–278

Ward JH (1963) Hierarchical grouping to optimize an objective function. Am Stat Assoc 58(301):236–244

Warner BD, Coomer BW (2003) Neighborhood drug arrest rates: are they a meaningful indicator of drug activity? A research note. J Res Crime Delinq 40(2):123–138

Weisburd D, Green L (1995) Policing drug hot spots: the Jersey City drug market analysis experiment. Justice Q 12(4):711–735

Part II
Crime Analysis

Chapter 3
Convicted Sex Offender Residential Movements

Alan T. Murray, Tony H. Grubesic, Elizabeth A. Mack, Ran Wei, Sergio J. Rey, Luc Anselin, and Marie L. Griffin

Abstract The social, economic, and cultural impacts of sex offender legislation are topics of considerable interest in recent years. Despite the number of studies evaluating the collateral consequences of these laws, the implications of spatial restrictions on housing availability and residential mobility for convicted sex offenders remain an empirical question. Because of the social implications, but also risks associated with recidivism, a better understanding of the spatio-temporal movements of sex offenders is critical for developing effective management policies and strategies aimed at promoting public safety. The purpose of this chapter is to analyze sex offender residential movement patterns over a 2.5 year period in Hamilton County, Ohio. Using geographic information systems and a developed exploratory system, SOSTAT, this study uncovers significant trends and behavioral patterns that shed light on offender reintegration, their residential mobility and the implications of residency restrictions on both offenders and community.

Keywords Spatial restriction zones • Sex offenders • GIS • Residential mobility

---

This material is based upon work supported by the National Science Foundation under grants 0924001 and 0922737. Any opinions, findings, conclusions or recommendations expressed in this material are those of the authors and do not necessarily reflect the views of the National Science Foundation.

A.T. Murray (✉) • E.A. Mack • R. Wei • S.J. Rey • L. Anselin
GeoDa Center for Geospatial Analysis and Computation, School of Geographical Science and Urban Planning, Arizona State University, Tempe, AZ 85287, USA
e-mail: atmurray@asu.edu

T.H. Grubesic
Geographic Information Systems and Spatial Analysis Laboratory, College of Information Science and Technology, Drexel University, Philadelphia, PA 19104, USA

M.L. Griffin
School of Criminology and Criminal Justice, Arizona State University, Phoenix, AZ 85004, USA

M. Leitner (ed.), *Crime Modeling and Mapping Using Geospatial Technologies*, Geotechnologies and the Environment 8, DOI 10.1007/978-94-007-4997-9_3,

3.1 Introduction

Sexual offenses, especially those committed against children, are of much concern to the general public and policy makers, primarily due to the long-lasting negative social and psychological effects on victims. These individuals are viewed as a threat to vulnerable populations, particularly children. In response to these concerns, local, state and federal legislators have passed a series of laws to govern aspects of post-release activities of sex offenders, largely with the intent of protecting the public. Successive changes to sex offender legislation at the federal, state and local levels have provided the general public unprecedented levels of information about convicted sex offenders, their crimes and where they reside. The information about offender residential locations, coupled with advances in geospatial technologies, enable law enforcement agencies to track the real-time movements of sex offenders and evaluate mobility trends over time. As of November, 2011 there were 747,408 sex offenders currently registered in local, state, and federal databases (NCMEC 2011).

The goal of these laws is to prevent the perpetration of future sex crimes by creating increased community awareness about this potentially dangerous segment of the population. The various state and local laws establishing sex offender residency restriction zones, such as Ohio's Megan's Law (Public Law 104–145 C.F.R. § 170101), are designed to reduce interaction between children and these potentially dangerous individuals. Although the general public frequently views all convicted sex offenders as violent, predatory, sexual criminals, this is not necessarily based on anything other than perception. Some studies have found that many sexual crime victims know their abuser (Groth et al. 1982), though some do not (WLWT 2009). The Bureau of Justice Statistics (2003) reports that just 5.3% of convicted sex offenders recidivated within 3 years after release from U.S. prisons while other studies have found significantly higher recidivism rates (Hanson and Morton-Bourgon 2004, 2005; Groth et. al. 1982).

Though there is some debate about the likelihood of recidivism among these individuals, the enactment of sex offender legislation in the last two decades has sparked considerable debate regarding their efficacy (Duwe et al. 2008; Minnesota Department of Corrections 2003) and unintended or collateral consequences (Burchfield 2011). For example, community notification and registration laws are thought to increase isolation and hinder re-integration into society (Tewksbury and Lees 2006; Levenson and Cotter 2005). Spatial restriction zones are also believed to eliminate housing options for offenders (Zandbergen and Hart 2006; Chajewski and Mercado 2009; Barnes et al. 2009) and force offenders to reside in socially disorganized areas (Socia 2011; Tewksbury 2011). Some studies also suggest that residence restrictions spawn a higher frequency of residency change for convicted offenders (Turley and Hutzel 2001; Mustaine et al. 2006).

In order to balance protection of the public, perceived or otherwise, and mitigate collateral consequences for offenders, accurate and effective measurement and analysis of sex offender residency issues is important. There is considerable work evaluating aspects of sex offender residency restrictions. A range of analytical techniques are employed to examine the implications of spatial restriction zones enacted in communities. Perhaps the most popular is the use of proximity analyses within geographic information systems (GIS) to determine the number of offender residences located within restricted areas of sensitive facilities (Tewksbury and Mustaine 2006; Zandbergen and Hart 2006; Grubesic et al. 2007; Barnes et al. 2009; Chajewski and Mercado 2009). Other techniques employed include clustering (Grubesic 2010), optimization models (Grubesic and Murray 2008, 2010), statistical measures (Tewksbury and Mustaine 2006; Mustaine et al. 2006; Grubesic et al. 2008; Youstin and Nobles 2009), and qualitative approaches (Tewksbury 2005; Levenson and Cotter 2005; Levenson and Hern 2007). While these studies have yielded a wealth of information about the impact of sex offender legislation on available housing for offenders, many questions about the collateral consequences of laws persist, particularly with respect to the residential mobility of convicted sex offenders. Obtaining a deeper understanding of policy related to offender residential mobility is critical not only because of the potential link between spatial restriction zones and recidivism, but also for effective offender management policies and strategies aimed at promoting public safety.

This chapter details a framework and developed system that combines geographic information systems (GIS) and exploratory methods, called SOSTAT (Sex Offender Spatio-Temporal Analytical Toolbox), for use in evaluating sex offender policies. It also demonstrates the utility of this tool in an analysis of sex offender movement patterns over a 2.5 year period in Hamilton County, Ohio.

3.2 Theories of Offender Residential Mobility

Residential mobility refers to a multifaceted process of urban demographic change. Typically, factors such as age, income levels, housing attributes, employment accessibility, social bonds, neighborhood satisfaction and mobility expectations all play a role in either encouraging or discouraging residential moves (Kan 2006; Quigley and Weinberg 1977; McHugh et al. 1990; Dieleman 2001; Boehm and Ihlanfield 1986). Not surprisingly, there are also links between residential mobility and crime. Recent work suggests that adolescents who moved recently experience a higher incidence of violence compared to adolescent non-movers (Haynie and South 2005).

3.2.1 Social Disorganization Theory

A more typical linkage made between residential mobility and crime is rooted in social disorganization theory (Shaw and McKay 1942). This theory suggests that variations in crime levels are strongly linked to the local ecological characteristics of a neighborhood: poverty, residential mobility, family disruption and ethnic heterogeneity (Sampson and Groves 1989). When high levels of each characteristic are found in a neighborhood, victimization rates often increase, especially when compared to neighborhoods with lower levels of disorganization (Sampson and Groves 1989; Bursik 1988). Studies of the collateral consequences of sex offender legislation intimate that these laws force offenders to reside in socially disorganized areas, which may negatively impact their prospects for rehabilitation and reintegration (Levenson and Cotter 2005; Mustaine et al. 2006). Socially disorganized communities have lower levels of social control, which contributes to a decrease in the monitoring of criminal activity by local residents (Sampson et al. 1997) and children (Putnam 2001). Studies suggest this breakdown in community ties decreases the likelihood that residents will have a vested interest in their community (Burchfield and Mingus 2008), and therefore, the link between social disorganization and sex crimes (e.g., Mustaine et al. 2006) is not surprising.

3.2.2 Routine Activities Theory

An increased number of offenders in socially disorganized communities, likely conducive to criminal activity, is a concern from a routine activities perspective as well. Routine activities theory identifies three things necessary for a crime: a motivated offender, a suitable target, and the absence of a capable guardian (Cohen and Felson 1979). If any of these are missing, a crime is far less likely to occur. Of course, common characteristics of socially disorganized neighborhoods are motivated offenders, suitable targets and few if any capable guardians, so there are serious implications of legislation that would relegate offenders to such areas.

Studies suggest the risk of recidivism increases for offenders that reside in target rich environments (Ouimet and Proulx 1994). In addition to a higher availability of targets, socially disorganized neighborhoods are also found to lack social support mechanisms and neighborhood organizations, the absence of which can fuel anti-social behavior, making recovery for offenders more difficult (Levenson and Cotter 2005; Levenson and Hern 2007; Burchfield and Mingus 2008). Rehabilitative prospects for offenders are also complicated by the location of support facilities, which are often found in dense urban areas, and are frequently proximal to sensitive facilities where children congregate (Grubesic 2010).

3.2.3 Rational Choice and Optimal Foraging Theory

Social disorganization and routine activities theories help explain the necessary conditions and characteristics of places for crime events to take place. They also highlight the concern surrounding the collateral consequences of current sex offender legislation. However, these theories of criminal activity do not deal with the decision making process of criminals (Clarke and Cornish 1985; Groff 2007).

Rational choice theory (Clarke and Cornish 1985) and optimal foraging theory (Krebs and Davies 1987) suggest several reasons for concern about the proximity of sex offender residences to areas where children congregate, particularly in socially disorganized areas. Proximity ensures a ready supply of potential targets and drastically reduces the friction of distance. Further, because potential victims are located in a familiar environment, albeit socially disorganized, they are less likely to be alerted to strangers with malicious intentions. The role of distance decay on human interaction and crime is well-documented (White 1932; Pyle 1974; Rhodes and Conly 1981; Brantingham and Brantingham 1984). This is highly related to the principle of least effort espoused by Zipf (1949), where people with a choice of possible actions will most often select the one that requires the least amount of effort. It is also illustrative of optimal foraging behavior in criminals. This theory of offender decision-making suggests that criminals attempt to target places that present the greatest opportunity for reward and minimize the risk of being apprehended (Krebs and Davies 1987). In fact, studies of the journey to crime find evidence of this risk- reward tradeoff as it relates to the friction of distance. These findings reflect the time, effort, and money required to overcome distance.

Criminals often select targets nearest to them, particularly if the level of target desirability is equal. Studies suggest that some criminals purposely avoid operating too close to home in order to reduce the risk of being recognized by neighbors or acquaintances (Brantingham and Brantingham 1984). This body of research also finds that criminal acts committed against people involve shorter journeys to crime than do crimes involving property (White 1932; Pyle 1974; Reppetto 1974; Rhodes and Conly 1981; Beauregard et al. 2005). In this context, socially disorganized neighborhoods are the places that likely offer the greatest potential for sex offenders to target victims and minimize the risks inherent to committing a sex crime. Further, Ouimet and Proulx (1994) argue that committing a sexual offense in one's home offers advantages over competing locations. Specifically, in a home, a child may feel more secure and comfortable than a strange location, helping facilitate a sexual encounter. In addition, if offenders are forced to hunt in more distant locations, it is less likely that an offender could find a child willing to take a car trip to his/her residence (Beauregard et al. 2005). Obviously, these are two compelling reasons that an offender may want to establish a permanent residence near socially disorganized areas and/or locations where children congregate.

3.3 Geographic Analyses of Collateral Consequences

As mentioned previously, the enactment of sex offender legislation has resulted in much interest in the impacts of these laws. Recent work suggests that management strategies are generating unintended social, cultural and economic effects, or collateral consequences (Lieb 2000; Petrunik 2003; Tewksbury 2005; Levenson and Cotter 2005). For example, in addition to the public shaming of offenders (Blair 2004), they often suffer from an increased sense of isolation (Tewksbury and Lees 2006). Research also points out that such formal responses may undermine an offender's ability to normalize social relationships, weakening bonds with existing family and friends, and leading to an increased likelihood of recidivism among some groups of offenders (Hepburn and Griffin 2004). Further, because spatial restriction zones limit housing availability (Zandbergen and Hart 2006; Chajewski and Calkins-Mercado 2009; Barnes et al. 2009), convicted offenders are often forced into socially disorganized areas (Levenson and Cotter 2005; Hipp et al. 2010). While this is not necessarily an outcome of residence restrictions (Grubesic et al. 2007), research suggests that offenders who suffer from housing instability are more likely to recidivate or abscond (Meredith et al. 2007; Willis and Grace 2008).

Much of the recent empirical work on sex offender public policy and legislation is oriented towards offender residency issues, the analysis of which is both important and potentially informative. A range of methods has been employed to evaluate the impact of spatial restriction zones on housing availability for offenders. These methods range from qualitative to quantitative.

Of these methods, several studies make extensive use of GIS. For example, Grubesic et al. (2007) and Barnes et al. (2009) utilized GIS based maps to visually depict the spatial relationship between offender residential locations and spatial restriction zones. Youstin and Nobles (2009) relied on a visually driven search for sex offender residence patterns using GIS mapping as well. A commonly used GIS technique in sex offender studies is proximity analysis. Buffer zones around sensitive facilities are created using GIS to analyze offender residence proximity to sensitive facilities, examples of which can be found in Tewksbury and Mustaine (2006), Zandbergen and Hart (2006), Grubesic et al. (2007) and Chajewski and Mercado (2009).

More advanced statistical and spatial analysis has been applied as well. For example, Youstin and Nobles (2009) compared and contrasted sex offender residential clusters in two time periods using local indicators of spatial autocorrelation. Grubesic (2010) investigated the application of several area-based clustering methods to evaluate offender residency issues. Regression models have also been structured to test for offender residency patterns, including the work of Mustaine et al. (2006), which employed a logistic regression model to examine the residential movement tendencies of sex offenders. Pope (2008) utilized hedonic models to estimate the impacts of established sex offender residences on neighborhood housing values. Grubesic et al. (2008) applied various regression models to test for demographic and socioeconomic differences inside and outside spatial restriction zones.

Finally, considerable work has employed qualitative methods, including interviews, surveys and case study histories. For instance, Tewksbury (2005) administered a questionnaire asking registered sex offenders to comment on negative experiences brought by offender registration laws. Levenson and Cotter (2005) surveyed sex offenders about whether spatial restriction zones exerted negative influences on re-integration or increased recidivism rates. Levenson and Hern (2007) surveyed residents to understand public perceptions of policies associated with sex offenders.

3.3.1 *Offender Mobility*

Even with all of this evidence in hand, there is no convergence in the literature on whether or not sex offender mobility and residential proximity to sensitive facilities impacts recidivism. For example, analysis by the Minnesota Department of Corrections (2003) reported that of the 224 recidivistic sexual offenses analyzed, none would have been prevented by residence restrictions. Duwe et al. (2008) suggest that the distance where offenders made first contact for predatory sexual assaults ranged between 2,500 ft. from the offenders' home (for minor victims) and approximately 5,280 ft. from their home (for adult victims). They found no systematic relationship between sensitive facilities and sexual assaults. That said, in 3.6% of the cases analyzed, the sex offender was a neighbor of the victim. Again, considering the zero-tolerance stance perpetuated via current public policies designed to manage offenders, many would find this statistic alarming. While predatory behavior and sexual recidivism are a concern for communities where sex offenders reside, an understanding of their behavior remains hampered by a lack of verified information regarding offender mobility. If, as noted previously, criminals and sexual offenders are a highly mobile population, many questions remain regarding the frequency, direction and neighborhood choices made by them.

Further evaluation of sex offender mobility is also important given the link between mobility and the mental stability of offenders. Literature evaluating mobility and criminal psychology suggests the most mobile offenders are also likely to be the most mentally unstable and dangerous. In a study of 311 male sexual offenders Hunter (2003) found that psychopaths displayed greater mobility than nonpsychopaths. Specifically, their levels of impulsiveness, unstable employment and a need for stimulation can predispose them to higher degrees of mobility (Cooke 1998). Further, Beauregard et al. (2005) suggest that psychopathic offenders frequently move because their ability to exploit others eventually becomes known and they are no longer able to take advantage of people in their local community. As a result, they move, directly increasing their levels of anonymity and establishing a new residence from which to exploit potential victims. Based on these studies, a high correlation between mobility and the type of offenses committed by an offender is likely.

Considering that so little is known about the geographic mobilities of convicted sex offenders, law enforcement officials, policy makers and analysts are in dire need of an analytical framework and associated methodologies that deepen their understanding offender movements. While rudimentary options for tracking mobility exist in many commercial GIS packages, these tools often lack sophistication, are difficult to access, require knowledge of programming or are only available in expensive, add-on extension modules. Exploratory methods that provide for integrated and interactive inquiry are needed to facilitate the analysis of offender movements. In the next section, we outline an example of an integrated framework for analyzing mobility, SOSTAT (Sex Offender Spatio-Temporal Analytical Toolbox).

3.4 Study Area and Data

The remainder of this study attempts to fill the void in better understanding one component of sex offender mobility: examining the residential movements of convicted offenders in Hamilton County, Ohio. Hamilton County is the central county of the Cincinnati-Middletown Metropolitan Statistical Area, located in southwestern Ohio with portions of the metropolitan area extending into both Kentucky and Indiana. Ohio's version of Megan's Law mandates that convicted offenders cannot establish a permanent residence within 1,000 ft. of schools, preschools or child day-care center premises.

Hamilton County was chosen for use in this analysis because of its prior use in studies of sex offender residency issues (Grubesic et al. 2007, 2008; Grubesic and Murray 2010; Mack and Grubesic 2010). This county was also chosen because of recent legal challenges related to the state's offender residence restrictions (e.g., Hyle v. Porter; 117 Ohio St.3d 165, 2008-Ohio-542), questioning the retroactive application of Megan's Law to offenders that owned their home and committed a sexual offense prior to the effective date of this law.

3.4.1 Data

Cadastral or parcel data and the associated land use information for the region were obtained for 2005 from the Cincinnati Area Geographic Information System (CAGIS). There are 406,527 parcels, 256,946 of which are classified as residential and 3,007 of which are classified as school-related. Information about school related parcels was acquired from the Ohio Department of Education (ODE 2005) and includes information about all high schools, junior high schools, middle schools, elementary schools, vocational schools, special schools, and non-public schools. Street data were obtained from the 2000 TIGER line file (U.S. Census Bureau 2011). Both the parcel and street data are the primary files used to spatially reference sex offender residential locations obtained from the Hamilton County Sheriff's Office.

3.4.2 Spatial and Temporal Matching of Offenders

Address information for registered sex offenders was obtained for four time periods: June 2005, December 2006, June 2007, and December 2007. After removing offenders without addresses and offenders with addresses outside of Hamilton County, there are a total of 1,098 offenders in June 2005; 1,198 in December 2006; 1,277 in June 2007; and 1,346 in December 2007. Geographic coordinates for offender addresses were assigned in one of two ways, parcel matching and TIGER-based geocoding. Parcel matching utilizes cadastral data, where surveyed property parcels display a high level of spatial accuracy. Cadastral data are the preferred geographic base files for sex offender studies (Murray and Grubesic 2012) and are used, when possible, for this study. The alternative, TIGER-based geocoding, generates the latitude and longitude coordinates of each address using reported address ranges, street segment length and pre-defined offset distances from street centerlines (Ratcliffe 2001). While this information is relatively good for general spatial analysis, results are subject to interpolation errors and related geographic innacuracies (Ratcliffe 2001; Grubesic and Murray 2004; Murray et al. 2011). TIGER-based geocodes are only used when parcel matching fails for this study.

As an intermediate step, between parcel matching and resorting to TIGER-based geocodes, problematic or initially undetermined addresses, were provided supplementary land use information. Specifically, offender addresses that could not be matched with the parcel database directly were matched to a parcel in one of two ways. Offenders were initially matched to the parcel with the most similar address. If this was not successful, an overlay procedure was used to determine the location of an address. In both cases, this process included manual verification involving the use of areal photographs, parcel data and local knowledge to assign locations to an address (see Murray et al. 2011). In sum, this multistep process resulted in the assignment of over 1,000 offender addresses in each time period, with 90% of the offenders assigned to a parcel in the CAGIS database.

In addition to matching offenders geographically, it was also necessary to match offenders temporally. This involved matching offenders across consecutive observation periods as well as non-consecutive observation periods. Details of the spatio-temporal data processing approach are given in Murray et al. (2011).

3.5 Methods

While considerable research has been dedicated to evaluating housing availability for offenders, the exploration of sex offender residency issues is distinct due to legislative questions and connections to sex offender residency, particularly spatial restriction zones. A combination of factors makes sex offender residency, movement and equity questions unique and challenging:

- Regions, communities and local neighborhoods are geographically complex, each displaying varying levels of housing availability, sensitive facilities, vulnerable populations and social support services. As a result, a one-size-fits-all analytical (or policy) based approach does not suffice.
- Contingency analysis (i.e., "what if" scenarios) of public policies are designed to highlight potential outcomes. To facilitate this, the analytical approaches utilized to evaluate sex offender policies need to represent a diverse array of complementary techniques that can illustrate the subtleties of proposed ordinances or laws. Communities do not exist in isolation, nor do sex offenders. Policies implemented in one jurisdiction can impact neighboring communities in many unanticipated ways, particularly where offender migration is concerned. As a result, the ability to evaluate both direct and indirect geographic impacts of sex offender policies

is essential, as is an ability to facilitate interregional or intercommunity coordination of management strategies.

- Communities do not exist in isolation, nor do sex offenders. Policies implemented in one jurisdiction can impact neighboring communities in many unanticipated ways, particularly where offender migration is concerned. As a result, the ability to evaluate both direct and indirect geographic impacts of sex offender policies is essential, as is an ability to facilitate interregional or intercommunity coordination of management strategies.

In order to deal with the complexities of analyzing sex offender residency issues, an exploratory spatial data analysis (ESDA) based system, SOSTAT, was developed. ESDA methods typically consist of non-confirmatory statistical measures, as well as graphics and map based displays (Anselin and Bao 1997; Rey and Janikas 2006; Murray 2010). ESDA has been used in prior crime analyses (Messner et al. 1999; Murray et al. 2001) and is ideal for exploring hypotheses about data and searching for relationships that require additional analysis (Messner et al. 1999).

SOSTAT was developed in Python as open source software that will ultimately be made available for public access, use and distribution. SOSTAT uses two sets of Python libraries, Shapely and PySAL, facilitating the construction of new methods, functions, graphics and an interface. Shapely facilitates the creation, manipulation, and non-statistical geometric analysis of spatial data. PySAL (Rey and Anselin 2007) provides access to a collection of modules capable of spatial analysis, ranging from data structures (e.g., for spatial weights), to geocomputation (e.g., convex hull), and specific statistical techniques (e.g., local Moran).

Key components of SOSTAT include a map based display, linked graphics (e.g., histograms, pie charts, etc.), statistical measures/methods, and optimization models. The system facilitates exploratory queries, enables the identification of offenders residing within spatial restriction zones, and is also capable of highlighting characteristics of offender groups. The system also supports classic geovisualization techniques that allow for the use of interactive graphics capabilities to create links between various information display sources, like plots, tables and maps, using linking and brushing techniques. This offers the capacity to dynamically extract and relate knowledge from data.

Figure 3.1 shows multiple display windows of the system. In this instance, a histogram and circular histogram summarizing sex offender residency movements are shown in the linked map, describing distance and direction attributes. Specifically, the distance histogram involves bins, 10 in this case, signifying the total number of movement vectors that fall within the distance range of each bin. Each bin is linked to the actual movement vectors in the map display, enabling interactive and dynamic exploration. Specifically, one or more bins in the histogram can be selected, initiating a real-time update of the map display to highlight all the vectors in the selected bin(s). This facilitates spatial analysis, providing the user with the ability to assess potential patterns based on distance, as is evident between the selected bin and the corresponding 36 vectors highlighted in the map display (blue vectors).

The circular histogram shown in Fig. 3.1 summarizes offender movement vectors by direction of residency change. In this case the circular histogram consists of

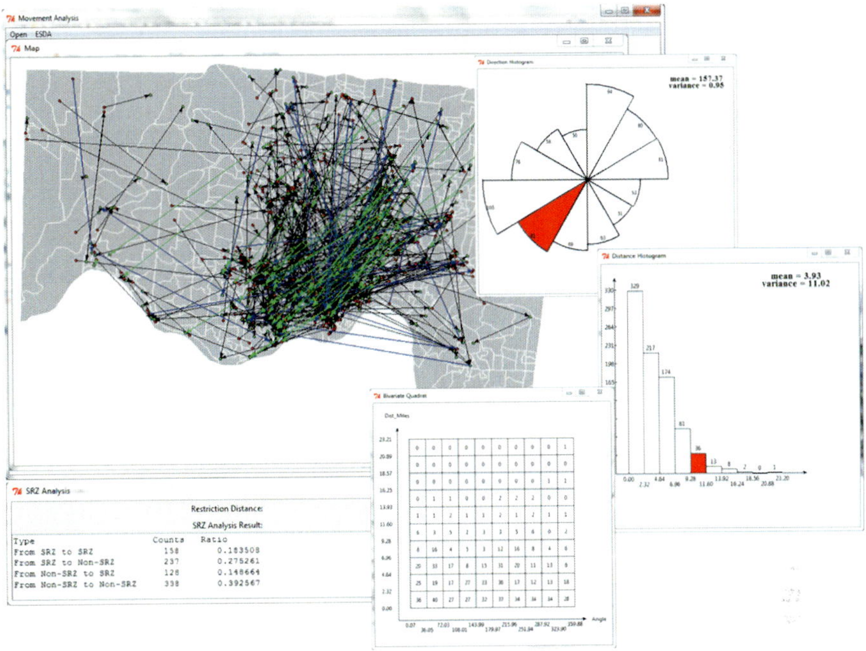

Fig. 3.1 SOSTAT

twelve fans (or bins), each spanning 30°, with the size of the fan representing the number of movement vectors in the associated direction. Each fan is linked to vector movement objects shown in the map based display as well. Selecting the fan in the circular histogram shown in Fig. 3.1 automatically highlights the associated 81 vectors (green) in the map display. This provides powerful and flexible geovisualization capabilities as one can select part of the histogram and see instantly where these directional movements are located in geographic space on the map display. Alternatively, one could select a subset of movement vectors using a "lasso" for summary using histogram or circular histogram displays. The two-way linkage between the different views provides for a more comprehensive visualization of these patterns than would be possible with traditional static graphics in existing software packages. The technical issues in system development require capabilities for tracking which objects are in which bins or fans. With this linkage, the system updates the map display to highlight those objects associated with interactively selected bins/fans.

Another important functionality of SOSTAT is its production of a bivariate matrix. Each interval of the bivariate matrix shows the number of observations whose attribute values fall into a range of values defined for each of the two variables of interest. An example of the bivariate matrix is displayed in Fig. 3.1, which shows the attributes of direction (x-axis) and distance (y-axis). The matrix therefore is a simultaneous summary of statistical graphics, the histogram (distance) and circular histogram (direction), which are shown in Fig. 3.1. The matrix is also linked to the map based display, which is important for data exploration. Selection of one or more interval cells in the bivariate matrix initiates the system to highlight the associated movement vectors in the map display, enabling an interactive query and investigation of offender movement.

3.6 Results

In order to evaluate the relative mobility of offenders through space and time, offenders that changed residences over the course of the study period were identified and matched across time periods, as described above. The matching process revealed several variations associated with potential residency change: (a) some individuals will not move; (b) a certain number of offenders are recently convicted and registering for the first time; (c) some offenders will fail to re-register; (d) some offenders will move into the county from other areas, states or countries; (e) offenders will move out of the county; and, (f) some offenders will move within the county. Although data limitations prevent a full description of each of these movement scenarios, the spatio-temporal resolution of the information was sufficient to identify a consistent set of offenders in the compiled offender database and track their movements through space and time (case f). In particular, 609 sex offenders did not move in any time period, 484 moved exactly once, 149 moved exactly twice and 26 moved exactly three times.

While there is some uncertainty about the number of newly convicted and recently registered offenders (case b) out of the 484 that registered only once, it is likely that most of these individuals reflect a residency move into or out of Hamilton County. Thus, if one combines these 484 offenders with the 660 that did move within Hamilton County, a total of 1,144 (65.2%) convicted sex offenders moved over a period of 2.5 years. Clearly, sex offenders appear to be a very mobile group, as suggested in the literature (Turley and Hutzel 2001), and have a greater propensity to move than a typical resident of the United States (13.6% national residential mobility rate for 2007).

3.6.1 *Offender Movements and Spatial Restriction Zones*

Aside from evaluating the relative mobility of offenders, the number of offender residences within spatial restriction zones was also tabulated. Spatial restriction zones were constructed by delineating a 1,000 ft Euclidean (straight-line) distance around the perimeter of each school parcel in accordance with the restriction distance specified in Ohio's Megan's Law; restriction zones account for approximately 18% of the area of Hamilton County (Fig. 3.2). Analysis of offender residence locations and movements with respect to these restriction zones reveals that for the June 2005 period, 41% of the convicted sex offenders resided in a spatial restriction zone. By December 2006 onward, only 30% of sex offenders resided in

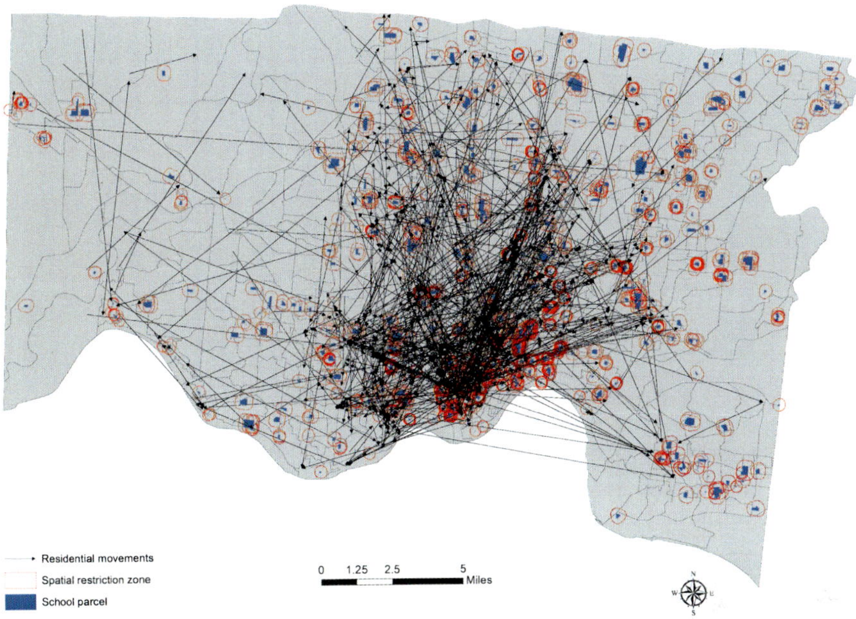

Fig. 3.2 Residency movement patterns of sex offenders

spatial restriction zones. This is a sizeable decrease, and one that appears to persist. There are two likely reasons for this. First, the decrease represents the outcome of a more stringent enforcement of residency restrictions at the local level by the Hamilton County Prosecutor's office and the county sheriff. Specifically, during the fall of 2005, these agencies began actively evicting offenders from residences within identified spatial restriction zones. Second, with the passage of the Adam Walsh Act (2006), it is possible that both offenders and local community members became more aware of the federal guidelines for stricter registration requirements, making the establishment of a residence by a convicted sex offender within a spatial restriction zone more difficult.

An interesting finding associated with the 484 sex offenders that registered only once, representing cases (b)–(d), is the propensity for them to reside within spatial restriction zones. Approximately 51% of these offenders live in a spatial restriction zone, whereas offenders that register more frequently are less likely to reside in restricted areas. In fact, offenders that register multiple times only violate spatial restriction zone laws in less than 30% of all cases. Based on this finding it appears that offenders that register less frequently are perhaps engaging in predatory behavior. Conversely, these findings could also indicate that single time registrants are simply returning to their initial residence upon release from prison. Of course, inadequate local knowledge, a poor understanding of residence restrictions and available housing options in Hamilton County remain

Table 3.1 Registered offender residency movement within county

Movement period	SRZ to SRZ (%)	SRZ to non-SRZ (%)	Non-SRZ to SRZ	Non-SRZ to non-SRZ (%)
June 2005 to later periods	18.97	31.62	13.11	36.30
December 2006 to later periods	17.33	23.76	18.32	40.59
June 2007 to December 2007	18.10	23.28	15.09	43.53

χ^2 test statistically significant with p-value $< 1 \times 10^{-8}$

relevant factors likely associated with residential housing decisions. Nevertheless, given that such individuals may also represent moves into or out of the region, this is a curious and noteworthy finding.

Examination of residency movements within Hamilton County in relation to spatial restriction zones is also particularly revealing. There are 861 changes in residence, as some individuals move multiple times over the study period. These spatial movement patterns are shown in Fig. 3.2, and summarized in Table 3.1. One important finding associated with spatial restriction zones is the residence movements occurring between June 2005 and December 2006. Nearly a third, 31.62%, of moves in this period were from a spatial restriction zone (SRZ) to a non-spatial restriction zone (non-SRZ). This figure declines to 23% in later periods, and perhaps reflects the increased enforcement of Megan's Law.

A more interesting finding from this analysis is summarized in Table 3.1 and Fig. 3.2 which illustrates that approximately 32.2% of offender residence moves within Hamilton County are moves into spatial restriction zones. Specifically, 18.35% of moves across all periods were from one spatial restriction zone (SRZ) to another SRZ. Further, 14.87% of offenders moved from a non-spatial restriction zone to a spatial restriction zone. Additional analysis of offender movements to restricted areas by offender sub-type provides additional resolution on offender movement trends.

This analysis by offender type uses the offender classifications from Ohio's classification system at that time, which tiers offenders by their potential dangerousness to society.[1] Three risk levels are reported, each imposing different registration requirements on offenders (Ohio Megan's Law 1996). Sexual predators are the most dangerous offenders in this tiered system, habitual sex offenders are moderately dangerous individuals, and sexually oriented offenders are the least dangerous offenders (Grubesic et al. 2007). A breakdown of offenders by this classification system reveals some slight differences in residential movements by offender type. Table 3.2 highlights that 31.86% of sexually oriented offender movements and 35.80% of sexual predator movements were movements into spatial restriction

[1] This was subsequently replaced with federal designations after the passage of the Adam Walsh Act.

Table 3.2 Offender sub-type movement within county

Offender classification	To SRZ (%)	To non-SRZ (%)
Sexual predator	35.80	64.20
Sexually oriented offender	31.86	68.14

χ^2 test statistically significant with p-value $< 1 \times 10^{-8}$

zones within Hamilton County. This consistent pattern across offender groups suggests a preference by offenders to reside in areas that have been deemed off limits by enacted residential laws. There are a variety of reasons for this apparent preference for restricted areas. First, there is a propensity for many rehabilitation and support facilities in urban environments to be located near sensitive facilities where children congregate (Grubesic 2010). Second, this preference may also indicate strategic movements to areas with a ready supply of available targets, as suggested by routine activities theory. Regardless of the underlying causes, which would require significantly more analysis to fully explain, the patterns exist.

3.6.2 *Distance and Direction of Offender Movements*

Given the evidence displayed in Fig. 3.2, which suggests a bias in the frequency of offender movements, the analysis of the 861 moves described previously suggests that there are some systematic distance and direction traits exhibited by offender residency change in Hamilton County. Most movements are relatively short, with an average distance of 3.93 miles. This average belies that fact that over one third (329/861) of these moves were less than 2.5 miles. Further analysis of these movements by their origin and destination uncovers a distance decay pattern from the city center. The circular histogram also suggests a northeast and southwest directional bias to these movements.

The use of SOSTAT to interactively select bins in the histogram associated with distances moved by offenders shows that many of the short-distance movements occur primarily in downtown Cincinnati. Interactive selection (lasso) and graphical summary of the selected areas, which is depicted in Fig. 3.3, illustrates that the distance moved by offenders exhibits regional variability within the county. Moves in the downtown area (selected area in map on the right in Fig. 3.3) largely conform to trends in the distance moved for the county as a whole, but have a slightly higher mean of 4.44 miles. Offender movements in the northeast area of the county however do not conform to the general county trend. The mean distance of movements in this portion of the county is larger than the county average at 7.35 miles. These trends appear to reflect the shorter distance between residences in downtown areas as opposed to more rural areas.

A combined analysis of trends in the distance and direction of offender residential movements is possible via the bivariate matrix, which as noted previously, may be used for any two attributes. The two attributes of interest in this portion of the analysis are the offender distance to a sensitive facility prior to a move and after a

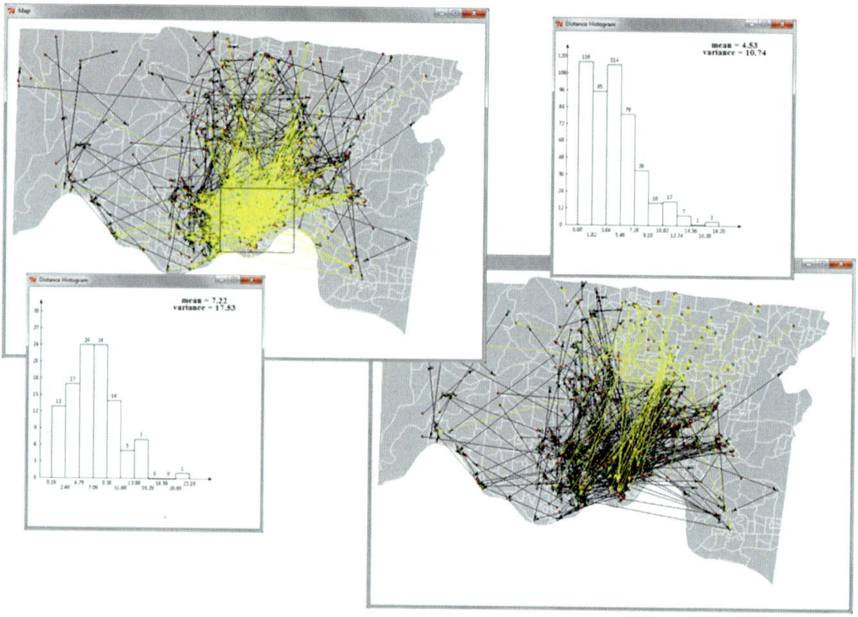

Fig. 3.3 Interactive map-based display environment

move. Figure 3.4 depicts an offender's distance from a school prior to a move (x-axis) and after a move (y-axis). Cells in the matrix are counts of offender origin and destination pairs at a particular distance from a school. The matrix is interpreted as follows: cells nearest the origin represent offenders whose origin and destination are very near a school and cells far from the origin represent offenders whose origin and destination are farther away from a school. The presence of large counts of offenders in cells nearest the origin suggests that many mobile offenders relocate from places relatively close to schools (less than 0.5 miles) to residences that are still relatively close to schools (less than 0.5 miles). There are few populated cells farther away from the origin, which signifies that there are few offenders whose origin residence and destination residence are farther away from sensitive facilities. An examination of these cells through interactive selection in Fig. 3.4 suggests that the majority of these moves occur in rural areas.

3.7 Discussion and Policy Implications

As noted previously, while the vast majority of research on sex offenders and residence restrictions deals with issues of housing availability, affordability or collateral consequences, very little research has focused on sex offender mobility, particularly as it relates to residential housing trends. One exception is the work of Turley and Hutzel (2001) examining the mobility of offenders post-release, finding them to be a highly mobile population. The analysis reported above supports these

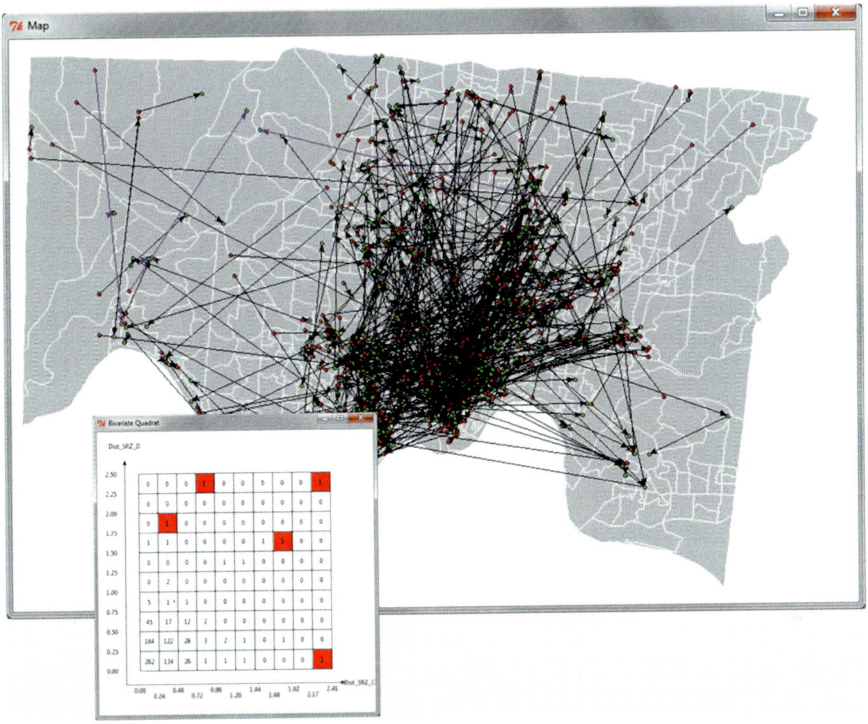

Fig. 3.4 Bivariate matrix

conclusions, but also uncovers some noteworthy mobility trends. First, over the 2.5 year period in this analysis, 65.2% of registered offenders changed residences. This is remarkable and highlights the logistical challenges that law enforcement officials face with respect to maintaining accurate and up-to-date registries, as well as notifying residents when/if offenders move into a neighborhood.

Second, sex offender residency movements are not random and exhibit distinct geographic patterns. The decline of offenders residing in spatial restriction zones after 2005 is likely attributable to increased law enforcement efforts in the region and more public awareness of sex offender regulations after the passage of the Adam Walsh Act in 2006. This suggests the combined passage and enforcement of these laws is effective in preventing these potentially dangerous individuals from residing in close proximity to places where children congregate. While this impact may vary in areas where housing has been demonstrated to be unavailable to offenders (Zandbergen and Hart 2006; Levenson and Hern 2007; Chajewski and Mercado 2009; Zgoba et al. 2009), it suggests that in Hamilton County, where these laws do not reduce housing availability for offenders (Grubesic et al. 2007), enforcement of these laws is correlated with achieving their stated objectives.

A third finding of note is the large proportion of offender movements (33.22%) into spatial restrictions zones. As noted above, there are two explanations for this pattern. First, offender rehabilitation facilities and social support services are often located in dense urban areas, which are proximal to sensitive facilities where children congregate (Grubesic 2010). Access to these facilities is necessary to receive treatment, which is a mandatory part of offender corrections and the parole process. Therefore, proximity to these important facilities may be a primary consideration in the residential search process of offenders. Second, these patterns may support the findings of Beauregard et al. (2005), which suggest that offenders locate and target victims near a permanent residence. Regardless of the motivation, the propensity of offenders to move into spatial restriction zones more than one third of the time in Hamilton County is a noteworthy, particularly given legislative and offender management efforts in the region.

Finally, the observed residency movement behavior of sex offenders appears to be significantly different from that of expected movement patterns. Though not reported here, Rey et al. (2012) compare movement patterns of sex offenders in Hamilton County to expected residency change patterns. Sex offender that currently reside in a spatial restriction zone tend to move shorter distances than expected, and have a higher propensity to move into another spatial restriction zone when they change residence. Overall, however, restriction zones do appear to change movement behavior as fewer sex offenders than expected move into these areas, but the intent of eliminating all offenders in restriction zones is not realized.

3.8 Conclusions

Although this study is subject to some limitations, it does effectively document the variability in residential mobility for Hamilton County, Ohio sex offenders. It also provides some insights about the mobility of these individuals and the challenges associated with the implementation of sex offender management policies. Residence restrictions continue to enjoy widespread public support and are unlikely to be repealed in the near future (Schiavone and Jeglic 2009). Of course, public perceptions of safety are predicated on the idea that laws restricting sex offender movement within the community are designed in such a way so as to achieve their goal of minimizing contact with vulnerable populations.

An interesting and important facet of this work is whether the analysis suggests anything about public safety or sex offender recidivism. With respect to recidivism, there is little that can be inferred as the analysis was not privy to any information on offender recidivism. Of course, the behavioral tendency of convicted sex offenders to move into a spatial restriction zone suggests an expectation of recidivism, but it is unknown whether this will prove to be true. Clearly this has implications for public safety.

Over the years, subsequent changes in laws governing the post-release activities of offenders are increasingly designed to monitor and track this group of individuals. The present study highlights however that despite these increasingly stringent laws, sex offenders move freely about communities and continue to reside in restricted residential areas. This mobility suggests that current policies may require modification in their design and implementation to achieve their intended goals.

References

Adam Walsh Child Protection Act (2006) Public Law 109–248. 42 U.S.C § 16901–16962

Anselin L, Bao S (1997) Exploratory spatial data analysis linking SpaceStat and ArcView. In: Fisher M, Getis A (eds) Recent developments in spatial analysis. Springer, Berlin, pp 35–59

Barnes JC, Dukes T, Tewksbury R, DevTroye TM (2009) Analyzing the impact of a Statewide residence restriction law on South Carolina Sex offenders. Crim Justice Policy Rev 20:21–43

Beauregard E, Proulx J, Rossmo K (2005) Spatial patterns of sex offenders: theoretical, empirical, and practical issues. Aggress Viol Behav 10:579–603

Blair M (2004) Wisconsin's sex offender registration and notification laws: has the Wisconsin Legislature left the criminals and the constitution behind? Marquette Law Rev 87:939–981

Boehm TP, Ihlanfeld KR (1986) Residential mobility and neighborhood quality. J Reg Sci 26(2):411–424

Brantingham P, Brantingham P (1984) Patterns in crime. Macmillan, New York

Burchfield K (2011) Residence restrictions. Criminol Pub Policy 10(2):411–419

Burchfield KB, Mingus W (2008) Not in my neighborhood: assessing registered sex offenders' experiences with local social capital and social control. Crim Justice Behav 35(3):356–374

Bureau of Justice Statistics (2003) Recidivism of sex offenders released from prison in 1994. Washington, DC: U.S. Department of Justice

Bursik RJ (1988) Social disorganization and theories of crime and delinquency: problems and prospects. Criminology 26(4):519–552

Chajewski M, Calkins-Mercado C (2009) An evaluation of sex offender residency restriction functioning in town, county, and city-wide jurisdictions. Crim Justice Policy Rev 20:44–61

Clarke R, Cornish D (1985) Modeling offender's decisions: a framework for research and policy. Crim Justice 6:147–185

Cohen LE, Felson M (1979) Social change and crime rate trends: a routine activity approach. Am Sociol Rev 44:588–608

Cooke DJ (1998) Psychopathy across cultures. In: Cornish DB, Clarke RV (eds) The reasoning criminal: rational choice perspectives on offending. Transaction, New Brunswick, pp 351–382

Dieleman FM (2001) Modelling residential mobility: a review of recent trends in research. J Hous Built Environ 16(3–4):249–265

Duwe G, Donnay W, Tewksbury R (2008) Does residential proximity matter? A geographic analysis of sex offender recidivism. Crim Justice Behav 35:484–504

Groff E (2007) Simulation for theory testing and experimentation: an example using routine activity theory and street robbery. J Quant Criminol 23(2):75–103

Groth AN, Longo RE, McFadin JB (1982) Undetected recidivism among rapists and child molesters. Crim Delinq 28:450–458

Grubesic TH (2010) Sex offender clusters. Appl Geogr 30:2–18

Grubesic TH, Murray AT (2004) Assessing the locational uncertainties of geocoded data. In: Proceedings from the 24th urban data management symposium, Chioggia, Italy

Grubesic TH, Murray AT (2008) Sex offender residency and spatial equity. Appl Sp Anal Policy 1:175–192

Grubesic TH, Murray AT (2010) Methods to support policy evaluation of sex offender laws. Pap Reg Sci 89(3):669–684

Grubesic TH, Murray AT, Mack EA (2007) Geographic exclusion: spatial analysis for evaluating the implications of Megan's Law. Soc Sci Comput Rev 25(2):143–162

Grubesic TH, Murray AT, Mack EA (2008) Sex offenders, housing and spatial restriction zones. GeoJournal 73:255–269

Hanson RK, Morton-Bourgon KE (2004) Predictors of sexual recidivism: an updated meta-analysis (User report 2004–02) (Public Safety and Emergency Preparedness Canada, Ottawa, Canada)

Hanson RK, Morton-Bourgon KE (2005) The characteristics of persistent sexual offenders: a meta-analysis of recidivism studies. J Consult Clin Psychol 73:1154–1163

Haynie DL, South SJ (2005) Residential mobility and adolescent violence. Soc Force 84(1):361–374

Hepburn JR, Griffin ML (2004) The effect of social bonds on successful adjustment to probation: an event history analysis. Crim Justice Rev 29:46–75
Hipp JR, Turner S, Jannetta J (2010) Are sex offenders moving into social disorganization? Analyzing the residential mobility of California Parolees. J Res Crime Delinq 47:558–590
Hunter S (2003) Geographical Mobility among male psychopaths. Paper presented at the western society of criminology, Vancouver, British Columbia
Hyle v. Porter (2008) Ohio Supreme Court ruling. URL: http://www.supremecourt.ohio.gov/rod/docs/pdf/0/2008/2008-Ohio-542.pdf. Accessed 28 June 2010
Kan K (2006) Residential mobility and social capital. J Urban Econ 61(3):436–457
Krebs J, Davies N (1987) An introduction to behavioural ecology. Blackwell, Oxford
Levenson JS, Cotter LP (2005) The impact of sex offender residence restrictions: 1,000 feet from danger or one step from absurd? Int J Offender Ther Comp Criminol 49:168–178
Levenson JS, Hern AL (2007) Sex offender residence restrictions: unintended consequences and community reentry. Justice Res Policy 9:59–73
Lieb R (2000) Social policy and sexual offenders: contrasting United States' and European policies. Eur J Crim Policy Res 8:423–440
Mack EA, Grubesic TH (2010) Sex offenders and residential location: a predictive analytical framework. Environ Plan A 42(8):1925–1942
McHugh KE, Gober P, Reid N (1990) Determinants of short- and long-term mobility expectations for home owners and renters. Demography 27(1):81–95
Megan's Law (1996) Public Law 104–145 C.F.R. § 170101 (d) of the Violent Crime Control and Law Enforcement Act of 1994
Meredith T, Speir J, Johnson S (2007) Developing and implementing automated risk assessments in parole. J Res Policy 9:1–21
Messner S, Anselin L, Baller R, Hawkins D, Deane G, Tolnay S (1999) The spatial patterning of county homicide rates: an application of exploratory spatial data analysis. J Quant Criminol 15:423–450
Minnesota Department of Corrections (2003) Level three sex offenders residential placement issues: 2003 report to the legislature. Minnesota Department of Corrections, St. Paul
Murray AT (2010) Quantitative geography. J Reg Sci 50:143–163
Murray AT, Grubesic TH (2012) Spatial optimization and geographic uncertainty: implications for sex offender management strategies. In Johnson M (ed) Community-Based Operations Research: Decision Modeling for Local Impact and Diverse Populations, (Berlin: Springer). pp 121–142
Murray AT, Grubesic TH, Wei R, Mack EA (2011) A hybrid geocoding methodology for spatio-temporal data. Transactions in GIS 15:795–809
Murray AT, McGuffog I, Western JS, Mullins P (2001) Exploratory spatial data analysis techniques for examining urban crime. Br J Criminol 41:309–329
Mustaine EE, Tewksbury R, Stengel KM (2006) Residential location and mobility of registered sex offenders. Am J Crim Justice 30:177
National Center for Missing and Exploited Children (NCMEC) (2011) Map of registered sex offenders in the United States. URL: http://www.missingkids.com/en_US/documents/sex-offender-map.pdf
Ohio Department of Education (ODE) (2005) ODE Interactive: extract Ohio educational information. URL: http://webapp2.ode.state.oh.us/data/extract_oed_addgrades.asp
Ouimet M, Proulx J (1994) Spatial and temporal behavior of pedophiles: their clinical usefulness as to the relapse prevention model. Paper presented at the meeting of the American Society of Criminology, Miami, Florida
Petrunik M (2003) The hare and the tortoise: dangerousness and sex offender policy in the United States and Canada. Can J Criminol Crim Justice 45:43–72
Pope JC (2008) Fear of crime and housing prices: household reactions to sex offender registries. J Urban Econ 64:601–614
Putnum R (2001) Bowling alone. Simon & Schuster, New York

Pyle GF (1974) The spatial dynamics of crime. Chicago, University of Chicago, Department of Geography Research Paper No. 159

Quigley JM, Weinberg DH (1977) Intra-urban residential mobility: a review and synthesis. Int Reg Sci Rev 2(1):41–66

Ratcliffe JH (2001) On the accuracy of TIGER-type geocoded address data in relation to cadastral and census areal units. Int J Geogr Inf Sci 15(5):473–485

Reppetto TA (1974) Residential crime. Ballinger, Cambridge

Rey SJ, Anselin L (2007) PySAL: a Python library of spatial analytical methods. Rev Reg Stud 37:5–27

Rey SJ, Janikas MV (2006) STARS: Space-Time Analysis of Regional Systems. Geogr Anal 38:67–86

Rey SJ, Murray AT, Grubesic TH, Mack EA, Wei R, Griffin M (2012) The impact of spatial restrictions on sexual offender residential movement patterns: a Markov chain analysis. Draft manuscript

Rhodes WM, Conly C (1981) Crime and mobility: an empirical study. In: Brantingham PJ, Brantingham PL (eds) Environmental criminology. Sage Publications, Beverly Hills

Sampson RJ, Groves WB (1989) Community structure and crime: testing social- disorganization theory. Am J Sociol 94:774–780

Sampson RJ, Raudenbush SW, Earls F (1997) Neighborhoods and violent crime: a multilevel study of collective efficacy. Science 277:918–924

Schiavone SK, Jeglic EL (2009) Public perception of sex offender social policies and the impact on sex offenders. Int J Offender Ther Comp Criminol 53:679

Shaw CR, McKay HD (1942) Juvenile delinquency in urban areas. University of Chicago Press, Chicago

Socia K (2011) The policy implications of residence restrictions on sex offender housing in Upstate NY. Criminol Pub Policy 10(2):351–389

Tewksbury R (2005) Collateral consequences of sex offender registration. J Contemp Crim Justice 21(1):67–81

Tewksbury R (2011) Policy implications of sex offender residence restrictions laws. Criminol Pub Policy 10(2):345–348

Tewksbury R, Lees M (2006) Perceptions of sex offender registration: collateral consequences and community experiences. Sociol Spectr 26:309–334

Tewksbury R, Mustaine E (2006) Where to find sex offenders: an examination of residential locations and neighborhood conditions. Crim Justice Stud 19(1):61–75

Turley E, Hutzel L (2001) West Virginia sex offender study. URL: http://www.dcjs.wv.gov/SAC/Documents/sexoffender%20report%2001.pdf

U.S. Census Bureau (2011) TIGER/Line Shapefiles. http://www.census.gov/geo/www/tiger/

White RC (1932) The relation of felonies to environmental factors in indianapolis. Soc Force 10:498–509

Willis GM, Grace RC (2008) The quality of community reintegration planning for child molesters: effects on sexual recidivism. Sex Abuse J Res Treat 20:218–240

WLWT (2009) Police: suspect in teen's death possible serial killer. URL: http://www.wlwt.com/news/18879648/detail.html. Accessed 28 June 2010

Youstin TJ, Nobles MR (2009) Residency estrictions: a geospatial analysis of sex offender movement over time. Crime Mapp 1:55–76

Zandbergen PA, Hart TC (2006) Reducing housing options for convicted sex offenders: investigating the impact of residency Restriction Laws using GIS. Justice Res Policy 8:1–24

Zgoba KM, Levenson J, McKee T (2009) Examining the impact of sex offender residence restrictions on housing availability. Crim Justice Policy Rev 20(1):91–110

Zipf GK (1949) Human behaviour and the principle of least effort. Addison-Wesley Press, Cambridge

Chapter 4
Street-Level Spatiotemporal Crime Analysis: Examples from Bronx County, NY (2006–2010)

Christopher R. Herrmann

Abstract The objective of this chapter is to illustrate how different geographic scales of violent crime analyses vary and can benefit from spatiotemporal analyses within a geographic information science framework. Geographic clusters of violent crime, typically referred to as 'hot spots', can be very difficult to interpret and address at small geographic scales. Incorporating various temporal resolutions to small-scale crime analysis, such as 'hot streets' of violent crime, provides law enforcement with a much more robust understanding of small-scale crime patterns. These small-scale street patterns can assist police departments in developing improved geospatial models for targeted police patrols and provide a more comprehensive understanding of the complex relationships between crime and place.

In the first part of this chapter, I illustrate several popular ways that 'hot spots' are typically generated and demonstrate how hot spots vary using several violent crime types and temporal analysis. In the second part of the chapter, I establish the importance of exploring crime hot spots at small geographic scales (e.g., streets) and demonstrate several spatiotemporal methods for 'hot streets'.

Keywords Crime mapping • Crime analysis • Micro-level • Hot spots • Hot streets • Small-scale

C.R. Herrmann, Ph.D. (✉)
Department of Professional Studies, Justice Studies Program, Berkeley College, New York, USA

John Jay College of Criminal Justice, City University of New York, New York, USA
e-mail: cph@berkeleycollege.edu

M. Leitner (ed.), *Crime Modeling and Mapping Using Geospatial Technologies*, Geotechnologies and the Environment 8, DOI 10.1007/978-94-007-4997-9_4,

4.1 Introduction

This chapter illustrates how complex issues in crime analysis can benefit from small-scale exploration within a geographical information system framework. Understanding both spatial and temporal variations in violent crime at the street level can have direct implications on apprehending criminals, police resource allocation & planning, crime modeling & forecasting, and evaluation of crime prevention & crime control programs (Ratcliffe 2004a; Boba 2001). In our current state of shrinking agency operating budgets, law enforcement (and other government agencies) needs to take the temporal dimensions of spatial crime patterns into consideration when identifying, exploring, and managing crime 'hot spots'. If Sherman's concept of 'wheredunit' (1989) for hot spots was the foundation for crime mapping and crime analysis between 1990 and 2010, I propose we consider a combination of 'whendunit' & 'wheredunit' at smaller geographic scales for 2011 and beyond.

4.2 Hot Spots

The crime analysis and crime mapping communities have become very proficient in locating, tracking, and managing 'hot spots'. This iterative crime analysis and crime control process has resulted in a steady ebb and flow of statistical and spatial crime patterns throughout many geographic levels (i.e., neighborhoods, police precincts, census tracts). Current research (Weisburd et al. 2009; Groff et al. 2010; Block 2011) indicates that as we drill down into the small-scales of geography (e.g., streets, tax lots, buildings), crime hot spots start to form new shapes (i.e., lines, points), sizes, and patterns.

In my experience as a crime analyst with the New York City Police Department, not all violent crime hot spots act the same and (almost) all hot spots have significant internal spatiotemporal variance – especially at small-scales (Ratcliffe 2004a, b, c, 2006; Groff et al. 2010). Crime analysts and researchers should not simply view hot spots as geographic polygons that become objectives for crime prevention, crime control, and targeted patrol efforts. Hot spots need to be examined from within. What (specifically) is generating the hot spot? On what days of the week and at what times of day are the problem(s) occurring within the hot spot? How many explicit problem properties ('hot points') and/or street segments ('hot streets') are there within the hot spot? Is the crime problem dispersing, clustering, or stationary? Are the problem areas diffused, focused, acute? Are the trends increasing, decreasing, remaining flat? (Ratcliffe 2004a, b, c).

The idea of hot spots (Sherman et al. 1989; Block RL and Block CR 1995; Levine 1999; Weisburd and Green 1995; Peuquet 1994; Ratcliffe 2002; Ratcliffe 2004a, b, c) has been the fuel for much of the interest in our current 'crime and place' research. Ever since the Sherman et al. article (1989), there has been a substantial body of literature that supports this concept of hot spots and crime concentrations. Hot spots can be calculated many different ways (Nearest Neighbor Hierarchical

Table 4.1 Review of hot spot policing programs

Hot spot study	Program elements
Minneapolis (MN) RECAP Sherman et al. (1989)	Problem-oriented policing to control crime at high-activity addresses; interventions comprised mostly traditional enforcement tactics with some situational responses
Minneapolis (MN) hot spots Sherman and Weisburd (1995)	Increased uniformed police patrol in crime hot spot areas; treatment group, on average, experienced twice as much patrol presence as the control group
Jersey City (NJ) DMAP Weisburd and Green (1995)	Well-planned crackdowns on street-level drug markets followed by preventive patrol to maintain crime control gains
Jersey City (NJ) POP at violent places Braga et al. (1999)	Problem-oriented policing to prevent crime at violent hot spot areas; interventions comprised mostly aggressive disorder enforcement tactics with some situational responses
St. Louis (MO) POP in three drug areas Hope (1994)	Problem-oriented policing to prevent crime at three high-drug activity addresses; interventions comprised mostly traditional enforcement tactics with some situational responses
Kansas City (MO) crack house raids Sherman and Rogan (1995a)	Court-authorized raids on crack houses conducted by uniformed police officers
Kansas City (MO) Gun project Sherman and Rogan (1995b)	Intensive enforcement of laws against illegally carrying concealed firearms in targeted beat through safety frisks during traffic stops, plain view, and searches incident to arrest on other charges
Houston (TX) targeted beat program Caeti (1999)	Patrol initiative designed to reduce Index crimes in seven beats: Three beats used "high visibility patrol" at hot spots Three beats used "zero tolerance" policing at hot spots One beat used a problem-oriented policing approach that comprised mostly traditional tactics to control hot spots
Beenleigh (AUS) Calls for service project Criminal Justice Commission (1998)	Problem-oriented policing to control crime at high-activity crime addresses; interventions comprised mostly traditional enforcement tactics with some situational responses

clusters, Getis-Ord Gi* statistics, Kernel Density Estimation, Standard Deviation Ellipses, K-Means Clustering, Local Moran's I statistics), however, none of these methods take the temporal aspect of crime into consideration during calculation.[1]

A recent Crime Prevention Research Review (Braga 2008) that was conducted for the Community Oriented Policing Services (COPS) office indicates that a majority of medium & large size police departments are using crime analysis and crime mapping to identify crime hot spots. Table 4.1 reports the cities where hot spot studies have been reported on and their associated program elements.

[1] For information on spatiotemporal clustering methods, see Kulldorff (1997) and Hardisty and Klippel (2010).

In his systematic review of hot spot interventions, Braga (2008) selected nine hot spot evaluations that were identified and reviewed for their effectiveness and impact on managing crime hot spots. He noted that seven of the nine selected studies contained significant crime reductions. Moreover, Clarke and Weisburd (1994) indicate that there is routinely a 'diffusion of benefits' that results from these types of police hot spot interventions. Not only does crime decrease throughout the targeted hot spot area as a result of the applied intervention, but the surrounding areas also typically experience a decrease in crime (even though they are not within the specified intervention boundaries). It should be noted that of the nine studies selected and reviewed, none of the studies focused specifically on spatiotemporal clusters of crime, but rather traditional (spatial) hot spots.

Throughout the study of crime and place, criminologists have examined the various relationships between crime and social forces at various geographic scales. There have been numerous studies of crime at higher level geographies; such as countries (Weir and Bang 2007; Gartner 1990), states (Rosenfeld et al. 2001; Faggiani et al. 2001), countries (Block and Perry 1993; Baller et al. 2001), cities (Martin et al. 1998; Cork 1999), and neighborhoods (Elffers 2003; Tita and Cohen 2004). Many of these studies have indicated various relationships between crime and socioeconomic factors (e.g., poverty, race, education, etc.). However, all of the socioeconomic and environmental criminology factors that are of interest to crime analysts and criminologists also vary 'beneath the surface' (Maantay et al. 2007).

In the past 30 years, there has been a renewal in interest in crime at smaller scales. Instead of looking at crime relationships at the county, city, and neighborhood levels – we are starting to recognize the value of studies of crime at smaller-scales (Taylor 1998; Weisburd et al. 2009; Groff et al. 2010). One of the current trends in crime and place research is small-scale geographies, where small-scale is defined as street segments, properties, or buildings. Most of this renewed interest is a result of small-scale research conducted in Minneapolis (Sherman et al. 1989), Baltimore (Taylor 2001), Seattle (Weisburd et al. 2004), and Jersey City (Weisburd and Green 1994).

Few of the previous macro-level (large scale) studies indicate that there is significant variation beneath the unit of analysis that is central to the research. However, as GIS analysts, we know when studying country level crime rates, we need to recognize that the entire country is not high crime or low crime, there is significant variation in crime at the state level within the country. When studying state-level crime rates, it is important to recognize that the entire state is not high crime or low crime, there is significant variation at the county level within each state. When studying county level crime rates, there is significant variation between cities/towns within each county. Lastly, within the cities and towns, there is significant variation at the neighborhood level. This research simply continues this process of 'zeroing in' on crime problems, beyond the neighborhood, census tract, and census block group levels.

4.3 Cones of Resolution

Brantingham et al. (1976) defined this process of zooming in to small-scale geographies as moving downward through different geographical 'cones of resolution'. "The mapping of variables into areal aggregates limits comparative analysis to variables which have been mapped into similar levels of aggregation and further limits the questions that can be validly asked of the data" (Brantingham et al. 1976, p.261).

Figure 4.1 shows the different spatial and temporal levels of analysis that are typically used in crime analysis research today. It is important to note that as we drill down to smaller spatial units of analysis, it also becomes essential to correspondingly drill down on the temporal units of analysis. Small-scale hot spots (i.e., streets, properties, buildings) vary by space and time much more so than large-scale hot spots (census tracts, neighborhoods, police precincts).

For example, residential buildings/streets that contain crime problems do not have high crime problems 24 h a day and 7 days a week. Residential buildings/streets have a unique temporal pattern, based on the occupancy of the residents in the building. If residents work a traditional work/school week (Mon-Fri), during the daytime (6 am – 6 pm), there is more likely to be criminal activity in the evening (6 pm–midnight) on 'workdays/schooldays' (Mon–Thur). However, we would expect to see a shift from evening (6 pm–midnight) to nighttime (midnight–4 am) on Fridays and Saturdays, since most of the working residents would not be going to work (or school) in the morning on the following day. This day of week and time of day pattern is very evident in the Bronx violent crime patterns (Figs. 4.2 and 4.3).

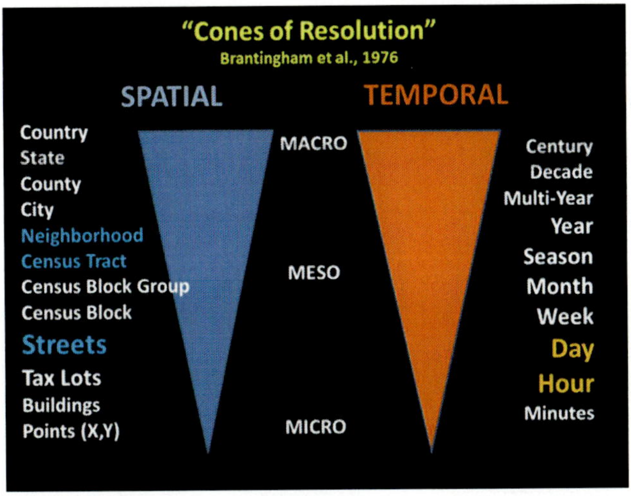

Fig. 4.1 Spatial and temporal units of crime analysis

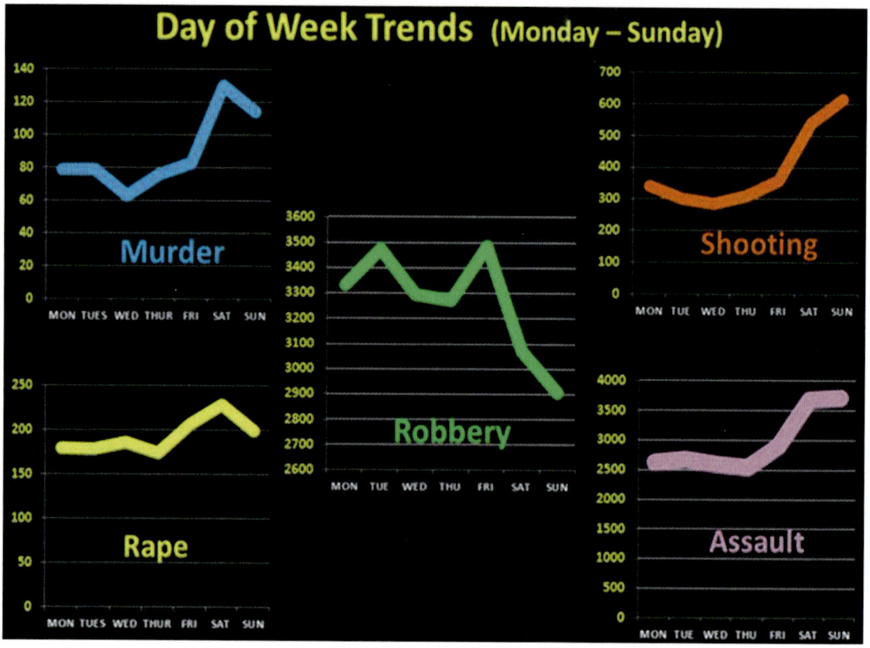

Fig. 4.2 Violent crime trends for the Bronx (2006–2010) by day of week

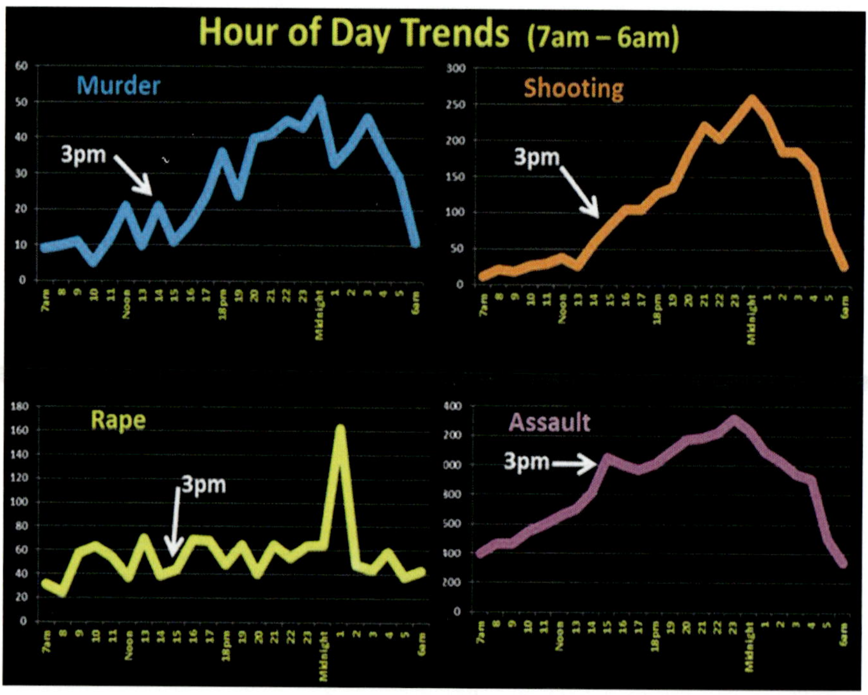

Fig. 4.3 Violent crime trends for the Bronx (2006–2010) by hour of day hour of day

If hot spot geographies 'move' based on the day of week and/or time of day, such as high crime Nearest Neighbor hierarchical (Nnh) clusters, they will also vary much more temporally when these Nnh clusters are constructed at smaller scales. If you query out crime data by hour of day and/or day of week and run the same Nnh clustering parameters, you will most likely obtain very different locations for the small-scale clusters. A very beneficial analysis is to do this over a 24-h period, to find out how each crime 'moves' over the course of one day. You can also query out crime by day of week to find out how the Nnh cluster locations vary day to day.

4.4 The Move to the Micro-level

One of the current trends in studying crime and place at smaller scales is simply a continuation of our historical interest in crime and place. If we continue to see clustering of crime at smaller geographic levels, then we need to recognize that there are significant benefits of studying crime and place at smaller scales. First and foremost, small-scale clusters provide easy 'targets' for directed police patrols. It is much easier for the police to target patrols at properties and street segments, than it is to target entire neighborhoods or police precincts. This is especially true when developing foot patrol strategies (Ratcliffe et al. 2011).

Moreover, if small-scale clusters of properties and street segments are responsible for a majority of the crime within an entire neighborhood, certainly a targeted patrol strategy would have a much more significant crime prevention/crime control benefit than police randomly patrolling entire neighborhoods. Problem-Oriented Policing (POP) strategies are much more effective when they target specific small-scale areas. Again, it is important to recognize that small-scare areas are not 24 × 7 problem areas. In order to effectively address small-scale crime hot spots, we must incorporate temporal trends into our small-scale crime prevention and control strategies. Not only would this small-scale place + time process maximize police impact and outcomes, it would also manage police resources much more effectively than 'random' 24 × 7 patrols. Furthermore, this type of small-scale research provides better understanding of the social, structural, and opportunity factors that are related to crime and small-scale places.

One of the current objectives in environmental criminology and crime analysis is drilling down/zooming in on typical hot spot geographies that are generated by density maps. Using longitudinal crime data, it is now possible to zoom in to the small-scale units of geography and determine the actual cause(s) of the hot spots. This is the reason we map crimes to begin with – to discover why crime patterns occur consistently at the same areas/places over time and to develop programs to intervene with these consistent crime patterns. However, when we analyze hot spots and disaggregate the data within, several unique patterns begin to develop. Every hot spot does not act the same way. In fact, few crime hot spots behave similarly.

4.5 Welcome to the Bronx

The research area and data for this study comprise of various Geographic Information Systems (GIS) datasets for Bronx County, New York; as well as several violent crime datasets for the Bronx from the New York City Police Department (NYPD). New York City is an ideal place to conduct geospatial research because New York City has been using GIS and collecting GIS data since the late 1970s (New York City Department of City Planning 2010). The GIS datasets include Bronx County (Borough), Bronx Neighborhoods (n=37), Bronx Census Tracts (n=355), Census Block Group (n=957), street segments (n=10,544), and property lot data (n=89,812) from the New York City Department of City Planning (NYC-DCP) and the New York City Department of Finance (NYC-DOF).

The neighborhood boundaries of the Bronx are defined by NYC-DCP to contain small area population projections of at least 15,000 people and the boundaries are designated according to historical geographic and sociocultural data (NYC-DCP, 2010). The resulting 37 neighborhoods shapefile contains entire census geographies that were subdivisions of New York City Public Use Microdata Area (PUMA) datasets. Within the 37 unique neighborhoods, Bronx County is further disaggregated into 355 census tracts, 957 census block groups, 10,097 street segments, 89,812 property lots, and 101,307 buildings.

While this chapter promotes advances in small-scale (geographical units below the block group level) crime analysis techniques, it does not intend to concentrate on the inherent problems that occur in most studies of crime and space; most notably the issues related to census unit boundaries (Rengert and Lockwood 2009; Hipp 2007), the modifiable areal unit problem (MAUP) (Openshaw 1984; Chainey and Ratcliffe 2005), and the issue(s) surrounding ecological fallacy (Robinson 1950; Subramanian 2009).

4.6 The New York City Police Department and GIS

The NYPD has been using GIS since the early 1990s, primarily for use in its innovative COMPSTAT process (Bratton 1996). The violent crime datasets for this research include traditional Uniform Crime Report (UCR) violent crime categories murder, rape, robbery, and assault records that were exported out of the NYPD Crime Data Warehouse (NYPD Computer File, 2011). In addition to UCR data, shooting incidents, where shooting locations are confirmed by evidence of a shooting (one or more credible witnesses, one or more shooting victims, gun shell casings, bullet holes, etc.) were also included in the violent crime dataset. All of the violent crime data were geocoded to the property lot level and then aggregated up to street segments and higher-level geographies (i.e., census tracts and neighborhoods).

This research takes place in Bronx county (shown in red, Fig. 4.4), the northernmost county of the five countries that make up New York City. The Bronx is 42 square

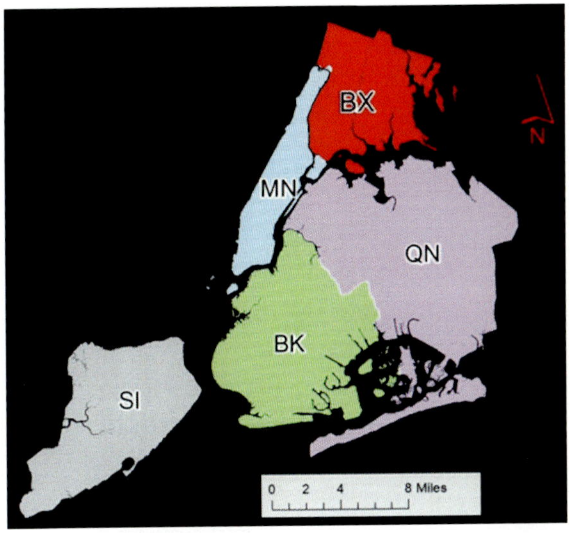

Fig. 4.4 New York city map

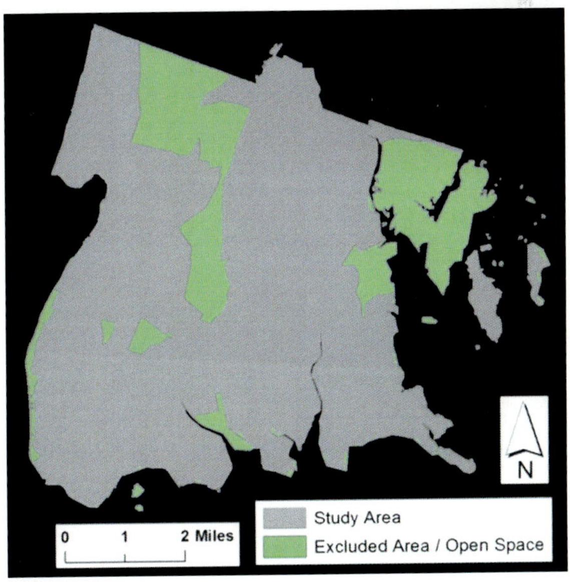

Fig. 4.5 Bronx study area and open space

miles in area, which makes it 14% of New York City's total geographical area. Even though the Bronx is the third most densely populated county in the United States (behind Manhattan & Brooklyn), about a quarter of its land area (shown in Fig. 4.5 in green) is uninhabited open spaces. These uninhabited open spaces include the largest park in New York City (Pelham Bay Park), the Bronx Zoo & Botanical Gardens, large cemeteries, industrial, and waterfront areas.

Table 4.2 New York City and Bronx violent crime between 2006 and 2010 (NYPD 2011)

Crime	NYC	Bronx	Bronx as % of NYC
Murder	2,622	657	25
Rape	6,944	1,510	23
Robbery	105,788	23,069	22
Assault	84,541	20,732	25
Shooting	7,998	2,256	28

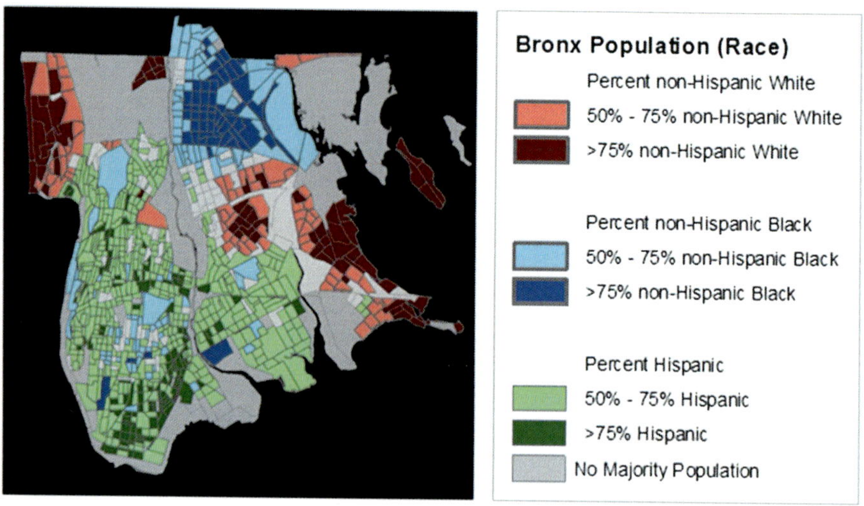

Fig. 4.6 Census tract by race

The Bronx is an ideal place to model human behavior (such as crime) because it is one of the smallest (in area), one of the highest in population density, it is the most diverse in ethnic/racial composition, and it has a substantial amount of violent crime (between 2006 and 2010). As Table 4.2 indicated, the Bronx contains a disproportionate amount of violent crime while considering its size (14% of NYC's total land area) and population (17% of NYC's total population).

The population of the Bronx is 1.4 million (U.S. Census, 2010). Figure 4.6 shows the population distribution by race throughout the Bronx. The Bronx River runs north/south thru the middle of the Bronx and separates the east/west parts of the Bronx. The US Census indicates that the Bronx is the most diverse county in the US: 15% Non-Hispanic White, 31% Non-Hispanic Black, 49% Hispanic, and 5% other. According to the US Census, if you randomly selected two Bronx residents, 90% of the time they would be of a different race or ethnicity (Newsweek 2009). Figure 4.7 shows the population distribution by race at the census tract level. As you can see, not only is the Bronx extremely diverse, but it is also very segregated by race.

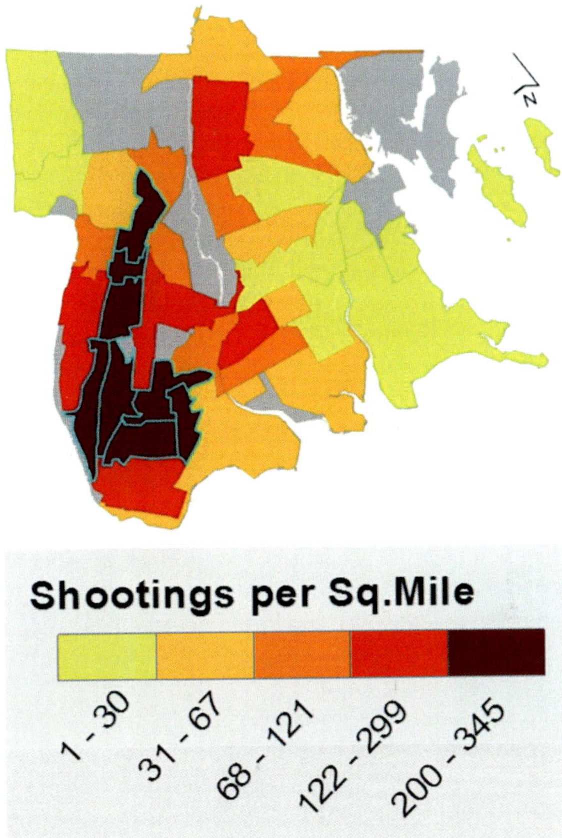

Fig. 4.7 Shooting density by neighborhood

4.7 Why Police Departments Must Focus on Small-Scale Crime Analysis

With police department budgets dwindling more and more during these difficult financial times, it is becoming vital for police departments to 'do more, with less'. New York City Mayor Michael Bloomberg eloquently stated this economic reality as the ability "to provide the service you need and then do it as efficiently as you can" (CBS Radio 2011). With estimates of a 2–4% NYPD budget cut looming in 2011–2012, now more than ever is it important for the NYPD (and any other police departments facing budget cuts) to efficiently analyze, model, and utilize geospatial technologies.

One way the NYPD achieves efficient crime prevention and crime control is by continuously analyzing crime and developing prevention and control strategies at both large-scales (county, precinct) and small-scales (police sectors, streets, properties). The NYPD COMPSTAT system was designed to analyze crime patterns at larger-scales – precincts, patrol boroughs (i.e. Manhattan, Brooklyn, and Queens

Counties are each split into two patrol boroughs based on north/south geography) and county levels on a weekly/monthly basis. The newer ‘Operation Impact’ system is a much more dynamic crime analysis management system, which continuously analyzes crime patterns and trends at the street and (police) sector levels on an hourly/day-to-day basis (police sectors at NYPD are very similar in size to census block groups). Under Operation Impact, hundreds of uniformed and plain-clothes police officers that are (foot) patrolling high crime areas one day can be redeployed to completely different small-scale areas the following day or week. Both COMPSTAT and Operation Impact are utilized by NYPD, but both operate at different spatiotemporal levels and have different goals/objectives.

There is (almost) always significant spatial clustering with violent crime data. Moreover, there is also (usually) significant temporal variation between and within violent crime data. This spatiotemporal realism is accentuated even more at smaller-scales. Not all violent crimes act the same way and even the same crime(s) have significant internal temporal variations.

We should consider the temporal variations of crime at higher spatial levels (neighborhood, tract, block group) primarily as a result of the dominant land uses (e.g., commercial, residential, recreational, transportation, vacant, etc.). According to the routine activities theory (Cohen and Felson 1979), we would expect to see more daytime violence patterns in geographical areas where large groups of people congregate (e.g., commercial, recreational, transportation) or where groups of people are intermingling (e.g., transportation hubs). Evening and nighttime violence patterns in geographical areas may be dominated by areas with higher percentages of vacant land, public transportation hubs, high-density residential areas, or commercial areas (especially those with late-night/24-h businesses serving alcohol) that lack effective place managers.

4.8 Bronx Shootings

Geospatial analysis was conducted on 2,752 shooting incidents throughout the Bronx between 2006 and 2010. Shooting locations were geocoded to the property lot level and then aggregated up to the street segment, census block group, census tract, and neighborhood levels. Shooting aggregates were divided by the polygon area (in square miles) which created a shooting density for each geographical unit. The shooting classes were then symbolized using a quintile classification method, where the dataset is split into five groups, each with an equal number (approximately 20%) of areal units. This quintile method works well with neighborhood and census level data in New York City because it allows comparison of geographical units that are similar in size and population. However, one of the weaknesses of quintile classification is that it masks some of the outlier classes because the values are grouped by an ordinal ranking system (Maroko et al. 2009).

The shooting maps (Figs. 4.7, 4.8, and 4.9) illustrate several problems that typically occur when analyzing crime desities at comparatively coarse geographies.

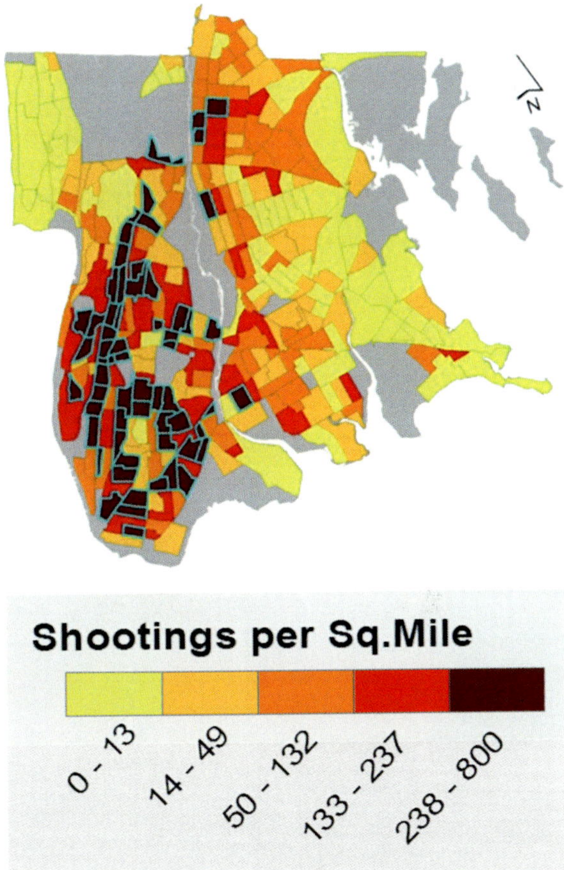

Fig. 4.8 Shooting density by census tract

The neighborhood (Fig. 4.7), census tract (Fig. 4.8), and census block group (Fig. 4.9) level maps all illustrate how shooting densities vary significantly at each geographic level. When starting at the neighborhood level (Fig. 4.7), each subsequent geographic level inidicates increased spatial variation.

Table 4.3 indicates the small-scale street variations within the higher level geographies – neighborhoods, census tracts, and census block groups. Overall, as we drill down through the spatial cone of resolution (from neighborhood to block group level); the number and percentage of shootings that occur within the top (20%) quintile (outlined in blue) increases and the number and percentage of street segments within the top quintile decreases. Likewise, the number and percentage of streets that have zero shootings on them decreases as we move from higher level to lower level geographies.

It is important to note that almost 80% of the street segments within the highest quintile shooting neighborhoods have zero reported shooting incidents on them. Therefore, even within the highest quintile shooting density neighborhoods, shooting incidents are highly clustered which indicates a need to examine shootings at smaller

Fig. 4.9 Shooting density by block group

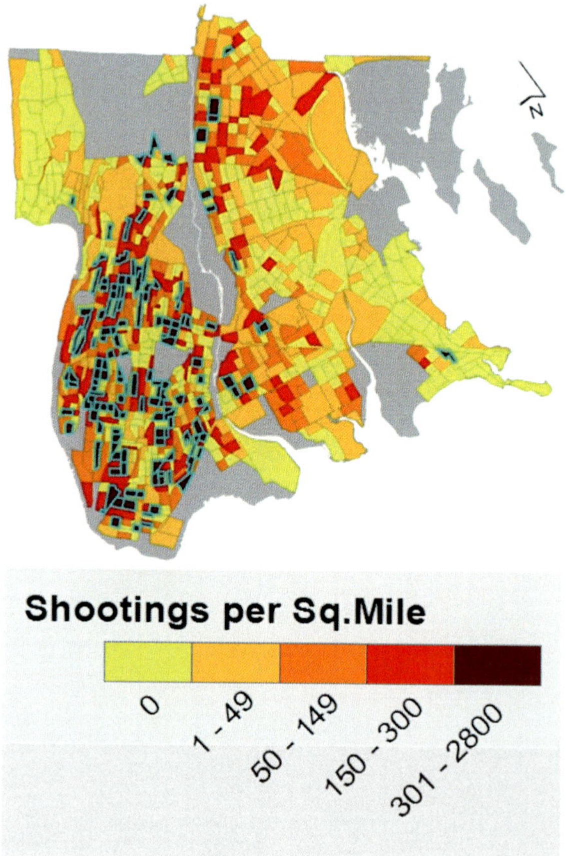

Table 4.3 Bronx shooting frequencies and percentages within top quintile class

Bronx Shootings 2006–2010 n=2,752	# of Total shootings in top quintile	% of Total shootings in top quintile	# of total streets in top quintile	% of total streets in top quintile	# of zero shooting streets in top quintile	% of zero shooting streets in top quintile
High shooting neighborhoods (8 out of 36)	1,006	37	1,820	18	1,430	79
High shooting Tracts (62 out of 329)	1,268	46	1,591	16	1,137	72
High shooting block groups (144 out of 924)	1,286	47	1,047	10	643	61

scales. Similarly, at the census tract level, almost half of the total reported shootings occur in the top 20% of tracts and the majority (72%) of street segments contain zero shootings. Even at the block group level, the majority (61%) of streets have zero reported shooting incidents within the highest (quintile) block groups.

4.9 Bronx Robbery

Robbery is the most common form of violent crime in New York City and one that many researchers consider to be the best indicator of street-level and neighborhood 'safety' (Kennedy and Baron 1993; Groff 2007; Block and Bernasco 2011). The New York City Police Department uses robbery as their primary violent crime indicator for the creation and development of its crime reduction "impact zone" program. Impact zones are small geographical areas, similar to clusters of street segments, that experience consistent high or sharply increasing rates of robbery (and/or other violent crimes). At a recent symposium on the "Understanding the Crime Decline in New York" (John Jay College 2011), Zimring noted that impact zones and destruction of outdoor drug markets were two NYPD initiatives that have helped reduce robbery 30% since 2000.

Spatiotemporal analysis was conducted on 22,824 robbery incidents that were reported in the Bronx between 2006 and 2010. Robbery locations were geocoded to the property lot level and then aggregated to the street segment, census block group, census tract, and neighborhood levels. Robbery aggregates were divided by the polygon area (in square miles) which created a robbery density for each geographical unit. The robbery classes were then symbolized using a quintile classification method, where the dataset is split into five groups, each with an equal number (approximately 20%) of areal units.

The robbery maps (Figs. 4.10, 4.11, and 4.12) illustrate many of the similar problems exhibited by the previous shooting maps. As was observed in the neighborhood (Fig. 4.10), census tract (Fig. 4.11), and census block group (Fig. 4.12) level maps – robbery densities vary significantly at each geographic level. When starting at the neighborhood level (Fig. 4.10), each subsequent geographic level indicates increased spatial variation.

Table 4.4 indicates similar micro-level street variations throughout the coarser level robbery geographies. Since robbery is more than nine times as prevalent and widespread (robbery is still clustered countywide, but it has a much larger spatial distribution) than shootings, the data showing the number/percentage of robberies in the highest quintile class does not suggest as much of a difference between the neighborhood/tract and tract/block group levels. The percentage of streets in the top quintile for robbery is almost the exact same as the percentage of streets in the top quintile for shootings. However, the differences in the percentage of top quintile streets demonstrates the primary difference between the two crimes. Again, it is important to note that even in the top quintile robbery areas, 26% of streets have fewer than 2 robberies and more than 20% of street segments contain zero robbery incidents

Fig. 4.10 Robbery denisty by neighborhood

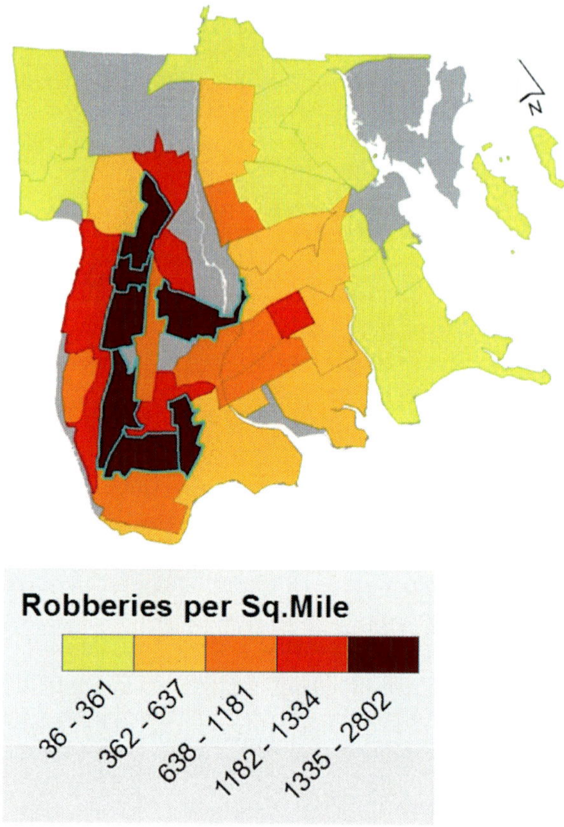

over the 5-year study period. Even with the most widespread violent crime in the Bronx, almost 50% of the streets contain less than the average number of robberies. Again, this suggests a need for analyzing crime at finer spatial resolutions.

4.10 Bronx Assault

Assault is the second most common form of violent crime in New York City. The New York City Police Department also uses assault as a secondary violent crime indicator for its crime reduction "impact zone" program.

Analysis was conducted on 20,726 assault incidents that were reported in the Bronx between 2006 and 2010. Similarly to shootings and robberies, assault points were geocoded to the property lot level and then aggregated to the street, census block group, census tract, and neighborhood levels. Assault aggregates were divided by the polygon area (in square miles) which created an assault density for each geographical unit. The assault classes were then symbolized using a quintile

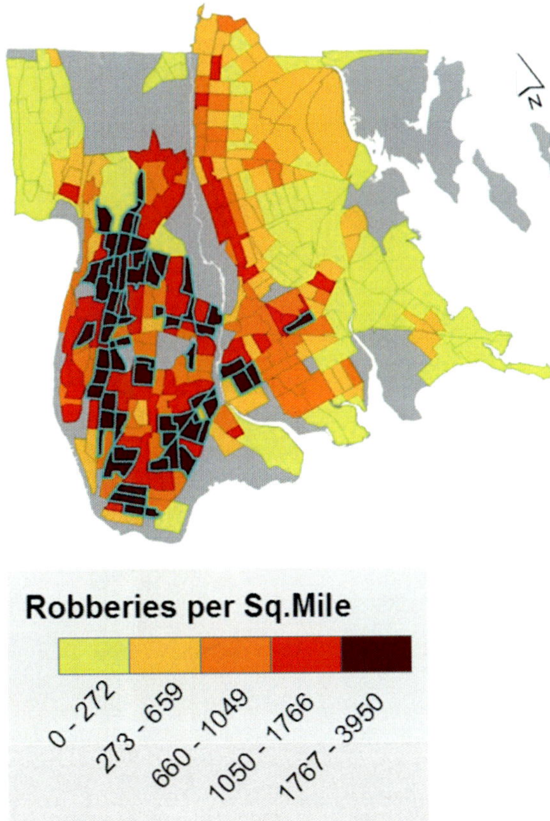

Fig. 4.11 Robbery by census tract

classification method, where the dataset is split into five groups, each with an equal number (approximately 20%) of areal units.

The assault maps (Figs. 4.13, 4.14 and, 4.15.) illustrate similar problems exhibited by the previous shooting and robbery maps. As you can see in the neighborhood (Fig. 4.13), census tract (Fig. 4.14), and census block group (Fig. 4.15) level maps – assault densities vary at each geographic level. When starting at the neighborhood level (Fig. 4.13), each subsequent geographic level indicates increased spatial variation.

Table 4.5 indicates similar micro-level street variations throughout the coarser level assault geographies. Since assault is also (7.5 times) more prevalent and widespread than shootings, the data showing the number/percentage of assaults in the highest quintile classes do not suggest as much of a difference between the neighborhood/tract and tract/block group levels. The percentage of streets in the top quintile for assault is similar to the percentage of streets in the top quintile for robberies. Again, it is important to note that even in the top quintile robbery areas; 22% of street segments at the block group level, 26% of street segments at the tract

Fig. 4.12 Robbery by census block group

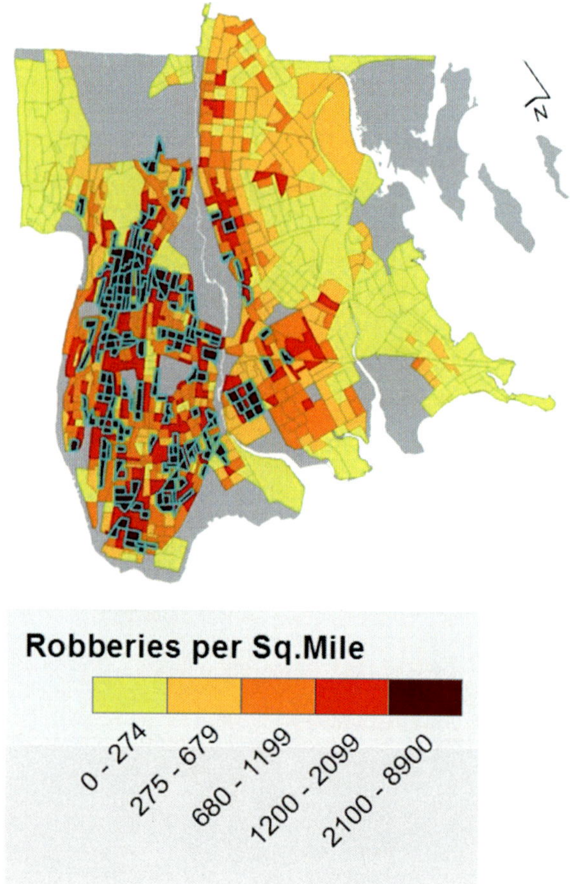

Table 4.4 Robbery frequencies and percentages within the top qunitile class

Bronx Robbery 2006–2010 n=22,824	# of Total robberies in top Quintile	% of total robberies in top quintile %	# of total streets in top quintile	% of total streets in top quintile	# of zero robbery streets in top quintile	% of zero robbery streets in top quintile
High robbery neighborhoods (7 out of 37)	6,630	29	1,656	16	446	27
High robbery tracts (64 out of 329)	8,495	37	1,635	16	345	21
High robbery block groups (169 out of 924)	6,953	31	1,047	10	229	22

Fig. 4.13 Assault Density by neighborhood

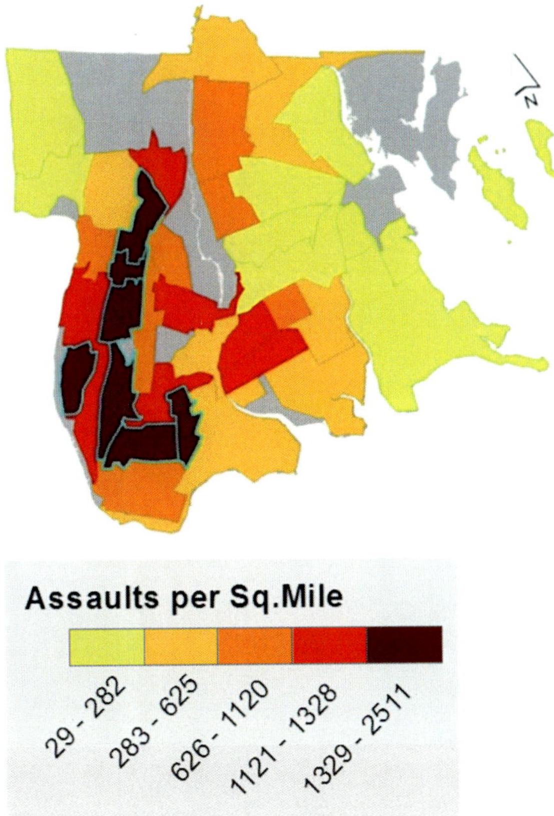

level, and 31% of street segments at the neighborhood level contain zero assault incidents over the 5-year study period. Again, this suggests a need for analyzing crime at finer spatial resolutions.

4.11 Crime Clusters and Crime Densities

There appears to be a continuously growing number of ways that police departments analyze crime clusters and crime densities today (especially if we consider a priori knowledge of analysts/officers). If we refer back to the cones of resolution (Fig. 4.1), we can see that there are at least 12 different geographic levels of resolution, almost all of which would produce different size, shape and strength crime clusters or crime densities. If we incorporated the concepts of scale, temporal trends, input parameters, and classification methods into this cluster/density analysis process, there would appear to be an exponential number of ways to analyze crime.

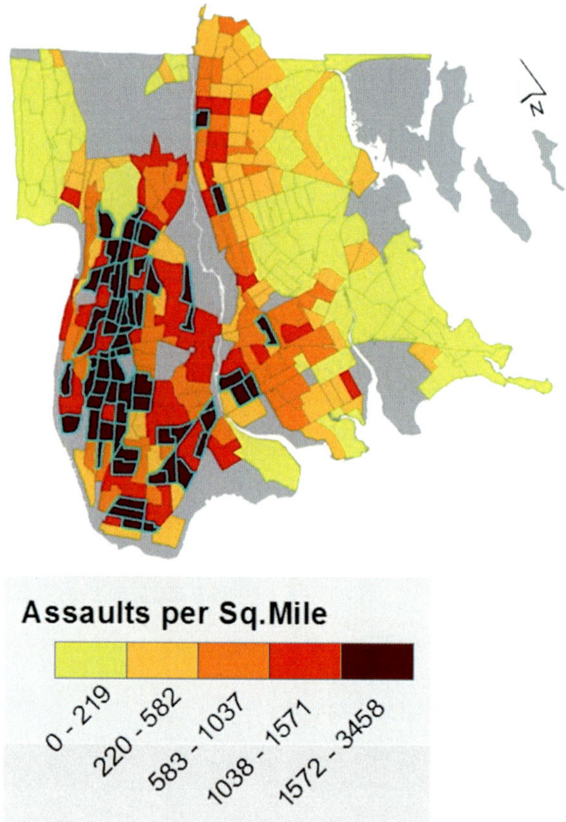

Fig. 4.14 Assault density by census tract

Furthermore, there also appears to be a growing number of geospatial analysis methods that are being used to analyze crime (ie. Nearest Neighbor Hierarchical clusters, Getis-Ord Gi* statistics, Ripley's 'K' Statistic, Single/Dual Kernel Density Estimation, LISA, Standard Deviation Ellipses, K-Means Clustering, Spatial and Temporal Analysis of Crime [STAC], Geary's 'C', Anselin's Local Moran's I statistics, SatScan, etc.). While each method/tool has advantages and disadvantages, several geospatial methods have become prevalent throughout the crime analysis community.

Perhaps the most common geospatial analysis methods used in crime analysis today are nearest neighbor hierarchical spatial clustering (using CrimeStat), Hot Spots/Getis-Ord Gi* (using the ArcGIS Hot Spot tool), and Single/Dual Kernel Density Estimation (using CrimeStat or ArcGIS Spatial Analyst) (McGuire and Williamson 1999; Chainey et al. 2002; Eck 2002a, b; Chainey and Ratcliffe 2005). While each of these popular geospatial methods does an excellent job of generating clusters/densities, none of these geospatial methods incorporate temporal trends into the analysis process. The primary objective of these clustering/density methods

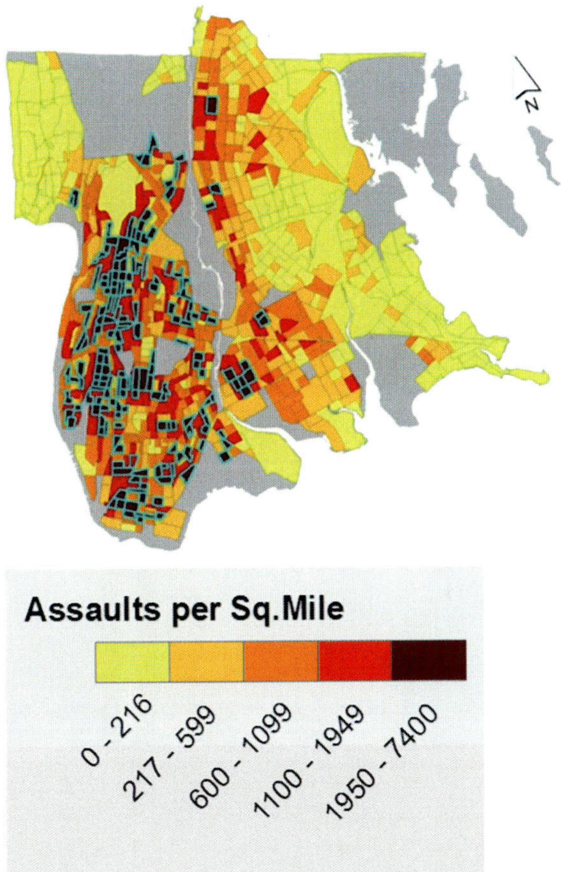

Fig. 4.15 Assault density by census block group

Table 4.5 Assault frequencies and percentages within top qunitile class

Bronx Assault 2006–2010 n=20,726	# of total assaults in top quintile	% of total assaults in top quintile	# of total streets in top quintile	% of total streets in top quintile	# of Zero assault streets in top quintile	% of zero assault streets in top quintile
High assault neighborhoods (7 out of 37)	5,643	27	1,507	15	460	31
High assault tracts (60 out of 329)	7,912	38	1,625	16	421	26
High assault block groups (167 out of 924)	7,737	37	1,335	13	293	22

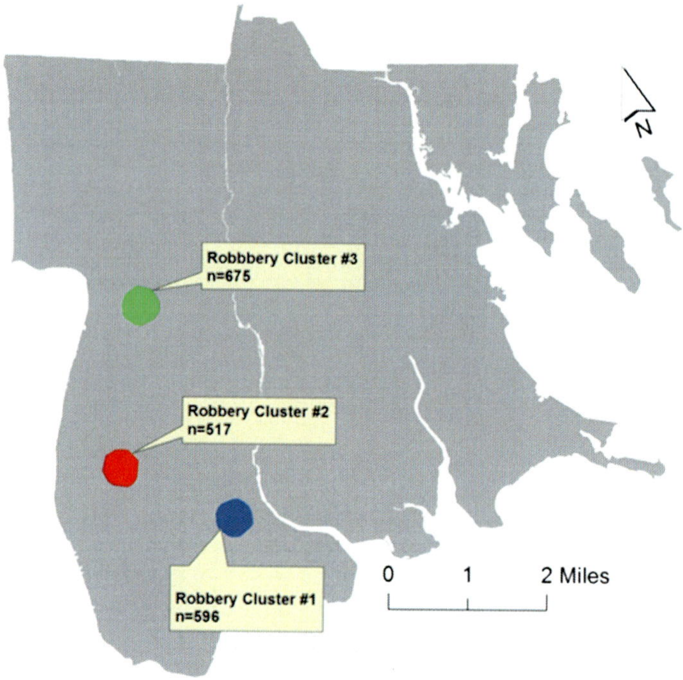

Fig. 4.16 Quarter-mile robbery clusters

is to isolate geographical areas of high/low concentration(s) so police departments can focus their resources on these areas and analysts can gain a better understanding of the complex relationship(s) between crime and place.

Figure 4.16 shows a typical nearest neighbor hierarchical cluster robbery map. The clustering routine was generated in CrimeStat III (Levine 1999) & ArcGIS using an iterative process with a fixed input parameter of a quarter mile and a minimum number of robbery points (greater than 500). My objective was to find the three 'highest' quarter mile robbery clusters. Figure 4.17 shows a single kernel density estimation map that was also generated in CrimeStat III & ArcGIS, using a quartic method of interpolation, fixed interval tenth-mile bandwidth, and relative densities output. While the two maps utilize the same exact robbery data, as you can see, there are several noteworthy differences between the two maps.

First, the clustering map (Fig. 4.16) provides a very easy, focused illustration of the areas that contain the highest number of robberies. The area of each cluster is approximately .18 miles, which makes it an ideal candidate for foot patrol or stationary foot post(s). The clustering routine can show you where the smallest areal units (which you can define) contain the highest number of points (which you can also define). On the other hand, the density map (Fig. 4.17) provides you with a much broader illustration of where robbery is/is not, since the method provides

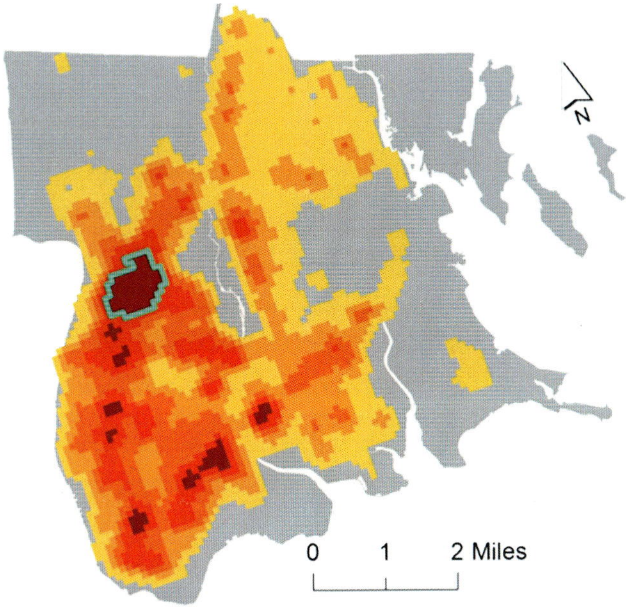

Fig. 4.17 Tenth- mile robbery densities with highest density area highlighted

an estimate for all parts of the study region. In this case, it highlights what areas of the Bronx have high/medium/low/no amounts of robbery. Density maps are great for providing a look at 'the big picture' of crime. (If you would like more detailed information and examples on clustering or density methods, please see Chaps. 6, 7, and 8 in the CrimeStat III manual).

4.12 Internal Temporal Variation Issues with Cluster and Density Methods

When we start to use cluster and density routines to 'zero in' on micro-level crime areas (streets, properties, buildings), we must also consider the temporal variation that will also occur at the micro-level. For example, when the robbery points that are used to generate the robbery cluster map (Fig. 4.16) are queried, exported, and analyzed in Microsoft Excel 3d contour charts, you can observe that there is significant variation by the day of the week (y-axis) and also by the time of day (hour of day, located on the x-axis) for the three distinct quarter-mile robbery clusters.

Robbery Cluster #1 (Fig. 4.16, blue cluster) contains 596 robberies. Table 4.6 illustrates the distinct day of week and hour of day temporal variations throughout this robbery cluster. The 3D contour chart is a very helpful illustration because it

Table 4.6 Robbery Cluster #1 – day of week and hour of day temporal variations

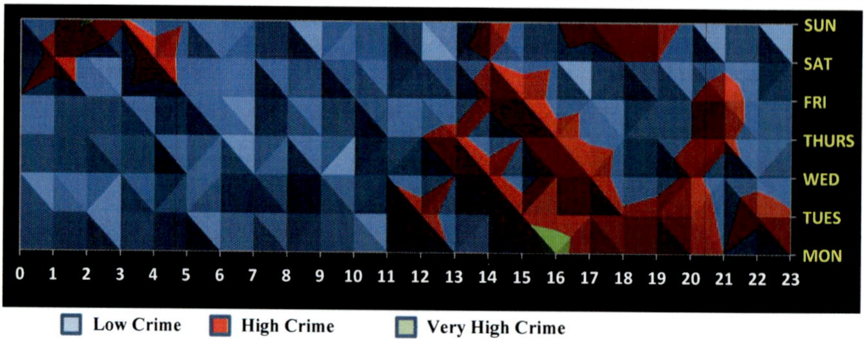

Table 4.7 Robbery cluster #2 – day of week and hour of day temporal variations

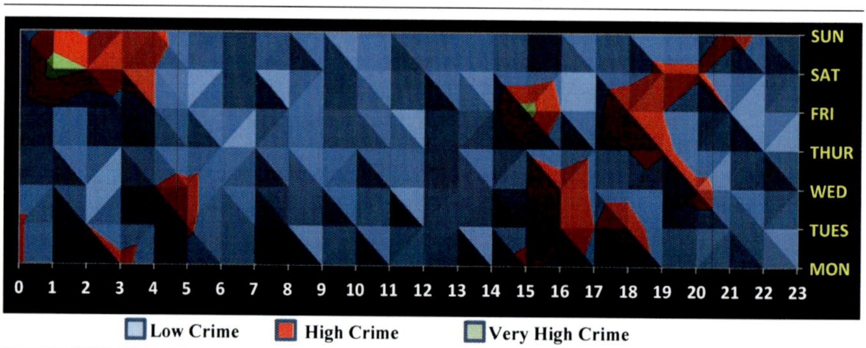

clearly identifies both day of week (y-axis) and hour of day (x-axis) patterns. As you can see, there are two robbery peaks (shown in green) – one on Mondays, around 1,600 h (4 pm) and another peak on Sundays at 0200 h (2 am).

Robbery Cluster #2 (Fig. 4.16, red cluster) contains 517 robberies. Table 4.7 illustrates the day of week and hour of day temporal variations throughout this robbery cluster. As you can see, there are two robbery peaks (shown in green) – one robbery peak on Fridays at 1,500 h (3 pm) and another robbery peak on Saturdays/Sundays between 0030 and 0200 h (12:30 am–2 am).

Robbery Cluster #3 (Fig. 4.16, green cluster) contains 675 robberies. Table 4.8 illustrates the distinct day of week and hour of day temporal variations throughout this robbery cluster. As you can see, this robbery cluster is more similar to robbery cluster #1, but very different from robbery cluster #2. There are two robbery peaks (shown in green) – one robbery peak on Fridays, between 1,300 h – 1,800 h (1 pm–6 pm) and another robbery peak on Tuesdays – Thursdays, between 1,500 h–1,700 h (3 pm–5 pm). Robbery clusters #1 and #3 are primarily weekday, afternoon patterns whereas robbery cluster #2 has distinct weekend, afternoon & late night temporal clustering.

Figure 4.17 shows the single kernel density estimation map that was generated in CrimeStat III & ArcGIS, using a quartic method of interpolation, fixed interval

Table 4.8 Robbery cluster #3 – day of week and hour of day temporal variations

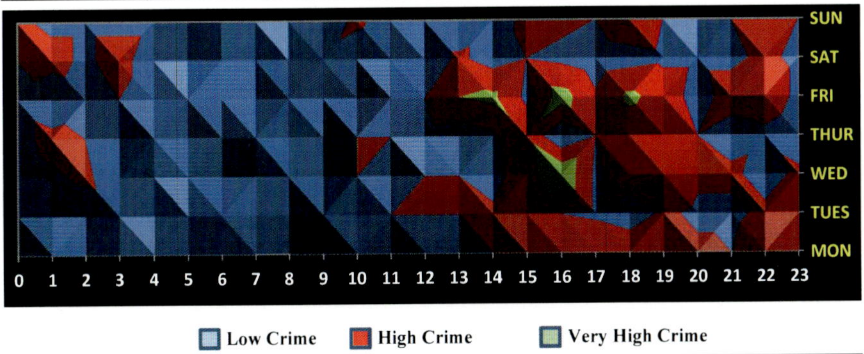

tenth-mile bandwidth, and relative densities output. The highest robbery density z-scores were queried using ArcGIS and a separate shapefile (polygon) was exported for the largest (and highest density) robbery area. This high density robbery zone has a geographical area of .51 square miles, which is almost three times as large as each of the clusters. There are 1,604 robbery points that fall within the high density robbery zone. The robbery zone contains 247 street segments, 51 of these street segments (21%) have no reported robberies on them. Even when using KDE, more than a fifth of the streets contain zero crime and the crime varies extensively based on time of day and day of week temporal trends.

4.13 Driving Crime Analysis Down to the Street (Level)

As this chapter has illustrated, crime continues to cluster as we move further down the cone of resolution to smaller and smaller geographic levels. This is great news for crime analysts, police officers, and police mangers because clustering of crime at smaller areal units makes increased accuracy of targeted police patrols and management of 'hot spots' much easier. However, as Fig. 4.1 illustrated, as we move down the spatial cone of resolution, we must also move down the temporal cone of resolution. We noted significant internal temporal variations within both the quarter-mile clusters and the high density zone(s).

Ratcliffe (2004a, b, c), recognizing both the importance of spatiotemporal clustering and the temporal variance within hot spots, developed a brilliant framework for evaluating and targeting hot spots and called it a "hot-spot matrix". The hot spot matrix incorporates both the spatial and temporal dynamics of hot spots into a manageable framework so police managers can optimize resource allocation and crime control strategies. Spatial events were classified into three spatial categories; dispersed, clustered, or hot points. Temporal events are also classified into three categories; *diffused*, *focused*, and *acute*. While this was originally developed for use with hot spots (polygons), I would like to propose that we transform this framework so it can also be used at the street segment level.

Dispersed spatial events are distributed throughout the hot street; there is no discernible intra hot street clustering. An example of a dispersed hot street pattern might be a street segment where there are a significant number of residential foreclosures. Each foreclosed (vacant) property attracts residential burglary, but no one individual property in particular is the cause of the burglary hot street. A hot street that is classified as *clustered* contains within hot street clustering at one or more points within the hot street. An example of this might be a strip mall parking lot, where cars are stolen at parking spaces near the entrances/exits at a much higher rate than other parking spaces in the parking lot. A *hot street* is exactly that, an individual street segment where crime consistently occurs over and over again. An example of a hot street might be parking spaces on a street near a bank (especially those with an outside ATM), where people get out of their car to take money out of the ATM and are robbed on the way back to their car.

Diffused temporal events have no discernible temporal pattern throughout the hot spot. There may be some temporal variation(s) within the hot spot, but nothing that creates a distinct temporal pattern. A *focused* temporal pattern may have one or more significant increasing trends throughout the day. These trends might require additional manpower, but are not quite considered an acute problem. *Acute* problems are confined to a much smaller period of time. If the majority of problems occur over a relatively short period of time, this is defined as an acute problem.

4.14 Creating Hot Streets

The street segment is becoming a more important unit of analysis in the crime analysis process as a result of the significant within-neighborhood crime variance. In addition, intra-hot spot temporal variation (as described in Figs. 4.16 and 4.17, Tables 4.6, 4.7, and 4.8) indicates significant temporal variation within crime clusters. The last part of this chapter will explain how hot streets are created, identified, analyzed and various methods to geovisualize hot streets.

Micro-level crime analysis begins like many other geospatial point pattern analyses, geocoding. The accuracy of geocoding becomes essential to any type of micro-level crime modeling and analysis. Crime locations that are unable to be geocoded or locations that are inaccurately geocoded can create significant problems for micro-level analyses by skewing statistical and/or spatiotemporal results.

While geocoding continues to be an issue (Ratcliffe 2004a, b, c) for some police departments, other departments have made significant improvements in the way that crime locations are assigned an X/Y coordinate on the map. The New York City Department of City Planning (2010) developed an innovative geocoding application, called 'GBAT', that allows New York City GIS analysts to batch geocode address files to the property lot level. Once crime locations are properly geocoded, they can be spatially joined to higher-level geographies. There are several ways to complete this process in ArcGIS.

The most effective way to begin this aggregation process is a 'bottoms-up approach', where the street (line) file is spatially joined to the crime (point) file (i.e. lines to points)

AND the crime (point) file is also spatially joined to the street (line) file (i.e. points to lines). This aggregation process will allow calculation of crime (points) per street and also permit calculation of temporal (or other) attributes of crime (assuming your crime points have temporal/other attributes) for each street segment.

4.15 Spatiotemporal Variations Within Neighborhoods

As was illustrated earlier in this chapter (Figs. 4.16 and 4.17, Tables 4.6, 4.7, and 4.8), there was distinct spatiotemporal variation within crime hot spots (both kernel densities and clusters). For this neighborhood level analysis example, I analyzed two different violent crimes (shootings and robbery) within the Mott Haven neighborhood in the south Bronx. Mott Haven was selected because it contained the highest number of shootings and robberies (combined over the 5-year study period) compared to the other 36 neighborhoods in the Bronx.

As was similar to the internal variation observed within the hot spots (densities and clusters), visual inspection of Fig. 4.18 indicates considerable spatial variation between streets containing robberies in this neighborhood. Since robberies are the most frequent violent crime in the Bronx, it is expected that robbery would also be the most widespread violent crime at the neighborhood level. Figure 4.18 indicates that 41% of the streets in the neighborhood have zero robberies; 67% of the streets have less than 3 robberies per segment; and 5% of streets contain 35% of the neighborhood robbery.

Table 4.9 indicates the day of week/time of day temporal patterns of robbery incidents within the Mott Haven neighborhood. There are two noticeable temporal

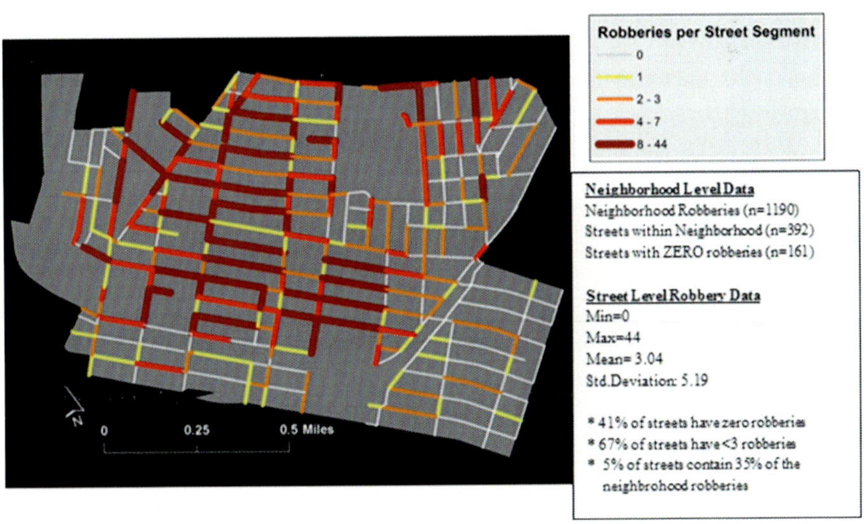

Fig. 4.18 Robbery streets in the neighborhood of Mott Haven

Table 4.9 Temporal patterns of robbery incidents in a Bronx neighborhood

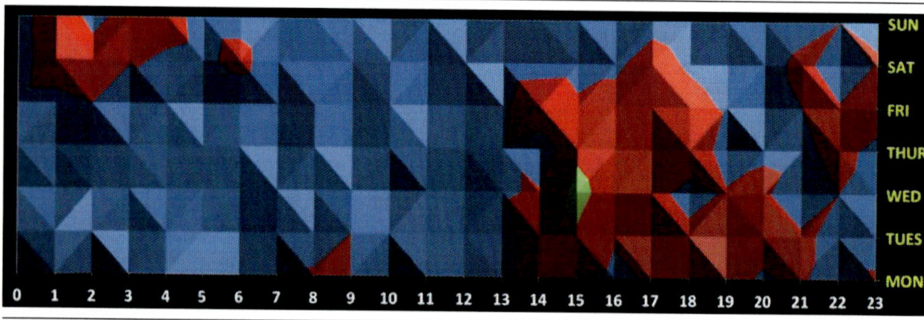

Table 4.10 Temporal patterns of shooting incidents in a Bronx neighborhood

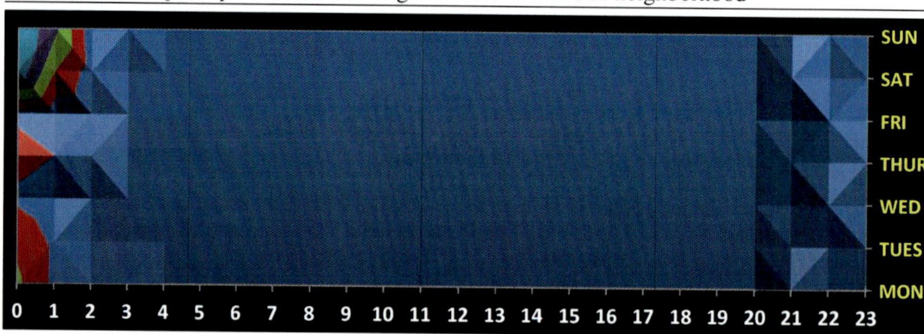

trends that can be observed within this robbery dataset. First, we can note that there are two (red) time periods highlighted on the chart, a weekday afternoon (2 pm–7 pm) trend and a weekend nighttime (10 pm–4 am) robbery trend. The highest temporal peak for robbery in Mott Haven is Wednesday at 3 pm.

Table 4.10 illustrates the temporal variation between the neighborhood shooting incidents. This is a very interesting temporal crime pattern, since there are no reported shootings that occur between 4 am and 8 pm on any day of the week. 52% of the shootings occur within a 1-h time frame, between the hours of midnight and 1 am. When the shooting data is disaggregated further (by hour of day and day of week), we can observe more specifically that 30% of shootings within this neighborhood occur on Saturdays & Sundays, between midnight and 1 am.

As was similar to the intra-neighborhood spatial variation observed within the robbery hot streets, visual inspection of the shooting hot streets (Fig. 4.19) indicates considerable spatial variation between streets containing shootings in this neighborhood. Since shootings occur much less frequently than robbery, it is expected that lower frequency crimes (like murder, rape, shootings) would cluster more at the different geographic levels. Table 4.10 indicates that 83% of the Mott Haven streets have zero shootings and 5% of streets contain 62% of the neighborhood shootings.

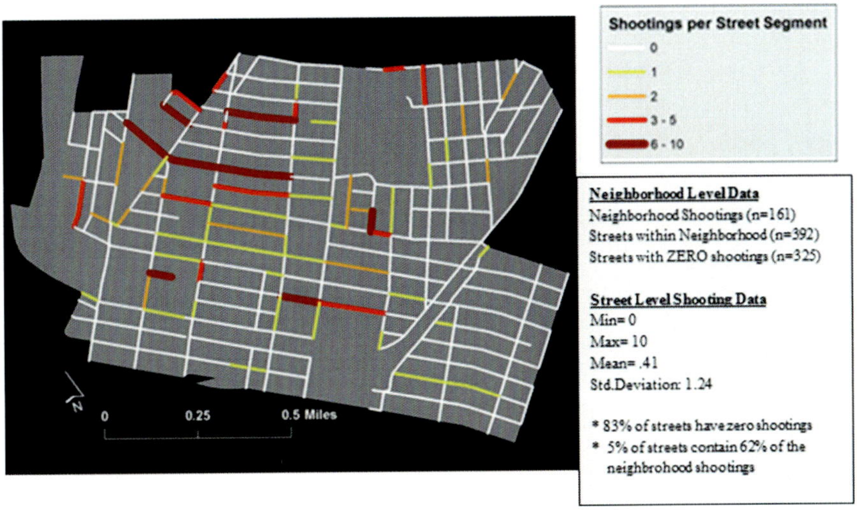

Fig. 4.19 Shooting streets in a Bronx neighborhood

4.16 Conclusion

This chapter identifies several unique advantages for using street segments as a small-scale unit of analysis when conducting geospatial modeling and mapping for crime analysis. As was illustrated earlier in the chapter, there is considerable internal spatiotemporal variation(s) when conducting traditional hot spot analyses and neighborhood level crime analyses. Understanding that crime is clustered in both space and time is not a new finding, however, this chapter highlights some of the benefits of utilizing street segments as units of analysis including identification of hot streets and detection of unique temporal patterns, both of which can assist police departments in crime prevention and control strategies.

It is important to note that the identification of spatiotemporal patterns of hot streets provides significant 'actionable intelligence' for police departments. Understanding that a small percentage of streets are responsible for a significant percentage of violent crime is an important finding of this research. Equally important, although often overlooked, is the number and percentage of streets with zero crime over the study period. Developing street level crime prevention and control strategies can save police departments considerable resources (manpower, time, money) and provide police with a much better understanding of the relationship between crime and opportunity at the street level.

Future research on hot streets should incorporate analysis of land-use and business types to determine what the spatiotemporal relationship(s) are between hot streets, violent crime types, and the smaller-scale units 'below' the street segment level.

References

Baller RD, Anselin L, Messner SF, Deane G, Hawkins DF, Baller RD, Anselin L, Messner SF, Deane G, Hawkins DF (2001) Structural covariates of U.S. county homicide rates: incorporating spatial effects. Criminology 39:561–590

Block R, Bernasco W (2011) Robberies in Chicago: a block-level analysis of the influence of crime generators, crime attractors, and offender anchor points. J Res Crime Delinq 48(1):33–57

Block RL, Block CR (1995) Criminal careers of public places. In: Eck JE, Weisburd D (eds) Crime and place. Criminal Justice Press, Monsey

Block RL, Block CR (1995) Space, place, and crime: hot spot areas and hot places of liquor-related crime. In: Eck JE, Weisburd D (eds) Crime and place, vol 4. Criminal Justice Press, Monsey, pp 145–183

Block R, Perry S (1993) STAC News. vol.1, No.1, Illinois Criminal Justice Information Authority: Statistical Analysis Center. Baltimore County Police fight crime with STAC. Available online at: www.icjia.state.il.us/public/index.cfm?metasection=publicationsandmetapage=STACNEWS_01_W9

Bloomberg M, CBS Radio (2011) State budget means NYPD must shrink

Boba R (2001, Nov). Introductory guide to crime analysis and mapping. Office of Community Oriented Policing Services, Washington, DC

Braga A (2008) Police enforcement strategies to prevent crime in hot spot areas. Crime prevention research reviews, No. 2. U.S. Department of Justice, Office of Community Oriented Policing Services, Washington, DC

Braga AA, Weisburd DL, Waring EJ, Mazerolle LG, Spelman W, Gajewski F (1999) Problem-oriented policing in violent crime places: a randomized controlled experiment. Criminology 37(3):541–580

Brantingham PJ, Dyreson DA, Brantingham PL (1976) Crime seen through a cone of resolution. Am Behav Sci 20:261–273

Bratton W (1996) Cutting crime and restoring order: what America can learn from New York's finest. Speech presented to the Heritage Foundation on 15 October 1996

Caeti T (1999) Houston's targeted beat program: a quasi-experimental test of police patrol strategies. Ph.D. dissertation, Sam Houston State University. University Microfilms International, Ann Arbor

Chainey S, Ratcliffe JH (2005) GIS and crime mapping. Wiley, Hoboken/Chichester

Chainey SP, Reid S, Stuart N (2002) When is a hotspot a hotspot? A procedure for creating statistically robust hotspot maps of crime. In: Kidner D, Higgs G, White S (eds) Innovations in GIS 9: socio-economic applications of geographic information science. Taylor and Francis, London

Clarke RV, Weisburd D (1994) Diffusion of crime control benefits: observations on the reverse of displacement. In: Clarke RV (ed) Crime prevention studies, vol 2. Criminal Justice Press, Monsey, pp 165–182

Cohen L, Felson M (1979) Social change and crime rate trends: a routine activity approach. Am Sociol Rev 44:588–605

Cork D (1999) Examining space-time interaction in city-level homicide data: crack markets and the diffusion of guns among youth. J Quant Criminol 15(4):379–406

Criminal Justice Commission (1998) Beenleigh calls for service project: evaluation report. Criminal Justice Commission, Brisbane

Eck JE (2002a) Crossing the borders of crime: factors influencing the utility and practicality of interjurisdictional crime mapping. Police Foundation and National Institute of Justice, Washington, DC

Eck J (2002b) Evidence-based crime prevention. Preventing crime at places. Routledge, New York

Elffers H (2003) Analyzing neighbourhood influence in criminology. J Neth Soc Stat Oper Res 57(3):347–367

Faggiani D, Bibel D, Brensilber D (2001) Regional problem solving using the national incident-based reporting system. In: Corina SB, Melissa R, Lisa C (eds) Solving crime and disorder problems. Police Executive Research Forum, Washington, DC

Gartner R (1990) The victims of homicide: a temporal and cross-national comparison. Am Sociol Rev 55:92–106

Groff ER (2007) Simulation for theory testing and experimentation: an example using routine activity theory and street robbery. J Quant Criminol 23:75–103

Groff ER, Weisburd D, Yang S-M (2010) Is it important to examine crime trends at a local "Micro" level? A longitudinal analysis of block to block variability in crime trajectories. J Quant Criminol 26:7–32

Hardisty F, Klippel A (2010) Analysing spatio-temporal autocorrelation with LISTA-Viz. Int J Geogr Inform Sci 24(10):1515–1526

Hipp JR (2007) Block. tract, and levels of aggregation: neighborhood structure and crime and disorder as a case in point. Am Sociol Rev 72(5):659–680

Hope T (1994) Problem-oriented policing and drug market locations: three case studies. Crime Prev Stud 2:5–32

Kennedy LW, Baron SW (1993) Routine activities and a subculture of violence: a study of violence on the street. J Res Crime Delinq 30(1):88–112

Kulldorff M (1997) A spatial scan statistic. Comm Stat: Theory Method 26:1481–1496

Levine N (1999) CrimeStat: a spatial statistics program for the analysis of crime incident locations (version 1.0). Ned Levine & Associates/National Institute of Justice, Annandale/Washington, DC

Maantay JA, Maroko A, Herrmann C (2007) Mapping population distribution in the urban environment: the Cadastral-based Expert Dasymetric System (CEDS), 2007 US National Report to the International Cartographic Association, Cartography and Geographic Information Science, vol 34, no 2, pp 77–102

Maroko AR, Maantay JA, Sohler NL, Grady K, Arno P (2009) The complexities of measuring access to parks and physical activity sites in New York City: a quantitative and qualitative approach. Int J Health Geogr 8(34):1–24

Martin D, Barnes E, Britt D (1998) The multiple impacts of mapping it out; police, Geographic Information Systems (GIS) and community mobilization during devil's night in detroit, Michigan. In: LaVigne N, Wartell J (eds) Crime mapping case studies: successes in the field. Police Executive Research Forum, Washington, DC

McGuire PG, Williamson D (1999) Mapping tools for management and accountability. Paper presented to the third international crime mapping research center conference, Orlando, Florida, 11–14 December 1999

Newsweek (2009) Retrieved at: http://www.thedailybeast.com/newsweek/galleries/2009/01/17/photos-bronx-residents-on-obama.html

New York City Department of City Planning (2010) Information retrieved from: http://www.nyc.gov/html/doitt/html/consumer/gis.shtml

New York City Police Department (2011) New York Police Department (NYPD) crime database [Computer file, 2011]

Openshaw S (1984) The modifiable areal unit problem. Concept Tech Mod Geogr 38:41

Peuquet DJ (1994) It's about time: a conceptual framework for the representation of temporal dynamics in geographic information systems. Ann Assoc Am Geogr 84(3):441–462

Ratcliffe JH (2002) Aoristic signatures and the temporal analysis of high volume crime patterns. J Quant Criminol 18(1):23–43

Ratcliffe JH (2004a) The hotspot matrix: a framework for the spatio-temporal targeting of crime reduction. Police Pract Res 5(1):5–23

Ratcliffe JH (2004b) Geocoding crime and a first estimate of a minimum acceptable hit rate. Int J Geogr Inf Sci 18(1):61–72

Ratcliffe JH (2004c) Geocoding crime and a first estimate of a minimum acceptable hit rate. Int J Geogr Inf Sci 18(1):61–72

Ratcliffe JH (2006) A temporal constraint theory to explain opportunity-based spatial offending patterns. J Res Crime Delinq 43(3):261–291

Ratcliffe JH, Taniguchi T, Groff E, Wood J (2011) The Philadelphia foot patrol experiment: a randomized controlled trial of police patrol effectiveness in violent crime hotspots. Criminology 49(3):795–831

Rengert G, Lockwood B (2009) Units of analysis and the analysis of crime. In: David W, Wim B, Gerben B (eds) Putting crime in its place: units of analysis in spatial crime research. Springer, New York

Rosenfeld R, Messner SF, Baumer EP (2001). Social capital and homicide. Soc Force 80(1):283–309

Sherman L, Weisburd D (1995) General deterrent effects of police patrol in crime 'Hot Spots': a randomized study. Justice Quart 12(4):625–648

Sherman L, Rogan D (1995a) Effects of gun seizures on gun violence: 'Hot Spots' patrol in Kansas City. Justice Quart 12:673–694

Sherman L, Rogan D (1995b) Deterrent effects of police raids on crack houses: a randomized controlled experiment. Justice Quart 12:755–782

Sherman L, Gartin P, Buerger M (1989) Hot spots of predatory crime: routine activities and the criminology of place. Criminology 27:27–56

Subramanian SV (2009) Revisiting Robinson: the perils of individualistic and ecological fallacy. Int J Epidemiol 38(2):342–360

Taylor RB (1998) Crime in small-scale places: what we know, what we can do about it. In: Crime and place: plenary papers of the 1997 conference on criminal justice research and evaluation. National Institute of Justice, Washington, DC, pp 1–22

Taylor RB (2001) Understanding the connections between physical environment, crime, fear and resident-based control. In: Wilson JQ, Petersilia J (eds) Crime: public policies and crime control. ICS Press, Oakland, pp 413–425

Tita G, Cohen J (2004) Measuring spatial diffusion of shots fired activity across city neighborhoods. In: Goodchild MF, Janelle DG (eds) Spatially integrated social science. Oxford Press, New York, pp 171–204

Weir R, Bangs M (2007) The use of geographic information systems by crime analysts in England and Wales, Home Office online report series. Home Office, London

Weisburd D, Green L (1994) Defining the street-level drug market. In: MacKenzie DL, Uchida CD (eds) Drugs and crime: evaluating public policy initiatives. Sage Publications, Newbury Park, pp 61–76

Weisburd D, Green L (1995) Policing drug hot spots: the Jersey City drug market analysis experiment. Justice Q 12:711–736

Weisburd D, Bushway S, Lum C, Yang S-M (2004) Crime trajectories at places: a longitudinal study of street segments in the City of Seattle. Criminology 42(2):283–322

Weisburd D, Bernasco W, Bruinsma G (eds) (2009) Putting crime in its place: units of analysis in spatial crime research. Springer, New York

Zimring F (2011) New York and the future of crime control. Paper presented at John Jay College of Criminal Justice – understanding the crime decline in New York

Chapter 5
Exploring Spatial Patterns of Crime Using Non-hierarchical Cluster Analysis

Alan T. Murray and Tony H. Grubesic

Abstract Exploratory spatial data analysis (ESDA) is a useful approach for detecting patterns of criminal activity. ESDA includes a number of quantitative techniques and statistical methods that are helpful for identifying significant clusters of crime, commonly referred to as hot spots. Perhaps the most popular hot spot detection methods, both in research and practice, are based on tests of spatial autocorrelation and kernel density. Non-hierarchical clustering methods, such as k-means, are less used in many contexts. There is a perception that these approaches are less definitive. This chapter reviews non-hierarchical cluster analysis for crime hot spot detection. We detail alternative non-hierarchical approaches for spatial clustering that can incorporate both event attributes and neighborhood characteristics (i.e., spatial lag) as a modeling parameter. Analysis of violent crime in the city of Lima, Ohio is presented to illustrate this for hot spot detection. We conclude with a discussion of practical considerations in identifying hot spots.

Keywords Clustering • Hot spots • Spatial patterns

A.T. Murray (✉)
GeoDa Center for Geospatial Analysis and Computation, School of Geographical Sciences and Urban Planning, Arizona State University, Tempe, AZ 85287-5302, USA
e-mail: atmurray@asu.edu

T.H. Grubesic
Geographic Information Systems and Spatial Analysis Laboratory, College of Information Science and Technology, Drexel University, Philadelphia, PA 19104, USA
e-mail: grubesic@drexel.edu

M. Leitner (ed.), *Crime Modeling and Mapping Using Geospatial Technologies*, Geotechnologies and the Environment 8, DOI 10.1007/978-94-007-4997-9_5,

5.1 Introduction

Cluster detection and hot spot mapping in criminology, geography and related socio-economic planning sciences has evolved significantly over the past decade (Eck et al. 2005; Chainey et al. 2008). While many of the most basic approaches remain popular, such as spatial autocorrelation, spatial ellipses, kernel density estimation and spatial scan statistics (Wang 2005; Eck et al. 2005; Kent and Leitner 2007; Chainey et al. 2008; Rogerson and Yamada 2009; Anselin et al. 2009), advanced approaches now include fuzzy clustering (Grubesic 2006), spatio-temporal modeling of crime (Ratcliffe 2002; Grubesic and Mack 2008; Leitner et al. 2011), geospatial visual analytics (Anselin and Kochinsky 2010), and agent-based simulation (Eck and Liu 2008). Further, the emergence of proactive policing, predictive hot spotting and crime forecasting strategies suggests a growing need for objective spatial pattern detection methods to establish a better understanding of the distributions and morphologies crime (Cohen et al. 2004; Gorr et al. 2003; Johnson and Bowers 2004; Wu and Grubesic 2010).

Broadly defined, a crime hot spot represents a grouping of incidents that are spatially and/or temporally clustered (Harries 1999; Eck et al. 2005; Grubesic 2006). The genesis of crime hot spots is often linked to environmental factors (Brantingham and Brantingham 1981), social disorganization (Shaw and McKay 1942; Sampson and Groves 1989; Morenoff et al. 2001) and opportunity (Cohen and Felson 1979). Regardless of the underlying factors that fuel the emergence of hot spots, law enforcement agencies recognize the importance (and benefits) of detection and intervention in these problematic areas (Harries 1999; Braga 2001; Ratcliffe 2004). However, the ability to identify hot spots is highly dependent on the capability to detect patterns, and this requires the selection of appropriate techniques for carrying out hot spot analyses. Such pattern detection is typically viewed as exploratory spatial data analysis (ESDA) (Murray and Estivill-Castro 1998; Anselin 1998; Wu and Grubesic 2010), but can be confirmatory in some contexts.

At the intersection of ESDA, GIS, and crime analysis is the use of ESDA for identifying significant patterns of criminal activity (Harries 1999; Anselin et al. 2000; Murray et al. 2001). Again, while local indicators of spatial association (Messner et al. 1999; Anselin et al. 2000, 2009) and kernel density mapping (McLafferty et al. 2000) are popular approaches for identifying hot spots, alternative techniques such as cluster analysis are less utilized in practice. Grubesic (2006) notes that there are three major problems associated with applying cluster analysis for crime hot spot detection:

1. The choice between non-hierarchical and hierarchical methods can be confusing[1];
2. There are problems regarding the manner in which some techniques treat geographic space (e.g., spatial bias);
3. There is relatively little guidance for determining the appropriate number of clusters in a study area.

[1] A discussion of hierarchical and non-hierarchical methods can be found in Hartigan (1975), Everitt (1993) and Kaufman and Rousseuw (2005), among others. Non-hierarchical, or partitioning, approaches identify a pre-specified number of clusters, k, such that each object is a member of exactly one cluster, where membership similarity is optimized. In contrast, hierarchical methods

While these challenges can be daunting, non-hierarchical cluster analysis is potentially useful for finding crime hot spots, reflected by its inclusion in the National Institute of Justice sponsored and supported crime analysis tool, CrimeStat (Levine 2010).

The non-hierarchical technique implemented in CrimeStat (version 3.3) is the *k*-means approach proposed by Fisher (1958). The *k*-means technique is based upon multivariate analysis of variance in the evaluation of homogeneity among entities (Estivill-Castro and Murray 2000). Specifically, the scatter matrix of similarity between entities may be evaluated by its trace (Aldenderfer and Blashfield 1984), and homogeneity is then measured for a grouping of events using the sum of squares loss function (Rousseeuw and Leroy 1987). The benefits of using *k*-means lie in its ability to handle extremely large numbers of observations and still generate clusters relatively quickly, although this is contingent on the number of iterations selected for the routine.

Other non-hierarchical clustering approaches have been developed and utilized. Some are detailed in Kaufman and Rousseuw (2005) In the context of geographic applications, a review of approaches is given in Murray and Estivill-Castro (1998), Murray (2000a, b) and Grubesic (2006). Clearly, if one is intent on identifying crime hot spots that are strongly related in some predefined sense (e.g. crime type), then multiple non-hierarchical clustering techniques may be useful. This is a subtle but important point. If an analyst is able to choose from a suite of alternative clustering approaches, a clearer picture of the spatial morphology of crime may emerge. However, it is also possible that the selection of an inappropriate technique may skew the identification and interpretation of crime hot spots, minimizing the usefulness of the approach. This is particularly true where non-hierarchical approaches are concerned because many analysts may not be aware of the biases and inaccuracies associated with a particular approach. Simply put, all clustering methods are not equivalent. Unfortunately, the overall body of research focusing on the subtle differences in the use and application of non-hierarchical techniques for geographic applications is rather limited (Murray 1999, 2000a; Murray and Grubesic 2002; Grubesic 2006). Empirical results suggest that substantial variation exists in the structure and quality of clusters, depending on the approach.

The purpose of this chapter is to review clustering approaches for identifying spatial patterns of crime, focusing on the basic tenets of crime mapping and analysis from a geographic perspective. This is followed by an examination of the statistical foundations of non-hierarchical cluster analysis, highlighting the strengths and weaknesses of the most widely utilized approaches. Section 5.4 introduces alternative approaches for non-hierarchical cluster analysis that incorporate additional geographic context through the use of spatial lags. Application results examine violent crime in Lima, Ohio. We conclude with a brief discussion and final remarks.

---

build clusters based on agglomeration (e.g., begin with *n* clusters and merge the two most similar groups to get *n*-1 clusters) or division (e.g., begin with one cluster and divide it into two most similar clusters), creating a decomposition hierarchy of clusters ranging from *n* to 1.

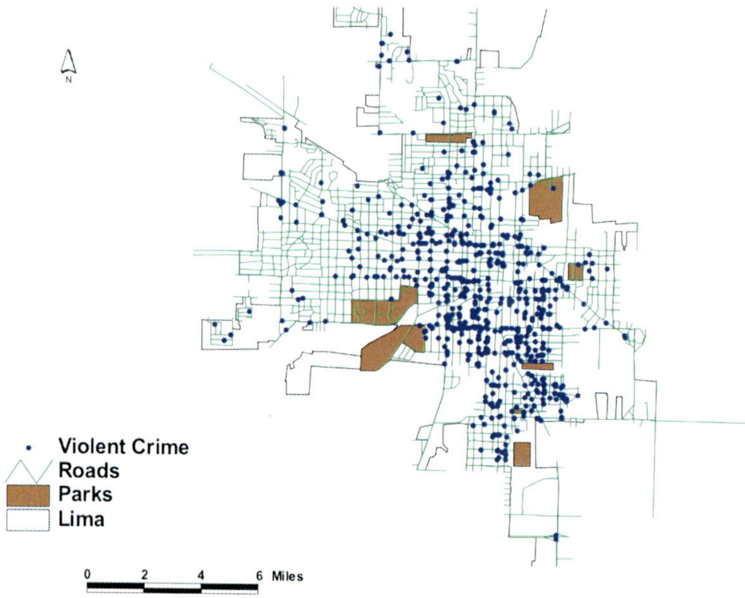

Fig. 5.1 Violent Crime in Lima, Ohio

5.2 Spatial Patterns of Crime

Identifying significant geographic relationships in the occurrence of criminal activity is, perhaps, the most fundamental component of crime mapping and analysis. Of course, the process is complicated by a vast array of techniques and methods available to analysts. In many instances, the first step in developing a better understanding of crime distributions and their contributing factors is to generate a map. This might involve plotting incident locations, differentiating them by crime type and adding topographic information for additional spatial context. For example, Fig. 5.1 illustrates 848 violent crimes (homicide, rape, robbery and assault) in the city of Lima, Ohio.[2] Alternatively, if the crime information is only recorded at a more aggregate level, such as census block groups, then a choropleth map of total crime or crime rates for a geographic area can be created. At this level of geographic detail, broader patterns of neighborhood distress and spatial inequity may become apparent. For instance, Fig. 5.2 depicts violent crime rates in Lima using a choropleth display of block group crime rates per 1,000 people. Ignoring the overlaid ellipses for the moment, this display emphasizes differences in the attribute of interest using seven unique classes. As with any choropleth display, the goal is to

[2] Lima, Ohio is a city of approximately 38,000 people and is located about 70 miles north of Dayton on the Interstate 75 corridor.

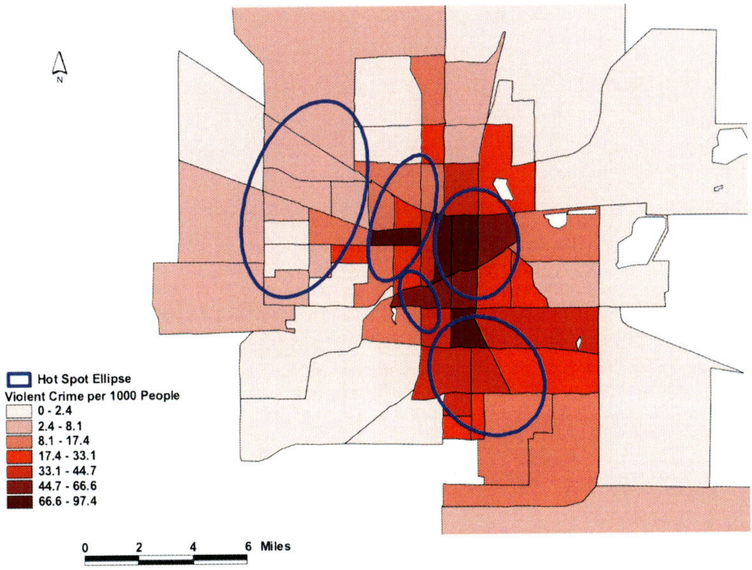

Fig. 5.2 Crime rates by block group in Lima

effectively show spatial variation in the variable's distribution. Creation of a traditional choropleth map involves deciding where to establish the class break/cutoff values (Dent 1999; Murray and Shyy 2000). In Fig. 5.2, class breaks of 2.4, 8.1, 17.4, 33.1, 44.7 and 66.6 (shown in the legend) are used, derived using the natural breaks options in ArcGIS. This classification helps communicate how violent crime rates vary spatially in Lima, but does so in a much different way than the point map displayed in Fig. 5.1.

Perhaps the most intriguing aspect of crime mapping and analysis is the subtle methodological overlap of choropleth mapping approaches, non-hierarchical cluster analysis and hot spot detection techniques. Choropleth mapping is an area of cartography and GIS that has received considerable interest over the past 50 years (Murray and Shyy 2000; Armstrong et al. 2003; Xiao and Armstrong 2005; Cromley and Cromley 2009). Numerous choropleth mapping approaches have been developed, most of which are accessible and readily available in commercial GIS and cartography software. As noted, the display shown in Fig. 5.2 was generated using the natural breaks option in ArcGIS (version 10.3), an approach that is also available in TransCad, MapInfo, Maptitude and many other GIS packages. Natural breaks is widely considered the standard/default choropleth mapping method. In brief, the natural breaks approach attempts to minimize the sum of variance in created classes (Dent 1999). This is identical to the goal of non-hierarchical clustering, such as k-means, a sum of squares approach.

By analyzing either Fig. 5.1 or Fig. 5.2, analysts could make inferences about the spatial distribution, and perhaps the potential impact, of violent crime in Lima. Clearly, the intent of crime analysis is that such displays are helpful for understanding crime trends and patterns so that appropriate law enforcement action can be prescribed.

The next step would typically involve assessment of spatial autocorrelation, at least for aggregate crime rates such as the block groups in Fig. 5.2, as this would help confirm whether clustering is occurring. Packages like as CrimeStat, GeoDa (Anselin et al. 2006) and ArcGIS allow analysts to derive such measures. In this instance, we find that Moran's I is 0.710 with a standard normal z-value of 11.43 ($p=0$), indicating spatial clustering of violent crime in Lima. Unfortunately, global metrics do not pinpoint where this clustering is taking place. As a result, if an analyst is interested in determining where hot spots exist, additional analysis is necessary. In many cases, local spatial statistics and non-hierarchical clustering approaches are advocated for identifying and assessing potential hot spots (Anselin 1995; Harries 1999; Messner et al. 1999; Levine 2006; Ratcliffe 2005). These approaches are typically coupled with standard deviation ellipses in an effort to represent the co-variation within a cluster group about the major and minor axes.

The ellipses associated with the k-means generated clusters using CrimeStat (version 3.3) are also shown in Fig. 5.2. Fundamentally, this shows the integration of non-spatial *and* spatial grouping processes. The ellipses represent the spatial grouping of the associated areas, whereas the choropleth classes reflect attribute (violent crime rate) variation. Furthermore, it is worth reiterating that the ellipses were generated in CrimeStat from spatial clusters identified using a k-means heuristic, although alternative options for summarizing distributions are also available. As noted previously, this is all the more interesting because the natural breaks choropleth classes shown in Fig. 5.2 are also identified using equivalent criteria.

There are a number of questions arising from this brief review on spatial aspects of crime hot spot detection. Is the sum of squares clustering approach and its most popular solution technique (k-means) viable for spatial data? If not, why? Are there feasible alternatives to these approaches that can either complement or improve upon the results generated through traditional solution techniques? In an effort to address these questions, the next section outlines the fundamental nature of non-hierarchical clustering, with a focus on the sum of squares approach.

5.3 Statistical Clustering

As noted previously, cluster analysis is a popular approach for developing classification systems and taxonomies. A simple search on the Social Sciences Citation Index reveals that nearly 130,098 entries have referenced "cluster analysis" since 1980, equating to approximately 6,195 per year (1980–2011). In crime analysis, as in other problem domains, the sum of squares variance minimization approach continues to be the dominant non-hierarchical partitioning technique (Levine 2010). In fact, most commercial statistical packages, including SPSS,

S-Plus, SAS, Stata and NCSS, provide capabilities for carrying out cluster analysis using the sum of squares approach (Murray and Grubesic 2002; Grubesic 2006). Consider the following notation:

$$\begin{aligned}
i &= \text{index of entities;}\\
k &= \text{index of clusters;}\\
p &= \text{total number of clusters;}\\
f_i &= \text{attribute measure;}\\
d_{ik} &= \text{measure of proximity between entity } i \text{ and cluster } k;\\
z_{ik} &= \begin{cases} 1 & \text{if entity } i \text{ is in cluster } k \\ 0 & \text{otherwise.} \end{cases}
\end{aligned}$$

Where crime analysis is concerned, entities correspond to the location of a crime(s). The variable f_i indicates the number of crimes occurring at a particular location i. If there is a need to attribute a measure of importance to particular crime types (e.g. severity), it is possible to extend the specification of f_i to reflect such differentiation.[3] The sum of squares approach is as follows:

Sum of Squares Clustering Model (SSCM)

$$Minimize \sum_i \sum_{k=1}^{p} f_i d_{ik}^2 z_{ik} \tag{5.1}$$

Subject to:

$$\sum_{k=1}^{p} z_{ik} = 1 \ \forall i \tag{5.2}$$

$$z_{ik} = (0,1) \ \forall i,k \tag{5.3}$$

The objective (5.1) of the SSCM is to minimize the total weighted squared difference in cluster group membership. This is equivalent to minimizing the within group sum of squares (Hartigan 1975; Kaufman and Rousseeuw 2005). Constraint (5.2) ensures that each entity is assigned to a group and Constraint (5.3) imposes integer restrictions on the decision variables.

The formulation of the sum of squares clustering model illustrates that this is an optimization problem. The overall goal of the SSCM is to identify the best, or optimal, partition of entities. One approach for solving the SSCM is the k-means heuristic developed by Fisher (1958) and MacQueen (1967), when Euclidean

---

[3] Details on multivariate integration for such purposes may be found in Kaufman and Rousseeuw (2005) as well as other clustering texts.

distance is considered. In vector quantization, this heuristic is also known as the generalized Lloyd algorithm (Estivill-Castro and Murray 2000). This optimization problem is recognized as being inherently difficult to solve optimally, so the application of heuristic techniques such as the *k*-means approach are considered a good option for obtaining a solution. The *k*-means heuristic has four main steps (Murray and Grubesic 2002):

1. generate *p* initial clusters
2. compute the center of each cluster
3. assign each entity to its closest cluster
4. if groupings have changed in step 3, return to step 2. If not, a local optima has been found.

A notable feature of the SSCM is that the center of each grouping is a centroid, reflecting the squared Euclidean proximity measure in the objective (5.1). In addition, the *k*-means heuristic is a popular approach for solving the SSCM for a number of reasons. First, it is statistically grounded and widely available in most commercial statistical software packages (Murray and Grubesic 2002). Second, it has the ability to handle relatively large data sets (Huang 1998). Third, it converges quickly to find a local optima (Murray and Grubsic 2002).

While these advantages are certainly appealing and have contributed to its widespread application, including the NIJ supported CrimeStat software package, there are questions pertaining to the appropriateness of the SSCM when applied to geographic data (Murray and Grubesic 2002; Grubesic 2006). Although many of the biases inherent in the SSCM are widely noted (see Murray and Estivill-Castro 1998; Kaufman and Rousseeuw 2005; among others), the SSCM continues to be relied upon in geographic and non-geographic inquiry.

What is wrong with the sum of squares approach, particularly with respect to the spatial analysis of urban crime? One major issue is the sub-optimality associated with the use of the *k*-means heuristic in solving the SSCM. Often, implementation of this heuristic provides analysts a solution based on one instance. In order for the *k*-means heuristic to be effective for solving the SSCM, it must be re-started hundreds or thousands of times (depending on problem size), using a different initial clustering in step 1 for each instance.[4] Standard practice, however, has been to use only one initial starting configuration. The result is that the identified cluster solutions are likely sub-optimal, which means that they may be of limited use for inferential analysis and policy making. The extent to which sub-optimality was an issues was examined in Murray and Grubesic (2002), who found that non-optimal solutions were generally identified using major statistical packages such as SPSS, S-Plus and SAS. In some instances, SSCM solutions were found to deviate more than 30% from the optimal solution, which means that subsequent analysis is being

[4] It is well known that any single application of the *k*-means heuristic is susceptible to becoming trapped in a local optima (Estivill-Castro and Murray 2000; Murray and Grubesic 2002), which prohibits the approach from identifying an optimal solution.

conducted on clusters that are not most similar. Further, limited testing of CrimeStat found instances where the identified solutions deviated more than 72% from the optimal solution.[5]

A second and more significant problem with the SSCM is that spatial clusters are biased by outliers. Although this bias is discussed by Kaufman and Rousseeuw (2005) and others, Murray and Grubesic (2002) demonstrated the influence of this bias using spatial information rather than non-spatial data. The SSCM is biased because of the use of the squared Euclidean distance measure in objective (5.1). The result in application is that outliers, or more distant events from others, have greater influence on the structure of the identified clusters, effectively distorting potential hot spots. One option is to identify and remove outliers using the approaches detailed in Messner et al. (1999) and Grubesic (2006). Alternatively, it may be preferable to utilize a modeling approach that does *not* spatially bias clusters.

Though not an issue with the SSCM generally, Murray and Grubesic (2002) note that most software packages do not provide the capability to include a f_i value in objective (5.1), rather this is assumed to equal 1.[6] Given this, it makes sense that statistical packages like CrimeStat would attempt to summarize *k*-means generated clustering results using standard deviation ellipses, because the clusters are identified on the basis of space alone.

Finally, the SSCM does not explicitly address attribute similarity, but rather focuses on spatial proximity. Integration of the choropleth display with the ellipses in Fig. 5.2 is an interesting approach for examining spatial and non-spatial patterning in this regard, but lacks direct examination of both issues. Murray and Shyy (2000) present a clustering based approach for choropleth mapping that considers attribute and spatial similarity simultaneously. Murray (2000b) details a spatial lag approach to integrate attribute and spatial proximity.

5.4 Spatial Lag in Cluster Analysis

Geographic analysis using spatial statistical techniques has been significantly enhanced when more is known about what is taking place near a particular entity of interest. The reason this has been the case is that the assumption of independence

[5] Analysis was carried out using 114 crime events in a neighborhood located in Akron, Ohio. The number of clusters obtained ranged from 4 to 11 (p=4–11). A separation value of 4 was utilized in the application of the *k*-means solution technique in CrimeStat for each value of p and the identified solution compared with the "optimal" solution using the approach reported in Murray and Grubesic (2002). For this range of clusters, the average sub-optimality of CrimeStat solutions was 38.28% (min = 12.01%; max = 72.19%). It should be noted that one can alter the separation distance in CrimeStat, in essence representing a pseudo-restart of the heuristic. Unfortunately, it is not possible to compare or assess cluster solution quality.

[6] An assumed value of $f_i = 1$ implies the occurrence of a single event, rather than reflecting the aggregate summary of areas like police beats, census blocks or alternative areal units.

between entities in statistical testing is known to be problematic for spatial data as the existence of spatial autocorrelation can alter significance levels and reduce interpretative capabilities (Griffith and Amrhein 1997). One approach for dealing with spatial autocorrelation involves the use of a spatial lag. A spatial lag represents an averaging process of an entity's neighbors. In most cases, neighbors represent other entities or areas next to a particularly entity. As a point of reference, consider the following notation:

i (and j) = index of entities;
l_i = spatial lag for entity i;
Ω_i = spatial neighbors of entity i

Neighbors are often defined as those entities sharing a common border or point and do not include the entity itself.[7] Using this notation, the spatial lag for entity i may be defined as follows:

$$l_i = \frac{\sum_{j \in \Omega_i} f_j}{|\Omega_i|} \tag{5.4}$$

Spatial lag enables one to summarize what is taking place in a neighborhood around a particular area. For example, one can compute the average number of crime events occurring in neighborhoods that are adjacent to a neighborhood of interest. This is an indirect spatial proximity metric. The integration of both space and attribute values is relatively straightforward:

$$\delta_{ik} = \sqrt{(f_i - \bar{f}_k)^2 + (l_i - \bar{l}_k)^2} \tag{5.5}$$

where $\bar{f}_k$ represents the average attribute value for cluster k and $\bar{l}_k$ indicates the average lag value for cluster k. With this, (5.5) represents an integration of attribute similarity with an indirect spatial proximity metric. Murray (2000b) introduced an alternative clustering model based on this:

Spatial Lag Cluster Model – Center 1 (SLCM-C1)

$$Minimize \sum_i \sum_{k=1}^{p} \delta_{ik} z_{ik} \tag{5.6}$$

Subject to:
(5.2)–(5.3)

Although the constraints for this model are the same as those in the SSCM, the objective of SLCM-C1 is much different. Objective (5.6) minimizes the total

[7] It is also common to view neighbors as being within a specified distance of a given location.

dissimilarity in selected clusters. This differs in three ways from objective (5.1) for the SSCM. First, there is no attribute (f_i) weighting. Second, there is no explicit representation of distance in (5.6) as there is in (5.1). Finally, the similarity measure, δ_{ik}, is not squared in (5.6), whereas it is in (5.1). The implication of this is that the cluster centers in SLCM-C1 are *not* centroids, in contrast to the SSCM. This general representational distinction is a subtle but exceptionally important point. Simply put, by avoiding the use of a centroid in (5.6), the biasing influence of outliers in the SLCM-C1 is minimized. That said, there are tradeoffs with this type of formulation; namely, solving the SLCM-C1 remains challenging due to its implied non-linear form. As a result, the alternating heuristic has generally been relied upon for solving the SLCM-C1 (Murray 2000b).

Clearly, one drawback of the SLCM-C1 is the inability to alter the importance of either attribute or spatial lag influence in the identification of clusters. The SLCM-C1 treats attribute and lag with equal importance. However, this may not necessarily be appropriate for exploratory analysis. For example, one might want to investigate the clusters associated with maximizing attribute similarity only (somewhat equivalent to classes created in choropleth maps using the natural breaks approach). Alternatively, one might wish to view the clusters where lag similarity is optimized. Given these two extremes, it is also possible that one might want to examine the clusters associated with slightly more importance on attribute similarity than lag – or something else in between. Unfortunately, it is not possible to structure the relative importance of variables using the SLCM-C1. In an effort to provide more flexibility, Murray (2000b) presented a modified interpretation of similarity:

$$a_{ik} = \left| f_i - \overline{f}_k \right| \tag{5.7}$$

$$s_{ik} = \left| l_i - \overline{l}_k \right| \tag{5.8}$$

Essentially, these measures track the similarity structured in (5.5), but do so separately. With this modified representation, it is now possible to alter how much significance the individual components have in structuring clusters. Incorporating them independently into a non-hierarchical clustering model may be accomplished by assigning weights to both attributes and lag:

w_a = weight for attribute similarity
w_s = weight for spatial lag similarity

Murray (2000b) derived a variant of the SLCM as follows:

Spatial Lag Clustering Model – Center 2 (SLCM-C2)

$$\mathit{Minimize}\ w_a \sum_i \sum_{k=1}^{p} a_{ik} z_{ik} + w_s \sum_i \sum_{k=1}^{p} s_{ik} z_{ik} \tag{5.9}$$

Subject to:

(5.2)-(5.3)

Objective (5.9) of the SLCM-C2 maximizes the total weighted attribute similarity and maximizes the total weighted spatial lag similarity in selected clusters. In this revised form, (5.9) is now a multi-objective optimization problem that may be used to identify a range of non-dominated clustering solutions (Cohon 1978), each potentially valuable in identifying crime hot spots. Unfortunately, the SLCM-C2 remains a difficult optimization problem to solve optimally, so a heuristic is necessary (Murray 2000b).

Finally, it is also possible to view the above lag models from the more traditional median perspective. Murray (2000b) proposed a multi-objective median based clustering model incorporating spatial lag. Using a median based approach, similarity may be defined as follows:

$$\hat{a}_{ik} = \left| f_i - \bar{f}_k \right| \tag{5.10}$$

$$\hat{s}_{ik} = \left| l_i - \bar{l}_k \right| \tag{5.11}$$

where j is the index of potential medians (same as the index i). This approach enables similarity to be defined a priori between entities, rather than being a function of identified clusters. In order to present the median clustering model, additional decision variables must first be defined:

$$x_j = \begin{cases} 1 & \text{if median } j \text{ is selected to facilitate cluster creation} \\ 0 & \text{otherwise} \end{cases}$$

$$y_{ij} = \begin{cases} 1 & \text{if entity } i \text{ is in cluster } j \\ 0 & \text{otherwise.} \end{cases}$$

With the above notation, it is possible to structure a median-based non-hierarchical clustering model with objectives for maximizing both attribute and spatial lag homogeneity.

Spatial Lag Clustering Model – Median (SLCM-M)

$$\textit{Minimize } w_a \sum_i \sum_j \hat{a}_{ij} y_{ij} + w_s \sum_i \sum_j \hat{s}_{ij} y_{ij} \tag{5.12}$$

Subject to:

$$\sum_j y_{ij} = 1 \quad \forall i \tag{5.13}$$

$$\sum_j x_j = p \tag{5.14}$$

$$y_{ij} \leq x_j \qquad \forall i, j \tag{5.15}$$

$$y_{ij} = (0,1) \qquad \forall i, j \tag{5.16}$$

$$x_j = (0,1) \qquad \forall j$$

Objective (5.12) of the SLCM-M minimizes the total weighted attribute dissimilarity and minimizes the total weighted spatial lag dissimilarity in selected clusters. This is equivalent to what is structured in objective (5.9) in the SLCM-C2. Constraint (5.13) ensures that each entity is included in a cluster. Constraints (5.14) and (5.15) require that only p clusters be generated. Constraints (5.16) impose integer restrictions on decision variables.

One of the most appealing features of the SLCM-M is that it is an integer programming problem that can be solved optimally for small and medium sized applications using commercial software and/or specialized techniques. This is a major departure from previously discussed models that rely on heuristic solution techniques and have the potential for getting "stuck" in a local optima. In addition, the multi-objective nature of this clustering model enables a number of things to be addressed. One important feature is that it simultaneously integrates both attribute similarity, as is done in choropleth mapping, and spatial proximity, as is done using standard deviational ellipses (along with the use of a k-means clustering heuristic). As with the other spatial lag models (SLCM-C1 and SLCM-C2), the SLCM-M avoids spatial bias inherent in the SSCM, but remains a within group variance minimization approach. One final feature is that the SLCM-M allows for non-dominated clustering solutions to be identified, an essential characteristic for ESDA and critically important for comparing alternative hot spot solutions.

5.5 Cluster Model Application for Hot Spot Detection

In an effort to illustrate the power and flexibility of the SLCM-M for exploratory analysis, the 62 block groups and violent crime rates for Lima, Ohio displayed in Fig. 5.2 will be used for analysis. Reported SLCM-M results are optimal to within 0.1% and the time required to solve associated problems was less than 1 s on an Intel Xeon quad core computer (2.27 GHz) with 8 gigabytes of RAM.

The first step in this exploratory analysis is deciding what number of clusters will be evaluated. Next, the associated non-inferior tradeoff curve must be generated using trial-and-error or techniques detailed in Cohon (1978). Considering that previous analyses in this chapter examined seven classes in Fig. 5.2, seven clusters will be evaluated using the SLCM-M. Figure 5.3 displays one non-dominated clustering

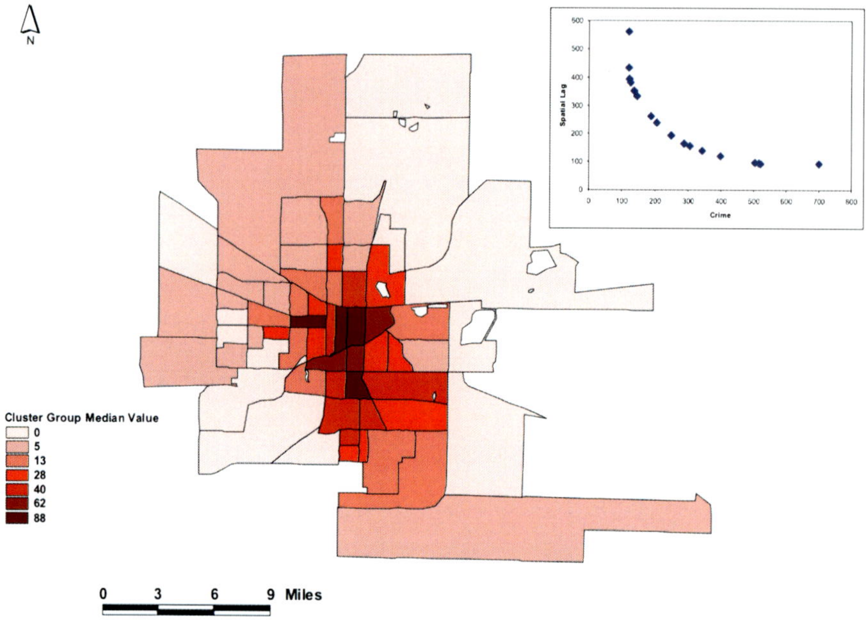

Fig. 5.3 Structured clusters using the SLCM-M ($w_a = 1$, $w_s = 0.01$)

solution using weights of $w_a = 1$ and $w_s = 0.01$.[8] In addition, Fig. 5.3 also shows the non-inferior tradeoff curve for the range of possible solutions that may be identified by varying the weights of importance for attribute similarity and lag similarity. Thus, plotted in this tradeoff curve is the total dissimilarity of violent crime against the total dissimilarity of spatial lag for the range of identified clustering solutions. The highlighted tradeoff point (*) corresponds to the displayed clustering solution. As a result, each point on the non-inferior tradeoff curve has an associated unique spatial clustering that may be analyzed and evaluated. For example, Fig. 5.4 depicts another tradeoff solution for weights of $w_a = 1$ and $w_s = 0.7$, which not only represents another point on the tradeoff curve but also has a unique corresponding spatial clustering pattern. Other tradeoff solutions could be shown as well. Comparing Figs. 5.3 and 5.4 (as well as 2), one can see subtle cluster changes as the influence of spatial lag is increased. The significance of this is that different spatial patterns emerge, patterns which may be more suggestive of underlying social and environmental characteristics or conditions for a region.

[8] The legend in this case does not have the same interval interpretation as that shown in Figure 5.2. Rather than depicting interval ranges, only the median group value is shown. Once spatial lag importance is increased, it is unlikely that groups will have non-overlapping values characteristic of choropleth maps. This point is discussed in Murray and Shyy (2000).

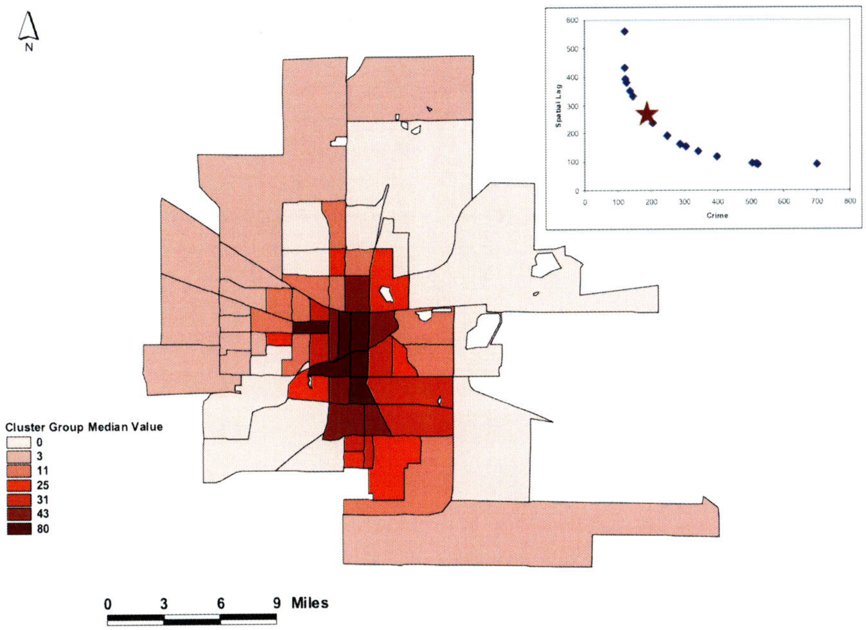

Fig. 5.4 SLCM-M clusters increasing spatial lag importance ($w_a = 1$, $w_s = 0.7$)

All of the figures suggest that there is a relative concentration of violent crime in the downtown area (center) of Lima. The highest crime rate areas in Fig. 5.3 correspond to lower income neighborhoods in the city. Further, these areas also have high minority concentrations, high unemployment, and a high percentage of households headed by single women. Thus, the choropleth displays (Figs. 5.2, 5.3, and 5.4) do a particularly good job highlighting higher violent crime rate areas and track well with the socio-economic factors likely to be influencing violent crime in Lima. Interestingly, as the weight for spatial lag is increased, the depicted geographic variation is less significant.

5.6 Discussion and Conclusion

The above analysis is insightful in many ways. There is a clear indication that downtown Lima represents one or more clusters in Figs. 5.1, 5.2, 5.3, and 5.4. However, point based displays (Fig. 5.1) are difficult to assess in a relative manner, ignoring background rates and activity. Ellipses (Fig. 5.2) are misleading, failing to adequately identify or delineate hot spot cluster. Figure 5.4, on the other hand, shows that there are actually spatial spillover effects that constitute a corridor area that is a hot spot (darkest units). This provides definitive instruction on where to allocate resources and personnel in order to combat violent crime in Lima.

There are a number of important issues associated with the detailed methods, and non-hierarchical clustering in particular. One important application issue remains identifying the appropriate number of clusters. There is actually little theoretical guidance for selecting the number of clusters to generate. In choropleth mapping, Dent (1999) suggests that 4–6 classes (clusters) should be selected (see also Harries 1999 as well with respect to crime analysis). Cromley (1995), also in the context of choropleth display, discusses the "elbow" in the curve approach. This is consistent with the rule of thumb well established in cluster analysis (Everitt 1993) as well as the economic interpretation found in location modeling (ReVelle 1987). However, this is less than definitive and certainly subjective, not unlike the criticisms of simple choropleth mapping and visual inspection (Messner et al. 1999). In the statistical literature additional methods for detecting the appropriate number of clusters have been proposed (Gordon 1996; Lozano et al. 1996; Podani 1996; Milligan and Cooper 1985). It is not clear, however, whether these alternatives might be useful in the analysis of crime. As a result, an important area for continued future research is exploring the applicability of these techniques for guiding users in the specification of the number of clusters to find.

Although there is significant flexibility and exploratory capabilities offered in the multi-objective structure and weighting in the SLCM, it does present a potential difficulty when carrying out analysis. Specifically, there is currently no theoretical basis for opting for a particular set of weights responsible for producing an associated non-dominated solution. In multi-objective modeling, the entire set of non-dominated solutions is considered potentially valuable (Cohon 1978). So, an analyst faces the question of addressing which ones are significant. This depends on external interpretation of the set of identified non-dominated solutions. It is unclear whether technical or theoretical approaches will be able to establish practical guidelines for analysts in the evaluation of alternative weightings.

One of the distinguishing features of non-hierarchical clustering is that of mutual exclusivity. In other words, entities are partitioned so that all of them are members of a cluster, but no two clusters share a common entity. As a result, the implication is that all of the identified clusters are significant. However, this is not well suited for hot spot detection in crime analysis. Rather, in hot spot detection it is recognized that crime events do and will happen, but it is when they localize and/or concentrate in some manner that a sub-area becomes a significant concern. This alternative interpretation of produced partitions leaves analysts to infer cluster significance using their own judgment. Given that hot spots represent areas in need of attention, this is obviously problematic. Potential approaches for addressing this issue may be found in the work of Arnold (1979) and Milligan and Mahajan (1980), which suggests Monte Carlo tests for examining partition validity and significance.

Aside from the detection of crime hot spots, the delineation of activity clusters does have a broader use. Clusters and their associated center locations may be important for finding criminals. In particular, the center of a cluster may correspond to where a perpetrator of certain crimes lives/works or where the next crime event may occur (LeBeau 1987; Rossmo 2000). Thus, the nature of the cluster (grouping of entities) and its subsequent interpretation (location of centers) is very spatial.

This suggests similarities with location modeling approaches, such as those discussed in ReVelle (1987) and Murray and Estivill-Castro (1998). More research is needed to establish the significance of cluster centers for this purpose as well as what interpretation of the "center" is most appropriate (e.g. mean, median).

A final point in the application of clustering models is the influence of scale variation. As an example, do clustering approaches produce equivalent results when using point based information as opposed to the use of area based aggregations of point information? In spatial analysis this line of inquiry is referred to as the modifiable areal unit problem (MAUP). Openshaw and Taylor (1981) note the possibility that analytical results may be altered by varying scale or modifying the boundaries of the reporting units. Criminology research has long been aware of scale and aggregation issues, and their implications in analysis (Parker 1985). Often times crime event locations are not made accessible for detailed analysis, making this concern a non-issue. However, when individual locations do exist, it is reasonable that clustering using these events be carried out. Another aspect of this issue is that a hot spot may exist in different ways and at different levels of spatial scale, as noted in Harries (1999) and Eck et al. (2005). At the individual crime incident level, hot spots may run along a particular street segment or route, rather than being circular (centered on a point) or elliptical. In such cases utilizing clustering models as currently specified may be problematic. Recent research has begun to deal with these spatial patterning issues (Yamada and Thill 2007; Shiode and Shiode 2009). Research examining scale and unit definition differences as well as patterning in clustering analysis is much needed.

This chapter has examined the statistical orientation of non-hierarchical clustering for assessing patterns of crime. Extensions and new approaches for this assessment were also reviewed and introduced. The use of spatial lag was shown to be an interesting way to incorporate geographic relationships and likely represents a promising avenue for relating non-hierarchical clustering to local spatial statistics. There are clearly unique and challenging aspects to the use of non-hierarchical clustering for identifying patterns of crime. Research examining these issues is necessary if clustering is to be effective tool in the exploratory analysis of crime activity.

Acknowledgement This material is based upon work supported by the National Science Foundation under grants SES-1154316 and SES-1154324. Any opinions, findings, conclusions or recommendations expressed in this material are those of the authors and do not necessarily reflect the views of the National Science Foundation.

References

Aldenderfer M, Blashfield R (1984) Cluster analysis. Sage Publications, Beverly Hills

Anselin L (1995) Local indicators of spatial association-LISA. Geogr Anal 27:93–115

Anselin L (1998) Exploratory spatial data analysis in a geocomputational environment. In: Longley PA, Brooks SM, McDonnell R, Macmillan B (eds) Geocomputation, a primer. Wiley, New York

Anselin L, Cohen J, Cook D, Gorr W, Tita G (2000) Spatial analysis of crime. In: Criminal justice volume 4. Measurement and analysis of crime and justice. National Institute of Justice, Washington, DC

Anselin L, Syabri I, Kho Y (2006) Geoda: an introduction to spatial data analysis. Geogr Anal 38:5–22

Anselin L, Meyer WD, Whalley LA, Savoie MJ (2009) Actionable cultural understanding for support to tactical operations (ACUSTO). US army corps of engineers. ERDC/CERL TR-09-13

Anselin L, Rey SJ, Kochinsky J (2010) Flexible geospatial visual analytics and simulation technologies to enhance criminal justice decision support systems. Crime Mapp 3(1).

Armstrong MP, Xiao N, Bennett DA (2003) Using genetic algorithms to create multicriteria class intervals for Choropleth maps. Ann Assoc Am Geogr 93(3):595–623.

Arnold FJ (1979) A test for clusters. J Mark Res 16:545–551

Braga AA (2001) The effects of hot spots policing on crime. Ann Am Acad Polit Soc Sci 578(1):104–125

Brantingham PJ, Brantingham PL (1981) Environmental criminology. Waveland Press, Prospect Heights

Chainey S, Tompson L, Uhlig S (2008) The utility of hotspot mapping for predicting spatial patterns of crime. Secur J 21:4–28

Cohen LE, Felson M (1979) Social change and crime rate trends: a routine activity approach. Am Sociol Rev 44(4):588–608

Cohen MA, Rust RT, Steen S, Tidd ST (2004) Willingness-to-pay for crime control programs. Criminology 42(1):89–110

Cohon J (1978) Multiobjective programming and planning. Academic, New York

Cromley R (1995) Classed versus unclassed choropleth maps: a question of how many classes. Cartographic 32(4):15–27

Cromley RG, Cromley EK (2009) Choropleth map legend design for visualizing community health disparities. Int J Health Geogr 8:52

Dent B (1999) Cartography: thematic map design, 5th edn. WCB/McGraw-Hill, Boston

Eck JE, Chainey S, Cameron JG, Leitner M, Wilson RE (2005) Mapping crime: understanding hot spots. U.S. Department of Justice. https://www.ncjrs.gov/pdffiles1/nij/209393.pdf

Eck J, Liu L (2008) Varieties of artificial crime analysis: purpose, structure, and evidence in crime simulations. In: Liu L, Eck J (eds) Artificial crime analysis systems: using computer simulations and geographic information systems. Information Science Reference, Hershey, pp 413–432

Estivill-Castro V, Murray A (2000) Hybrid optimization for clustering in data mining. CLAIO 2000, on CD-ROM. IMSIO, Mexico

Everitt B (1993) Cluster analysis. Halsted, New York

Fisher W (1958) On grouping for maximum homogeneity. J Am Stat Assoc 53:789–798

Gordon AD (1996) How many clusters? An investigation of five procedures for detecting nested cluster structure. In: Hayashi C, Ohsumi N, Yajima K, Tanaka Y, Bock H, Baba Y (eds) Data science, classification, and related methods. Springer, Tokyo, pp 109–116

Gorr W, Olligschlaeger A, Thompson Y (2003) Short-term forecasting of crime. Int J Forecast 19(4):579–594

Griffith D, Amrhein C (1997) Multivariate statistical analysis for geographers. Prentice-Hall, Upper Saddle River

Grubesic TH (2006) On the application of fuzzy clustering for crime hot spot detection. J Quant Criminol 22(1):77–105

Grubesic TH, Mack EA (2008) Spatio-temporal interaction of urban crime. J Quant Criminol 24(3):285–306

Harries K (1999) Mapping crime: principle and practice. National Institute of Justice, Washington, DC

Hartigan J (1975) Clustering algorithms. Wiley, New York

Huang Z (1998) Extensions to the k-means algorithm for clustering large data sets with categorical values. Data Min Knowl Discov 2:283–304.

Johnson SD, Bowers KJ (2004) The burglary as clue to the future: the beginnings of prospective hot-spotting. Eur J Criminol 1(2):237–255
Kaufman L, Rousseeuw PJ (2005) Finding groups in data. Wiley, New York.
Kent J, Leitner M (2007) Efficacy of standard deviational ellipses in the application of criminal geographic profiling. J Investig Psychol Offender Profil 4:147–165
LeBeau JL (1987) The methods and measures of centrography and the spatial dynamics of rape. J Quant Criminol 3:125–141
Leitner M, Barnett M, Kent J, Barnett T (2011) The impact of Hurricane Katrina on reported crimes in Louisiana: a spatial and temporal analysis. Prof Geogr 63(2):244–261
Levine N (2010) CrimeStat: A spatial statistics program for the analysis of crime incident locations, version 3.3. Ned Levine & Associates/National Institute of Justice, Washington, DC
Levine N (2006) Crime mapping and the crimestate program. Geogr Anal 38(1):41–56.
Lozano JA, Larranaga P, Grana M (1996) Partitional cluster analysis with genetic algorithms: searching for the number of clusters. In: Hayashi C, Ohsumi N, Yajima K, Tanaka Y, Bock H, Baba Y (eds) Data science, classification, and related methods. Springer, Tokyo, pp 117–124
MacQueen J (1967) Some methods for classification and analysis of multivariate observations. In: Le Cam L, Neyman J (eds) Proceedings of the fifth Berkeley symposium on mathematical statistics and probability, vol I. University of California Press, Berkeley.
McLafferty S, Williamson D, McGuire PG (2000) Identifying crime hot spots using kernel smoothing. In: Goldsmith V, McGuire PG, Mollenkopf JH, Ross TA (eds) Analyzing crime patterns: frontiers of practice. Sage, Thousand Oaks, pp 77–85.
Messner SF, Anselin L, Baller RD, Hawkins DF, Deane J, Tolnay SE (1999) The spatial patterning of county homicide rates: an application of exploratory spatial data analysis. J Quant Criminol 15:423–450
Milligan GW, Cooper MC (1985) An examination of procedures for determining the number of clusters in a data set. Psychometrika 50:159–179
Milligan GW, Mahajan V (1980) A note on procedures for testing the quality of a clustering of a set of objects. Decis Sci 11:669–677
Morenoff JD, Sampson RJ, Raudenbush SW (2001) Neighborhood inequality, collective efficacy, and the spatial dynamics of urban violence. Criminology 39:517–559
Murray A (1999) Spatial analysis using clustering methods: evaluating the use of central point and median approaches. J Geogr Syst 1:367–383
Murray A (2000a) Spatial characteristics and comparisons of interaction and median clustering models. Geogr Anal 32:1–19
Murray A (2000b) Spatially lagged choropleth display. In: Forer P, Yeh A, He J (eds) In: Proceedings of 9th international symposium on spatial data handling, 1a40–49. International Geographical Union, Beijing
Murray A, Estivill-Castro V (1998) Cluster discovery techniques for exploratory spatial data analysis. Int J Geogr Inf Sci 12:431–443
Murray AT, Grubesic TH (2002) Identifying non-hierarchical spatial clusters. Int J Ind Eng 9:86–95.
Murray A, Shyy T (2000) Integrating attribute and space characteristics in choropleth display and spatial data mining. Int J Geogr Inf Sci 14:649–667
Murray A, McGuffog I, Western J, Mullins P (2001) Exploratory spatial data analysis techniques for examining urban crime. Br J Criminol 41:309–329
Openshaw S, Taylor P (1981) The modifiable areal unit problem. In: Wrigley N, Bennett R (eds) Quantitative geography: a British view. Routledge and Kegan Paul, London, pp 60–69
Parker RN (1985) Aggregation, ratio variables, and measurement problems in criminological research. J Quant Criminol 1:269–280
Podani J (1996) Explanatory variables in classifications and the detection of the optimum number of clusters. In: Hayashi C, Ohsumi N, Yajima K, Tanaka Y, Bock H, Baba Y (eds) Data science, classification, and related methods. Springer, Tokyo, pp 125–132
Ratcliffe JH (2002) Aoristic signatures and the spatio-temporal analysis of high volume crime patterns. J Quant Criminol 18(1):23–43

Ratcliffe JH (2004) The hotspot matrix: a framework for the spatio-temporal targeting of crime reduction. Police Pract Res 5(1):5–23
Ratcliffe JH (2005) Detecting spatial movement of intra-region crime patterns over time. J Quant Criminol 21(1):103–123
ReVelle C (1987) Urban public facility location. In: Mills E (ed) Handbook of regional and urban economics. Elsevier Science, New York
Rogerson P, Yamada I (2009) Statistical detection and surveillance of geographic clusters. CRC Press, New York
Rossmo DK (2000) Geographic profiling. CRC Press, Boca Raton
Rousseeuw P, Leroy A (1987) Robust regression and outlier detection. Wiley, New York
Sampson RJ, Groves WB (1989) Community structure and crime: testing social-disorganization theory. Am J Sociol 94:774–802
Shaw CR, McKay HD (1942) Juvenile delinquency and urban areas: a study of rates of delinquency in relation to differential characteristics of local communities in American cities. University of Chicago Press, Chicago
Shiode S, Shiode N (2009) Detection of multi-scale clusters in network space. Int J Geogr Inf Sci 23(1):75–92
Wang F (2005) Geographic information systems and crime analysis. Idea Group Publishing, Hershey
Wu X, Grubesic TH (2010) Identifying irregularly shaped crime hot-spots using a multiobjective evolutionary algorithm. J Geogr Syst 12:409–433
Xiao N, Armstrong MP (2005) ChoroWare: a software toolkit for Choropleth map classification. Geogr Anal 38(1):102–121.
Yamada I, Thill J-C (2007) Local indicators of network-constrained clusters in spatial point patterns. Geogr Anal 39(3):268–292.

Chapter 6
Reconstitution of the Journeys to Crime and Location of Their Origin in the Context of a Crime Series. A Raster Solution for a Real Case Study

Jean-Paul Kasprzyk, Marie Trotta, Kenneth Broxham, and Jean-Paul Donnay

Abstract In the region of Charleroi (Belgium), a series of criminal acts were committed by the same group, using the same vehicle. The events were located in space and time. The car used during these criminal activities was stolen (first event) and was later retrieved (last event) after a period of 4 days of offences. Police recorded a crucial clue: the total mileage covered by the vehicle between the first and the last event was estimated with an admissible approximation. Thanks to this information, we were able to choose the most probable journey-to-crime among several scenarios. These depended on the combination of cost surfaces built with distance propagation algorithms starting from each criminal event in raster mode. The distance propagations were limited to the road network and the combinations of the cost surfaces had to respect the chronology of the facts. The most plausible scenario suggested that the criminals hided the car into a withdrawal site between their activities. In order to improve the precision of the location of this withdrawal site, we used a multi-criteria analysis taking account of the journey of the vehicle and other environment variables. At the end of these treatments, the small stretch of road that we isolated actually included the withdrawal site, as confirmed by the police later.

Keywords Geographic profiling • Road network • Cost surfaces modelling • Multi-criteria analysis • Withdrawal area delineation

J.-P. Kasprzyk (✉) • M. Trotta • K. Broxham • J.-P. Donnay
Department of Geography (Geomatics Unit), Faculty of Sciences,
University of Liège, Liège, Belgium
e-mail: Jp.Kasprzyk@ulg.ac.be; Marie.Trotta@ulg.ac.be; Jp.Donnay@ulg.ac.be

M. Leitner (ed.), *Crime Modeling and Mapping Using Geospatial Technologies*, Geotechnologies and the Environment 8, DOI 10.1007/978-94-007-4997-9_6,

6.1 Introduction

6.1.1 Generalities

Journey-to-crime (JTC) has significantly increased and broadened its scope over recent decades. The scientific researches are led by many diverse disciplines such as criminology, geography or psychology with as many different concerns. While some of them study the behavioural or socio-economic offender's characteristics influencing the traveled distances (Lundrigan and Czarnomski 2006; Lundrigan et al. 2010; Snook et al. 2005), others, like geographers, develop new techniques, governed by spatial principles, in order to delineate prior search areas for the offender. It is what Rossmo (2000) defined as geographic profiling and several software packages, such as CrimeStat (Levine 2007), Rigel (Rossmo 2000), Dragnet (Canter et al. 2000), have been developed to automatically delimit this search area.

This book chapter proposes a new methodology structured around three original aspects for geographic profiling: the environmental influence on the choice of crime sites, a raster-based methodology and an operational-oriented solution.

6.1.2 Environmental Influence

Environmental criminology demonstrated the influence of the environment on criminal behaviours (Brantingham and Brantingham 1981). Somes studies considered environmental factors in geographic profiling (Kent and Leitner 2009; Snook et al. 2005; Mohler and Short 2012).

> Operationally the crux of geographic profiling involves combining geographical theory and research with our experience of offender behaviour, while considering relative environmental factors and forensic issues,... (Daniell 2008).

The influence of the environment interfering with the shape of the distance decay function is widely recognised. The new Bayesian approaches introduce matrix of origin–destination in order to integrate the influence of several, combined, unknown features on the chosen journeys (Levine and Lee 2009; Kent and Leitner 2009; Leitner and Kent 2009).

Many aspects of the environment could nevertheless be directly integrated into the modelling of JTC. First of all, the road network poses constraints to the offender's movements and directions. It is often admitted that Euclidean or Manhattan distances provide accurate geoprofiles (Kent et al. 2006; Canter 2003). However, Alston (1994) mentions that micro-spatial behaviours for sexual rapists in England were completely different from those in North-American cities because the organisation of cities influence the offender's mobility. By contrast to the North-American cities that present a quite regular grid organisation, European cities have been modelled by successive, often uncontrolled developments making their structure very complex. Their irregular networks challenge the classical use of direct and indirect distances. "Besides, transportation modellers usually conceptualize distance not as an independent variable but as the result of predisposition, attraction and networks" (Levine and Block 2011).

Identifying all the physical or social barriers of the JTC is impossible (Bernasco and Block 2009). Some rough aspects like the exclusion of uninhabited areas (water bodies, woodlands, *etc.*) or of specific route categories, avoided because of police monitoring, could nonetheless restrict the search area.

6.1.3 Crime Mapping Research and Raster Analysis

Geospatial analysis can be done in two ways, depending on how the spatial primitives are defined and structured. The point defined in 2 or 3 dimensions is the geometric primitive of the vector mode. It allows the use of spatial entities such as segments, poly-lines, polygons and volumes. The raster mode considers a grid of cells covering the territory. Its geometric primitive, the cell, has a surface which introduces a bias in the analysis of linear or point features. However, the contiguity of grid cells allows treating spatially continuous phenomena. Indeed, in raster mode, the thematic dimension is given by the numeric values of the grid while its spatial dimension is computed thanks to the relative position of the cell in the grid. This format allows computing quick treatments on large amounts of data like those describing continuous phenomena.

In crime analysis, the vector approach has been preferred to the raster approach. Centrographic statistics, such as centroids and ellipses of dispersion (well described in Chainey and Ratcliffe (2005)) and applications of graph theory such as searching the shortest path (e.g. Qian et al. 2011), all emerge from the vector mode.

However, the ability to process spatial continuity offered by the raster mode allows the integration of environmental phenomena in the analysis. In addition, the matrix structure of information in raster mode allows the modelling of surfaces and facilitates their combination with simple operators in the so-called "map algebra".

Literature shows some uses of the raster approach in crime mapping and geographic profiling. Groff and La Vigne (2001) build an overall risk index surface for residential burglaries. Eck et al. (2005) estimate the density of crimes thanks to a continuous surface obtained by interpolation in order to highlight hot spots. Kent and Leitner (2009) integrate rasterized land cover features to enhance JTC models. In identifying the JTC, the solutions in raster mode typically use the construction of cost surfaces, possibly restricted to the road networks. Thus Trotta et al. (2011) generate all possible JTC from the various crime sites to determine the starting point of the criminal with a time constraint. The raster mode was chosen as it gives the possibility to consider each cell as a potential anchor point by contrast to the vector mode which is restricted to the only nodes of the graph.

6.1.4 Operational Aspect of Geographic Profiling Studies

In the current techniques of geographical profiling, the distance from the anchor location to each crime site is unknown. It is often estimated thanks to the analysis of similarly solved crimes in the study area by a classical calibration of a distance decay function or by a Bayesian approach.

But an important limitation of geographic profiling studies is the reliance on solved crime series (Tonkin et al. 2010). First, many authors adopt an inductive approach. They compare the JTCs of several crimes series to compute a distance-decay function instead of performing a proactive deductive analysis. For example, they do not specify how crimes have been connected to the same series.

Secondly, the form of the distance decay function is dependent on the studied area and varies with the scale of analysis (Canter and Hammond 2006). If crime mapping is considered as an essential tool to fight criminality in the United States, most of European polices, with the exclusion of the United Kingdom, are only in the early stages of transposing American methodologies on their own spatial patterns. They do not possess long solved geo-coded crime series for this calibration.

Another criticism made to the calibrated distance function is that it still provides a global approach that neglects the specificities of the individual behaviour whereas mobility characteristics such as the mode of transportation or some environmental preferences may be crucial in delineating a prior search area.

Indeed, studies demonstrated that the variability of intra-offender crime trips is smaller than the inter-offender one, offenders being more consistent with their own journeys than with others for the same crime type (Smith et al. 2009; Lundrigan et al. 2010).

For these reasons, we would like to highlight that every operational investigation can have its own specificities and its own spatial and temporal constraints finding use in geographical profiling.

Behavioural hypotheses can be built on those indications. For example, similar time slots for crimes occurring at different days suggest a constant departure time. In the same vein, successive offences committed the same day suppose a unique JTC. Moreover, all the distances to the unknown anchor point can also be combined as it will be illustrated in this chapter.

This literature review identifies three aspects that have been understudied in geographic profiling: the integration of environmental influence, the opportunities provided by a raster approach and the use of the specificities of an isolated crime series. These aspects are addressed in the study of a real and a-priori unsolved case of criminal investigation.

6.2 Context

Data provided by the police refer to a series of offenses committed by a group of criminals with a single stolen car in the urban area of Charleroi (Belgium) within a short period of time (4 days). The first offence is the theft of the car and the last one

Table 6.1 Summary of the event date/time as given by the Police

Event	Date	Time	Event
1	05/05/30	9:00 am	Car-jacking of the first car
2	05/05/31	4:00 pm	Hold-up at a supermarket
3	05/06/02	7:00 am	Suspicious behaviour in the first car
4	05/06/03	2:00 am	Car-jacking of a second car
5	05/06/03	3:00 pm	Discovering of the first car

is its abandonment. Between these two events, three criminal activities were committed with the same vehicle. Each activity is located in space (mailing address) and time, except the abandonment of the car for which we only know where and when it was discovered by the police (Table 6.1). Thanks to the car owner who remembered the mileage, we are able to estimate the total distance traveled by criminals between the first and the last event (approximately 100 km). This information, that is rare in such cases, is a key element of our methodology.

The purpose of this study is to delineate a withdrawal area where the criminals could hide the car between their activities, by using the chronology and the locations of the five events with the constraint of the known total mileage. The withdrawal area will be based on a subset of road segments selected from all the paths traveled with the vehicle on the road network of the urban region under study.

Belgian police gave us this real case study without informing us of the actual location of the withdrawal site. This allowed us to apply our methodology in real field conditions and to check its accuracy by comparing our results to the reality at the end of the study.

6.3 Data

As the research method is based on the movements of the vehicle, it is necessary to collect and evaluate the information about the road network of the study area at the time of the events. This information is both geometric (axe and width of the road) and semantic (state of the road, traffic direction, *etc.*).

We can obtain the geometry of the network with a sufficient precision for this application, thanks to the different vector spatial databases of the institutional providers of spatial data (e.g. IGN in Belgium).

The geometry is fairly stable in time but the semantic information evolves much faster, especially in urban spaces, without any archive of past modifications (e.g. inaccessible segment because of road-building). Therefore, the earlier the criminal analysis is achieved, the more relevant is the semantic information of the road network.

Belgian Police provided the location of the events with postal addresses. In order to integrate the data of the investigation on the road network, the postal addresses must be transformed in (X, Y) coordinates. Indeed, the localization of the events and the geometrical description of the road network must share the same coordinate system for the analysis (in our case: Belgian Lambert Coordinate System). An automatic geo-codification of postal addresses is possible tanks to the address files provided by private or public institutions. However an interactive localization can be necessary if an address is not available or invalid.

6.4 Methodology

6.4.1 Choice of the Method

When the entire road network is extracted and every event located, the data is ready for the first step of the study: the computation of the shortest path from each event site through the network. In this case, a traditional Dijkstra algorithm would not be pertinent because it would give the shortest path from an event only to the nodes of the graph formed by the network. What we are looking for is the shortest path to every point of the network.

Therefore, we need to use a distance propagation algorithm limited to the road network. This treatment requires working in raster mode by building a cost surface limited to the cells belonging to the network. This implies a preliminary rasterization of the network which is only available in vector mode. The parameters of the transformation "vector to raster" (the resolution of the cells and the extension of the study area) have an important influence on the final results of the procedure. On the one hand, a high resolution and/or a large study area increase the number of cells so that they lead to heavy treatments for the computer. On the other hand, a low resolution diverges from the real dimensions of the studied spatial entities. Moreover the extension of the study area must include the five criminal events and the most probable localizations of the withdrawal site (see Sect. 6.4.4). Therefore, we had to find the best compromise between these parameters. We first tested the treatments explained in the following sections (see Sects. 6.4.2, 6.4.3, and 6.4.4) with a low resolution on a large territory to be able to isolate the most adapted study area. After fixing this first parameter, we tested the treatments with different resolutions on the chosen study area so that we decided to keep a resolution of 20 m on the ground per cell. Indeed, it corresponded to the precision of the initial vector network and computations with a higher resolution risked to become too heavy for the computer.

6.4.2 Building a Cost Surface

We build a cost surface, based on the propagation of distances, for each event considered as a start cell. The principle of the propagation algorithm in raster mode is the following: we propagate a crossing "cost" of the cells, from neighbour to neighbour, from a start cell (a located criminal event of the data) to the borders of the study area. The spread is not isotropic since it is hampered by a friction factor

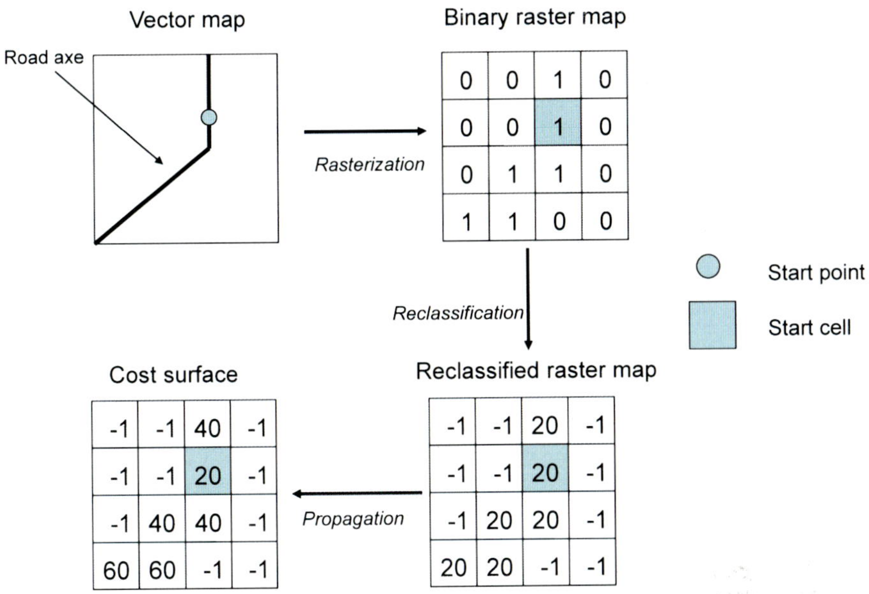

Fig. 6.1 Building a cost surface

in each cell. This factor reflects the difficulty of crossing the cell and, ultimately, the cell can be described as impassable. In the first analysis, the friction of a cell can be considered equal to its length field (or resolution).

Technically, we start with the binary map of the road network that we created by superposing a grid of cells (the size of a cell is 20 m) to the vector data (rasterization). The value of a cell is "1" if it belongs to the road network and "0" if it does not (like every binary image). We reclassify the image to prepare it for the propagation algorithm: the values of the cells become "20" (which corresponds to the resolution of a cell) when they are on the road and they become "–1" when they are off the road. Now we are able to build the cost surface by using the propagation algorithm: from the start cell, each cell's value is cumulatively summed from neighbour to neighbour except the "–1" cells that are considered as impassable by the algorithm (Fig. 6.1).

At the end of the procedure, the cost surface informs us about the cumulated distance required to reach every cell of the network from the starting point. In other words, the value of each cell is the distance between it and the source by following the shortest path along the road network.

This computation is only a first approximation. Indeed, we should take into account imprecision caused by a small map scale (the size of the cells is superior to the width of the road) and the resolution of the cells as a constant crossing friction. Indeed, the precision could be improved with a friction depending on the direction of the crossing. For example, the resolution of the cell is geometrically correct when it is crossed perpendicularly but is equal to $\sqrt{2}*r$ (with r = resolution = cell size) when it is crossed along the diagonal. One way to improve the estimation of friction

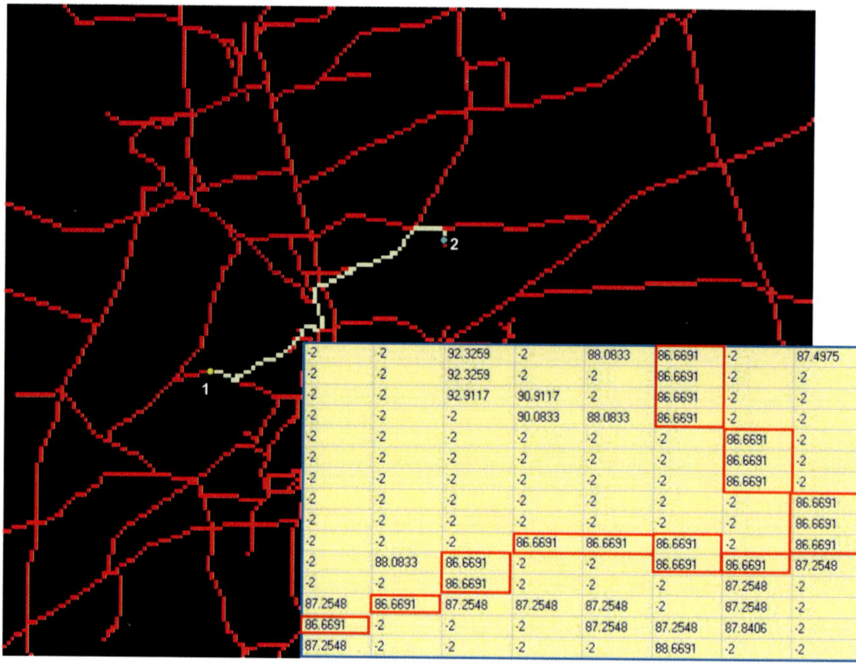

Fig. 6.2 Shortest path between two events by summing two cost surfaces

per cell is to calculate the average crossing length of the cells belonging to one edge of a road network by dividing the total length of this edge, calculated from the vector data, by the number of cells used for the rasterization of the edge.

The next step is the computation of the total shortest distance to travel to every criminal event in chronological order. If we add two cost surfaces cell by cell, the resulting map informs about the shortest cumulated distance from each cell to the two events. Therefore, the group of cells with the minimum value is logically the shortest path between the two events assigned to the two cost surfaces, and this value is the length of this shortest path (see Fig. 6.2). We repeat this operation with every couple of successive events so that we obtain every shortest path between the criminal events in chronological order. When we sum every value of these shortest paths, the result is the minimum theoretical distance traveled by the stolen car. It cannot be superior to the total mileage traveled by the car (100 km, as provided by the police). The positive difference between this total mileage and the total distance of the shortest paths supports the different scenarios elaborated to locate the possible withdrawal area.

6.4.3 Scenarios Development

Geographic profiling is always based on several assumptions with regard to the offender's spatial behaviour. When multiple hypotheses could describe the events, geographic profiling must be seen as a tool to choose among those possibilities.

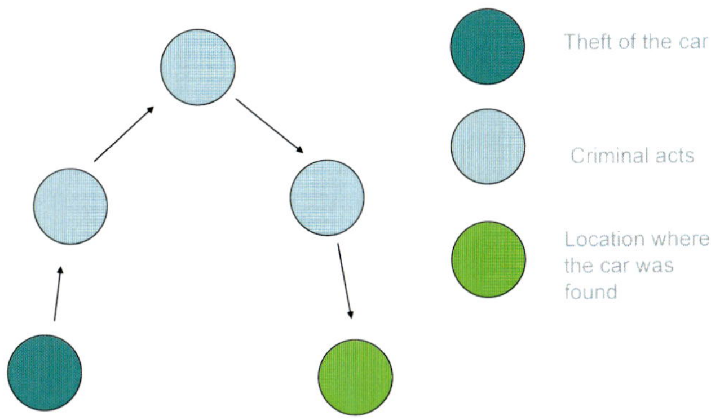

Fig. 6.3 First scenario

In this criminal investigation, several scenarios respecting the total traveled distance have been proposed. They are discussed in detail in this section and the credibility of each of them is estimated thanks to spatial information brought by raster computation and temporal constraints.

The chronology of the events indicates that they did not successively happen during the same day. As soon as we notice a time interval of several hours or even several days between two successive crimes, we can imagine that the criminals came back to the withdrawal site between each couple of consecutive events. Different scenarios can be considered according to the chronology of the facts.

In the sum of the cost surfaces, different weights can be assigned to each event, depending on the number of associated travels. For example, giving a double weight to a specific event means the criminals did a round trip from the withdrawal site to this event instead of a simple one-way trip. Technically, the choice of a scenario can be summarised in the assignation of the weights to the terms of the sum of the cost surfaces.

In the first scenario, which is the simplest one, offenders travel successively from one crime site to another, simply stopping between them for the night when events occur at different dates. Such a hypothesis corresponds to a kind of loop journey that can be easily tested with a geographic information system. In such a situation, the sum of the shortest paths joining each couple of events according to the chronology gives a result of 86 km, which is a too small (<100 km) to validate this hypothesis. A central withdrawal site is then considered as a better scenario for the series (Fig. 6.3).

In the second scenario, the offenders stole the car, came back to the withdrawal site and committed each crime from this site. If two events are close in time, we can imagine that they are successively committed without a ride back to the anchor point. However, it is unclear whether the offenders returned to the withdrawal site after the last criminal event (second scenario). It is plausible that the offenders traveled directly from the final criminal event to the location where the car was found (third scenario). As previously stated, these scenarios condition the way to combine the cost surfaces of each crime site. We attribute a double weight to the

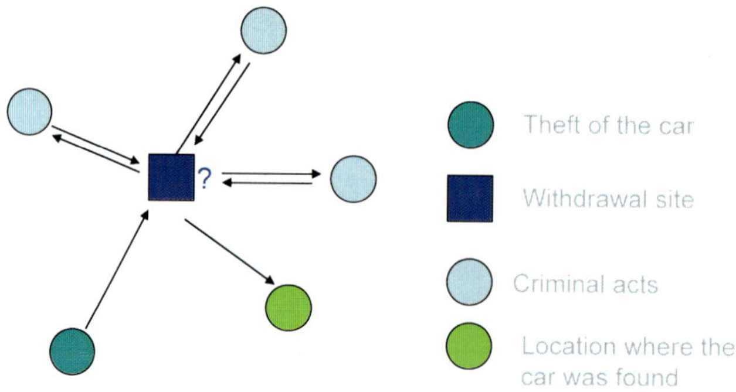

Fig. 6.4 Second scenario

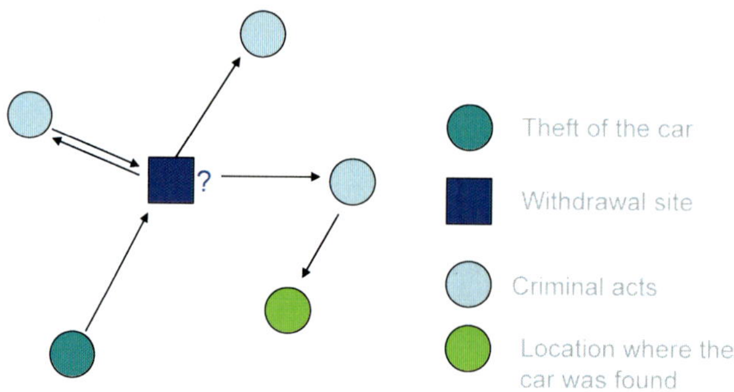

Fig. 6.5 Third scenario

round trips and a single for the others. In the case of the third scenario, the constant distance of the shortest path from the last event to the abandoning of the car is added to every cell of the network (Figs. 6.4 and 6.5).

A fourth scenario can be imagined if we formulate the hypothesis that the criminals did a reconnaissance before the hold-up at the supermarket (second event). The reconnaissance implies a new round trip associated to the second event that the offenders made before the hold-up in order to prepare it. Therefore, we give it a quadruple weight: one round trip for the tracking, and one round trip for the hold-up. When we add the cost surfaces in this way, we notice that the minimum value of the cells of the resulting map is slightly superior to 100 km. That means that with this scenario, we overvalue the total distance traveled by the car. Therefore, we reject the hypothesis of a reconnaissance (Fig. 6.6)

Four scenarios have been imagined based on the distance data and the chronology of events. We already said that the first and the fourth one are not plausible due to the estimation of the total distance traveled by the car (too short for the first scenario, and too long for the fourth one). Therefore, we have to choose between the second and the third one. The only difference between them is an eventual journey to the withdrawal site between the last event and the abandonment of the car. As the police

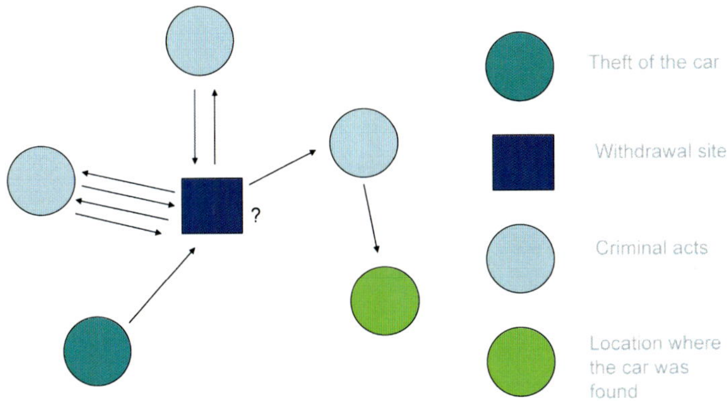

Fig. 6.6 Fourth scenario

told us they found the car only a few hours after the last event (car jacking of a second car), we can therefore consider that the criminals abandoned the car directly after the car-jacking which actually corresponds to the third scenario.

6.4.4 *Classification*

For the resulting cost surface corresponding to the chosen scenario, we isolate the cells having a value close to 100 km. Indeed, we have to take into account various errors caused by the imprecision of the mileage of the stolen car: the rasterization of the network, the variable accuracy of the geometric and semantic data, the geo-codification of the postal addresses, *etc*. Therefore, for a better representation, we classify the map using classes of 5 km around the central value of 100 km. This map gives an idea of the probability of finding the withdrawal site in a cell: the closer to 100 km we get, the more chance we have of finding the withdrawal site (Fig. 6.7).

6.4.5 *Multi-criteria Analysis*

The search area delimited by the sum of the cost surfaces is still too large to conduct an effective police raid. However, taking into account other factors than the traveled distances will help to reduce this area. This is the aim of the multi-criteria analysis.

6.4.5.1 Generalities

The multi-criteria analysis is a "decision support" method involving various criteria, including the "criterion of scenario" in the final map. The criteria are often

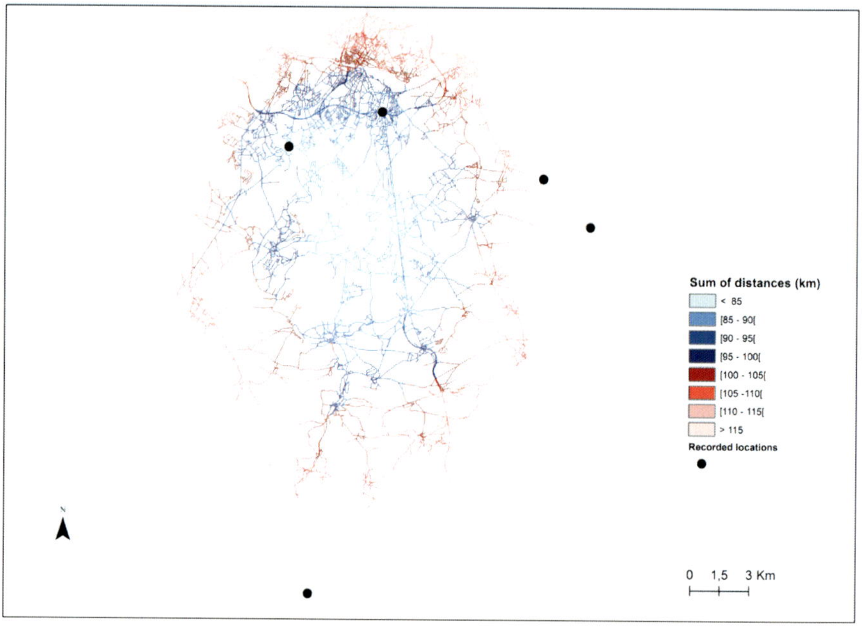

Fig. 6.7 Reclassified sum of the cost surfaces based on the third scenario

heterogeneous, conflicting and with unequal importance. Their combination leads to a performance index associated to each cell which, in our case, is related to the probability of finding the withdrawal site.

To build the index, we use two categories of criteria: judgement and admissibility. A judgement criterion allows us to measure and evaluate an action in regards to its function. This kind of criterion integrates the decision-maker's preferences. It can be positive or negative in the analysis. An admissibility criterion is a constraint limiting the number of actions taken into account. It can allow or forbid a given action (Eastman et al. 1995). First, we calibrate the factors of judgment on the same scale and we give each a relative weight according to the importance given to them by the decision maker. The sum of the weights must be equal to unity and the coherence of relative weight is controlled by the method proposed by Saati (1980). Then we perform the weighted sum of factors between homologous cells of different images containing each judgment criterion. Finally we take the admissibility criteria into account by multiplying the resulting judgement image by the different binary images assigned to the admissibility criteria (Fig. 6.8).

In this present case, we introduce these four criteria:

- A sociological factor taking account of the urban/rural character of the area (judgement criterion);
- A technical factor taking account of the distance to the main roads (judgment criterion);

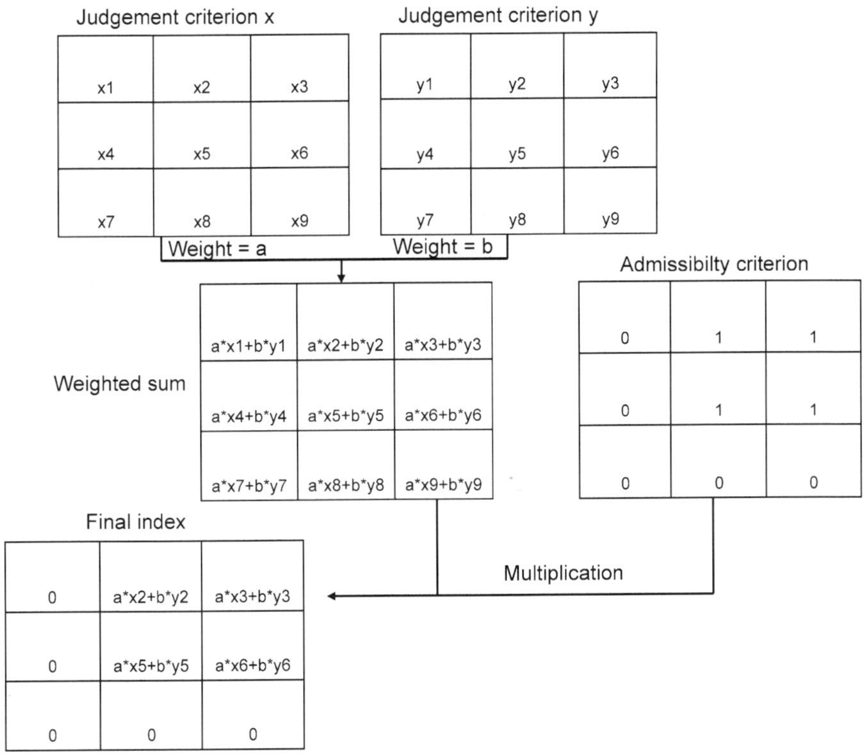

Fig. 6.8 Multi-criteria analysis

- A factor taking account of the chosen scenario, which is actually the result of the summed cost surfaces of the third scenario (judgment criterion);
- A built-up constraint: the withdrawal site must be located in a built-up area (admissibility criterion).

6.4.5.2 Factor 1: Integration of the Rural Character

The first factor gives a score to each commune (the smallest administrative entity in Belgium) relating to its urban/rural character. This character was evaluated with a principal component analysis (PCA). The input variables, provided by the Belgian National Institute of Statistics for the year 2005 for each commune, were the followings:

- percentage of strangers;
- percentage of urban spaces;
- percentage of rural spaces;
- population density;
- average income.

The PCA shows that the first four variables are correlated and can be reduced to a single factorial score explaining the rural character of the communes (the higher the score, the less rural the commune).

We noticed that the PCA was applied on 30 communes in the neighbourhood of the criminal events. This cross-section could seem too small to have reliable results. However, it did not make any sense to extend the territory of this analysis due to the characteristics of rural areas that vary depending on the regions of Belgium. For example, the density of population in rural areas is higher in the north than in the south. We also noticed that due to the commune being the smallest administrative entity of Belgium, it proved impossible to collect more data for the same territory.

Nevertheless, the risk of a PCA based on a small population is mainly influenced by a few individuals that could have very different characteristics from the rest of the group. Therefore, we applied a Jacknife process: the PCA is computed 30 times in total on 29 communes by dropping a commune from the PCA in turn. We noticed that the factorial scores remained relatively identical from one PCA to another. This means that no individual in the population was unique enough to have a major influence on the final result. Thus, despite the small population, we consider the PCA as relevant.

As we formulate the hypothesis that the withdrawal site should be located in a rural space, we create a raster map (still with the same usual resolution) and we allocate the value of the factorial score resulting from the PCA to each cell of every commune.

The only way to compare several criteria is to calibrate them with the same scale. In other words, the minimum and the maximum values of each criterion have to be the same. We arbitrarily chose to encode them in one byte, which means a range of 256 different integer values comprised between 0 and 255. The procedure used to obtain this new range of value is a simple linear transformation described in the following equation.

$$X' = Round\left[\frac{(X - X_{\min})}{(X_{\max} - X_{\min})} * 255\right] \tag{6.1}$$

Where X: original value
X': calibrated value

We observed that the factorial score was decreasing when the rural character was increasing. Keeping in mind that the withdrawal site should be located in a rural area, we have to reverse the range of the rural criteria according to:

$$X'' = 255 - X' \tag{6.2}$$

Where X'': reversed value.

6.4.5.3 Factor 2: Proximity of the Main Roads

We formulate the hypothesis that the withdrawal has to be close to a main road to facilitate an eventual escape of the criminals. Therefore, we have to create a new raster map with a score to each cell increasing with the proximity of the main road.

We start with a binary raster map representing the main roads (the value of a cell is 1 if it covers the road, 0 if not). For each cell of the map, we calculate the distance to the closest main road. The distance is computed by counting the number of cells separating the present cell to the closest road and by multiplying the result by the resolution (20 m). The algorithm returns a new map with a real value for each cell corresponding to the distance to the closest main road. We calibrate it to one byte and reverse the range as we did for the rural factor so that the longer the distance, the lower the chance of finding the withdrawal.

6.4.5.4 Factor 3: Integration of the Scenario

The integration of the scenario implies a particular treatment of the cost surface. The procedure that implements the scenario limits the cost surface to the sole road network. We can admit that the withdrawal site is not located on the road and we must therefore expand the cost surface in the entire territory.

We consider the space outside the network as isotropic so that we can apply the same propagation algorithm (used in Sect. 6.4.2). The propagation starts from the cells located on the network to the cells located perpendicularly of it. The value assigned to a cell located off the road is obtained by adding the value of its closest neighbour belonging to the network to the distance separating the cell from this neighbour.

In order to build the criterion, we reclassify the map resulting from the cost surface. This map represents a range of values comprised between 85 and 115 km. As previously stated, the withdrawal should be located on a cell having a value of 100 km which corresponds to the theoretical distance traveled by the criminals. The values not equal to 100 km were kept to take into account the errors (see Sect. 6.4.4). Therefore, the closer a cell is to 100 km, the greater the chance of finding the withdrawal. The absolute value of the difference to 100 km is calibrated on one byte and reversed to maintain the consistency of the relationship.

6.4.5.5 Constraint: The Withdrawal Must Be in a Built-up Area

The criminals must hide the car inside a garage or a warehouse. Therefore, we use the data from "CORINE Land Cover" (CCE 1993) to extract the built-up area. This vector database stores land use information for the whole of Europe. The different polygons are attributed by a code linked to the "CORINE Land Cover" nomenclature. When we rasterize the data for the studied area, we only keep the polygons which the code refers to a built-up area. The result is a simple binary map: the value of the cells is 1 if they are included inside a built-up area, 0 if they are not.

6.4.5.6 Construction of the Multi-criteria Index

The construction of the final map is realized in two steps. We first combine the three factors and then we apply the constraint.

Table 6.2 Weights assigned to the different factors

Factor	Weight
Factor 1: rural character	0.1
Factor 2: main road proximity	0.2
Factor 3: integration of the scenario	0.7

The combination of the factors is a weighted sum. The weights of each factor, which mean their relative importance in the probability of finding the withdrawal, were arbitrarily fixed with the help of the Police. In other words, the larger the weight, the larger the influence of the factor. Every cell of the three maps is then summed to create a new map representing the combination of these factors. The range covered by the different values of the cells is still the same: integers from 0 to 255 as the sum of weights equals unity (Table 6.2).

We finally apply the constraint to the last map by multiplying it by the binary constraint map. Consequently, the resulting map keeps the values of the combined factors when the cells are included in the built-up areas (the constraint), and displays 0 for the cells outside the built-up area (Fig. 6.9).

6.5 Validation

After presenting these results to the Police, it appeared that the actual withdrawal site was located on a cell of the final map having the value 252 (the maximum value of the cells is 255). If we classify the map only to keep the cells having a value superior to 251, we can see that only a small stretch of road remains on the map (Fig. 6.10).

6.6 Conclusion

This study referred to a real case provided by the Belgian Police. A series of five events were committed by criminals with a stolen car for which we knew the traveled distance. GIS techniques combined with raster maps allowed us to test the hypothesis that the criminals actually used a withdrawal site during the short time period covered by the events. Indeed, the theoretical distance using the shortest paths along the road network that links each event in accordance with the timeline was actually too short compared to the real one (100 km). Then, several scenarios concerning the movements of the vehicle were devised and evaluated using the total distance factor.

After choosing the most probable one, we delineated the first area for the withdrawal site by selecting the group of cells for which the distance value was close to 100 km in the weighted sum of the cost surfaces of each event. Those cost surfaces resulted from a distance propagation algorithm and the weights of the sum depended

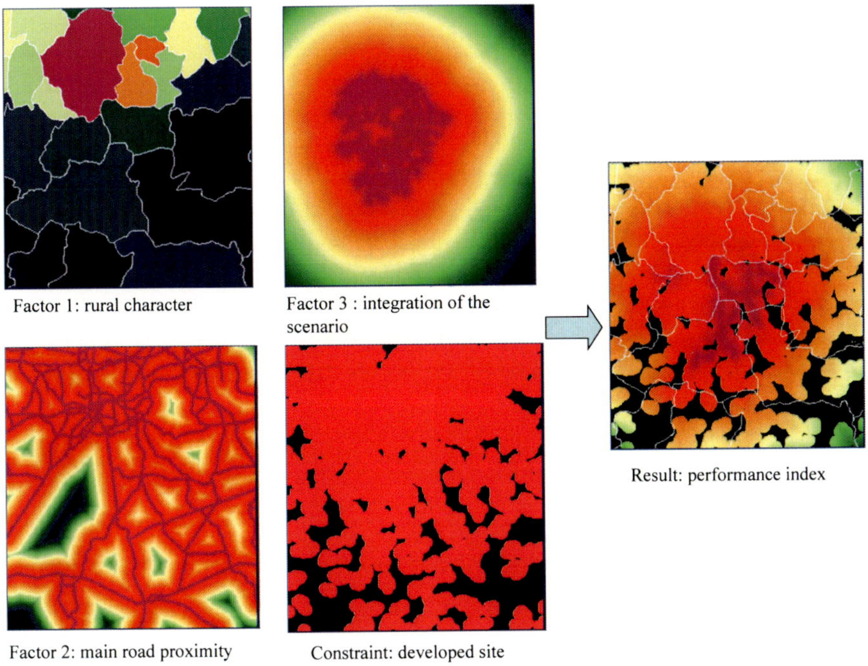

Fig. 6.9 Combination of the different criteria

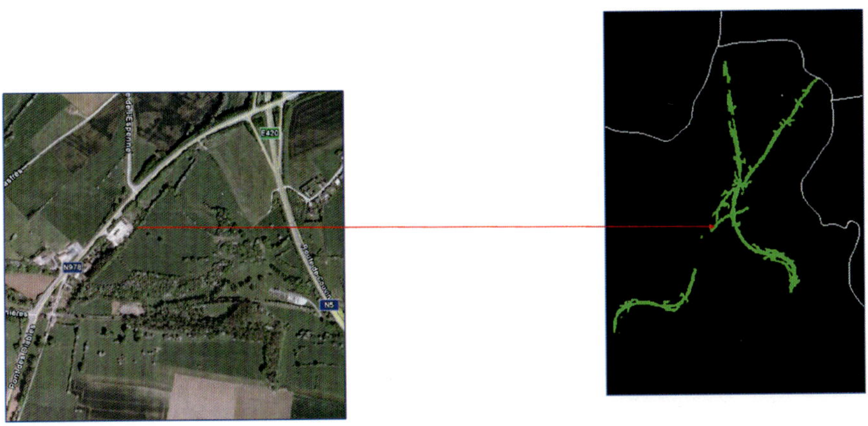

Fig. 6.10 Stretch of road resulting from the reclassification of the final map

on the chosen scenario. Unfortunately, the result covered an area too extensive for a police investigation. In order to reduce it, a multi-criteria analysis was conducted by taking into account various field elements (rural spaces, built-up areas and proximity of the main roads). We finally obtained an area corresponding to a small stretch of road where the criminals actually hid the vehicle, as later confirmed by the police.

Multi-criteria analysis should interest other scientists or police investigators that would like to synthesize several different environmental influences in one probable surface. However, the difficulty remains in properly identifying and weighting those influences. For this reason, working in collaboration with psychologists remains unavoidable.

Moreover, most studies in geographic profiling use behavioural models calibrated with other solved cases of similar crimes. Nevertheless, consistency is more an issue of individual than crime type. This work shows that models built only with data coming from the investigation (especially distance data) can be very useful, even if we must admit that some of them are rarely present (like the mileage of a stolen car).

Another originality of this study is the use of distance along a road network. Indeed, the Euclidian and even better the Manhattan distance is very popular in journey-to-crime, especially in USA. The road networks in American cities follow a very regular structure and can easily be considered as an isotropic space, which is simply incorrect in Europe. If we want to properly understand and model the offender's mobility characteristics, working with street networks is essential.

References

Alston J (1994) The serial rapist's spatial pattern of target selection. Master thesis, Simon Fraser University

Alston J (2001) The serial rapist's spatial pattern of victim selection. In: Godwin MG (ed) Criminal psychology and forensic technology: a collaborative approach to effective profiling. CRC Press, Boca Raton

Bernasco W, Block R (2009) Where offenders choose to attack: a discrete choice model of robberies in Chicago. Criminol Interdiscip J 47(1):93–130

Brantingham PL, Brantingham PJ (1981) Notes on the geometry of crime. In: Brantingham PL, Brantingham PJ (eds) Environmental criminology. Sage, Beverly Hills

Canter D (2003) Mapping murder: secrets of geographic profiling. Virgin, London

Canter D, Hammond L (2006) A comparison of the efficacy of different decay functions in geographical profiling for a sample of US serial killers. J Investig Psychol Offender Profil 3:91–103

Canter D, Coffey T, Huntley M, Missen C (2000) Predicting serial killers' home base using a decision support system. J Quant Criminol 16:457–478

CCE (1993) CORINE land cover. Guide technique, Office des publications officielles des Communautés Européennes, 144 p

Chainey S, Ratcliffe J (2005) GIS and crime mapping. Wiley, West Sussex

Daniell C (2008) Geographic profiling in an operational setting: the challenges and practical considerations, with reference to a series of sexual assaults in Bath, England. In: Spencer Chainey LT (ed) Crime mapping case studies. Wiley, West Sussex

Eastman JR, Jin W, Kyem PAK, Toledano J (1995) Raster procedures for multi-criteria/multi-objective decisions. Photogramm Eng Remote Sens 61(5):539–547

Eck J, Chainey S, Cameron J, Leitner M, Wilson RE (2005) Mapping crime: understanding hot spots. In: Justice UDO (ed). National Institute of Justice, Washington, DC

Groff ER, La Vigne NG (2001) Mapping an opportunity surface of residential burglary. J Res Crime Delinq 38:257–278

Kent J, Leitner M (2009) Utilizing land cover characteristics to enhance journey-to-crime estimation models. Cime mapping. J Res Pract 1:33–54

Kent J, Leitner M, Curtis A (2006) Evaluating the usefulness of functional distance measures when calibrating journey-to-crime distance decay functions. Comput Environ Urban Syst 30:181–200

Leitner M, Kent J (2009) Bayesian journey-to-crime modelling of single and multiple crime-type series in Baltimore County MD. J Investig Psychol Offender Profil 6(3):213–236

Levine N (2007) CrimeStat: a spatial statistics program for the analysis of crime incident locations 3.2a ed. Ned Levine & Associates/National Institute of Justice, Houston/Washington, DC

Levine N, Block R (2011) Bayesian journey-to-crime estimation: an improvement in geographic profiling methodology. Prof Geogr 63:213–229

Levine N, Lee P (2009) Bayesian journey-to-crime modelling of juvenile and adult offenders by gender in Manchester. J Investig Psychol Offender Profil 6(3):237–251

Lundrigan S, Czarnomski S (2006) Spatial characteristics of serial sexual assault in New Zealand. Aust N Z J Criminol 39:218–231

Lundrigan S, Czarnomski S, Wilson M (2010) Spatial and environmental consistency in serial sexual assault. J Investig Psychol Offender Profil 7:15–30

Mohler G, Short M (2012) Geographic profiling from kinetic models of criminal behavior. SIAM J Appl Math 72:163–180

Qian C, Wang, Y, Cao J, Lu J, Kurths J (2011) Weighted-traffic-network–based geographic profiling for serial crime location prediction. EPL 93:6

Rossmo K (2000) Geographic profiling. CRC Press, Boca Raton

Saati TL (1980) The analytical hierarchy process for decision in a complex world. RWS Publications, Pittsburg

Smith W, Bond JW, Townsley M (2009) Determining how journeys-to-crime vary: measuring inter- and intra-offender crime trip distributions. In: Weisburd D, Bernasco W, Bruinsma GJN (eds) Putting crime in its place. Springer, New York

Snook B, Cullen RM, Mokros A, Harbort S (2005) Serial murderers' spatial decisions: factors that influence crime location choice. J Investig Psychol Offender Profil 2:147–164

Tonkin M, Woodhams J, Bond JW, Loe T (2010) A theoretical and practical test of geographical profiling with serial vehicle theft in a U.K. context. Behav Sci Law 28:442–460

Trotta M, Bidaine B, Donnay JP (2011) Determining the geographical origin of a serial offender considering the temporal uncertainty of the recorded crime data. In: Iaria (ed) GEOProcessing 2011: the third international conference on Advanced Geographic Information Systems, Applications, and Services, Gosier, pp 40–45

Chapter 7
Journey-to-Crime by Gender and Age Group in Manchester, England

Ned Levine and Patsy Lee

Abstract This study examines journey-to-crime trips by gender and by age group for offenders who committed crimes in Manchester, England. The data are 97,429 crimes committed in 2006 by 56,368 offenders in which both the residence location of the offender and the crime location were known. Approximately one in six crimes was committed by women and by juveniles. The analysis showed gender differences in crime travel with interactions by age group, location of the crime, the presence of co-offenders, and ethnicity. Juvenile males had the shortest average trip lengths while adult males had the longest. Female offenders, both juveniles and adults, had crime trips of intermediate length but with a higher percentage being committed in major commercial centres. Around one-quarter of the trips were committed in conjunction with co-offenders, who generally lived quite close to the offender.

A negative binomial regression model showed that multiple factors contribute to the journey-to-crime distance traveled including type of crime, age and ethnicity of the offender, crime prolificacy of the offender, presence of co-offenders, and location and land use where crimes occurred. Controlling for these factors, with the exception of shoplifting, female offenders traveled shorter distances in committing their crimes, on average, than male offenders. For shoplifting, female offenders

The authors wish to thank the National Institute of Justice for funding (grant 2005-IJ-CX-K037) and the Greater Manchester Police for access to crime and offender data, compliant with the 1998 U.K. Data Protection Act. The authors also wish to thank Dr. Richard Block of Loyola University, Chicago, for suggestions on improving the paper. An earlier version of this paper was presented at the 4th International Conference on Women's Issues in Transportation sponsored Transportation Research Board, Newport Beach, CA., 2009.

N. Levine (✉)
Ned Levine and Associates, Houston, 8422 Bluegate St 77025-3212, TX, USA
e-mail: Ned@nedlevine.com

P. Lee
Greater Manchester Police,
Manchester, UK

M. Leitner (ed.), *Crime Modeling and Mapping Using Geospatial Technologies*, Geotechnologies and the Environment 8, DOI 10.1007/978-94-007-4997-9_7,

traveled longer distances because a higher proportion of those crimes were committed in the central retail core or in town centres.

The results indicate that simple generalisations about criminal travel are suspect. Instead, crime travel must be understood as reflecting the interaction of the type of crime, the characteristics of the metropolitan structure, the presence of accomplices, and offender characteristics, particularly gender and age.

Keywords Journey-to-crime • Gender • Age • Spatial • Opportunities

7.1 Introduction

In this paper, a journey-to-crime analysis was conducted in Manchester, England to understand how the interaction of gender and age affect the crime trip made by offenders. The aim is to show that the journey-to-crime trip is an interaction of individual characteristics of the offender along with the crime type, the role of co-offenders, and the metropolitan organisation. First, we discuss the journey-to-crime literature, the nature of crimes committed by women and by juveniles and how their crime trips differ from male and adult offenders respectively. Second, we examine variables affecting journey-to-crime and conduct a multivariate analysis of crime trips by offenders in Manchester.

7.1.1 Journey-to-Crime

There is substantial literature on the journey-to-crime. Many studies have demonstrated that most offenders live within a short distance from the location where they commit crimes and that offenders are less likely to commit crimes as the distance from their home increases (Lottier 1938; Turner 1969; Pyle 1974; Capone and Nichols 1975; Smith 1976; Phillips 1980; Rhodes and Conly 1981; LeBeau 1987; Canter and Larkin 1993; Rengert et al. 1999; Wiles and Costello 2000). Some crimes may occur in the residence of the offender and be domestic in nature or be in the same multi-unit building. At the same time, there are crime trips that are quite long (Paulsen 2007; Levine 2005, Ch. 11; Canter and Larkin 1993).

In more recent years, there have been more focused studies of travel behaviour by type of crime: commercial robberies in the Netherlands (Van Koppen and Jansen 1998); vehicle thefts in Baltimore County (Levine 2005, Ch. 11); robberies in Chicago and confrontations, burglaries, and vehicle thefts in Las Vegas (Block and Helms 2005); residential burglaries in The Hague (Bernasco and Nieuwbeerta 2005); homicides in Washington, DC (Groff and McEwen 2005); bank robberies in Baltimore County (Levine 2007); robberies in Chicago (Bernasco and Block 2009); and the trips of drunk drivers involved in crashes in Baltimore County (Levine and Canter 2011). These studies show substantial variability in crime trip lengths with many trips being long.

7.1.2 Importance of the Issue

Understanding the travel patterns of offenders is important for two major reasons. First, crime patterns have changed as the population has redistributed. Since World War II, society has become very mobile on a daily basis, especially in the United States but also in other countries. The almost universal use of personal automobiles has increased daily mobility. For example, in the United States, more than 90% of households owned or had regular access to at least one motor vehicle (NHMC 2012; U.S. Census Bureau 2002). In the United Kingdom, the percentage was 73% (U.K. National Statistics 2000). Since there is no data available that could test whether offenders are less likely to own an automobile than non-offenders, it has to be assumed that many offenders have access to an automobile for the use of committing a crime. While offenders will probably commit crimes in locales for which they are familiar, there is no reason to think that those locales will necessarily be the communities in which they live.

The widespread availability of motor vehicles has allowed major shifts in intra-urban travel patterns. In the last census (2000), approximately half the U.S population lived in areas that would normally be called suburbs, even though the U.S. Census Bureau does not use this nomenclature (non-central city, metropolitan population; U.S. Census Bureau 2000; Demographia 1998). Within metropolitan areas, approximately two-thirds of the U.S. population lived in suburban areas. In Britain, the definition of a suburb is less clear but approximately 80% of the British population lived within an urban area in 2001 (Pointer 2005). Suburbanisation has been occurring in many, if not most, metropolitan areas for a very long time and this has shifted the distribution of the population and of commercial activities.

In addition, most metropolitan areas have redeveloped their central cities, the effect of which has been to shift low income populations towards the suburbs. Much of the community-oriented crime patterns that were described by the so-called Chicago School of Criminology in the 1920s and 1930s are no longer true (Burgess 1925; Thrasher 1927). Over the last 20 years, crime in the U.S. has decreased substantially in the suburbs of many metropolitan areas, though not as fast as in the central cities, and the differences in crime between the central city and suburb is decreasing (Kneebone and Raphael 2011). Some emerging suburbs and exurban areas have shown increases in violent crime. Thus, those offenders that exist would be expected to show more mobility than in past years.

Second, it appears that many, if not most, crimes, are committed in non-residential locations, both commercial and other non-residential uses. Some of the non-residential areas are adjacent to residential communities so that crimes committed in those locations can be seen as being related to the generating neighbourhoods. Yet, as we shall illustrate in this paper, a sizeable number of crimes occur in major commercial centres, far away from the residential communities where offenders live.

In short, understanding the links between the residences of the offenders and the locations where they commit crimes is important for crime analysis. In this case, the crime journey becomes the unit of analysis rather than just the locations where the crimes occurred or where the offenders lived.

7.2 Factors Affecting Journey-to-crime

There are a number of factors that influence the journey-to-crime trip.

7.2.1 *Environmental Characteristics*

7.2.1.1 Crime Type

First, there is the crime type. It has generally been accepted that property crime trips are longer than violent crime trips (Block et al. 2007; Levine 2005; LeBeau 1987; White 1932), though exceptions have been noted (Turner 1969). This implies different behavioural strategies by offenders with property crimes involving some element of planning while violent crimes generally involve momentary reactions to events.

7.2.1.2 Familiar Settings

Second, familiarity with the attractions appears to be a necessary condition for offenders seeking out those locations. Routine activity theory states that crimes occur in the process of daily activities when there is an opportunity and a lack of guardianship (Cohen and Felson 1979) while crime pattern theory extends this by showing that crimes occur in the areas where the offender is familiar with the environment, both the nodes where activities occur and the immediate awareness space around those activities (Bernasco 2010; Brantingham and Brantingham 2008). This occurs in both the neighbourhoods where offenders live and commit crimes as well as in other areas that they visit.

7.2.1.3 Metropolitan Organisation

Third, there is the metropolitan organisation. The activities that attract offenders and the land uses supporting those activities are structured by the organisation of the metropolitan area. In most metropolitan areas, residential neighbourhoods tend to be separated from commercial areas, though they are often adjacent to smaller (local) shopping areas. However, major commercial stores and office space tend to be concentrated in specialised areas which are typically more central in the metropolitan area. Industrial activities are generally separated from both commercial and residential activities. This formal separation of land uses is often reinforced through zoning and land use restrictions by local governments, though market conditions will lead to separation as well. While there are an increasing number of mixed use developments that are occurring, these still represent a very small percentage of both residential and commercial uses. Separate land use is still the norm in metropolitan areas and is expected to remain that way for the foreseeable future.

7.2.1.4 Transportation Networks

Fourth, transportation networks tie the different areas together, particularly the commercial areas. Roadways link every type of land use together while public transportation typically link dense residential neighbourhoods with major commercial areas and with the dense central city. The transportation system facilitates mobility and thereby allows offenders to easily travel outside their immediate neighbourhoods. Ease of access to both automobiles and to regional roads facilitates the diffusion of crime from high risk neighbourhoods to many other areas of a city.

7.2.2 Individual Characteristics

These factors are environmental and influence all crime trips irrespective of who commits them. However, they cannot explain why there are individual differences in crime travel. Some offenders travel farther than others in committing crimes even if they come from the same neighbourhoods. In other words, there are characteristics of the offender that affect crime trips, particularly gender and age.

7.2.2.1 Gender

There are substantial differences between men and women in the commission of crimes, factors that influence their crime travel. Males commit the vast majority of crimes. In the United States, the FBI reported that in 2008, 25% of arrests made were of females (FBI 2008). However, in the U.S., the percentage of females arrested has increased quite substantially. For example, in 1991, only 18% of arrestees were female (FBI 1997). The proportional shift over the 17 years in the U.S. involved an absolute decline in the number of males being arrested but an absolute increase in the number of females being arrested. This is different from the trend from 1960 through 1990 where arrests by both males and females increased substantially (Steffensmeier and Allan 1996).

Schwartz and Steffensmeier (2007) reviewed offending patterns by males and females and concluded that, while there were many similarities with both being heavily involved in minor property and substance abuse offenses, there were also substantial differences. Men committed more serious crimes than women and at much higher rates for all crime types except prostitution. Broidy and Cauffman (2006) noted that the percentage of property crimes, assault crimes, and drinking-related crimes committed by females has been rising since 1960. However, they noted that there are still substantial differences in crime rates between male and female offenders that persist over time.

In England and Wales, somewhat similar results are found. The Ministry of Justice reported that in 2005–2006, 18% of arrests made were of females (Ministry of Justice 2008). Between 1999–2000 and 2005–2006, there was a slight increase in the proportion of female arrests (18% compared to 16% in 1999–2000). However, there was a decline in the absolute number of females arrested (Home Office 2001).

7.2.3 Journey-to-Crime by Gender

Several studies have examined crime travel by gender. Rengert (1975) found that female offenders were more likely to commit crimes within their own residential area than male offenders, hence making shorter trips, a result supported by Pettiway (1995) and by Groff and McEwen (2005). However, Phillips (1980) found that female offenders traveled longer distances, on average, than male offenders, a result supported by Fritzon (2001) who studied female arsonists. Conclusions about female crime travel need to be tempered by the knowledge that these studies were conducted over a span of more than 25 years. None controlled for a myriad of factors associated with journey-to-crime trips or changes in the role of women over time.

7.2.3.1 Expected Gender Differences in Crime Travel

Gender differences in crime travel patterns are expected for several reasons. First, the somewhat different distribution of crime types committed by females and males would suggest differential target location and opportunities. The higher proportion of simple assaults committed by females suggests more localised locations since assaults in general tend to occur closer to the offender's home. Also, the higher proportion of less serious property crimes in itself would suggest a tendency to travel to commercial areas where many property crimes are committed. Therefore, there should be greater variability among female offenders compared to male offenders with most crime trips being short but a small proportion being quite long.

A second factor influencing gender differences in crime travel is the gendered context of crimes. Fighting is considered a masculine behaviour among certain subgroups of the population, typically in stable, low income neighbourhoods. At bars and night clubs, the use of alcohol tends to exacerbate fighting, particularly among males (Mulvey et al. 2010; Rand et al. 2010, Table 20; Newton and Hirschfield 2009; Roncek and Maier 1991). While females do fight too, it tends to be less frequent and less serious (Rand et al. 2010). Similarly, sexuality as an attraction in a crime scene differs substantially between female and male offenders. Miller (1998), for example, reports how female robbers use sexual attraction to lure male victims into compromising situations where they can be robbed, often through prostitution. While male prostitutes could do the same, by most accounts there are far fewer of them than of female prostitutes.

Third, there are gender differences in taste that affect locational choices. Men are more likely to go to sporting events. Thus, crimes occurring at or near sports venues are disproportionately more likely to be committed by males. Conversely, women are more likely to shop and travel to commercial shopping areas (Wharton and Verde 2007). Thus, crimes committed in shopping areas are disproportionately more likely to be committed by females, though the majority of crimes committed in shopping areas are still by men. In short, there are a number of reasons why gender differences in journey-to-crime are expected.

7.2.4 *Crime Travel by Juveniles*

In the U. S., about one in six offenders are juveniles, age 17 or younger. In 2006, for example, 85% of arrested individuals were adults, age 18 or older, a percentage barely changed from 1991 (84%; FBI 1997). This percentage has remained stable over time. In 2006, the percentage of arrested females who were juveniles (younger than age 18) has also remained the same at 19%. Juveniles typically commit a higher proportion of property crimes than adults (Crowe 2000). In 2008, for example, juvenile offenders in the United States were arrested for 16% of all violent crime incidents but 26% of all property crime incidents (Puzzanchera 2009).

It is difficult to find comparable statistics for the United Kingdom due to the handling of juvenile offenders. For example, in 2008 juveniles, ages 10–17, accounted for approximately 30% of cautions given for offences but only 14% of persons convicted of indictable offences (Ministry of Justice 2010), a lower percentage than in the United States. Like the U. S., however, crimes committed by juveniles tend to be less severe, on average, than those committed by adults (Berman 2010).

Regarding travel behaviour by juveniles, several studies (Groff and McEwen 2005; Snook et al. 2005; Bernasco and Nieuwbeerta 2005; Snook 2004; Warren et al. 1998) have shown that generally juveniles make shorter trips. This would be expected given that juveniles are less likely to be employed, more likely to have peers who live in or near their neighbourhood, and less likely to have access to an automobile.

However, there are most likely interactions with gender since juvenile females are more likely to seek social contacts outside their neighbourhood than juvenile males. Also, as with gender, the relationship of age to crime travel has not been studied while controlling for other factors that affect travel such as the type of crime committing and the location where those crimes were committed.

7.3 Data Sources

The data used in this study came primarily from Greater Manchester Police (GMP), which is the third largest police force in England and Wales. Manchester is the third largest urban conurbation in the United Kingdom covering an area of 1284 km^2 (496 square miles) and with a population of around 2.5 million in 2006 representing about 4.2% of the UK population (Manchester City Council 2012; Wikipedia 2009). Manchester also has an extensive bus system and a metropolitan-wide light rail (tram) system that links the older and denser residential neighbourhoods to the town centres and to the central city. Transit ridership is very high in the region with more than 220 million bus trips and 21 million rail trips being taken every year (TGM 2012). There is also an extensive roadway system with expressways that can allow cross-town travel fairly quickly.

7.3.1 Arrest Records with Known Coordinates for 2006

Data on 97,429 arrest records for crimes committed in 2006 in which both the crime location and the offender's residence location were known were examined. These records represented 56,368 different individuals. Only offenders whose residence was confirmed and who lived within the Manchester region were analysed, though they represented the vast majority of arrestees by the GMP.[1]

The journey-to-crime analysis was conducted on the individual crimes. Of the 97,429 crimes, 16,871 (or 17%) were committed by females and 23,128 (or 24%) were committed by juveniles. While the time period is narrow (1 year), the dataset is very large. This allows in-depth analysis of travel relationships in a way that could not be easily done with a smaller dataset. We have no reason to believe that 2006 was an unusual year with respect to crime.[2]

We extracted seven pieces of information from the records aside from the gender and the age of the offender.

---

[1] These records provided information on the offenders and their crimes. The address where the offender was living at the time of his or her arrest was taken as the residence address. We recognise, of course, that some offenders may have moved since the time they committed their crime, a well known phenomenon with police records (Bernasco 2010). Change of address details are not indicated on crime reports, but as intelligence on an individual tracking system. Upon investigation, any change of address is dealt with according to what is happening to the offender. For instance the offender may have been bailed from one offence, and have moved since. The individual could then be arrested whilst on bail, in which case the new address would be recorded on the second crime record. Bail would then be opposed, citing the fact that the offender moved whilst on bail for the previous offence. Analysts on the other hand would refer to the intelligence about offenders' addresses to assess the geography of their criminal activities; attribute other crimes to them (comparative case analysis) or suggest locations for finding them if they had gone missing.

[2] For example, in the previous year (2004/05) there were 85,816 persons arrested in the Greater Manchester region and in the subsequent year (2006/07) there were 89,510 arrested compared to 87,858 for 2005/06 (Ministry of Justice 2008). The trend over the three years represents an average annual increase of 2%.

7.3.1.1 Distance from Offender Residence to Crime Location

First, we defined the journey-to-crime distance as that between the offender's residence and the crime location. This was defined in kilometers. We understand that a trip may not have gone directly between the home and crime locations and that there may be intermediate locations. The lack of data about intermediate trips precludes this analysis. Nevertheless, at some point, the vast majority of offenders return home after committing crimes. Thus, the linkage between home and crime location is the unit of analysis (See Levine 2005, Ch. 14).

7.3.1.2 Crime Type

Second, the crimes were divided into three major categories – violent (42%), property (48%), and other (10%), and nine minor categories for which there was sufficient data – simple assault (35%), aggravated assault (non-sexual; 4%), rape and sexual assault (1%), burglary (8%), robbery (4%), shoplifting (9%), vandalism (11%), vehicle theft (3%), and drug dealing (8%).[3]

There were some substantial differences between male and female offenders. For all types of crime except cruelty and neglect, men committed more offences than women. Overall, the ratio of male-to-female offences was 4.8 to 1 but this ratio varied from 0.4 to 1 (for cruelty and neglect, mostly to children) to 143 to 1 (for rape). These results are similar to the general trends found elsewhere (Steffensmeier and Allan 1996).

The most common *specific* crime type committed were simple assaults (less serious wounding) accounting for 41% of the crimes committed by women and 34% of those by men. Other frequent specific crimes committed by both female and male offenders were shoplifting (22% for women; 7% for men), criminal damage/vandalism (9% for women; 12% for men), and the supplying or passing of drugs (5% for women; 9% for men).

Similarly, there were substantial differences between juvenile and adult offenders. For crimes committed by juveniles, 33% were against property, 18% were criminal damage/vandalism, 44% were violent (including robbery), 3% were drug-related, 0.2% were sexual, and about 5% were other types. For crimes by adult offenders, 27% were against property, 10% were criminal damage/vandalism, 48% were violent crimes, 9% were drug-related, 0.4% were sexual, and about 6% were other types. In other words, juveniles were more likely to commit property crimes and criminal damage than adults who, in turn, were more likely to commit violent and drug-related crimes.

---

[3] We did not examine the remaining crimes as a minor category (17%) because they represented many different types.

7.3.1.3 Repeat Offenders

Third, the database included information on prior offending. In the British system, an individual who comes in contact with the criminal justice system is assigned an ID that stays with them for life. This standardised identification system allows tracking of individual offenders within the entire United Kingdom.

Of the 56,368 unique individuals in our database, 56% had been arrested prior to 2006. Further, of the remaining offenders with no prior arrest history, 4% committed multiple offences in 2006 bringing the total with multiple criminal offences to 60%. Keep in mind that these are crimes for which the offender has been arrested and charged. The 'true' number of offenders who committed multiple offences is most likely higher.

Females were less likely to be repeat offenders than males with 43% having either prior arrest histories or committing multiple crimes in 2006 or later compared to 64% for male offenders. However, it is very clear that the majority of male offenders and close to half of the female offenders committed multiple crimes. Similarly, juveniles, of course, were less likely to be repeat offenders than adults. Of the juvenile offenders, 41% had been charged with prior offences or committed multiple offences in 2006 compared to 65% for adult offenders.

A similar variable to prior offending was the number of crimes committed in 2006. Smith et al. (2009) and Townsley and Sidebottom (2010) have argued that a typical journey-to-crime data base is hierarchical with some offenders being represented multiple times. Further, the travel distance of offenders with multiple representations (trips) in a data base may be different than for those offenders represented only once. In the Manchester data base, there were 40,755 individuals who were listed only once compared to 15,613 individuals who were listed two or more times. To account for such an effect, the number of 2006 crimes associated with each offender was identified for each crime record.

7.3.1.4 Ethnicity

Fourth, aside from gender and age, the arrest records also provided information about ethnicity. Manchester has both a large African-Caribbean population and a large Asian population. About 6% of the offenders were of African-Caribbean ethnicity, 7% were of Asian ethnicity, and 74% were of White ethnicity and British born (as opposed to being White ethnicity but being born in a non-British country).

7.3.2 Co-offenders

Fifth, about a quarter of the crimes (26%) involved two or more offenders and were usually identified by the charges brought against the individuals. The database was cleaned so as not to duplicate events. When there were more than two offenders (rare), we defined the co-offender as the oldest. Thus, our database has a slight bias towards older offenders. The analysis was done by the offender trips only with the presence of an accomplice (a co-offender) being considered a contextual variable.

7.3.2.1 Land Use Associated with Crimes

Sixth, we extracted information about the land use associated with the crime location. Each crime record has a *premise code*, which is a categorisation of the type of premise where the crime occurred and is, essentially, a land use categorisation of the building or location where the crime occurred. We allocated the premise code to six general land use categories:

1. Crimes committed on a street (37% of all the crimes in the database)[4];
2. Crimes committed at residential locations (29%);
3. Crimes committed at commercial locations (19%);
4. Crimes committed at transport-related facilities (4%);
5. Crimes committed at schools (2%); and
6. Crimes committed at other uses (9%).

To compare this, in most metropolitan areas the majority of land parcels are residential and the majority of street segments are in residential neighbourhoods, consistent with the general land use pattern. Thus, it appears that crimes are disproportionately committed in non-residential locations and, in particular, commercial locations relative to the distribution of parcels.

7.3.2.2 Town Centres

Seventh, in addition, we coded whether the crimes occurred in a town centre. In the Greater Manchester region, there are 48 town centres that are located within surrounding areas. They were once separate market towns but were eventually absorbed by metropolitan growth. Over time, the market areas evolved into commercial and business centres surrounding local government functions (e.g., town halls). The residential neighbourhoods radiate outward from these town centres.[5] In the central city of Manchester, there is a much larger commercial and business centre that dominates the central city; at its centre is a *central retail core* and, in particular, the Arndale Centre of central Manchester.[6]

---

[4] Unfortunately, as is well known in police departments, crimes committed 'on the street' is a miscellaneous category that is non-specific as to the land use relationship. In an earlier study (Levine and Wachs 1986), it was shown that a sizeable proportion of crimes that occurred in central Los Angeles were categorised as being committed 'on the street', many of which were related to transit use by the victims.

[5] Note that this organisation is different than in most U.S. cities where large suburban shopping centres compete with those in the central city. Many cities in the U.S. have reduced commercial activities in the central city.

[6] The Arndale Centre was redeveloped after the bombing of the earlier centre by the Provisional Irish Republican Army in 1996 (Arndale 2012; Manchester 2011; Wikipedia 2010). The bombing became a rallying point for redeveloping the entire central city.

7.3.3 Limitations to the Data

Greater Manchester Police, in common with other forces, evaluate all recorded crime as soon as offences are reported by complainants. One of the main factors considered is the likelihood of detection of each specific crime. A proportion of all crime is deemed, at evaluation, to be highly unlikely to be detected and is therefore not prioritised for investigation. The crimes are filed with a 'no further action' marker. At a later date, upon the arrest of an offender for another offence a proportion of these filed crimes may be looked at by crime pattern analysts and any bearing similarities to the offence committed by the arrestee are put to him/her at interview. If the offender admits these crimes they are taken into consideration at a later prosecution. These individual crimes are then reclassified as 'detected' and included in later detected crime statistics. Thus, the accuracy of the data does vary by crime type.

Also, since these data are from arrest records where the offender's residence could be identified in addition to the crime location, there is always the possibility that they do not represent all offenders and that the conclusions are biased. In 2006, for example, the GMP recorded 335,942 crimes within the region. Thus, the 97,429 crimes in our database represent only 29% of all GMP crimes committed that year. We compared the distribution of crime locations in our database by the ten police districts with the distribution of crimes committed by those offenders who were not arrested.[7] Given the sample size, a Chi-square test showed statistically significant differences. However, the differences were not particularly large. For example, 27.2% of crimes committed by offenders who were not arrested were committed in the Manchester district compared to 27.1% of crimes committed by the offenders who were arrested. For the Bolton district, the corresponding percentages were 9.1% and 9.4% and for the Stockport district the percentages were 9.3% and 9.6% respectively. The biggest difference occurred in the Wigan district where 9.5% of the non-arrested offender crimes were committed compared to 5.2% of the crimes committed by the arrested offenders and in the Tameside district where 6.2% of the non-arrested offender crimes were committed compared to 9.5% for the arrested offender crimes.

In other words, while there are statistically significant differences in the distribution of crimes by police districts for the offenders who were caught compared to the offenders who were not caught, those differences are not very large nor does there appear to be an obvious biasing pattern (e.g., arrested offender crimes would occur in one part of the region whereas the crimes committed by non-arrested offenders would occur in other parts).

[7] This was estimated by the difference between the total number of crimes per district and the number of crimes in our database per district. We recognise that some of the crimes attributed to non-arrested offenders may actually have been committed by the arrested offenders. However, this seemed a better test than comparing the distribution of crimes by arrested offenders to the total distribution of crimes.

Nevertheless, the differences do suggest conclusions based on arrested offenders need to be tempered, at least until replication of results can be shown. There are undoubtedly other discrepancies in the distribution of crimes committed by arrested offenders compared to those who were not arrested by gender, age, type of crime, weapon use and other variables so there is always the potential for biased conclusions based on a journey-to-crime analysis of arrested offenders. With this type of research, replication is the only way to support the accuracy of conclusions.

7.4 Results

The entire data base was analysed for crime trip length and crime travel to commercial areas (town centres and the city centre).

7.4.1 Crime Trip Length

We analysed the 97,429 trips that the 56,368 individual offenders made. That is, each crime committed by an offender was analysed as a single crime trip. Even though the records represent multiple crimes committed by some individuals, it was important to understand the travel dimensions of their offending behaviour. A trip was assumed to occur between the offender's home address (the origin) and the crime location (destination).

For the sample as a whole, the range varied from 0 km (i.e., the crime was committed in the same building where the offender lived) to almost 40 km. The average trip was 2.6 km from the home location to the crime location with the median being 1.3 km. However, 16% of the trips were longer than 5km and 7% were longer than 8 km.

7.5 Crime Trip Length by Gender and Type of Crime

Figure 7.1 shows an interpolated graph of the travel distances made by female and male offenders. The two categories have been standardised by converting them into proportions.[8] Notice that the proportion of both sets of trips show a rapid decay by distance. The distribution of crime by distance is extremely skewed for both sexes and there was not a meaningful difference in distances traveled. The Mann–Whitney 'U' test of the rank order of the two distributions did find significant statistical

[8] The function was estimated with the *CrimeStat* journey-to-crime calibration routine that interpolates travel distances to a distance scale using a smoothing function, called a *kernel*, which is then re-scaled as a proportion (Levine 2005, Ch. 10).

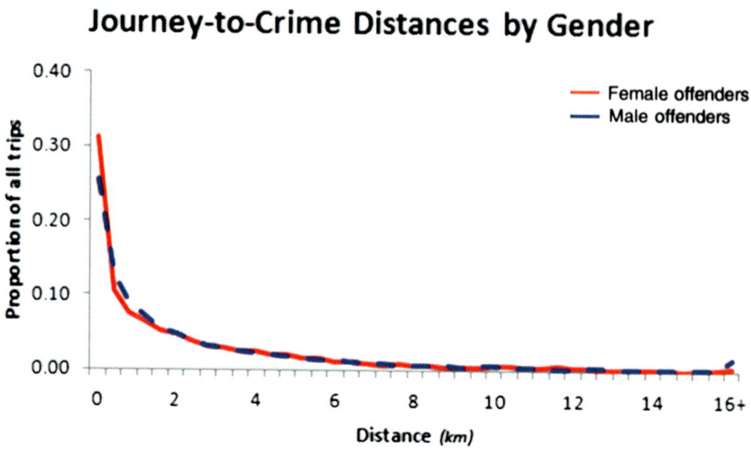

Fig. 7.1 Journey-to-crime by gender

differences in the distributions.[9] But, this is primarily due to the very large sample size of the database and subtle differences in the distribution of the trip lengths.

The trip length of females and males need to be related to crime type, however, as trip length varies by the type of crime. On average, property crime trips were longer than crime trips ending in violent incidents (with an average that was 29% longer –2.9 km compared to 2.3, and a median that was 67% longer – 1.6 km compared to 1.0 km). Other types of crime, mostly drug related, were intermediate in length (an average of 2.6 km and a median of 1.1 km). These differences were highly statistically significant.

There were meaningful differences by gender in the distances traveled for different crime types (Table 7.1). Comparisons were made for the three major crime groupings – violent, property and 'other', and the nine minor crime groupings for which adequate sample sizes were available– simple assault, aggravated assault (non-sexual), rape and sexual assault, burglary, robbery, shoplifting, vandalism, vehicle theft, and drug dealing. All of the comparisons except robbery and vehicle theft were statistically significant. Overall, with the exception of vandalism/criminal damage, violent crime trips were shorter than property crime trips for both sexes.

For the nine minor crime groupings and comparing the mean only, males made substantially longer crime trips than females for drug dealing (53% longer), rape and sexual assaults (35% longer), and simple assaults (21% longer). Conversely, female offenders made longer shoplifting trips (21% longer) than male offenders. There were no substantial differences in trip length for robbery and vehicle theft trips and small differences for aggravated assault trips and vandalism trips.

[9] Even though the Mann–Whitney 'U' test is used as a non-parametric version of the t-test, it actually is a distributional test and not just of the means relative to their standard deviations.

Table 7.1 Trip length by gender for different crime types: mean and median distance (km)

Type of Crime[a]	Females			Males		
	N	Mean	Median	Mean	Median	U[a]
All crimes	**97,429**	**2.56**	**1.22**	**2.56**	**1.22**	**10.6******
Major crime category						
Minor crime category						
Violent/Person	**41,370**	**1.87**	**0.69**	**2.30**	**0.97**	**150.7*** **
Simple assault	*34,381*	*1.87*	*0.72*	*2.27*	*0.97*	*82.29****
Aggravated Non-sexual Assault	*4,322*	*1.96*	*0.56*	*2.19*	*0.97*	*35.09****
Rape/Sexual Assault	*1,080*	*1.90*	*0.23*	*2.56*	*0.87*	*4.01**
Property	**46,539**	**3.32**	**2.00**	**2.78**	**1.45**	**208.6*** **
Burglary	*7,535*	*2.54*	*1.09*	*2.88*	*1.48*	*18.1****
Robbery	*4,070*	*3.36*	*1.92*	*2.98*	*1.92*	*0.4 n.s*
Shoplifting	*8,898*	*4.25*	*3.01*	*3.51*	*2.20*	*153.4****
Vandalism	*11,135*	*1.71*	*0.68*	*1.87*	*0.76*	*9.8***
Vehicle theft	*3,366*	*3.17*	*1.64*	*2.74*	*1.50*	*0.4 n.s*
Other	**9,520**	**1.80**	**0.26**	**2.56**	**1.22**	**204.8*** **
Drug dealing	*7,762*	*1.67*	*0.24*	*2.56*	*1.21*	*177.1****

<none> Not significant
* Significant at p≤.05
**Significant at p≤.01
***Significant at p≤.001
[a]Mann-Whitney 'U' test of difference between two populations with same mean, chi-square approximation (Kanji 1993, 89)

Therefore, it is very clear that there are gender differences in crime travel even when controlling for the type of crime. But, the differences are not consistent; female offenders travel shorter distances for most crimes but not all, especially shoplifting.

7.5.1 *Crime Trip Length by Gender and Age Group*

There were differences in crime travel by age group. Figure 7.2 shows an interpolated graph of the travel distances made by juveniles and adults. In general, juvenile offenders traveled shorter distances, an average of 1.9km and a median of 1.0 km compared to an average of 2.7 km and a median of 1.3km for adult offenders. Overall, 52% percent of the juvenile crime trips were less than 1km in length compared to 43% for the adults. On the other hand, only 6% of the juvenile crime trips occurred at the same address as where the offender lived compared to 11% for adult offender trips. Juveniles are more likely to commit crimes near where they live compared to adults but are less likely to commit them at their residence address.[10]

[10] This analysis is based on a crime location that has the same coordinates as the residence location. We cannot determine whether the crime occurred within the residence of the offender (domestic crime) or at another unit within the same building.

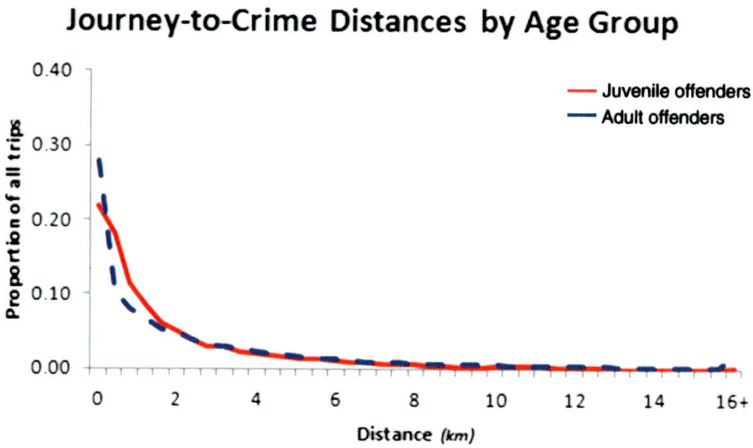

Fig. 7.2 Journey-to-crime by violent and property crimes

Table 7.2 Crime trip length by gender-age group interaction mean and median distance (km) (N=97,429 crime trips)

Sub-group	N	Mean	Median	Standard deviation
Juvenile males	18,430	1.87	0.87	2.69
Adult males	62,128	2.78	1.38	3.81
Juvenile females	4,698	2.58	1.34	3.35
Adult females	12,173	2.56	1.19	3.70
K-W[a] = 711.2 ($p \leq .001$)				

[a]Kruskall-Wallis rank sum test of difference between populations with same mean (Kanj, 1993, 89)

However, there is a distinct gender-age interaction in crime trip length that is substantially more differentiating than the simple gender and age group comparisons (Table 7.2). Adult males made the longest crime trips while juvenile males had the shortest crime trip lengths. These differences were statistically significant as well as meaningful. Adult males traveled 53% longer for the average and 59% longer for the median than juvenile males while there was no real difference between adult females and juvenile females. Adult males also traveled slightly longer than adult and juvenile females. Finally, juvenile females traveled 42% longer for the average and 54% longer for the median than juvenile males. In short, juvenile males traveled significantly shorter, on average, in committing their crimes than the other three groups.

7.5.2 Crimes Committed in Town Centres

These trip length differences partially reflect the locations chosen for the crimes (Table 7.3a). For the female offenders, 16% of the crime trips occurred at a town centre with another 7% being committed in the central retail core compared to 10%

Table 7.3 Percent of crime trips to commercial centres

	% to Town centres	% to Central retail core	% to Commercial centres
Total sample	**11.3%**	**4.4%**	**15.7%**
A. By gender[a]			
Males	10.2%	**3.8%**	14.0%
Females	16.3% $p \leq .001$	7.1% $p \leq .001$	23.4% $p \leq .001$
B. By gender-age group interaction			
Juvenile males	6.7%	1.5%	8.2%
Adult males	11.2%	4.5%	15.8%
Juvenile females	16.9%	8.5%	25.4%
Adult females	16.1%	6.6%	22.7%
K-W[b] = 711.2 ($p \leq .001$)			

[a]Kruskall-Wallis rank sum test of difference between populations with same mean (Kanji 1993, 89)
[b]Z-test of difference between proportions with unequal variances (Blalock 1979, 232–234)

and 4% respectively for males, a difference that is highly significant. In other words, female offenders were more likely to commit crimes in a commercial centre than male offenders.

There were also interactions with age group (Table 7.3b). Juvenile males were less likely to commit crimes in a town centre or in the central retail core than adult males, who were, in turn, less likely to commit their crimes in the central retail core than females, both juveniles and adults. All of these differences were highly significant.

The type of crime committed can explain some of these differences. Females were more likely to commit shoplifting offences than males (22% compared to 7%), and retail stores tend to be in town centres across the conurbation. Figure 7.3 illustrates major inter-zonal origin–destination links for crime trips by female offenders (both juveniles and adults) in central Manchester while Fig. 7.4 illustrates major inter-zonal origin–destination links for crime trips by juvenile males.[11] The town centres and central city are shown in the maps as small polygons. As can be seen, the spatial patterns of the juvenile males are more localised and diffuse than those of the female offenders, both juveniles and adults. The patterns of the adult males are closer to those of the females (not shown).

7.5.3 *Multivariate Predictors of Crime Trip Length*

To examine the interaction of factors predicting crime trip length, a multivariate negative binomial regression model was run. The dependent variable was the trip length from the offender's residence to the crime location in kilometers. In order to

[11] The maps were created with *CrimeStat* using the trip distribution module tools (Levine 2004, Ch. 14). Essentially, a fine grid was overlaid on to a map of Manchester and the routine assigned the crime events to both an origin cell and a destination cell. The number of links between the origin and the destinations are calculated and output as a line.

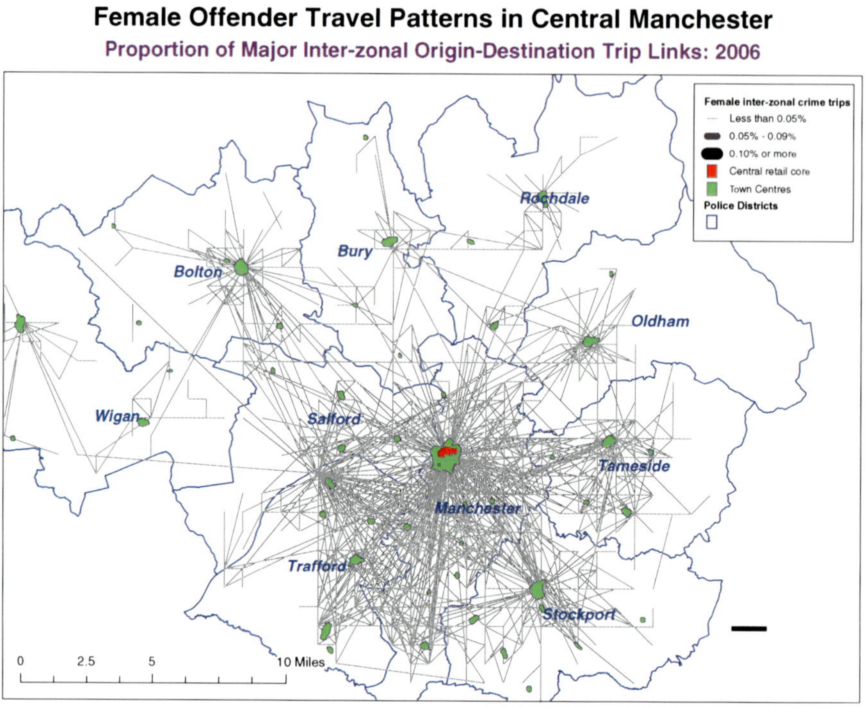

Fig. 7.3 Major O-D links for female crime trip in central Manchester

properly test the effect on gender and age group on crime travel distance, a number of contextual and statistical control variables were used. There were five categories of independent variables (Table 7.4).

The variables were defined in the data sources section.

7.5.3.1 Negative Binomial Model of Crime Trip Length

Because the data are highly skewed, a negative binomial regression function was used (Cameron and Trivedi 1998; Lord 2006). This model can be applied either to counts or to an extremely skewed continuous variable with a zero minimum (Hilbe 2008). The distance traveled for a crime trip is assumed to be Poisson distributed and independent over all offender trips and has the form:

$$Y|_i\ \mu_i \sim \text{Poisson}(\mu_i) \tag{7.1}$$

The mean of Poisson is organised as:

$$\mu = e^{(X_{ik}\beta_{ik})}e^{\varepsilon_i} \tag{7.2}$$

where $e^{(X_{ik}\beta_{ik})}$ is an exponential function of the k linear covariates (X_k), β_k is a vector of unknown coefficients for the k covariates, and ε_i is the model error independent

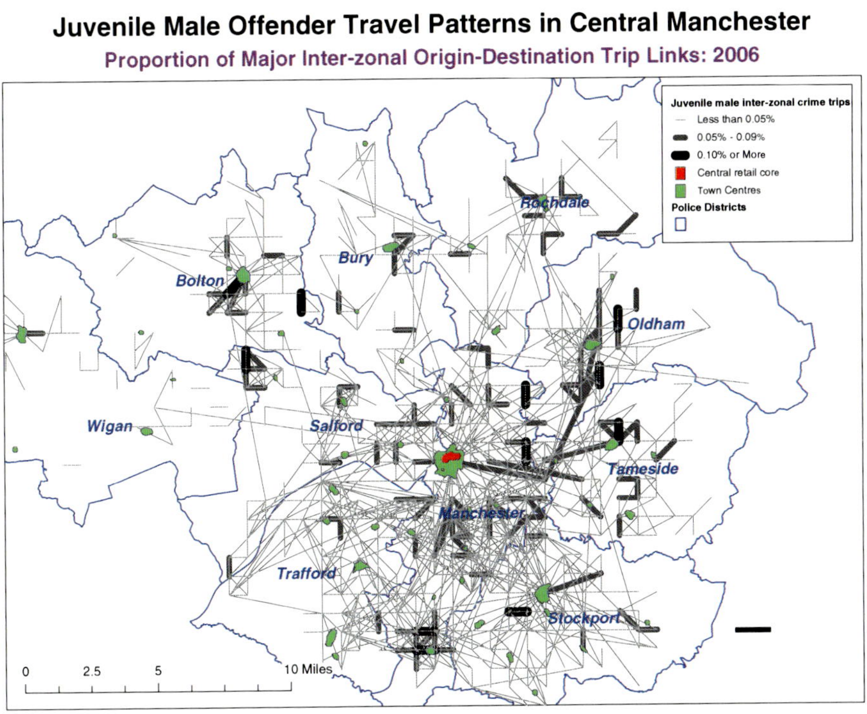

Fig. 7.4 Major O-D links for juvenile male crime trips in central Manchester

Table 7.4 Multivariate Predictors of Crime Trip Length

Offender characteristics
1. African-Caribbean ethnicity
2. Asian ethnicity
3. White, British-born
4. Age
5. Female (v. male)
6. Juvenile male
7. Arrested prior to 2006
8. Number of crimes committed by offender in data base

Type of crime
9. Violent (v. property and other type)

Effect of co-offenders
10. Presence of one or more co-offenders

Land use of the crime location
11. Residential land use
12. Commercial land use
13. Transport facility
14. On street/road

Metropolitan Characteristics
15. Crime occurred in the city centre
16. Crime occurred in a town centre
17. Distance from offender's residence to city centre (km)

of all covariates. Depending on the amount of dispersion, the error term can be modeled as a true Poisson function (when the dispersion is equal to the mean), a linear-corrected Poisson function for under-dispersion (when the variance is less than the mean, called an NB1 model), or a Gamma function for over-dispersion (when the variance is greater than the mean, called a negative binomial or NB2 model). In the negative binomial, which is a mixed function model, it is usually assumed that e^{ε_i} is independent and Gamma distributed with a mean equal to 1 and a variance equal to $\alpha = 1/\psi$ where both α (the dispersion parameter) and ψ (the inverse dispersion parameter) are greater than 0 (Lord 2006; Cameron and Trivedi 1998).

The model is tested with a link function that relates the natural log of expected mean to a linear combination of independent variables:

$$Ln(Y_i) = Ln(\mu_i) = \beta_0 + \beta_1 X_1 + \beta_2 X_2 + \cdots + \beta_k X_k + \varepsilon_i \tag{7.3}$$

The interpretation of this is that the trip length (the distance traveled) is an exponential function of the linear predictors. The distance traveled increases exponentially with a unit change in one of the predictors, controlling for the others.

Because it is a non-linear model, the model is estimated with the maximum likelihood approach. Also, again because it is non-linear, the goodness of fit of the model cannot be evaluated with R^2 but instead with a log-likelihood statistic (Hilbe 2008; Cameron and Trivedi 1998; Miaou 1996); higher log-likelihood values are best. The *Aikaike Information Criterion* (AIC) and the *Bayesian Information Criterion* (BIC) adjust the log likelihood for degrees of freedom (Hilbe 2008, 27). They are used to compare models; smaller values are best. The *deviance* statistic compares the model to that which would be achieved if there was perfect fit (Cameron and Trivedi 1998, 152–53). It is approximately Chi-square distributed with degrees of freedom equal to N-K-1 where *K* is the number of parameters estimated (including the constant). If the deviance value greater exceeds N-K-1, this suggests the model is over-dispersed due to missing values or a non-Poisson form.[12]

Two other summary statistics are presented. First, the Mean Absolute Deviation (MAD) is the average of the absolute value of the residual error between the observed and predicted values. It is a similar measure to the mean squared error but is more intuitive. A high MAD value is indicative of a poorly fit model. Second, the dispersion parameter, α, and its inverse, ψ, indicate the extent to which conditional variance (the remaining variance after accounting for all the independent predictors) exceeds the conditional mean (the expected mean taking into account all the independent predictors).

Table 7.5 presents the results of the full model and the best reduced form model that has eliminated non-statistically significant variables and multi-collinear inde-

[12] Over-dispersion occurs when the expected (predicted) variance is greater than the expected mean and is most likely the result of combining different underlying Poisson distributions. The Poisson distribution assumes that the expected variance will equal the expected mean. However, most count distributions show over-dispersion and are the primary rationale for using models like the negative binomial (Cameron and Trivedi 1998).

Table 7.5 Multivariate predictors of crime trip length: poisson-Gamma model: coefficients and standard errors (N=97,429 trips)

DepVar: Log of distance to crime location (km) for crimes committed in 2006				
	Full Model		Reduced Model	
N:	97,429		97,429	
Df:	97,410		97,415	
Log likelihood:	−189,337.0		−189,347.5	
AIC:	378,712.0		378,724.9	
BIC:	378,892.3		378,867.2	
Deviance:	52,649.5***		52,657***	
Mean absolute deviation	2.29		2.29	
Dispersion multiplier:	2.59		2.59	
Inverse dispersion multiplier:	0.39		0.39	
	Full Model Coefficient	(StdErr)	Reduced Model Coefficient	(StdErr)
Constant	**1.026*****	(0.033)	**1.120*****	(0.017)
Offender characteristics				
African-Caribbean	**0.165*****	(0.022)	**0.164*****	(0.022)
Asian	**0.161*****	(0.024)	**0.157*****	(0.024)
White and British-born	0.023	(0.016)	–	–
Age	0.0003	(0.001)	–	–
Female	**−0.198*****	(0.016)	**-0.198*****	(0.016)
Juvenile Male	**−0.438*****	(0.017)	**-0.452*****	(0.015)
Arrested prior to 2006	0.014	(0.014)	–	–
Number of crimes committed in 2006	**0.006*****	(0.001)	**0.006*****	(0.001)
Violent	**−0.198*****	(0.012)	**-0.205*****	(0.012)
Effect of co-offenders				
Co-offender	**0.091*****	(0.013)	**0.090*****	(0.013)
Land use of crime location				
Residential	**−0.445*****	(0.020)	**−0.496*****	(0.015)
Commercial	**0.088*****	(0.022)	–	–
Transport	**0.177*****	(0.032)	**0.123*****	(0.029)
Street/road	**−0.165*****	(0.019)		
Metropolitan characteristics				
City centre	**0.773*****	(0.027)	**0.782*****	(0.027)
Town centre	**0.309*****	(0.018)	**0.322*****	(0.018)
Distance from offender residence to city centre (km)	**0.012*****	(0.003)	**0.012*****	(0.003)

<none>Not significant
* Significant at p≤.05
**Significant at p≤.01
***Significant at p≤.001

pendent variables.[13] The full model included all the variables listed in Table 7.6 while the reduced model excluded the commercial land use, White and British-born, age, prior arrest, and street/road location variables. Note that the dependent variable is the

[13] The model was tested with negative binomial regression model in version 3.3 of *CrimeStat* (Levine 2010).

Table 7.6 Multivariate predictors of crime trip length by crime type: poisson-gamma model: coefficients of reduced model

DepVar: Log of distance to crime location (km) for crimes committed in 2006					
	All crimes	Simple assault	Aggravated assault	Rape/sexual assault	Burglary
N:	97,429	34,381	4,322	1,080	7,535
Df:	97,416	34,368	4,309	1,067	7,522
Log likelihood:	−189,699.5	−60,699.7	−7,649.7	−2,004.0	−15,644.3
AIC:	379,424.9	121,425.3	15,325.5	4,034.0	31,314.6
BIC:	379,548.2	121,535.2	15,408.3	4,098.8	31,404.7
Deviance:	52,916.2***	17,831.8**	2,071.4***	652.3***	3,757.0***
Mean absolute deviation:	2.30	2.04	2.02	2.60	2.44
Dispersion multiplier:	2.62	3.14	3.10	3.58	2.30
Inverse dispersion multiplier:	0.38	0.32	0.32	0.28	0.43
	Coeff.	Coeff	Coeff.	Coeff.	Coeff.
Constant	0.890***	**0.758*****	**0.758*****	**1.056*****	**1.067*****
Offender characteristics					
African-Caribbean	**0.160*****	**0.269*****	**0.238***	−0.099	**0.172***
Asian	**0.164*****	**0.147***	−0.169	0.261	0.200
Female	**−0.152*****	**−0.271*****	**−0.344*****	−0.174	**−0.191***
Juvenile male	**−0.438*****	**−0.418*****	**−0.427 *****	**-0.484****	**−0.548*****
Number of crimes committed in 2006	**0.010*****	0.002	**−0.016****	0.015	**0.009*****
Number of crimes committed in 2006					
Co-offender	**0.125*****	−0.010	0.077	0.321	−0.010
Land use of crime location					
Residential	**−0.373*****	**−0.494*****	**−0.305*****	−0.208	0.067
Transport	**0.251*****	**0.351*****	**0.459****	0.428	0.328
Metropolitan characteristics					
City centre	**0.785*****	**1.007*****	**0.772*****	0.699	**0.434****
Town centre	**0.331*****	**0.457*****	**0.432*****	0.398	−0.009
Distance from offender residence to city centre (km)	**0.010****	**0.019****	0.032	−0.037	−0.008

(continued)

Table 7.6 (continued)

DepVar: Log of distance to crime location (km) for crimes committed in 2006					
	Robbery	Shoplifting	Vandalism	**Vehicle** Theft	**Drug** Dealing
N:	4,070	8,898	11,135	3,366	7,762
Df:	4,057	8,885	11,122	3,353	7,749
Log likelihood	−8,905.9	−21,641.3	−17,971.2	−6,944.1	−14,431.5
AIC	17,837.7	43,308.6	35,968.5	13,914.3	28,889.0
BIC	17,919.8	43,400.8	36,063.6	13,993.9	28,979.4
Deviance::	1,954.3***	4,832.1***	4,790.3***	1,702.2***	4,161.5***
Mean absolute deviation:	2.25	2.69	1.80	2.37	2.18
Dispersion multiplier	1.69	1.28	3.72	2.27	2.53
Inverse dispersion multiplier	0.59	0.78	0.27	0.44	0.39
	Coeff.	Coeff.	Coeff.	Coeff.	Coeff.
Constant	**1.035*****	**1.076*****	**0.809*****	**1.165*****	**0.785*****
Offender characteristics					
African-Caribbean	−0.002	0.039	0.138	0.077	0.028
Asian	**−0.179***	−0.029	0.037	**−0.298***	**0.150****
Female	−0.076	**0.118*****	**−0.180****	0.040	−0.130
Juvenile Male	**−0.331*****	**−0.137****	**−0.418*****	**−0.431*****	**-0.194****
Number of crimes committed in 2006	**0.013****	**0.011****	0.005	−0.001	**-0.028*****
Effect of co-offenders					
Co-offender	**0.167****	**0.100****	−0.060	−0.002	**0.190****
Land use of crime location					
Residential	−0.036	0.197	**−0.277*****	0.066	**−1.401*****
Transport	**0.248****	0.281	**0.370****	0.075	**0.190***
Metropolitan characteristics					
City centre	**0.398*****	**0.439*****	**1.111*****	0.342	**0.860*****
Town centre	**0.272*****	−0.027	**0.373*****	**0.405***	**0.546*****
Distance from offender residence to city centre (km)	**−0.038****	**0.029*****	−0.009	−0.021	**0.078*****

<none>Not significant
* Significant at p ≤ .05
** Significant at p ≤ .01
*** Significant at p ≤ .001

natural logarithm of the trip length and the independent variables are linear. By convention, the results are presented in this log-linear form. However, in the final interpretation, the trip length is an exponential function of the linear predictors.

The summary statistics show that both models are properly fit, as indicated by the deviance statistic (Cameron and Trivedi 1998, 152–53). Note that the AIC of the full model is smaller than that of the reduced model (and that the log likelihood of the reduced model is greater than that of the full model), but the differences are not great. These indicate that the reduced model maximised the likelihood almost as well as the full model. By eliminating non-statistically significant and correlated (multicolinear) independent variables, it is a more stable model. The Mean Absolute Deviation of the two models is identical indicating that the reduced model predicted as well.

The significance of individual coefficients was tested with a Z-test, estimated by the coefficient divided by the standard error. The Z-values are not shown due to space limitations but significant coefficients are indicated by asterisks.

7.5.4 Longer and Shorter Crime Trips

Longer crime trips were associated with persons of African-Caribbean and Asian ethnicity and for those offenders living farther from the city centre, committing crimes with one or more co-offenders, at transport facilities, in the city centre, or at town centres. Shorter crimes trips were associated with juvenile males, and females (both juveniles and adults), and for crimes that were violent and committed in residential buildings.

The number of 2006 crimes listed in the data base, was associated with longer trips, consistent with the results of Smith et al. (2009) and Townsley and Sidebottom (2010). That is, offenders who are represented multiple times in the data base tend to make longer crime trips. However, the strength of the relationship was not particularly strong, as evidenced by the coefficient (and Z-value, which is not presented).

Overall, the strongest relationship with distance traveled is for crime trips to the city centre, crime trips made by juvenile males, and crimes committed in residential locations. In terms of gender, we see that females generally make shorter crime trips than adult males while juvenile males generally make the shortest crime trips. The general pattern of women making shorter trips than men holds for most crime trips.

7.5.5 Multivariate Predictors of Crime Trip Length by Crime Type

The model was then tested for each of the nine minor crime categories (Table 7.6). The model was the same as that of the reduced model except that the dummy crime type variable for violent crimes was dropped. For comparison, the reduced model was re-run on all the crime types (first model column).

There is both consistency and uniqueness among the nine specific models. The deviance values were smaller than the degrees of freedom for all nine models indicating that the models were properly fit. Also, the Mean Absolute Deviation values varied but were quite small for the nine models. In other words, the overall fit of the model was good for each of the nine crime types. Further, each crime type has its own unique combination of variables that were statistically significant. No two models were the same. Nevertheless, the coefficients were quite similar across the nine models.

Table 7.7 Signs of coefficients and significance levels for multivariate: predictors of crime trip length by 9 crime types

Coefficient				
	Positive/ Longer	(significant)	Negative/ Shorter	(significant)
Constant	9	(9)	**0**	(0)
Offender characteristics				
African-Caribbean	**7**	(3)	**2**	(0)
Asian	**5**	(2)	**4**	(2)
Female	**2**	(1)	**7**	(4)
Juvenile Male	**0**	(0)	**9**	(9)
Number of crimes committed in 2006	**6**	(3)	**3**	(2)
Effect of co-offenders				
Co-offender	**5**	(3)	**4**	(0)
Land use of crime location				
Residential	**3**	(0)	**6**	(4)
Transport	**9**	(5)	**0**	(0)
Metropolitan characteristics				
City centre	**9**	(7)	**0**	(0)
Town centre	**7**	(6)	**2**	(0)
Distance from offender residence to city centre	4	(3)	**5**	(1)

7.5.5.1 Common Factors Affecting Trip Length for All Crime Types

Table 7.7 presents the number of coefficients that had the same sign for each of the independent variables as well as the number that were statistically significant. Aside from the intercept, which was significantly positive for all nine models (essentially meaning that there is a crime-specific positive distance used as a reference point), the strongest variables overall were juvenile males, which was negative (shorter) and statistically significant for all nine models, crimes committed in the city centre, which was positive (longer) for all nine models and statistically significant for seven, and crimes committed at transport facilities, which was positive (longer) for all nine models and statistically significant for five.

Variables that generally had the same sign were individuals of African-Caribbean ethnicity (seven positive with three being statistically significant), females (seven negative with four being statistically significant), and crimes committed in town centres (seven positive with six being statistically significant). The remaining variables showed mixed tendencies including the number of crimes committed by the offender in 2006 (three significantly positive and two significantly negative).

In other words, over all nine crime categories, trips involving juvenile males and females (both juveniles and adults), and crimes committed in residential buildings were generally shorter. Conversely, trips by offenders of African-Caribbean ethnicity, by offenders who committed more crimes in 2006, and for crimes committed in transport facilities, the city centre, or the town centres were generally longer in length.

These variables reflect an interaction between the crime type, the presence of co-offenders, the organisation of the metropolitan area, and the personal characteristics of individuals – gender, age group and ethnicity. The shorter crime trips by female offenders are consistent with the hypothesis presented earlier. On the other hand, juvenile females did not particularly make short crime trips, certainly not compared to juvenile males. There clearly is an interaction between gender and age group.

In terms of the ethnicity variables, which were used as statistical controls, the longer trips by offenders of African-Caribbean ethnicity may reflect the concentration of immigrant housing towards the periphery of the Manchester area. Asian and African-Caribbean communities are generally very tight-knit in Manchester. It is possible that any involvement in criminal behaviour would likely lead to their recognition by family members or leaders of the small communities.

7.5.5.2 Unique Factors Affecting Trip Length for Crime Types

At the same time, there were meaningful differences among the predictors for several crime types. Among the offender characteristics, females made significantly longer trips in committing shoplifting than males in spite of generally making shorter trips for other types of crime. This is most likely the result of a large number of shoplifting offences being committed in the central retail core. For shoplifting, 15% of the events occurred within the central retail core and another 34% occurred in other town centres compared to 3% and 9% respectively for other types of crime (or about four times the relative likelihood). Since females commit shoplifting crimes at more than three times the proportion for that of males, it is not surprising that their trips are longer for this type of offence.

Another unique interaction appears to be the role of an offender's residence relative to the central city. In general, offenders who lived farther away from the city centre traveled farther, on average, than those who lived closer. However, for robberies, the reverse effect was found with offenders living farther away making shorter robbery trips. This may suggest some interaction with nearby neighbourhoods and commercial areas for robbery offenders living in the farther suburbs.

A third unique interaction was the number of crimes committed in 2006, a proxy variable for prolific offenders. While this variable was generally positive (longer) for six of the nine models with three being statistically significant (robbery, burglary, and shoplifting), it was significantly negative (shorter) for two crime types – assaults and drug dealing. One hypothesis is that prolific offenders may be more cautious in committing property crimes than less prolific offenders. But, those same persons may also be more disturbed and prone to reacting more quickly in an emotional situation than an offender who does not commit as many crimes.

These variables had a statistically significant effect for a variable that was opposite to its general pattern. But, there may certainly be other unique interactions. As with most social phenomena, there are always unique factors in spite of general trends in variables.

7.6 Conclusions

Overall, the crime trip length was a function of the general factors outlined above. First, consistent with other research, property crime trips were longer, on average, than violent crime trips.

Second, crime trip length was a function of the location where crimes occurred. Crimes committed in residential locations were generally shorter than those committed in non-residential locations with those committed in commercial areas or in transit station locations being the longest. More critically, crimes committed at one of the town centres or in the central retail core were generally much longer than other kinds of crimes. This is a function both of the centrality of those locations compared to residential neighbourhoods as well as to an extensive public transportation system that links residential communities with the town centres. While poverty still underlies most crime, the patterns seen are related as much to the transportation network as to the distribution of poor neighbourhoods. Further, the location of commercial centres is an intermediating variable in effecting crime trips, at least for a substantial number of crimes.

Third, crime trip length was a function of interaction with other offenders. Overall, approximately one-quarter of the offences involved a co-offender. Offenders who had accomplices were more likely to commit property crimes and more likely to travel farther than those who committed the crimes by themselves. This suggests that the crime emerged either out of deliberate planning or out of a decision to go to a specific location with the crime being the result of an opportunity occurring at that location.

Fourth, crime trip length was a function of the characteristics of the offender, in particular gender and age. With the exception of shoplifting, female offenders traveled shorter distances than male offenders for most types of crime. However, juvenile male offenders made the shortest crime trips, on average, while adult male offenders made the longest crime trips, on average. Further, female offenders, both juveniles and adults, were much more likely to commit crimes in one of the town centres or the central retail core. Part of this is due to a higher percentage of females committing shoplifting offences than males but females also commit other crimes in the commercial areas proportionately more than males.

Simple generalisations do not account for all of the variance. The crime trip length of females is mediated by the crime type. For four of the crime types, female offenders commit more crimes closer to their homes than male offenders. But, for shoplifting, the crime trips are longer. Since these are 22% of all their committed crimes, the exception is considerable. Further, for four other types, there is no relationship with distance for females offenders. Juvenile males have very short trips but juvenile females do not. Violent crimes tend to be closer to home than property crimes except for those committed in the city centre. The multivariate analysis shows these interactions whereas univariate analysis would cover up much of it.

7.6.1 Explanations for Shorter Female Crime Trips

The result for juvenile male crime trips is more consistent with the expectations of juveniles being more neighbourhood-based and less likely to be employed and less likely to have access to an automobile. On the other hand, the result for female crimes trips is complicated. There is not a single explanation that can account for it. While economic marginalisation is a factor contributing to crimes for both genders (Morash 2006), there are gender specific factors that interact with economic marginalisation.

Steffensmeier and Allan (1996) outlined five factors that can explain differential access to criminal opportunities by gender: (1) strong gender norms against female involvement in crime; (2) greater amenability to affiliation by females; (3) greater social control on women that limits their ability to commit crime; (4) less physical strength by females which limits the amount of violent crime; and (5) reproductive-sexual differences that limit female opportunities for sexual deviance. These factors explain why females commit fewer and generally less serious crimes than males.

To this list can be added greater acceptance of parental and household responsibilities by women and gender norms that shape preferences (Rosenbloom 2005; Hanson and Hanson 1980). We do not have information on whether the female offenders had children, but it is likely that many did as well as many who were tied to parents. This could explain both why females commit fewer crimes than males as well as the shorter journey-to-crime trips for most crimes with the exception of shoplifting.

Increasing gender equality over time certainly has increased female participation in public activities, such as work, shopping, or driving. But, whether this trend can explain increasing crime involvement by women is questionable given that public participation is associated with higher income levels which, in turn, are negatively correlated with crime levels. At best, increased gender equality might explain the increase in petty offenses committed by economically marginal women (Chesney-Lind 1989).

Part of the reason for shorter crime trips by females may also be due to a higher proportion of their offences being simple assaults (41% compared to 34% for male offenders). Since assault 'trips' are generally short, the proportional mix is a small factor. Domestic offences, which we defined as crimes committed in the residence location of the offender accounted for 12% of the crimes committed by females and 10% of the crimes committed by males. A slightly higher proportion of domestic offences committed by females than males contributed to a slight reduction in the typical crime trips.

Nevertheless, these effects are very small and not sufficient to explain the generally shorter trips of female offenders. Further, the hypothesis of dependence by female offenders on males in committing their crimes was not strongly supported by our data. Only about one-quarter of the offenders committed crimes with others. Females were slightly more likely than males to have had an accomplice though, with the exception of shoplifting, the vast majority of crimes were committed without co-offenders. Even with shoplifting, the majority of the co-offenders were also female. Age group was more critical than gender in explaining this as juveniles, both female and male, were far more likely to commit crimes with others than adults.

7.6.2 Gendered Opportunities for Crime Travel by Females

The greatest difference between females and males in their crime trips, controlling for age, appears to be in their choice of locations where they commit their crimes, what might be called *gendered opportunities*. The higher frequency of crimes committed in the commercial areas by female offenders, particularly petty theft and shoplifting, appears to reflect gender norms. Generally, women shop in retail locations more than men, both in terms of frequency as well as duration, though there is enormous variability in the behaviour (ICSC 2009; Wharton and Verde 2007; Ditmar et al. 2004; Otnes and McGrath 2001).[14] That female offenders follow similar patterns is not surprising. This may also explain why juvenile females made much longer trips than juvenile males as gender norms may attract them disproportionately to commercial areas. There is easy transit access through an extensive light rail and bus system and is linked up to the town centres and central retail core.

7.6.3 Alternative Models of Crime Travel

In this study, we modeled the distance traveled by offenders in committing their crimes. The results show that the distance traveled is a function of multiple factors, involving opportunities, personal characteristics, and metropolitan structure. However, this is only a limited view of crime travel. An offender does not start out by saying that he/she will travel only, say, two kilometers and no further. Instead, an offender is attracted to a location because of the potential targets available, whether consciously or not. Thus, we find that more than half of the crimes committed by these offenders are in non-residential locations. The idea of a marauding offender who commits crimes in his or her neighbourhood is not a particularly accurate description of the search and travel pattern. While it may have once been true, given the early research on journey-to-crime by researchers like Lottier (1938) or the Chicago-school of criminology (Burgess 1925; Thrasher 1927), today's offenders are much more mobile.

[14] This is true for conventional shopping, not online shopping where men exceed women in purchases; Ditmar et al. 2004.

Thus, other approaches are needed to understand the decision structure of offenders. The use of the conditional logit model to describe the decision structure of offenders places an emphasis on the factors that predict the decision to commit a crime in a particular location. Several interesting studies have been conducted that use this framework and which provide insights into the decision factors (Bernasco and Block 2009; Bernasco and Nieuwbeerta 2005).

Another approach is the modeling of crime trips over an entire metropolitan area using the Crime Travel Demand framework (Levine 2005). This framework models the crime trip in stages but is applied to aggregate totals by zone and separately predicts the number of crime trips that will emanate from each zone, the number of crimes that will be attracted to each zone, the number of trip links from each origin zone to each destination zone, the travel mode used for each of these links, and the likely routes taken. This approach can provide a model of where crime trips typically occur in space, at least for the higher volume crime types (Levine and Canter 2011; Levine 2007).

In short, understanding the dynamics of crime trips is an important goal for predicting how crime will change as metropolitan areas change and that a variety of different approaches will be necessary.

7.6.4 Practical Implications

From a practical viewpoint, understanding the relationship between crime origins and crime destinations can aid police in their enforcement efforts. In looking for specific offenders, knowing the link between the origin and the destination can decrease the search area that has to be examined and increase the accuracy of the search (Levine and Block 2011). Differentiating serial offenders by gender and by age might lead to further improvement in the methodology.

From a crime prevention perspective, understanding the areas where offenders commit crime can allow local government to target both origin and destination locations with targeted enforcement, selected social services, and educational efforts (Block and Bernasco, 2009; Leitner and Kent, 2009; Levine and Lee, 2009). Such a perspective gives police and other agencies increased flexibility in responding to crime (Levine and Canter 2011; Levine 2007). Since crimes committed by females appear to be increasing, there is a need for police to focus more on commercial areas where a sizeable proportion of crimes committed by females do occur. These commercial, retail areas tend to have a predominance of private security, especially in the larger stores. Coordination between police and private security might offer one avenue for improved enforcement of the retail area in general.

These results question the simple premise that crime trips are short and that offenders adopt a home-based search strategy in committing their crimes. While most crime trips are relatively short, many are not. More importantly, what we have shown is that the length of the crime trip is a function of the gender and age characteristics of the offender as well as the type of crime committed, the interaction of the offender with co-offenders, and the organisation of the metropolitan area including its transportation facilities. Crime travel is complex and is a function not only of the opportunities available to offenders but also their personal predispositions, defined in part by their gender norms and stage in the life cycle.

Finally, a word of caution is needed. These results are based on the crimes committed in one metropolitan area during a single year. Clearly, more research is needed to show whether these results hold over time in Manchester and for other metropolitan areas, too. Replication is essential for confirming any complex research.

References

Arndale (2012). History of Manchester Arndale. http://www.manchesterarndale.com/history_of_manchester_arndale.aspx. Accessed 12 Jan 2012

Berman G (2010) IV. Youth crime and punishment in England and Wales. Research paper 10/02. Parliament, London. http://www.parliament.uk/briefingpapers/commons/lib/research/briefings/snsg-05277.pdf

Bernasco W (2010) A sentimental journey to crime: effects of residential history on crime location choice. Criminology 48(2):389–416

Bernasco W, Block R (2009) Where offenders choose to attack: a discrete choice model of robberies in Chicago. Criminology 47(1):93–130

Bernasco W, Nieuwbeerta P (2005) How do residential burglars select target areas? Br J Criminol 44:296–315

Blalock HM Jr (1979) Social statistics, Rev. 2nd edn. McGraw-Hill, New York

Block R, Bernasco W (2009) Finding a serial Burglar's home using distance decay and origin–destination patterns: a test of empirical Bayes journey-to-crime estimation in The Hague. J Investig Psychol Offender Profil 6:187–211

Block R, Helms D (2005) Chapter 17: Case studies in crime travel demand modeling. In Levine N (ed) CrimeStat: a spatial statistics program for the analysis of crime incident locations (version 3.0). Ned Levine and Associates, Houston TX, National Institute of Justice, Washington, DC. November. http://www.icpsr.umich.edu/crimestat.

Block R, Galary A, Brice D (2007) The journey to crime: victims and offenders converge in violent crime offences in Chicago. Secur J 20:123–137

Brantingham PJ, Brantingham PL (2008) Crime pattern theory. In: Wortley R, Mazerolle L (eds) Environmental criminology and crime analysis. Willan, Devon

Broidy LM, Cauffman EE (2006) Understanding the female offender. Final report 2001-IJ-CX-0034. National Institute of Justice, U.S. Department of Justice, Washington, DC. http://www.ncjrs.gov/pdffiles1/nij/grants/216615.pdf

Burgess EW (1925) The growth of the city: an introduction to a research project. In: Park RE, Burgess EW, Mackensie RD (eds) The city. University of Chicago Press, Chicago, pp 47–62

Cameron AC, Trivedi PK (1998) Regression analysis of count data. Cambridge University Press, Cambridge

Canter D, Larkin P (1993) The environmental range of serial rapists. J Environ Psychol 13:63–69

Capone DL, Nichols WW Jr (1975) Crime and distance: an analysis of offender behaviour in space, Proc Assoc Am Geogr 7:45–49

Chesney-Lind, M (1989) Girls' crime and woman's place: awards a feminist model of female delinquency. Crime Delinq 35(1):5–29

Cohen LE, Felson M (1979) Social change and crime rate trends: a routine activity approach. Am Sociol Rev 44:588–608

Crowe A (2000) "Chapter 2: Jurisdictional and program self-assessment", Jurisdictional technical assistance package for juvenile corrections. Office of Juvenile Justice and Delinquency

Prevention, Office of Justice Programs, U.S. Department of Justice, Washington, DC. http://www.ncjrs.gov/html/ojjdp/juris_tap_report/ch2_06.html.

Demographia (1998) U. S. Metropolitan Areas: 1998 central city and suburban population. Wendell Cox Consultancy, Belleville, IL. http://www.demographia.com/db-usmsacc98.htm

Ditmar H, Long K, Meek R (2004) Buying on the internet: gender differences in on-line and conventional buying motivations. Sex Roles 50(5/6):423–443

FBI (1997) Table 35 Total arrest trends, Sex, 1991–1995, In Federal Bureau of Investigation Crime in the United States 1997, U.S. Department of Justice, Washington, DC. http://www.fbi.gov/ucr/Cius_97/95CRIME/95crime4.pdf.

FBI (2008) Table 33 Ten-year arrest trends by sex 1999–2008, In Federal Bureau of Investigation, Crime in the United States 2008, U.S. Department of Justice, Washington, DC. http://www.fbi.gov/ucr/cius2008/data/table_33.html.

Fritzon K (2001) An examination of the relationship between distance travelled and motivational aspects of firesetting behaviour. J Environ Psychol 21:45–60

Groff ER, McEwen JT (2005) Disaggregating the journey to homicide. In: Wang F (ed) Geographic information systems and crime analysis. Idea Group Publishing, Hershey

Hanson S, Hanson P (1980) The impact of women's employment on household travel patterns: a Swedish example. Rosenbloom S (ed), Women's travel issues: research needs and priorities: conference proceedings and papers, U.S. Department of Transportation. Washington, D.C.: U.S. Government Printing Office

Hilbe JM (2008) Negative binomial regression (with corrections). Cambridge University Press, Cambridge

Home Office (2001) Arrests for notifiable offences and the operation of certain police powers Under PACE England and Wales 2000/01 19/01 http://rds.homeoffice.gov.uk/rds/pdfs/hosb1901.pdf

ICSC (2009). How the recession has impacted consumer shopping habits. International Council of Shopping Centers, New York. http://www.icsc.org/web/RecessionBooklet_lores.pdf

Kanji GK (1993) 100 statistical tests. Sage Publications, Thousand Oaks

Kneebone E, Raphael S (2011) City and suburban crime trends in Metropolitan America. Metropolitan opportunity series, Metropolitan Policy Program, Brookings Institution, Washington, DC. http://www.brookings.edu/papers/2011/0526_metropolitan_crime_kneebone_raphael.aspx. 28 May

LeBeau JL (1987) The journey to rape: geographic distance and the rapist's method of approaching the victim. J Police Sci and Adm 15(2):129–136

Leitner M, Kent JK (2009) A Bayesian journey to crime modeling of single- and multiple crime type series in Baltimore County, MD. J Investig Psychol Offender Profil 6(3):213–236

Levine N (2005) CrimeStat: a spatial statistics program for the analysis of crime incident locations (version 3.0). Ned Levine and Associates/National Institute of Justice, , Houston, TX/Washington, DC. November. http://www.icpsr.umich.edu/crimestat. Chapters 6, 10–16

Levine N (2007) Crime travel demand and bank robberies: using CrimeStat III to model bank robbery trips. Soc Sci Comput Rev 25(2):239–258

Levine N (2010) *CrimeStat III:* A Spatial Statistics Program for the Analysis of Crime Incident Locations (version 3.3). Ned Levine & Associates, Houston, TX.; National Institute of Justice, Washington, DC. July

Levine N, Block R (2011) Bayesian journey-to-crime estimation: an improvement in geographic profiling methodology. Prof Geogr 63(2):213-229

Levine N, Canter P (2011) Linking origins with destinations for DWI motor vehicle crashes: an application of crime travel demand modeling. Crime Mapp 3:7–41

Levine N, Lee P (2009) Bayesian journey-to-crime modeling of juvenile and adult offenders by gender in Manchester. J Investig Psychol Offender Profil 6:237–251

Levine N, Wachs M (1986) Bus crime in Los Angeles: I - Measuring the incidence. *Trans Res 20*(4): 273–284

Lord D (2006) Modeling motor vehicle crashes using Poisson-gamma models: examining the effects of low sample mean values and small sample size on the estimation of the fixed dispersion parameter. Accid Anal Prev 38:751–766

Lottier S (1938) Distribution of criminal offences in metropolitan regions. J Crim Law Criminol Police Sci 29:37–50
Manchester City Council (2012) The council and democracy. List of metropolitan areas in Greater Manchester. Opening page. http://www.manchester.gov.uk/info/100004/the_council_and_democracy/66/councils_in_greater_manchester/1. Last accessed 12 Feb 2012
Manchester UK (2011) The IRA Bombing of City Centre Manchester – June 1996. http://www.manchester2002-uk.com/buildings/bombing.html. Last accessed 12 Feb 2012
Miaou S-P (1996) Measuring the goodness-of-fit of accident prediction models. FHWA-RD-96-040. Federal Highway Administration, U.S. Department of Transportation, Washington, DC
Miller J (1998) Up it up: gender and the accomplishment of street robbery. Criminology 36(1):37–65
Ministry of Justice (2008) Arrests for recorded crime (Notifiable Offences) and the operation of certain police powers under PACE England and Wales 2006/07. http://www.justice.gov.uk/publications/arrestsfornotibleoffences.htm. Accessed 26 July 2010
Ministry of Justice (2010) Criminal statistics: England and Wales 2008. Ministry of Justice: London. http://webarchive.nationalarchives.gov.uk/+/http://www.justice.gov.uk/about/docs/criminal-stats-2008.pdf. Accessed 13 May 2011
Morash M (2006) Understanding gender, crime and justice. Sage, Thousand Oaks
Mulvey EP, Schubert CA, Chassin L (2010) Substance use and delinquent behaviour among serious adolescent offenders. NCJ 232790, Office of Juvenile Justice and Delinquency Prevention, Office of Justice Programs, U.S. Department of Justice, Washington, DC. http://www.ojjdp.gov/publications/PubAbstract.asp?pubi=254883.
Newton A, Hirschfield A (2009) Measuring violence in and around licensed premises: the need for a better evidence base. Crime Prev Commun Saf 11(3):171–188
NHMC (2012) Quick facts: resident demographics. National Multi Housing Council: Washington, DC. http://www.nmhc.org/Content.cfm?ItemNumber=55508. Accessed 28 Apr 2012
Otnes C, McGrath MA (2001) Perceptions and realities of male shopping behaviour. J Retail 77(1):111–137
Paulsen D (2007) Improving geographic profiling through commuter/marauder prediction. Police Pract Res 8:347–357
Pettiway LE (1995) Copping crack: the travel behaviour of crack users. Justice Q 12(3):499–524
Phillips PD (1980) Characteristics and typology of the journey-to-crime. In: Georges-Abeyie D, Harries KD (eds) Crime: a spatial perspective. Columbia University Press, New York, pp 156–166
Pointer G (2005) The U.K.'s major urban areas. Focus on people and migration. U.K. National Statistics, Office for National Statistics. Department of the Environment, Transport and the Regions, United Kingdom. http://www.statistics.gov.uk/downloads/theme_compendia/fom2005/03_fopm_urbanareas.pdf
Puzzanchera C (2009) Juvenile Arrests 2008. Juvenile Justice Bulletin. Office of Juvenile Justice and Delinquency Prevention, Office of Justice Programs, U.S. Department of Justice, Washington, DC, December. http://www.ncjrs.gov/pdffiles1/ojjdp/228479.pdf
Pyle GF (1974) The spatial dynamics of crime. Department of Geography Research Paper No. 159, University of Chicago, Chicago
Rand MR, SabolWJ, Sinclair M, Snyder HN (2010) Table 20: Location of alcohol-involved incidents known to law enforcement, 2007. Alcohol and crime: data from 2002 to 2008. Bureau of Justice Statistics, U.S. Department of Justice: Washington, DC. http://bjs.ojp.usdoj.gov/content/acf/20_location_by_offender.cfm. Accessed 30 Nov 2010
Rengert GF (1975) Some effects of being female on criminal spatial behaviour. PA Geogr 13(2):10–18
Rengert GF, Piquero AR, Jones PR (1999) Distance decay re-examined. Criminology 37(2):427–445
Rhodes WM, Conly C (1981) Crime and mobility: an empirical study. In: Brantingham PJ, Brantingham PL (eds) Environmental criminology. Waveland Press, Inc, Prospect Heights, pp 167–188
Roncek DW, Maier PA (1991) Bars, blocks and crimes revisited: linking the theory of routine activities to the empiricisms of 'Hot Spots'. Criminology 29(4):725–753
Rosenbloom S (2005) Understanding women's and men's travel patterns: the research challenge. In Federal Highway Administration, Women's travel issues: proceedings of the second national

conference, U.S. Department of Transportation, Washington, DC. 7–28. http://onlinepubs.trb.org/Onlinepubs/conf/CP35v1.pdf

Schwartz J, Steffensmeier D (2007) The nature of female offending: patterns and explanation. In: Zaplin RT (ed) Female offenders: critical perspectives and effective interventions, 2nd edn. Jones and Bartlett Publishers, Sudbury, pp 43–75

Smith TS (1976) Inverse distance variations for the flow of crime in urban areas. Soc Force 25(4):804–815

Smith W, Bond JW, Townsley M (2009) Determining how journeys-to-crime vary: measuring inter- and intra-offender crime trip distributions. In: Weisburd D, Bernasco W, Bruinsma G (eds) Putting crime in its place: units of analysis in spatial crime research. Springer, New York

Snook B (2004) Individual differences in distance travelled by serial burglars. J Investig Psychol Offender Profil 1:53–66

Snook B, Taylor PJ, Bennell C (2004) Geographic profiling: the fast, frugal and accurate way. Appl Cogn Psychol 18:105–121

Snook B, Cullen RM, Mokros A, Harbort S (2005) Serial murderers' spatial decisions: factors that influence crime location choice. J Investig Psychol Offender Profil 2:147–164

Steffensmeier D, Allan E (1996) Gender and crime: toward a gendered theory of female offending. Annu Rev Sociol 22:459–487

Thrasher FM (1927) The gang. University of Chicago Press, Chicago

TGM (2012) Transport for greater Manchester. Manchester, England. http://www.tfgm.com/Pages/default.aspx

Townsley M, Sidebottom A (2010) All offenders are equal, but some are more equal than others: variations in journeys to crime between offenders. Criminology 48(3):897–917

Turner S (1969) Delinquency and distance. In: Wolfgang ME, Sellin T (eds) Delinquency: selected studies. Wiley, New York, pp 11–26

U.K. National Statistics (2000). Table 10.2. Households with regular use of car: 2000. General household survey and family expenditure survey, Office for National Statistics National Travel Survey, Department of the Environment, Transport and the Regions, United Kingdom. http://www.statistics.gov.uk/STATBASE/ssdataset.asp?vlnk=6067

U.S. Census Bureau (2000) All across the USA: Population distribution, 1999. In Population profile of the United States: 1999. Bureau of the Census, U. S. Department of Commerce, Washington, DC, chapter 2

U.S. Census Bureau (2002) "Vehicles available and household income in 1999: 2000". Table QT-H11, Census 2000 Summary File 3 (SF 3) – Sample Data, Bureau of the Census, U. S. Department of Commerce: Washington, DC. http://factfinder.census.gov/servlet/QTTable?_bm=yand-geo_id=Dand-qr_name=DEC_2000_SFAIAN_QTH11and-ds_name=Dand-_lang=en.

Van Koppen PJ, Jansen RWJ (1998) The road to the robbery: travel patterns in commercial robberies. Br J Criminol 38:230–246

Warren J, Reboussin R, Hazelwood RR, Cummings A, Gibbs N, Trumbetta S (1998) Crime scene and distance correlates of serial rape. J Quant Criminol 14(1):35–59

Wharton and Verde (2007) 'Men buy, women shop': the sexes have different priorities when walking down the aisles. Jay H. Baker Retail Initiative, Wharton School, University of Pennsylvania and Verde Group: Philadelphia and Toronto. http://knowledge.wharton.upenn.edu/article.cfm;jsessionid=9a30c1e100f3523587b6?articleid=1848

White RC (1932) The relationship of felonies to environmental factors in Indianapolis. Social Force 10(4):488–509

Wikipedia (2009) List of metropolitan areas in the United Kingdom. Wikipedia. http://wapedia.mobi/en/List_of_metropolitan_areas_in_the_United_Kingdom#1. Accessed 8 Jan 2009

Wikipedia (2010) Manchester Arndale. *Wikipedia.* http://en.wikipedia.org/wiki/Manchester_Arndale_Centre. Accessed 20 July 2010

Wiles P, Costello A (2000) the 'road to nowhere': the evidence for travelling criminals. Home Office Research Study, No. 207. Research, Development and Statistics Directorate, London. http://www.homeoffice.gov.uk/rds/prgpdfs/brf400.pdf

Part III
Crime Modeling

Chapter 8
Crime Scene Locations in Criminal Homicides: A Spatial Crime Analysis in a GIS Environment

Hyun Kim, Yongwan Chun, and Casey Anderson Gould

Abstract Law enforcement agencies recently have recognized the significance and impact that comes from a better understanding of the dynamics of crime in a geographical perspective. In particular, due to its complex nature, analyzing criminal homicide is often challenging when connecting multiple locations to persons involved in crimes in order to solve cases. The main purpose of this study is to provide an analytical framework to understand criminal homicide from a socio-geographical perspective using GIS. This chapter explores how social relationships between victims and offenders, and the spatial characteristic of a community, affect the geographical pattern of underlying locations in criminal homicides. We propose a conceptual model named *Spatial Configurations of Homicide Crime* (SCHC), which categorizes sequences of homicides from a geographical perspective. The spatial configurations of criminal homicides is defined by combinations of locations including: Offender's residence (O), Victim's residence (V), Murder location (M), and Disposal location of victim (D), all of which are expressed as a set of (O, V, M, D). The relationships between SCHC and (1) social relationships among victims and offenders, (2) the context of the crimes, and (3) the ethnic composition of people within a community, are analyzed with *Multinomial Logit Models* (MNL). Based on

H. Kim (✉)
Department of Geography, University of Tennessee, 311 Burchfiel Geography Building, 1000 Philip Fulmer Way, TN 37996-0925, USA
e-mail: hyunkim@utk.edu

Y. Chun
School of Economic, Political and Policy Sciences, University of Texas at Dallas, 800 W. Campbell Rd, Richardson, 75080-3021, USA
e-mail: ywchun@utdallas.edu

C.A. Gould
Southwestern Institute of Forensic Sciences, 2355 North Stemmons Freeway, Dallas, TX, USA 75207
e-mail: Casey.Anderson@dallascounty.org

M. Leitner (ed.), *Crime Modeling and Mapping Using Geospatial Technologies*, Geotechnologies and the Environment 8, DOI 10.1007/978-94-007-4997-9_8,

geospatial analysis using GIS, this chapter identifies the critical socio-geographic factors and reveals the structure of locational components in homicidal crimes. A case study with more than 300 homicide incidents in Hillsborough County, Florida from 1997 to 2007 demonstrates social relationships among victims and offenders in their geographical locations.

Keywords Criminal homicide • Socio-geographic context • Social relationship • GIS • Multinomial Logit Models (MNL)

8.1 Introduction

The causes and amount of violence in today's society have been the primary focus of analysis in many fields, including prevention and intervention for law enforcement agencies across the country. These agencies are frequently charged with the task of uncovering evidence and the circumstances surrounding criminal activities. The duty ranges from revealing associations between multiple locations to the social relationships of the individuals involved in crimes (Federal Bureau of Investigation 2004:41). Among violent crimes, criminal homicide is defined as an exchange between a victim and an offender, within a context, that results in the demise of the victim (Swatt and He 2006:279). In general criminal homicide cases, the social relationships between the people involved have the potential to provide important information about the connection of scenes (Silverman and Kennedy 1987). For example, the victim-offender relationship is crucial to unlocking the mystery of why people kill each other (Avakame 1998). The locations of the murder site, residences, and the bond between the victim and killer can play a vital role in recovering the body and solving the case (Gagne 1992).

The main differences between homicide and other types of crimes are highlighted in two points. First, the criminal homicides involve the serious social costs by an exchange between a victim and an offender, resulting in the demise of the victim. Second, locations involved in homicides include four clearly distinctive sites, ranging from the victim and the offenders' residences, to the place where the crime occurred, and a site of deposition, which is defined as the location of victim's body after murder. Other types of crime, for example, theft from a motor vehicle or theft of a motor vehicle, may involve more diverse locations and are often difficult to capture the characteristics. Moreover, the crime becomes more complicated based on the arranged sequence among multiple locations with interactions between offender and victim. In other words, by constructing a dynamic spatial configuration from combinations of the multiple locations, how these scenes and people connect and interact can play a vital role in solving a case in a timely fashion. Many criminal homicide studies have focused on the dynamics of social relationships among offenders and victims involved in criminal homicides (Wolfgang 1958, 1967; Daly and Wilson 1988; Polk 1993; Avakame 1998). However, some studies use spatial analyses which simply use distance measures between relevant locations for certain crimes (for example, Brantingham and Brantingham 1984; Messner et al. 1999; Canter 2000; Van Patten and Delhauer 2007; Grubesic and Mack 2008). Ideally, law enforcement agencies would use social relationships and spatial configurations to deduce conclusions for homicide analyses. It is obvious that there is a paucity of research integrating and supporting the two foci using Geographical Information System methods. Specifically, a geographical spatial analysis, paired with a social relationship as well as geographical context, has not yet been attempted in the context of an area where the urban society is characterized as a place with a large, diverse population with a great deal of immigration and residential movement (Frey and Zimmer 2001:25).

In this context, the main purpose of this chapter is to provide an analytical framework to understand criminal homicide from a socio-geographical perspective using Geographic Information System (GIS) for urban areas as exemplified by Hillsborough County, Florida. Using a socio-geographical perspective is important because the social relationship can be manifested in space with significance. For example, it is expected that the relationship of the offender to the victim will significantly affect the spatial distribution of a murder. As implied by previous literature, some socio-demographic characteristics can be related systematically to the geographical proximity of homicides (Cohen and Felson 1979; Messner and Tardiff 1985). If both offender and victim are in close proximity due to social relationship, then the place of murder is much more likely to occur in proximity between their residences. An offender has more opportunity to commit a homicide in this instance, while the reverse is true for stranger homicides in which that victim and offender have never met before. In other words, locations of domestic homicides, which involve intimate partners and family, are hypothesized to occur in a smaller spatial area than locations of stranger homicides, particularly in the situation of co-habitation between the offender and victim (Silverman and Kennedy 1987). This is due to the social relationships occurring within the context of the crimes. When a homicide transpires between strangers, it is hypothesized that the spatial distribution of the different locations will be more distant from each other; this is particularly true for distance between the body disposal site and the murder scene. This expectation is based on the assumption that the offender would not want to leave behind evidence of his offense. According to Silverman and Kennedy (1987:274), homicides that occur in rural areas are expected to have a higher rate of domestic murder, whereas urban area homicides are expected to have a higher prevalence of stranger killings. Therefore, the geographical patterns of criminal homicides can be recognized as an outcome of complex social relationships, the context of crime, and characteristics of the demographics in a community.

To better understand the phenomena of homicide, we take three steps. First, we identify the hot-spots of criminal homicides in terms of the locations of offenders, victims, murders, and the deposition of the victim. The significance of spatial concentrations is evaluated with Nearest Neighbor Statistics for all location components of homicides. As implicitly noted by Rossmo (1999), the significance of spatial concentrations of crime may imply that homicides occur in targeted communities with similar socio-economic relationships between victims and offenders, the context of crimes, and the geographical characteristics of those communities. The second is to analyze the complicated relationships among these factors; we categorize the spatial configurations considering the locations of offenders, victims, murder, and deposition in a geo-spatial context. Then what is the best representation using GIS to exploit the relationship of the involved individuals? To address this problem, we provide a conceptual model, named the Spatial Configurations of Homicide Crime (hereafter, SCHC). According to this framework, criminal homicide incidents are categorized with a set of variations and geographically represented in a GIS environment, which enable us to understand the characteristics of homicide in a spatio-time as well as socio-economic context. The relationship of the offender to the victim affects the spatial configuration of a murder and its deposition. The final step is to focus on answering these questions raised by Wolfgang (1958): What are potential variables causing complexity involving social relationships among victims and offenders, the context of the crimes, and the geographical characteristics within a community? In detail, what variables are significantly related to the geographical configurations of homicide scenes? To reveal the relationships between SCHC and the homicide cases, this study designs three Multinomial Logit Models (MNL) with different sets of independent variables. The main goal of this step is to reveal how the characteristics of criminal homicides are related to spatial configuration as well as underlying socio-economic factors. The chapter is organized as follows. The next section presents relevant literature, and the third section provides analytical frameworks for the analysis, followed by results in the fourth section. Concluding remarks will be provided in the final section.

8.2 Background: Social Relationship and Spatial Configuration in Criminal Homicide

A majority of criminal homicide research has focused on the social relationship between offenders and victims in order to understand the dynamics between society and crime (Bridges et al. 1987; Broidy et al. 2006; Decker 1996). Wolfgang (1958, 1967), one pioneer of homicide studies, established a generalization of involved individuals in criminal homicides based on demographics and victim-offender relationships with the case study of Philadelphia, Pennsylvania in the late 1940s. The analysis focuses on what relationships among European and African males and females are found in criminal homicides in terms of specific victim-offender relationships and patterns among ancestry, sex, and age. In detail, he categorized social relationships in homicides into 11 different groups, including close friend, family, acquaintance, stranger, paramour, sex rival, enemy, a lover of offender's mate, felon or police officer, innocent bystander, and homosexual partner. The study concludes that the most common form of victim-offender relationships are close friends and family and European males were more commonly killed during the commission of a robbery than other ethnic group in Philadelphia.

Following Wolfgang's work, demographic statistics of victims and offenders have been an interest in many homicide studies because the implications inferred from demographic relationships may drive a generalized pattern among social relationships to be used as a strong predictor (Curtis 1974; Harries 1997). Main findings suggest that: (1) the most disproportionate offenders are young adult males around age 20 (Curtis 1974); (2) younger women are exposed to a higher risk of being victimized by strangers and intimate partners than older women (Lauritsen and Schaum 2004); and (3) females were more likely killed by a male while males were more likely to kill other males (Silverman and Kennedy 1987). The ethnicities of victims and offenders in homicides have been used to uncover intra- or inter-ethnic correlations. Van Patten and Delhauer (2007) found most sexual homicides to be intra-ethnic in the city of Los Angeles, California. In Websdale's (1999) analysis of 78 male–female domestic homicides in Florida, results showed that an equal proportion of homicides were committed by people of both European and African ancestries, followed by Hispanic. In a similar vein, Harries (1997) ascertained homicide to be the leading cause of death for African-Americans, and the fact that they were 6.7 times more likely to be involved in a homicide than European-Americans were.

These studies highlight demographic representations among criminal homicides are important factors in criminal homicides; however, many law enforcement agencies have recently recognized the significance and impact of geographical characteristics to better understand the dynamics of criminal homicide. With circumstantial variables related to the probability of becoming involved in homicides, the spatial analysis using GIS has been stressed in recent research. Canter (2000) explains the advantages of GIS for tactical crime analysis, introducing pattern detection and linkage analysis, which contribute to a better understanding of offender patterns in a geographical context. Messner et al. (1999) and Santtila et al. (2007) analyzed homicide patterns through spatial and temporal analyses surrounding similar communities. These studies focus on detecting the spatial clustering of homicides based on victim and offender residences, and crime scene locations. Specifically, Messner et al. (1999) concludes that spatial randomness does not occur among sites and there is the suggestion of a diffusion process within the homicides. In agreement, Santtila et al. (2007:12–14) discovered the distances between sites differed significantly and it is possible to identify crime features that are correlated with distances. However, both studies are at a disadvantage due to the lack of acknowledgement toward the social relationships between the victim and offender, as well as pertinent case information. The case information and social variables may have explained the non-random movement. Recently, Van Patten and Delhauer (2007) used GIS with police and medical examiners' records to visualize the geometric distributions of homicides. Interestingly, the geometric realities of the spatial areas were used to determine the probabilistic significance for a case to be closed, assuming that simpler geometries had a higher probability of being solved than more complex spatial geometries. The study concludes that a vast majority of crime trips involved neighborhood trips of less than half a mile, and stresses that such case specifics as spatial geometry, manner of death, and offender motivation must be considered in determining how to best improve investigative approaches.

In a geo-spatial approach, Ye and Wu (2011) use spatial panel regression to estimate homicide rates across a community area in Chicago. Although their work does not directly reveal social and demographic characteristics in homicides cases, the models show that geographical analysis using GIS is an effective tool to detect hot-spots of homicides and visualize the surface of homicide rates over time. Considering the fact that criminal homicides result from the combinations of locations of a victim, an offender, the murder and deposition sites, a spatial configuration among them can be drawn in a geographical perspective. Using the concepts of mobility triangles and polygons, the pattern of criminal homicides is categorized by classes, which suggest that geographic profiling is necessary to resolve the crimes (Tita and Griffiths 2005; Groff and McEwen 2006; Frank et al. 2012).

Although two different venues in criminal homicide studies have been developed, research considering both social relationship dynamics and spatial distribution in homicides is rarely found (Sacco et al. 1993; Snook et al. 2005). A spatial analysis of criminal homicide is often challenging considering the complex social relationships between victim-offender in nature, multiple locations which are involved in a criminal homicide incident, and its dynamic sequence within space and time (McCall et al. 2008; Broidy et al. 2006). In other words, the key to a homicide study is to construct a new view, named socio-geographic perspective in this research, where the geographical sequence of multiple locations (victim and offender's residences, place of murder, and disposal location) and social relationships are reflected within an analytical framework. The following section provides an analytical framework to understand criminal homicide from a socio-geographical perspective using GIS.

8.3 Analytical Frameworks

8.3.1 Data and Construction of Variables

In our study, we consider only closed cases of criminal homicides where the offender is known and has been arrested, died, or sentenced. Non-criminal homicide cases that were excluded involved cases of self-defense and manslaughter. Under this protocol, we collected 368 criminal homicide cases in Hillsborough County, Florida from 1997 through 2007. The data were provided by Tampa Police Department, Hillsborough County Sheriff's office and Temple Terrace Police Department. This dataset contains not only geographical locations but also social relationships among victims and offenders. All of the locations of offenders, victims, murder, and deposition are geo-coded in a GIS environment with the address in the police reports. If the time of murder and deposition is known, that information is also reflected in constructing a spatial configuration of homicide. To construct the spatial analytical framework for criminal homicides, a geographic profiling of social relationship and locations for each case is necessary (Rossmo 1999). In this study, variables causing complexity of criminal homicide in terms of socio-demographic dimensions are prepared in three aspects, involving: (1) social relationships among victims and offenders; (2) the context of the crimes; and (3) the composition of people within a community. Each aspect is critical in current criminal homicide research. Note that the categories of each relationship are made based on the data collection protocol that the police report and autopsy records provide.

First, the victim-offender relationships form the basis for fundamentally distinct homicide situations (Miethe and Regoeczi 2004). Further, relational distances among individuals can have a great influence on predicting elements of homicide (Silverman and Kennedy 1987). In this study, we categorize the social relationships between victim-offender into five categories: (1) Stranger relationship: offenders who had no directly known relationship with the victim; (2) Romantic relationship: the offender and victim have a relationship including marriage, boy/girlfriend, or any sexual relationship; (3) Blood-related: living in the same domicile and co-existence (e.g., parent/child); (4) Acquaintance: the victim and offender have met at least once, or have been friends for any amount of time; and (5) other relationships which are not categorized into (1) through (4).

Second, the victim-offender relationships can be classified in the context of crime, which has five discrete situations: (1) Victim-precipitated: a situation in which the actions or words of a victim prompt an offender to commit homicide; (2) Domestic-dispute: a situation where homicide occurred between intimate partners and family; (3) Robbery-related: a homicide associated with robbery; and (4) Rape-related: a homicide associated with rape, (5) Sexual-homicide: a homicide involved a sexual relationship between the offender and victim. Specifically, the category of 'Victim-precipitated' is important because many homicides occur because of the victim's actions which resulted in their demise. In this form of murder, the deceased may have made a menacing gesture, was first to pull a weapon, or used words to elicit a deadly response from the killer (Felson and Messner 1998).

Finally, population composition within a community is related to the ethnic composition or race of victims and offenders. Many researchers in sociology and anthropology have studied that the potential racial and/or ethnic stratification can affect the level of crime rates due to the community structures (see Wolfgang 1967; Bridges et al. 1987). As summarized in Table 8.1, the study area, Hillsborough County, has a substantially large population totaling 998,948 residents. When compared to the national average, female population (51.1%) is slightly larger than male (49.1%). Interestingly, the ratio of Hispanics (18%) and Blacks (14.4%) groups are higher than the national average, although the White group is recognized as its major ethnic group (63.3%).

In detail, in terms of homicide cases, Table 8.2 summarizes the demographic profile of victims and offenders for 368 homicide cases. Obviously, there are underlying differences in demographic structures between victims and offenders. First, in terms of sex, more than 90% of offenders are male but male victims only appear 67%. Second, in terms of ethnicity, three groups (White, Black, and Hispanic) are

Table 8.1 Population composition in the study area (as of 2000)

	Hillsborough country (%)	United States (%)
Population	998,948 (100.0)	281,421,906(100.0)
Sex		
Males	488,772 (48.9)	138,053,563 (49.1)
Females	510,176 (51.1)	143,368,343 (50.9)
Ethnicity		
White	632,605 (63.3)	211,460,626 (75.1)
Hispanic	179,692 (18.0)	35,305,818 (12.5)
Black	144,259 (14.4)	34,658,190 (12.3)
Asian	21,585 (2.2)	10,242,998 (3.6)
American-Indian	2,991 (0.3)	2,475,956 (0.9)
Others	47,266 (1.8)	398,835 (0.1)

Source: U.S. Census Bureau (2000)

Table 8.2 Demographic profiles of victims and offenders

	Victims' demography			Offenders' demography		
Ethnicity	Male (%)	Female (%)	Total (%)	Male (%)	Female (%)	Total
White	111	65	176 (47.8)	138	13	151(41.0)
Black	88	37	125 (34.0)	134	17	151(41.0)
Hispanic	42	16	58 (15.8)	59	4	63(17.2)
American-Indian	1	0	1(0.3)	1	0	1(0.3)
Asian	1	1	2(0.5)	0	0	0(0.0)
Others	5	1	6(1.6)	1	1	2(0.5)
Total	248(67.4)	120(32.6)	368(100)	333(90.5)	35(9.5)	368(100)

Source: Tampa Police Department (2007), Hillsborough County Sheriff's Office (2007), and Temple Terrace Police Department (2007)

dominant for victims (97.6%) and offenders (99.2%). Other ethnic groups consist of very small portions in both demography (2.4% and 0.8%, respectively). Considering the ethnic structure in Hillsborough County, the proportion of Black or Hispanic groups, in both victims and offenders, may be stressed in homicide incidents.

8.3.2 Spatial Configurations of Homicide Crime (SCHC)

Conventional spatial crime analyses are heavily focused on detecting meaningful patterns behind crime scenes (for example, hot-spots). These studies, however, may not reveal the spatial relationships between offender and victims, which construct a more complex spatial context and geometries. Specifically, any criminal homicide involves a sequence of four locations which include Offender's residence (O),

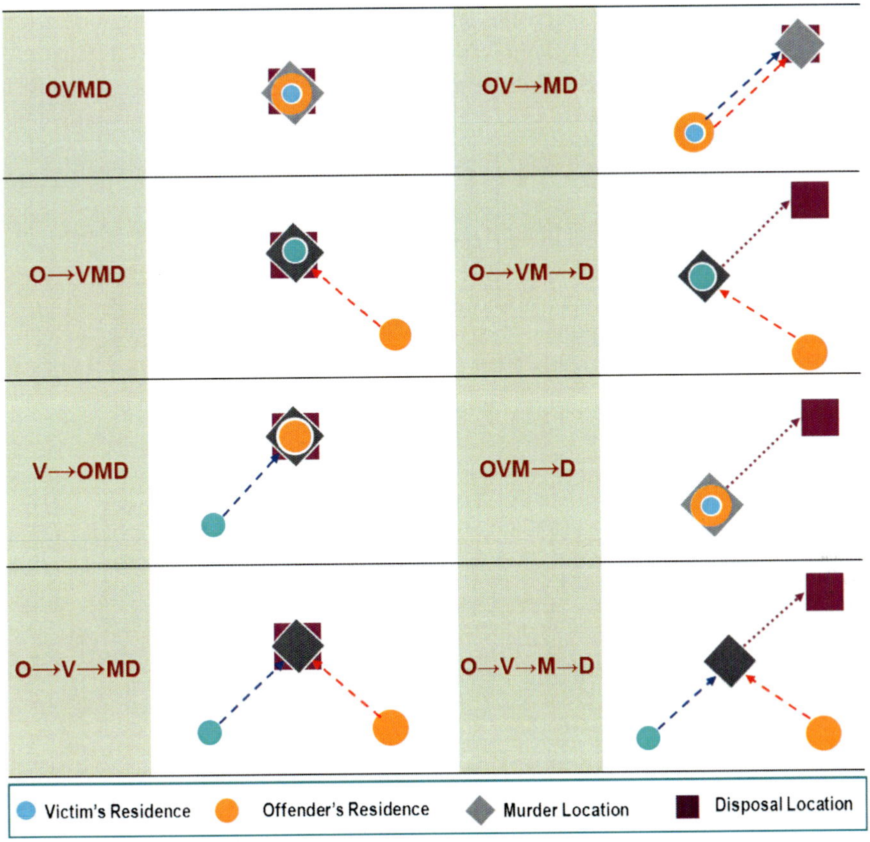

Fig. 8.1 Variation of spatial configurations of homicide crime (SCHC)

Victim's residence (V), Murder location (M), and Disposal location of victim (D). These four components are symbolized as a set of spatial configurations (O, V, M, D) of homicide crime in order to identify the different geographic sequences between offender and victim. Possible combinations of these four components are constrained by two phases. The first phase is made in terms of the residential location of offender (O) and victim (V) before the murder occurs. Offender and victim are cohabited (OV) or have a separate residence (O→V or V→O). The second phase refers to the geographic sequence after the murder. For example, the sequence (M→D) represents the body of a victim which was moved to a different location than the murder site while (MD) means the location of the murder and disposal is same. In this context, a conceptual model, named Spatial Configuration of Homicide Crime (hereafter, SCHC), is proposed. As illustrated in Fig. 8.1, criminal homicide incidents are categorized with 8 variations of (O, V, M, D). In detail, each SCHC set (O, V, M, D) defines a distinct situation of criminal homicide as follows.

- (OVMD): An offender and a victim reside together, a murder occurred in the shared household, and the victim's body was not transported to another location.
- (O→VMD): An offender and a victim lived in separate residences. A murder occurred at the victim's residence. In this situation, no movement of the victim occurs after the homicide.
- (V→OMD): An offender and a victim lived in separate residences. A murder occurred at the offender's residence. In this situation, the deposition of the victim's body remains at the homicide location.
- (O→V→MD): An offender and a victim resided at different locations, and a murder occurred in a location other than their homes. No movement of the victim's body occurs after homicide.
- (OV→MD): An offender and a victim reside together but a murder occurred elsewhere and the victim's body was not transported to another location.
- (O→VM→D): An offender and a victim lived in separate residences. A murder occurred at the victim's residence, but the victim's body was transported to another location after the homicide and deposited.
- (OVM→D): An offender and a victim reside together and a murder occurred in the shared household, but the victim's body was transported to another location and deposited.
- (O→V→M→D): An offender and a victim lived in separate residences and a murder occurred in a location other than their homes. In addition, the victim's body was transported to another location other than the residences of the offender, victim, and the place of murder.

8.3.3 *Relationships Between SCHC and Complexity of Homicides*

Of major concern in this chapter is to understand what potential variables causing complexity are associated with social relationships among victims and offenders, the context of the crimes, and the ethnic composition of victims and offenders within a community. Specifically, what variables are significantly related to the geographical configurations of homicide scenes? To reveal the relationships between SCHC and the homicide cases, this study designs three Multinomial Logit Models (MNL).

MNL is a special type of regression model, which is used to find the best fit and a parsimonious but reasonable model to describe the relationship between a dependent and a set of explanatory variables. What differentiates a MNL from the other regression models is that the dependent variable in the model is assumed categorical data or dichotomies for polytomous variables (Hosmer and Lemeshow 2000). In this sense, the MNL models in our study define dependent variables as the categories of spatial configuration of SCHC, and employ explanatory variables for three different aspects (social relationship, context of crime, and ethnic composition of victims and offenders). The main goal of the constructed models is to identify significant relationships of the independent variables in each model to the categories in SCHC. In detail, three MNL models are constructed as shown in Table 8.3. Each

Table 8.3 Design of MNL Models

Type of model	Independent variables and categories
Model I	Five categories in one independent variable are used in the model I
SCHC vs. Social relationship in crime	(Stranger), (romantic), (blood-related), (acquaintance), (other relationships)
Model II	Four discrete variables are used in the model II
SCHC vs. Context of crime	(Victim-precipitated), (domestic-dispute), (robbery-related), (sexual-homicide)
Model III	Six categories in two independent variables are used in the model III
SCHC vs. Ethnic composition of people: victim and offender	(Victim.European), (Victim.African-American), (Victim.Hispanic), (Offender.European), (Offender.African-American), (Offender.Hispanic)

has different sets of independent variables with a same dependent variable, which are the eight variations of SCHC.

In Model I, the social relationship is defined as a victim-offender's relationship and categories including (1) Stranger relationship, (2) Romantic relationship, (3) Blood-related, (4) Acquaintance, and (5) other relationships which are not categorized into (1) through (4). The definitions of each category used in Model I are slightly different than those described in previous section. The Romantic relationship includes the victim-offender relationship of spouse (married, separated, divorced) and boy/girlfriend, and the Blood-related includes the relationship between parents and child. The homicide between partners in marriage is counted as the Romantic relationship. Acquaintance includes coworker and neighbor. In Model II, the context of crime has four discrete situations categorized as (1) Victim-precipitated, (2) Domestic-dispute, (3) Robbery-related, (4) and Sexual-homicide which include the category of Rape-related. In Model III, the ethnic composition of people involves seven discrete variables for seven ethnic groups. These variables are the combination of victim-offender and three major ethnic types (European, African-American, Hispanic), and one variable for other ethnic groups.

8.4 Analysis

8.4.1 Spatial Patterns and Nearest Neighbor Statistics for Homicide Scenes

Crime opportunities are not randomly organized in space and time, so that identifying hot-spots from a crime pattern is meaningful for spatial understanding (Ratcliffe 2010). In particular, many offenders and victims are frequently observed in small areas where the place has an environment susceptible to crime activities, often called 'hunting grounds' or 'target-backcloth' (Rossmo 1999).

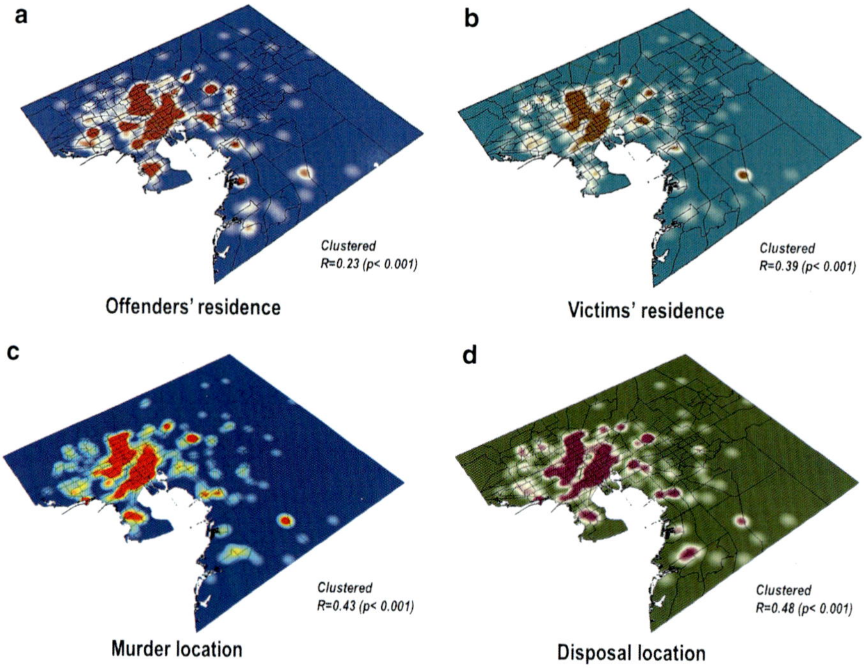

Fig. 8.2 Spatial patterns and Nearest Neighbor Statistics of homicide scenes

As shown in Fig. 8.2, the patterns of each scene location are clustered according to the Nearest Neighbor Statistics. The R scales, the ratio of the observed average distance between nearest neighbors of a point location and the expected average nearest neighbor distance, are measured as 0.23 (Offenders), 0.39 (Victims), Murders (0.43), and 0.48 (Disposal) for $p<0.001$, indicating that homicides are concentrated in particular areas, although the disposal locations are relatively dispersed compared to the other patterns. In Hillsborough County, highly clustered areas are found in highly populated tracts around the cities of Tampa, Lutz, and Brandon along with Interstate Highways I-275 and I-4. Notice that simply comparing the patterns among scenes often fails to capture the complexity of multiple scenes. For example, the patterns of murders (Fig. 8.2c) and disposal (Fig. 8.2d) look similar; however, 13.3% of homicides are the cases that the locations of disposals are not identical to that of murder location. In other words, the victim's body was moved to another location after the murder. Conventional crime analyses can detect the patterns, but may not reveal the relationships among O, V, M, and D, which construct a more complex spatial context and comprehensive geometries.

Figure 8.3 illustrates how the conceptual design for the eight variations in SCHC is represented in a space-sequence context and can be structured in a GIS environment (Fig. 8.3a). Note that the third dimension represents a temporal component if the timeline is clearly known with the locations for O, V, M, and D. As shown in Fig. 8.3b, our GIS SCHC model generates a three-dimensional space

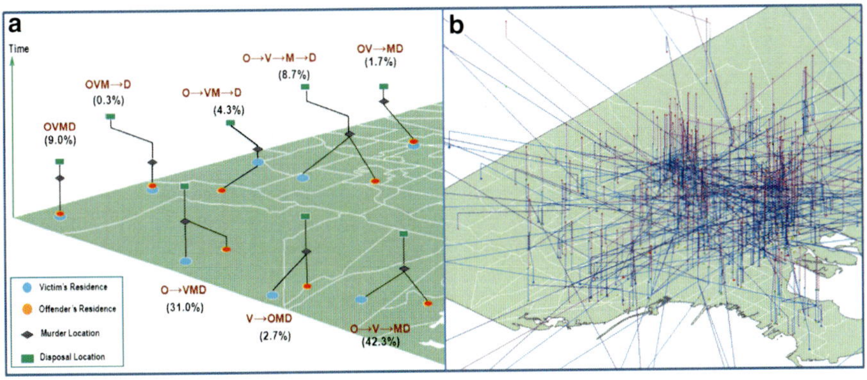

Fig. 8.3 Conceptual model of SCHC (**a**) and geographical representation in a GIS environment (**b**)

of SCHC based on the different geographic sequences for each event with O, V, M, and D which are conceptualized in Fig. 8.3a. All locations of O, V, M, and D are geocoded and paths are generated with ArcGIS 3D Analyst by the authors. When compared to the spatial patterns shown in Fig. 8.2, visualizing the pattern of SCHC in a three-dimensional framework has advantages for understanding the characteristic of SCHC because the spatial extent for each type of homicide can be evaluated with a geometric form as well as purely visualizing the complex geographic phenomena (Miller 2005). Not surprisingly, the activities of each variation of SCHC may form different geographic patterns. In our study, the most dominant SCHC is (O→V→MD: 42.3%) where the victim and offender live at different places but they met in a location other than their homes, followed by the murder where the victim was left on the spot without being transported to another place. In this case the major social relationships between the victims and offenders are strangers (35.4%), followed by romantics (19.7%) and acquaintances (10.2%). In the case of the SCHC (O→VMD), where offender and victim lived in separate residences but the murder and deposition are made in the victim's house, nearly half of the cases (48.2%) are characterized as romantic relationships. Interestingly, blood-relationship is only dominant in this SCHC sequence (15.7% within the sequence O→VMD) but is very rarely found in other SCHCs. Our interest is to uncover the significant relationship between SCHCs and particular variables in social relationships, the context of crime, and ethnic groups.

8.4.2 MNL Models: Identifying Significant Types of SCHC in Socio-Geographic Context

Since the dependent variable is in our MNL models is a categorical variable, the MNL models are estimated with a reference level for dependent variable. When any discrete variable or category is used for a reference given *J* variables or categories

Table 8.4 Result of Model I for a reference (O→V→MD) in SCHC

Reference	Social relationship			
O→V→MD	Stranger	Romantic	Blood-related	Acquaintance
O→V→M→D	−.762 (1.96)	−.174 (.097)	.560 (.20)	−1.312 (1.45)
O→VM→D	−1.455 (1.63)	.924 (1.78)	2.35 (4.34)*	−16.38 (.00)
O→VMD	−1.342 (7.87)**	.989 (7.59)**	2.39 (8.92)**	.033 (.00)
OV→MD	−.806 (.00)	17.14(.00)	18.56 (.00)	16.69 (.00)
OVM→D	−.806 (.00)	17.65 (.00)	1.21 (.00)	−.435(.00)
OVMD	−17.41 (.00)	−.174(.12)	1.37 (2.04)	−17.04(.00)
V→OMD	−1.86(2.86)	−1.27(1.32)	−15.64(.00)	−17.28(.00)

The value in parenthesis represents the coefficient of test (Wald-statistics) at $p<0.05$ (*), and $p<0.01$(**). Likelihood Ratio Test (−2 Log Likelihood = 72.780) is significant at $p<0.001$

($j=1,\ldots,J$) in the dependent variable (SCHC), the probability (π_{ij}) that the observation i falls in response category j, is defined as below.

$$\pi_{ij} = \begin{cases} \dfrac{\exp(x_i'\beta_j)}{1+\sum_{j=1}^{J-1}\exp(x_i'\beta_j)}, & j = 1,\ldots J-1 \\[2ex] \dfrac{1}{1+\sum_{j=1}^{J-1}\exp(x_i'\beta_j)}, & j = J \end{cases}$$

Where

β_j = a vector of the MNL coefficients to be estimated

x_i = a vector of the ith observation for all variables.

Note that, in these MNL models, any category in SCHC can be used as a reference level for the dependent variable. Technically, the selected reference level is suppressed as zero in a coding scheme so that other levels are tested to determine if each is statistically significant compared to the specified level, in the types of SCHC. In the Model I, a specific category, 'Other relationships' among five categories of the independent variable is selected as reference so that the MNLs are estimated with reference levels. The reason to choose the category is that the category is a less-important category in interpretation against other 4 categories in the MNL model. In detail, Table 8.4 summarizes the result of Model I when one SCHC (O→V→MD) is selected as the reference in the model. According to this model design, the coefficients β are estimated and the significance of each coefficient is tested based on Wald-statistics.

Three findings are worthy to note from this result. First, the coefficients for Stranger, Romantic, and Blood-related social relationships for O→VMD have been found to be statistically different from those for the reference 'O→V→MD'. Specifically, Stranger relationship is a negative sign. This indicates that the probability that O→VMD is committed by a stranger is significantly less than that of O→V→MD. But Romantic and Blood-related relationship variables have a positive sign, indicating that in Romantic and Blood-related relationship, O→VMD

Table 8.5 Result of model I for all references in SCHC

Reference	Significant categories	Social relationship			
		Stranger	Romantic	Blood-related	Acquaintance
O→V→MD	O→VM→D			2.35 (4.34)*	
O→VM→D	O→V→MD			−2.35 (4.34)*	
O→V→MD	O→VMD	−1.34 (7.87)**	.989 (7.59)**	2.39 (8.92)**	
O→VMD	O→V→MD	1.34 (7.87)**	−.989 (7.59)**	−2.39 (8.92)**	
O→V→M→D	O→VMD		1.163 (4.30)*		
O→VMD	O→V→ M→D		−1.163 (4.30)*		
OVMD	O→VMD		1.163 (5.36)*		
O→VMD	OVMD		−1.163 (5.36)*		

All of model Likelihood Ratio Test (−2 Log Likelihood) are significant at $p<0.001$ level. The value in parenthesis represents the coefficient of Wald-statistics at p<0.05(*) and p<0.01(**). The category 'Other relationships' was selected for all references in social relationship variables

happens more likely than O→V→MD. Second, the coefficient for Blood-related social relationship is also identified as statistically significant for the category of O→VM→D with a positive sign to the reference level. This implies that, in a blood-related case, O→VM→D happens more than O→V→MD. Finally, other categories of SCHC are not shown as significantly important to the reference level (O→V→MD). As noted, to draw significant relationships between categories, it is necessary to test the Model I by changing the reference level for the other variations of SCHC. As results, Table 8.5 summarizes the statistically significant relationships among the categories at $p<0.05$ or $p<0.01$ level.

First, the results reveal the model does not have any significant relationships among categories to the particular references in the SCHC, OV→MD, OVM→D, and V→OMD, which means that those variations in the SCHC are not likely to have statistically differences with other SCHC categories. More importantly, notice that the Blood-related homicide cases are significantly associated with the type of O→VM→D and O→VMD to the particular reference O→V→MD, interpreted as for Blood-related homicide incidents, a murder is more likely to occur at the victim's residence when an offender and a victim do not live together. O→VMD is significantly and positively associated with Romantic-relationship to the O→V→MD and O→V→M→D. This result may imply a tendency that an offender with romantic relationship usually visits his/her partner's house and leaves the decedent on site after the murder. The result reveals that the category 'Acquaintance' does not have a significant difference to any SCHC for all reference levels. Three categories, OV→MD, OVM→D, and V→OMD, do not find any significant differences in social relationships. However, it should be careful when we interpret the test result of the Model I because the reference variable in social relationship is set with the category 'Other relationships', which is the largest portion (35%) among all categories. In other words, if the cases are clarified with known social relationships and classified into an appropriate category, and then the result may change.

Table 8.6 Result of model II: SCHC vs. context of crime

Reference	Significant categories	Context of crime: Victim precipitated	Domestic dispute	Robbery related	Sexual relationship
OVMD	O→V→M→D		−3.13(10.09)		
	O→V→MD		−3.73(19.03)		
	O→VM→D		−2.95 (8.09)		
	O→VMD		−2.26 (7.08)		
	V→OMD		−3.10 (7.91)		
V→OMD	O→V→M→D				17.24(113.08)
	O→V→MD				15.97(124.41)
	O→VM→D				18.13(124.06)
	O→VMD				17.89(197.62)
	OV→MD				19.78(155.60)
	OVMD		3.10 (7.91)		
O→V→MD	O→VMD		1.47(18.21)		1.92(4.79)
	OV→MD		2.63(4.54)		3.81(6.22)
	OVMD		3.73(19.03)		
	V→OMD				−15.97(124.41)
O→VM→D	OVMD		2.95(8.09)		
	V→OMD				−18.13(124.06)
O→V→M→D	OVMD		3.13(10.09)		
	V→OMD				−17.24(113.08)
O→VMD					
O→VMD	OVMD		2.26(7.08)		
	V→OMD				−17.89(197.62)
	O→V→MD		−1.47(18.21)		−1.92(4.79)
OV→MD	O→V→MD		−2.63(4.54)		−3.81(6.22)
	V→OMD				−19.78(155.60)

All the coefficients by Wald Stat are significant at $p<0.01$ level in Model II. All of model Likelihood Ratio Test (−2 Log Likelihood) are significant at $p<0.001$ level

As shown in Table 8.6, in the Model II between SCHC and context of crime, 'Domestic-dispute' related murders and 'Sexual relationship' homicide are highlighted. First of all, Domestic-dispute homicide is most likely to occur in the sequence of OVMD. When OVMD is the reference, all SCHC categories except OV→MD are significant with a negative sign for Domestic-dispute cases. Conversely, this result indicates that Domestic-dispute homicide is significantly less related to O→V→MD, O→VM→D, O→V→M→D, O→VMD, and V→OMD than OVMD. Intuitively, the result implies Domestic-dispute homicide may depend on whether offender and victim are cohabitating or not (OV vs. O→V or V→O). Rather, the Domestic-dispute related homicide is more linked to the situation in which an offender and a victim live together than the situation that their residences are separated. Second, one interesting result is found in the case that the sexual relationship is involved in the context of criminal homicide. As noticed in the final column, all types of SCHC except OVMD are significantly different with a positive sign when the type of V→OMD is the reference. This indicates that homicide in sexual relationship is not likely to happen in the location sequence of V→OMD,

Table 8.7 Model significance test for Model III

Model fitting information		Pseudo R^2	
−2 Log likelihood	Chi-square (df = 20)	Cox and Snell	McFadden
115.45 (Sig. $p < 0.292$)	22.93	0.077	0.029

Several similar models with different specification also provide the results, supporting that any ethnic composition is not associated with particular types of SCHC

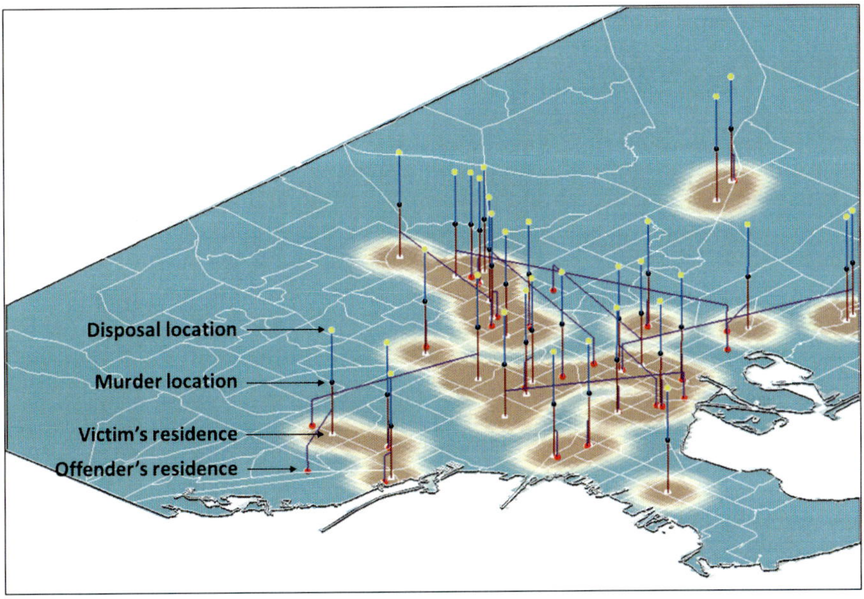

Fig. 8.4 Reconstructing the hot-spots of significant criminal homicide: a case of SCHC (O → VMD) positively related to Romantic relationship and Domestic-dispute

implying that sexual relationship homicides does not tend to happen at an offender's residence. In other words, an offender (within the sexual relationship component) commits homicide while visiting a victim's house or other locations. Finally, the results reveal no statistically meaningful categories in the SCHC occur for cases of Victim-precipitated or Robbery-related homicides. Therefore, it is difficult to tell that such criminal contexts as Victim-precipitated or Robbery-related homicides have distinctively different geographic sequences from other contexts of crime.

In the final model, the relationship between SCHC and ethnical composition of people (victim and offender) is tested. As summarized in Table 8.7, the Likelihood Ratio Tests (−2 Log Likelihood) of all models are 115.45 and shown as significant at $p < 0.292$. This result implies that there is no statistically significant relationship between any variations in SCHC and the independent variables of ethnicity. The values of Pseudo R^2 are also very low, supporting that no meaningful statistical relationships are expected. The model concludes that any racial types are not significantly associated with specific types of SCHC, at least in the cases of homicide incidents in Hillsborough County, Florida during 1997–2007.

8.4.3 Reconstructing Hot-Spots in Socio-Geographic Context

As discussed by Keppel and Weis (1994), the more information regarding locations of the crimes, the greater the chance that the case will be solved. The types of homicides associated with a specific social relationship or criminal context is highly likely to be identified using a specific spatial sequence of homicide (O, V, M, D) and the MNL model results. In the case of Hillsborough County, several socio-geographical contexts are highlighted. For example, the Romantic-relationship category between victim and offender is likely to manifest in the SCHC of O→VMD. The geographic sequence of these corresponding cases can be restructured in a GIS environment. As illustrated in Fig. 8.4, 39 cases are re-selected according to the result of Model I, and hot-spots from the cases are re-captured. When compared the map Figs. 8.3b and 8.4 more clearly shows where romantic-relationship associated homicides are clustered, which is statistically significant to the other types of SCHC in the County. Two things are worthy to note from the results in Fig. 8.4. First, when the pattern is compared to Fig. 8.3b, the hot-spots from Model I in Fig. 8.4 are more narrowed in terms of geographical space, which may provide a new tool to help solve homicide cases. Nearest Neighbor Statistics for this pattern confirms that the spatial patterns of (O→VMD) also clustered in the County ($R=0.109$, $p<0.01$). The results indicate homicides associated with Romantic-relationship are not formed randomly.

To support this statistical result, GIS easily reports useful proximity (distance) information between an offender's residence and a victim's residence, after identifying the significant type in SCHC. In the example shown in Fig. 8.4, the average distance between victim and offender residences is 2.517 km for 30 cases. Furthermore, 95% of offender's residence are located within 2.06 km (Std = 3.8 km) from the victim's house. This information supports that with most O→VMD cases, if the case suspects Romantic-relationship between the victim and offender, it is highly probable to find the offender within a small geographic range. Similarly, as tested in both the result Model I and II, if the case (O→VMD) is associated with Romantic-relationship, then it is highly probable that the case is caused by a Domestic-dispute situation. In summary, the classical hot-spot or cold-spot detection approach using GIS can be enhanced by employing such a design of geographic sequence, SCHC with MNL models.

8.5 Concluding Remarks

Criminal homicides are a significant part of organized societies, and although many efforts have been made to prevent them, success has not been achieved. The majority of previous research for criminal homicides considerably focuses on the social factors in victim-offender relationships. However, any criminal homicide can be characterized by space where the sequences of multiple locations of victims, offenders, murder, and a body deposition are made, which would be a product of the social relationships between victim and offender. This study highlights that a detailed and advanced geographic profiling is possible by categorizing geographic sequences, which is named SCHC. This categorization, as an advanced technique of geographic profiling, helps to clearly identify the relationship between the geographic properties of criminal homicides and other important variables, including (1) social relationship, (2) context of crime, and (3) composition of people using the MNL models.

Based on our conceptualized model and the geospatial analysis using GIS, this study restructures the hot-spots within a socio-geographic context. As indicated by results from the models, some social relationships are considerably related to specific types of SCHC. These results imply homicide cases can be effectively resolved if we know the social relationship between victim and offender and the context of the crime. Then the location of murder or disposal can be inferred, or particular sequences of locations (O, V, M, D) can give law enforcement agencies a clue to solve cases. Furthermore, detecting hot-spots of criminal homicide can be sophisticated by employing the statistical models with SCHC because the hot-spots are formed differently in space according to the type of homicide. For example, in our results, homicide incidents with Romantic-relationship categorization form a clustered hot-spot where virtually all murders occur in the victim's home and the offender lives within 2 km. To build a solid foundation of geographic profiling with detected geographic hot-spots or cold-spots and significant categories in the SCHC, GIS can be used to further explore the interactions of multiple locations involved in criminal homicides (Grubesic and Mack 2008; Snook et al. 2005).

Suggestions for similar studies involve the inclusion of more spatial variables of the individuals and communities. For example, the actual socio-economic status of both the victim and offender could shed light on the social factors that influence homicide, such as social segregation and economic impacts on crime rates. On a technical side, more cases should be considered in the model to construct a more reliable one. Second, it is obvious that more accurate time of death and deposition is necessary to construct three dimensional sequences SCHC, which improve the quality of the analysis.

References

Avakame EF (1998) How different is violence in the home? An examination of some correlates of stranger and intimate homicide. Criminology 36(3):601–632

Brantingham P, Brantingham P (1984) Patterns in crime. Macmillan Publishing Company, New York

Bridges GS, Crutchfield RD, Simpson EE (1987) Crime, social structure, and criminal punishment: white and non-white rates of imprisonment. Soc Probl 34(4):345–361

Broidy LM, Daday JK, Crandall CS, Sklar DP, Frost PF (2006) Exploring demographic, structural, and behavioral overlap among homicide offenders and victims. Homicide Stud 10(3):155–180

Canter P (2000) Using a geographic information system for tactical crime analysis. In: Goldsmith V, McGuire PG, Mollenkopf JH, Ross TA (eds) Analyzing crime patterns: frontiers of practice. Sage Publications, London, pp 3–10

Cohen L, Felson M (1979) Social change and crime rate trends: a routine activity approach. Am Sociol Rev 44(4):588–608

Curtis LA (1974) Criminal violence: national patterns and behavior. Lexington Books, Toronto

Daly M, Wilson M (1988) Homicide. Aldine De Gruyter, New York

Decker SH (1996) Deviant homicide: a new look at the role of motives and victim-offender relationships. J Res Crime Delinq 33(4):427–449

Federal Bureau of Investigation (2004) Uniform crime reporting handbook. United States Department of Justice, U.S. Government Printing Office, Washington, DC

Felson RB, Messner SF (1998) Disentangling the effects of gender and intimacy on victim precipitation in homicide. Criminology 36(2):405–423

Frank R, Andresen MA, Felson M (2012) The geodiversity of crime: evidence from British Columbia. Appl Geogr 34:180–188

Frey WH, Zimmer Z (2001) Defining the city. In: Paddison R (ed) Handbook of urban studies. Sage Publications, London, pp 14–35

Gagne P (1992) Appalachian women: violence and social control. J Contemp Ethnogr 20(4):387–415

Groff E, McEwen T (2006) Exploring the spatial configuration of places related to homicide events: final report. Institute for Law and Justice, Alexandria, VA

Grubesic TH, Mack EA (2008) Spatio-temporal interaction of urban crime. J Quant Criminol 24(3):285–306

Harries KD (1997) Serious violence: patterns of homicide and assault in America. Charles C Thomas Pub Ltd, Springfield

Hillsborough County Sheriff's Office (2007) Police report of criminal homicides 1997–2007, Tampa

Hosmer DW, Lemeshow S (2000) Applied logistic regression. Wiley-Interscience, New York

Keppel R, Weis J (1994) Time and distance as solvability factors in murder cases. J Forensic Sci JFSCA 39(2):386–401

Lauritsen JL, Schaum RJ (2004) The social ecology of violence against women. Criminology 42(2):323–357

McCall PL, Parker KF, MacDonald JM (2008) The dynamic relationship between homicide rates and social, economic, and political factors from 1970 to 2000. Soc Sci Res 37(3):721–735

Messner SF, Tardiff K (1985) The social ecology of urban homicide: an application of the "routine activities" approach. Criminology 23(2):241–267

Messner SF, Anselin L, Baller RD, Hawkins DF, Deane G, Tolnay SE (1999) The spatial patterning of county homicide rates: an application of exploratory spatial data analysis. J Quant Criminol 15(4):423–450

Miethe TD, Regoeczi WC (2004) Rethinking homicide: exploring the structure and process underlying deadly situations. Cambridge University Press, New York

Miller HJ (2005) A measurement theory for time geography. Geogr Anal 37:17–45

Polk K (1993) Observations on stranger homicide. J Crim Justice 21(6):573–582

Ratcliffe J (2010) Crime mapping: spatial and temporal challenges. In: Piquero AR, Weisburd D (eds) Handbook of quantitative criminology. Springer Science+Business Media, LLC, New York, pp 5–24

Rossmo DK (1999) Geographic profiling. CRC Press, Boca Raton

Sacco VF, Johnson H, Arnold R (1993) Urban-rural residence and criminal victimization. Can J Sociol 18(4):431–451

Santtila P, Laukkanen M, Zappala A (2007) Crime behaviours and distance travelled in homicides and rapes. J Investig Psychol Offender Profil 4(1):1–15

Silverman RA, Kennedy LW (1987) Relational distance and homicide: the roles of the stranger. J Crim Law Criminol 78(2):272–308

Snook B, Cullen RM, Mokros A, Harbort S (2005) Serial murderer's spatial decisions: factors that influence crime location choice. J Investig Psychol Offender Profil 2:147–164

Swatt ML, He N (2006) Exploring the difference between male and female intimate partner homicides: revisiting the concept of situated transactions. Homicide Stud 10(4):279–292

Tampa Police Department (2007) Police report of criminal homicides 1997–2007, Tampa

Temple Terrace Police Department (2007) Police report of criminal homicides 1997–2007, Tampa

Tita G, Griffiths E (2005) Traveling to violence: the case for a mobility-based spatial typology of homicide. J Res Crime Delinq 42(3):275–308

U.S. Census Bureau (2000) Census of population and housing supplement. U.S. Census Bureau, U.S. Government Printing Office, Washington, DC

Van Patten IT, Delhauer PQ (2007) Sexual homicide: a spatial analysis of 25 years of deaths in Los Angeles. J Forensic Sci 52(5):1129–1141

Websdale N (1999) Understanding domestic homicide. Northeastern University Press, Boston

Wolfgang ME (1958) Patterns in criminal homicide. University of Pennsylvania, Philadelphia

Wolfgang ME (1967) Studies in homicide. Harper and Row Publishers, New York

Ye X, Wu L (2011) Analyzing the dynamics of homicide patterns in Chicago: ESDA and spatial panel approaches. Appl Geogr 31(2):800–807

Chapter 9
A Methodology for Assessing Dynamic Fine Scale Built Environments and Crime: A Case Study of the Lower 9th Ward After Hurricane Katrina

Andrew Curtis, Jacqueline W. Curtis, S. Wright Kennedy, Amit Kulkarni, and Traci Auer

Abstract This chapter will present a method of data collection and analysis for fine scale environments experiencing change. The setting for this work is the Lower 9th Ward of New Orleans for the period 2010 and 2011. The approach described here utilizes a low cost mobile data collection strategy involving a spatial video, a built environment coding scheme, and fine scale spatial analysis using a spatial filter that creates a surface of abandonment/blight/returnee rates linked to individual crimes. This chapter will also address the need for longitudinal analysis beyond simply considering changes in crime events by framing crimes between two data collection periods. Although this chapter should be viewed as a methodological example, including the importance of primary data collection and spatial investigation at the street segment scale, one interesting result is that crimes in association with abandonment and blight only became statistically significant for the 2011 landscape. The chapter concludes with several examples of spatial video derived fine scale maps that can be used to advance current spatial crime theories.

A. Curtis, Ph.D. (✉) • J.W. Curtis, Ph.D.
GIS Health & Hazards Lab, Department of Geography, Kent State University, Kent, OH, USA
e-mail: acurti13@kent.edu; jmills30@kent.edu

S.W. Kennedy, M.A.
Department of History, Rice University, Houston, USA
e-mail: swk@rice.edu

A. Kulkarni
Department of Geography and Anthropology, Louisiana State University, Louisiana, USA
e-mail: akulka1@tigers.lsu.edu

T. Auer
Department of Economics, University of Southern California, Los Angeles, CA, USA
e-mail: tauer@usc.edu

M. Leitner (ed.), *Crime Modeling and Mapping Using Geospatial Technologies*, Geotechnologies and the Environment 8, DOI 10.1007/978-94-007-4997-9_9,

Keywords Hurricane Katrina • Lower 9th Ward • Spatial video • Disaster recovery and crime

9.1 Introduction

The link between ecological processes and crime is frequently made in both the academic and popular literature. Simply put, some spaces are perceived to be, or indeed are, associated with or encouraging of criminal activity. Unfortunately, data to capture these associations, especially at the finest of sub-neighborhood scales involving the interrelationship between buildings and streets, is hard to acquire. Attempts to make such fine scale comparisons often involve considerable effort (and expense) and as a result tend to be cross-sectional in nature. This is particularly problematic for landscapes of change; the two most obvious examples being foreclosed neighborhoods and post-disaster landscapes. This chapter will present a methodological approach that can be used to extract patterns in such dynamic conditions; including data collection, a method of coding and analysis, and the importance of temporal bracketing for this type of ecological analysis. The neighborhood chosen, the Lower 9th Ward in New Orleans, does not have particularly high crime levels, but was chosen as an excellent example of neighborhood extremes, where home owners (potential victims) reside next to broken and unsecured homes (potential crime spaces). In addition some streets are impassable while many blocks contain vegetation that is now at roof level. Within this extreme landscape, change is still occurring. New homes are constructed, both systematically and in isolation as "pioneers" within the sea of devastation and blight, while other buildings become progressively more distressed. The Lower 9th Ward presents the extremes of what might happen in an American neighborhood, and this chapter shows how this dynamism can be captured and linked to crime.

9.2 Fine Scale Built Environments and Crime

9.2.1 The Link Between Geography and Crime

Across several academic disciplines there has been a shift towards more ecological assessments of the built environment in association with different social outcomes, with arguably the most interesting work now involving the finest of geographic scales. This makes intuitive sense especially if the goal is to understand patterns and processes with a view to intervention. For example, play spaces may provide positive development opportunities *but what of their quality, including the perceived attributes of both the space itself and the path to that space from the residence*? Elevated levels of diabetes may present as hotspots on a map, *but what about the streets, living environment, and access to food opportunities for each patient within that area of elevated disease*? And with regard to crime, *how will the configuration of individual buildings, the condition of those buildings, and the surrounding micro-environment either promote or hinder crime*? Answers to such fine scale questions with regard to crime can frame policy suggestions (Loukaitou-Sideris 1999) and lead to successful crime reduction interventions (Dalton et al. 2008; Peed et al. 2008) including location denial strategies (Taniguchi et al. 2011).

The link between geography and crime is well established in criminology (Brantingham and Brantingham 1995), following a larger trend in academia embracing an ecological approach to investigating the patterns and processes of social problems within the built environment, where the relationship is arguably both contextual and compositional; the landscape is both an indicator of activity and may also attract/repel people who participate in that activity (Shaw et al. 1929; Bursik and Grasmick 1993; Ackerman and Murray 2004). The classic example often used in criminology classrooms is the theory of broken windows (Wilson and Kelling 1982), whereby the visual presence of socially unacceptable elements on the landscape (such as graffiti) may indicate that an environment contains crime, while also signaling that the lack of social oversight allowing these visual cues to occur will not hinder future criminal activity (Wilson and Kelling 1982; Skogan 1990). In terms of the post-disaster landscape described in this chapter, vacant and abandoned properties provide hidden spaces (Ackerman and Murray 2004), while dumping and overgrown vegetation might indicate a lack of localized activity making such areas ideal for criminal participation (Wilson and Kelling 1982; Skogan 1986, 1990; Perkins et al. 1993). At the same time these visual cues obviously drive perception and fears (Skogan 1986; LaGrange et al. 1992; Perkins et al. 1992; Perkins and Taylor 1996) that in turn might impede the desire to change the neighborhood for the better (to recover), and may perpetuate or even further develop a landscape suitable for crime. A linked ecological concept is routine activity theory, that the presence of oversight and guardianship will have a deterring effect on crime (Reynald 2011). In this regard criminals will be less attracted to locations where there are obvious signs of inhabitation, which in terms of the post-disaster landscape include rebuilt/refurbished buildings and signs of ongoing occupancy.

Ecologically based theories explaining crime locations also generate discussion with regard to an appropriate spatial scale for research. Although the physical condition of a neighborhood may contain visible signs of disrepair, crime does not occur homogeneously throughout the area. Crime is more spatially localized, with particular buildings, block faces or street corners being important (see Sherman et al. 1989; Taylor 1997; Smith et al. 2000; Kinney et al. 2008). These fine-scale geographies can play different roles, though the concept most relevant to the discussion of post-disaster landscapes is of crime attractors (Brantingham and Brantingham 1995); a location where a specific need is met, whether that is the presence of a victim or concealment.

There are several crime-related examples where the importance of spatial specificity is obvious, such as the built environment around transportation hubs. For example in Loukaitou-Sideris's (1999) work in Los Angeles, micro-geographies around bus stops were mapped identifying various negative visual cues (graffiti) or locations (such as alleyways) and then overlaid with crime locations. Street corners are important territories for gangs so Taniguchi et al. (2011) focused their research on crime counts around them. Both these papers are examples of how a small area is the only relevant space for analysis and any broader scale is problematic as it may either smooth out these patterns or mislead the reader towards conclusions of a more geographically extensive problem (Tita et al. 2005; Pitcher and Johnson 2011).[1] Therefore the units of geographically focused research should be based around theoretical places rather than convenient aggregations of data (Bernasco and Block 2011).

However, data are not readily available for these small spaces. In the previously mentioned Taniguchi study, census aggregations were manipulated using Thiessen polygons to create more appropriate geographies representative of street-corner areas of influence. In another census based manipulation, Bernasco and Block (2011) used mattress style configuration of blocks to investigate crime attractors and spillover effects for street robberies in Chicago between 1996 and 1998. If primary data are collected then it usually involves labor intensive surveys and assessments of the local environment, that can be vulnerable to fieldworker error.[2] For example Reynald (2011) used house-to-house surveys to assess both visual signs of guardianship and actual resident reaction to strangers in the street. Although data collected in this way provide valuable insight, in this case identifying the "direct active" role of guardianship in limiting crime for fine scale spaces, it is not surprising that this effort is usually only applied to one area for one time period.

9.3 Longitudinal Crime Analyses

9.3.1 The Lack of Available Data

An analysis for one time period is problematic for different reasons. It is acknowledged that crimes cluster in space and time as it is likely that the same combination of environment and activity make a target attractive to repeated crimes, especially with initial success (Felson 2006; Braga et al. 2011). If these environmental conditions persist over time, then clusters will be both spatially and temporally stable at that same location or in close proximity where a successful experience makes nearby victims more susceptible in future time periods (Townsley et al. 2003; Pitcher and Johnson 2011). However, fine scale spaces are also dynamic and the characteristic

[1] There are other methodological advantages of working with fine-scale geographies, such as a street segment, as coding errors are minimized (Brantingham et al. 2009).

[2] For other examples see Taylor et al. (1985), Perkins et al. (1992, 1993), Perkins and Taylor (1996), Loukaitou-Sideris (1999), Sampson and Raudenbush (1999), and Loukaitou-Sideris et al. (2002).

of a building, or adding or subtracting light, or the removal of a trash pile from the street can effect immediate change in patterns resulting in shifting crime geographies. Unfortunately, the number of studies that include fine-scale analysis for multiple time periods is sparse (see Weisburd et al. 2004; Braga et al. 2010; Groff et al. 2010), and even more so for those that include environmental primary data (Taylor 2001), such as monitoring physical change or capturing residents' shifting perception of impacts to crime (Robinson et al. 2003).

One reason for this paucity in primary data driven research is the logistical challenges it poses, especially for multiple time periods (Sampson et al. 1997). However, recent developments in individual mobile assessment tools, volunteered geographic information, and as will be discussed in this chapter, mobile mapping techniques, have opened the way for more dynamic street-level analyses allowing researchers to unshackle themselves from secondary data especially census enumerations at inappropriate scales (Immergluck and Smith 2006).

9.3.2 *Fine Scale Dynamic Landscapes*

One of the more visually recognized outcomes of the economic downturn in the United States since 2008 has been the rise in foreclosed neighborhoods (Bess 2008; Dalton et al. 2008). Even before these events, criminologists had realized that there might be a link between such dynamic landscapes and crime (Immergluck and Smith 2006). Prior to the foreclosure crisis, vacant and abandoned structures were primarily viewed as the loss of capital from neighborhoods experiencing economic downturn, usually in the inner cities (Accordino and Johnson 2000), but now residential foreclosures associated with global economic shifts have emerged as a factor in vacancy and abandonment; and what was once a concern in poor, inner-city neighborhoods, now extends to suburbs, middle class neighborhoods and satellite cities (Wilson and Paulsen 2008). Although it is often suggested that the resulting decline of buildings, streets and neighborhoods may be linked to fine-scale spatial and temporal patterns of crime (Rogers and Winter 2009), the lack of available data makes robust analyses problematic (Schuetz et al. 2008). Post-disaster landscapes offer similar dynamic street and building environments where an externality and not internal neighborhood social process have changed the normal pattern of living. Just as with foreclosed neighborhoods, but often rendered in even more extreme terms, the neighborhoods of New Orleans after Hurricane Katrina have left open spaces, abandoned buildings, trash (debris), and severely reduced social cohesion and service infrastructure. It is not hard to imagine how landscapes harbor or encourage crime if the same patterns and processes associated with "Broken Windows" or "Routine Activity Theory" (Cohen and Felson 1979) is relevant irrespective of genesis. After all, abandoned buildings and vegetation overgrowth provide concealment for criminal activities, a situation reinforced by reduced city services (policing routes) and fewer occupied homes providing oversight. A further component in the post-disaster landscape is a novel mix of attractors including crime-susceptible spaces and victims. Victims may live next to abandoned homes or vegetative tracts that provide the mix of both concealment and opportunity, or they can live in isolation, where the unique setting of an absence of community might result in severe stress and domestic violence. In addition, "pioneer" homes, those buildings that remain isolated in blocks of overgrown vegetation, meet the lack of supervision criteria identified by Felson (2006). These landscapes suggest a hybrid "holistic" theory including both ecological theories might be appropriate (Reynald and Elffers 2009; Reynald 2011).

Given such a potentially interesting mix of factors it is a shame so much post-disaster crime research usually focuses on the immediate aftermath of the event, including looting, fraud or violent crime, or the movements of crimes by those who have been displaced (Decker et al. 2007; Frailing and Harper 2007, 2010; Cromwell et al. 1995; Tierney et al. 2006; LeBeau 2002; Davila et al. 2005; Brezina and Kaufman 2008; Varano et al. 2010). Research on the post disaster recovery phase frequently involves stresses and coping mechanisms, such as substance use, or crime resulting from psychopathology and domestic violence (Weisler et al.2006; Adams and Adams 1984; Foa et al. 2006; Fothergill 1996, 1999; Morrow and Enarson 1996; Enarson 1999). Even here the research landscape tends to be short in duration, often within a year of the disaster, not the 5 and 6 years later after the event as is the setting for this chapter. As a result, there is little written on the relationship of long term recovery and crime as seen through the lens of a lack of collective efficacy and routine activity.

If it is possible to remove oneself from the emotional aspects of these landscapes, and the suffering that created them, an opportunity is presented to observe fine scale ecological processes on social outcomes. For example, it is well accepted that there is a bidirectional observation between landscape and crime – it is both reflective of local social processes while also being affective of social events. Translating this to the Lower 9th Ward (though obviously without the typical causative pathway), blighted and abandoned buildings, trash or graffiti are indicative of the lack of local social cohesive bonds (returnees) while also providing potential crime locations. Whereas normally these physical aspects occur through a process of social degradation and loss of control, in the Lower 9thWard they were imposed through an exogenous process; buildings lie abandoned and decaying because of flood waters, and graffiti initially covered this landscape because of search and rescue teams. The one post-disaster feature more commonly associated with a lack of typical social cohesion is dumping, though in this case performed by outsiders seeing opportunity for a free way to dispose of trash.[3] The lack of social cohesion that "allows" such negative visual evidence is based on a lack of ability to return and rebuild, and not disinterest in the neighborhood.

[3] A New York Times magazine article "Jungleland: the Lower Ninth Ward in New Orleans gives new meaning to 'urban growth'" by Nathaniel Rich describes in a general sense many of the visual vegetative, returnee and blight problems faced by the neighborhood (see http://www.nytimes.com/2012/03/25/magazine/the-lower-ninth-ward-new-orleans.html?_r=1&ref=magazine).

9.4 A New Method for Fine-Scale Data Collection

9.4.1 The Spatial Video

In order to collect data at the street scale for a changing landscape of this type, a cost effective surveying system is needed. The spatial video approach described in this chapter has been successfully used by the authors in a variety of situations to collect building-scale data that can be combined in ecological analyses. These applications have included post-disaster neighborhood recovery, damage assessment, and linking crimes to different built environment features (Curtis et al. 2007; Mills et al. 2008; Curtis et al. 2010a, b; Mills et al. 2010; Curtis and Mills 2011).

Although video has previously been used for the assessment of the built environment especially with regard to visual cues of decline (Sampson and Raudenbush 1999), the spatial video used here is different in that location information linked to the image allows for viewing within a GIS, that in turn improves the ability to perform a spatial analysis of the image content. The spatial video consists of two (or more) cameras mounted to the side of a vehicle on window clamps. The only requirement for the camera type is that an audio input socket is available as a global positioning system (GPS) encodes vehicle location as sounds on one of the audio tracks. The GPS receiver is attached to a roof mounted aerial. The entire setup: two cameras, window mounts, GPS receiver and aerial, all easily fit into a small back pack. The vehicle drives along a neighborhood street at no more than 15 mph. Once finished, the tapes are digitized and processed using a GIS extension that creates an XML file associated with the video format mpg. Using the same software (GeoVideo by Red Hen Systems), the path of the vehicle is displayed onscreen in the GIS as a cursor moving along a path of points. In an inset window the video plays allowing the user to match images to the exact position on the map. Although environmental conditions can vary the precision along the route, it is still relatively easy to match the video image with a building outline in a GIS once the GPS path has been overlaid on high resolution aerial photography. Various other tools are available within this GIS extension, such as extracting video segments, or single images, that remain linked to the path allowing the user to return to any image's original location on the map. This chapter illustrates this primary data collection approach and extends the work of Curtis and Mills (2011) who had previously analyzed crime in another post-Katrina neighborhood. This earlier paper focused on a smaller neighborhood, Holy Cross, that had received less hurricane damage and more closely resembles typical urban decline in the United States. The neighborhood being analyzed in this chapter is larger and more extreme in all senses. The Lower 9th Ward is a post-disaster urban wilderness within which pockets of recovery are occurring.

9.5 Research Methods and Findings

9.5.1 Data Collection and Coding

Two spatial video trips were performed in August 2010 and 2011 to cover the majority of the Lower 9th Ward (approximately one block north of Clairborne Ave and above) in New Orleans. Once the video had been collected and processed, all buildings on the routes were heads-up digitized in ArcMap 10 using a combination of the available online imagery option with that GIS, and building locations seen on the video itself. This process was aided by a previous layer of buildings digitized for the Lower 9th Ward in the months following Hurricane Katrina using imagery acquired from the Louisiana State University Katrina Clearinghouse (Mills et al. 2008). In addition, a layer of parcels (owned property boundaries) was also overlaid on the available imagery.

Each building was coded using the same recovery score previously employed by the authors in the Holy Cross neighborhood of New Orleans to monitor recovery and crime (Curtis et al. 2010a, b; Curtis and Mills 2011). Buildings that were not occupied scored 1, that were cleared to the ground scored 2, that were being rebuilt scored 3, and were occupied scored 4. Those scoring 1 also received a second score of 2 if the property was sound but empty, 5 if it was boarded up, and 10 if it showed multiple aspects of severe blight including being unsecured, damaged, and with vegetation overgrowth. For the purposes of this chapter scores of 5 and 10 were combined to identify locations of "severe" abandonment that could present opportunities for crime and would indicate a lack of social oversight. The coding scheme has evolved from several years of recovery work in New Orleans and has proven to be suitable in terms of consistency in coding between different people, capturing important aspects of the recovering landscape, and allowing for easy GIS manipulation. Although originally designed as a recovery metric, the categories are consistent with Reynald's (2011) approach to capture local area guardianship especially occupancy (and visible occupancy).

In Fig. 9.1 three images illustrate this coding system with example scores of 1.10 (A), meaning the building scored a 1 and a second score of 10, a score 4 and 1.10 neighboring each other (B) and a 4 and 1.5 also as neighbors (C). One difference between the coding employed here and in previous papers is that no "clearing" (code 2), was used during 2010 as so much of the Lower 9th Ward had returned to open land. However, "2" was utilized in 2011 as this would capture properties that had been bulldozed since 2010. The video was progressed to each property for both years, with the code being added into the attribute table of the digitized building. During the encoding process, images of interest with respect to the purpose of this chapter were extracted from the video.

Fig. 9.1 Three examples of the recovery coding score; $A = 1.10$, $B = 4$ and 1.10, $C = 1.5$ and 4 (Source: Spatial video collected by S. Wright Kennedy August 2010)

9.5.2 Spatial Filter Analysis

After coding, the latitude and longitude center of each building was calculated and used as input into a spatial filter analysis more commonly used in epidemiology studies. The software employed, DMAP (Rushton and Lolonis 1996), calculates a user specified grid over the study area. A rate is calculated at a user specified distance (called a filter) around each node. These filters usually overlap to create a smoothed rate surface. In keeping with the justification for performing fine scale analyses of changing landscapes described at the beginning of this chapter, that social oversight and blight might act as ecological determinants on different types of crime, rates were generated for reoccupied homes and blighted buildings. Therefore, for every node, the rate of blight (scores of 5 or 10 as the numerator, and all buildings as the denominator) and occupied home (scores of 4 as the numerator) were calculated. Rate surfaces were generated for both years (2010 and 2011), remembering that both the numerators and denominators changed between these time periods. Further rates were calculated for the same variables and the same time periods where the denominator was the number of parcels. The difference in using these two denominators is that the first captures the visual built environment while the second is a more standardized measure of *what was* the built environment space (capturing cleared but now overgrown parcels). The results reported in this chapter are for the finest scale (a 0.025 mile filter) capturing the immediate neighbor effect on the landscape, though more coarse filters were also calculated (0.05 miles). Previous work of the authors has analyzed recovery at multiple scales (different filter sizes) in order to eliminate the potential for analysis bias based on any one bandwidth. A minimum denominator was established of five buildings so as not to capture isolated houses, but also having a threshold low enough not to miss the small clusters of pioneer returnees in the neighborhood.

The final rate surface is presented as a series of nodes overlaid onto the Lower 9th Ward map. The locations of crimes[4] proximate to the spatial video path were added to the GIS along with crime type and date. Only crimes for the period July 2010 to September 2011 were included in the analysis. Each crime location[5] was linked in the GIS to the closest spatial filter node. In this way for each crime the neighborhood (here defined as the same street, series of buildings or possibly block) rate of blight and occupied homes could be determined. From the total list of crimes,

[4] Crimes were extracted from the Metro New Orleans Crime Map (http://www.nola.com/crime/nolasearchresults.ssf) which repurposes crime data from a variety of sources including the New Orleans Police Department.

[5] Crimes were reported by block segment, which even with digitizing using the associated map as guide, still leaves a degree of uncertainty as to exactly where the event took place. For several crimes it was easy to determine the location on the sparsely populated landscape, though for the purpose of maps presented in this chapter, the vagueness is preserved for ethical reasons. However, other studies are confident at working with just the aggregation of the street segment or block face which then limits errors associated with geocoding accuracy (see Weisburd et al. 2004; Braga et al. 2011).

Table 9.1 Summary of coding for 2010 and 2011

	2005	2010	2011	Parcels
All buildings	3508	904	1010	4290
Blighted		397 (43.9 %)	307 (30.4 %)	
Returned		365 (40.4 %)	463 (45.8 %)	

A summary of all buildings within the Lower 9th Ward

four were chosen to illustrate either the importance of occupied homes (domestic violence and burglary) or with little oversight often characterized by blight or open space (incivility or auto[6]).

9.6 Results

9.6.1 Changes in the Lower 9th Ward Between 2010 and 2011

A total of 6 hours of driving time video were captured for the Lower 9th Ward for 2010 and 2011. The digitizing and coding of the spatial video took one person approximately 20 h for 2010, and then 10 h for 2011, a shorter period as the majority of the buildings only had to be populated with new scores. Table 9.1 presents an overview of the total buildings, both blighted and occupied (called "return") extracted for both years. The total number of buildings digitized in the Lower 9th Ward immediately after Hurricane Katrina is also provided for comparison. There are no blighted or returned numbers for 2005 as the entire neighborhood was under water.

What is evident from Table 9.1 is that, although slow, progress between 2010 and 2011 has been made with regard to clearing old blighted property (a drop of 90 buildings) and with more homes being occupied (an increase of over 90). It is also poignant to see that the number of occupied homes in 2011 is still less than 20% of the buildings in the neighborhood from before Katrina.

Visual clues on the landscape

> A body was discovered in a burned car Monday afternoon in the Lower 9th Ward. The car -- a white Dodge Charger with a Memphis, Tenn., license plate -- was driven into towering bushes on Law Street between Flood and Choctaw streets. Much of the vegetation in the area has grown more than 12 ft high. "We call the city often," said Richelle Jackson, who could not see the Dodge Charger from the porch of her elevated home a stone's throw from where the car and body were found . "But they don't care". She and Sylvester and Molly O'Neal, who rebuilt in the Lower 9th Ward after their home was destroyed by floodwaters from broken levees during Hurricane Katrina, said this is the second burned car within

[6] For this analysis, crimes were collapsed into convenient categories thought to be of similar types with regards to their relationship with the landscape. In this way "incivility" combines simple drug arrests, simple arson and vandalism; "auto" combines car theft and theft from a car.

Fig. 9.2 Example images extracted from the spatial video showing crime-associated aspects of the landscape (Source: Spatial video collected by S. Wright Kennedy August 2010)

> weeks to appear in their neighborhood. The cars were located within a block of each other. The Charger was less than half a block from the O'Neal home, that has a manicured lawn adorned with lots of roses. But they could not see it from their porch either because of the neglect of public and private property around their home.
>
> *Burned body discovered in Lower 9th Ward neighborhood of New Orleans* by Leslie Williams, The Times-Picayune, Monday August 29, 2011

During the coding of all the buildings on the spatial video routes, various images were extracted and maps generated to illustrate landscape features relevant to the general theoretical discussions described earlier in this chapter. Figure 9.2 displays six of these images. Both A and B are typical of the Katrina-devastated buildings that are still found throughout the Lower 9th Ward. Many of these are damaged, sometimes in danger of collapse, and often overgrown with vegetation. In addition, many are "unsecured" meaning doors and windows are open. As with the opening quote of this section, it is not surprising to find that crimes and evidence of crimes are continually linked in the media to these spaces in post-Katrina New Orleans. Similarly, many old lots are overgrown with vegetation to the point of being impenetrable, presenting further opportunity for activities beyond normal neighborhood oversight. B and C also display aspects of neighborhood decline more commonly associated with crime in non-disaster landscapes; graffiti and trash (dumping).

Although there is evidence of graffiti on several structures in the Lower 9th Ward, it is not as widespread as one might have thought given the number of abandoned properties. Of more concern is the illegal dumping, which ranges from vehicles (burned out joy rides), to multiple piles of tires and black bags of trash left by the road side. One poignant homemade sign on a street with both blighted and occupied homes states "this is a neighborhood not a dump". A further visual clue to a lack of normal social activity is the many abandoned churches and stores, locations that would have previously played an important role in the social fabric of the community. In addition, many roads are in a terrible condition, even to the point of being impassable, with D showing a bollard placed in the middle of the road to indicate a severe pot-hole. These roads limit access throughout the neighborhood, especially at night. Finally, E and F display aspects of gang activity in the neighborhood, both being memorials to someone who has died. In the first a t-shirt has been nailed to a telegraph pole, while in the second an "RIP" message is spray painted on the sidewalk. These originated from different locations within the neighborhood.

9.6.2 Results of the Spatial Filter Analysis

Once all the buildings had been digitized, eight different spatial filter maps were constructed; for the years 2010 and 2011, for blight and return, and with both buildings and parcels as denominators. Figure 9.3 displays the locations of the four crime categories overlaid on all nodes meeting a minimum of five buildings in the filter threshold. The mix of blighted and occupied homes is so interwoven that the node surface output for each rate surface in 2010 and 2011 is almost identical (only 2010 is shown here). The one difference is in the top left of the map for 2010 where a high rate of rebuilt homes can be seen. Each node is buffered by 41 m (approximately 0.025 miles) and these are merged to show the geographical extent covered by the filters in the rate calculation. In addition, a further buffer extends this area by 59 m to show all crimes that are within at least 100 m of a rate node. It is interesting to note that only ten crimes fall outside of the buffered areas, but of these, four are "incivility" and three are "domestic". One interpretation might be that the minimum of five homes in the denominator does not capture isolated "pioneering" homes where a high stress burden may lead to domestic violence. The map for 2011 is similar to 2010 with only a few subtle shifts in coverage, though the number of crimes outside of the buffer area now falls to six. It is important to note that this also captures an important aspect of a dynamic landscape, that reliance

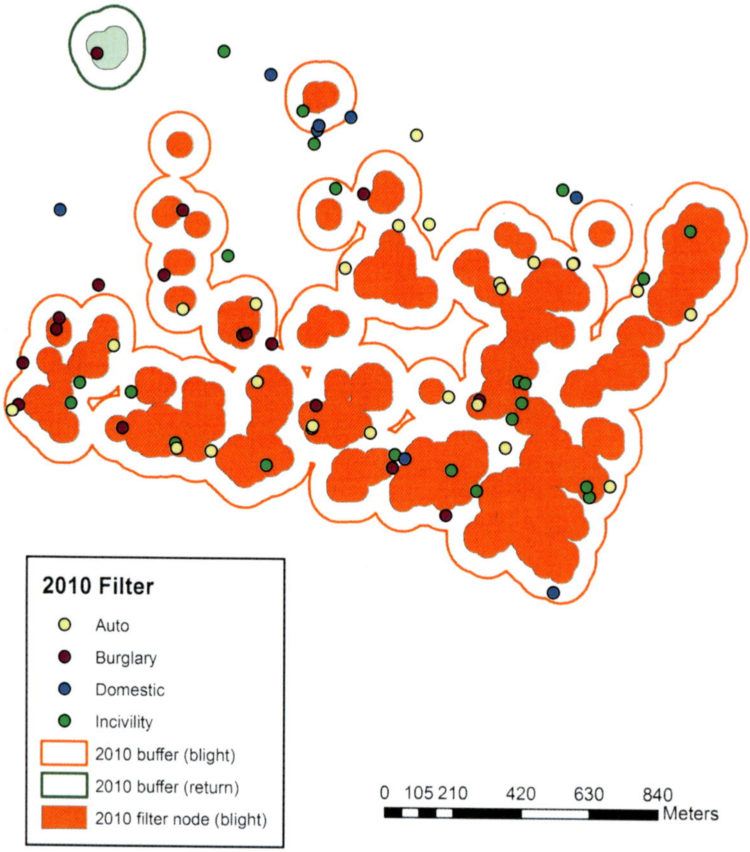

Fig. 9.3 The spatial filter map for 2010 showing the location of all buffered nodes and crimes by type

on cross-sectional data has limitations and the two bounded ecological landscapes presented here at least capture aspects of that change.

As each crime was spatially linked to the closest node, the rates attached to that node could be extracted to form a summary of the average percentage of blighted or returned homes at the node closest to each crime in that category (Table 9.2).

Although the numbers of crimes is too low by category to perform a difference of proportions t test, the overall pattern is still interesting in terms of what it suggests. As an example, there is little variation across all four crime categories using buildings as denominator in 2010 in the blight calculation. The biggest differences occur when the rate calculation uses parcels instead of all buildings as the denominator as this would capture smaller groupings of homes. Although there is relatively little difference between all four crime types and the blight rate during 2010, both auto and incivility have markedly higher average blight rates in 2011. Interestingly, the reverse is true for proximity to occupied homes, with little difference between

Table 9.2 Average blight and return rates for each crime type for 2010 and 2011

Denominator	2010 Blight Building	2010 Blight Parcel	2010 Return Building	2010 Return Parcel	2011 Blight Building	2011 Blight Parcel	2011 Return Building	2011 Return Parcel
Auto (25)	43.1	8.7	38.1	15	30.4	50.8	50.8	9.1
Burglary (17)	33.3	9.5	51.6	14.4	20	7.7	63.5	64
Domestic (8)	40	9.4	50	16	40	5.7	50	50
Incivility (22)	44.8	14	43.7	16.7	27.6	46.6	46.6	9.7
Auto and Incivility	43.95	11.35	40.9	15.85	29	48.7*	48.7	9.4

Where * means p=0.01 and the number in parenthesis is the total for that crime type

all four crime types in 2010, but now a markedly lower rate for auto and incivility in 2011. One possible interpretation of these findings is that when considering the entire landscape (the pre-Katrina footprint and not just the remaining buildings), auto crimes and incivility crimes are more linked to a lack of activity, while the reverse is true for crimes of the home such as domestic disturbances and burglaries. By combining auto and incivility together based on these results (and a belief that this ecological finding makes sense), a difference of proportions *t* test was used to determine that the average percentage of blighted homes (using parcels as the denominator), was the only statistically significant difference from the background population. In other words, the rate of blighted properties around these crimes was higher than the percentage of blighted properties spread across the neighborhood. What is interesting from this finding is that the relationship only holds for 2011. Why? Is it possible that in the slow move to normalcy with the removal of blighted properties and a returning population, a tipping point has been reached whereby a devastated landscape begins to function more similarly to other urban landscapes?

9.7 Discussions

9.7.1 The Purpose of the Chapter

This chapter has presented a methodological approach to collecting fine scale built environment data that can be used to investigate street level patterns of crime. The particular focus was on a more extreme landscape than has previously been analyzed, in this case the Lower 9th Ward of New Orleans after Hurricane Katrina. Although the results presented here are interesting, the take-home message of this chapter is that a technology and technique exists to facilitate more fine-scale built

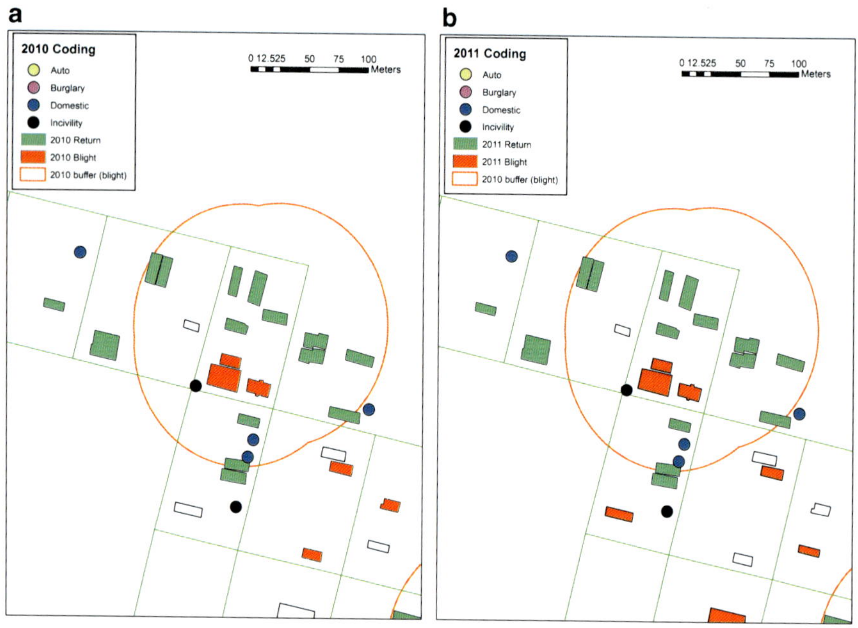

Fig. 9.4 (**a**) and (**b**) A relatively isolated recovery area of the Lower 9th Ward with a high number of domestic violence arrests

environment research suitable for many dynamic environments. Not only can the spatial video be used to capture the variables described here, it also provides an archival source that can be returned to for different investigations, or to follow new lines of inquiry that might emerge. This would be difficult from "walking" survey responses alone, as even if a photograph was taken, these are more likely to be seen in spatial isolation, whereas the video allows for more spatial context. As previously mentioned, images and maps have been generated from the spatial video source to illustrate different fine-scale discussions common in the crime literature, and this chapter will end with a series of block scale maps generated after the analysis had been completed.

9.7.2 Returning to the Spatial Video for Visual Evidence

Relatively little has been written about the space-time evolution of crimes, yet time is considered implicit in most theories (Pitcher and Johnson 2011). While the spatial video approach utilized here can be used to advance general theories, there are specific temporal questions that arise in a post-disaster landscape. For example, Fig. 9.4 a, b map the highest fine-scale concentration of domestic disturbances in the Lower 9th Ward. By comparing the area between 2010 and 2011 we can suggest

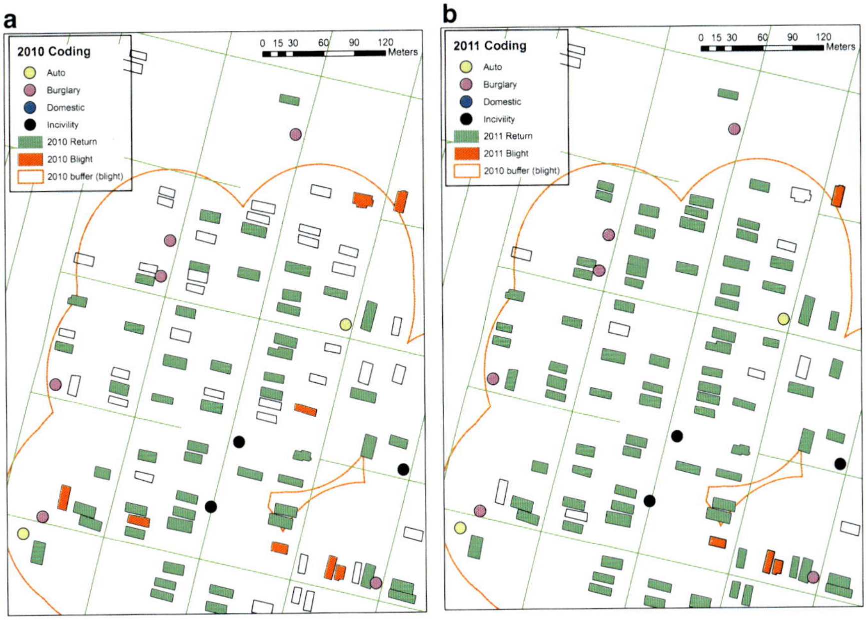

Fig. 9.5 (**a**) and (**b**) The recovery growth pole generated by "Make it Right"

a temporal explanation behind this spatial pattern. Over this period there has been no further rebuilding, and while two of the blighted properties have been cleared, two others have become blighted. By returning to the video, it is evident that a home being rebuilt in 2010 was subsequently abandoned. The question this map raises is, does the relative isolation of these occupied homes, and the ongoing presence of blighted buildings (in combination with severe local vegetation overgrowth) provide the kind of increasing stressful environment that leads to domestic violence?

Previous work has considered the length of residential tenancy to crime (e.g., Kubrin 2003). For example Taniguchi et al. (2011) linked street corners with high percentages of people who had lived there for less than 5 years, with violent and property crime (see also Xie and McDowall 2008). This is also supportive of the idea that neighbor-to-neighbor surveillance is a product of how long residents know each other (Freudenburg 1986). For some post disaster neighborhoods in New Orleans not only has there been a significant reduction in overall population levels, but there has also been an influx of individuals from outside the area. Alternatively, new developments such as "Make It Right", a nonprofit that is relentlessly building new homes from one corner of the neighborhood outwards, are being built systematically to develop "community". Areas of new construction could be investigated in terms of place-based routine activities (Sherman et al. 1989) as this section of the neighborhood, at least during the day, contains both builders and disaster tourists. Figure 9.5a, b show the increase in occupied homes around "Make It Right", and the decrease in blighted homes. Of the ten crimes, five are burglaries; the impressive new

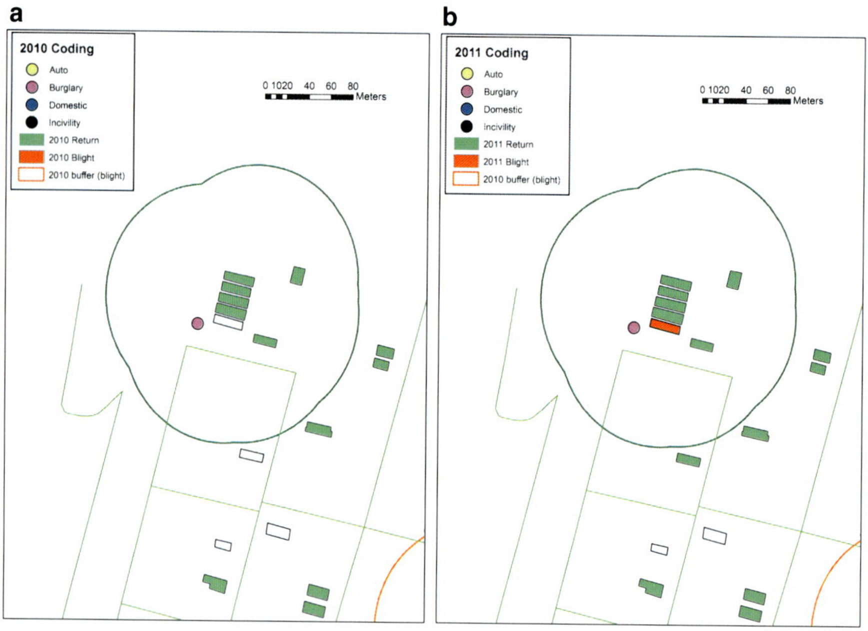

Fig. 9.6 (**a**) and (**b**) An example of "pioneer" returnees who are surrounded by open land

homes and easy access to a main road (North Claiborne) provide a possible crime attractor space. Further questions to consider could be, has this "constructed" community led to a degree of neighbor oversight one wouldn't normally expect given the length of residential tenancy?

A further aspect of post-disaster neighbor oversight is illustrated in Fig. 9.6a, b where the majority of homes are occupied, though still in relative isolation and providing an example of a pioneering mentality. However one home that was being rebuilt in 2010 has since been abandoned and moved into the blighted category. This provides an excellent example of how temporal framing captures more of the story than any cross sectional analysis. Interestingly, in this landscape of relative isolation, a burglary has still occurred.

Another common theme in fine scale crime research is the effect of spatial dependency. For example, Bernasco and Block (2011) found that robbery related crime attractors "radiate" out risk to neighboring census blocks. Similar questions could be posed here, for example, is the presence of blighted buildings in proximate blocks more likely to cause crime? One difference between studies in post-disaster landscapes such as the Lower 9th Ward is that crime is probably not a zero sum game, in other words, crime is attracted into the neighborhood from outside whereas other spillover studies suggest a set amount of crime only shifts geographically. A further difference is with the scale of movement. Although Bernasco and Block (2011) comment on the importance of neighboring blocks in terms of spillover

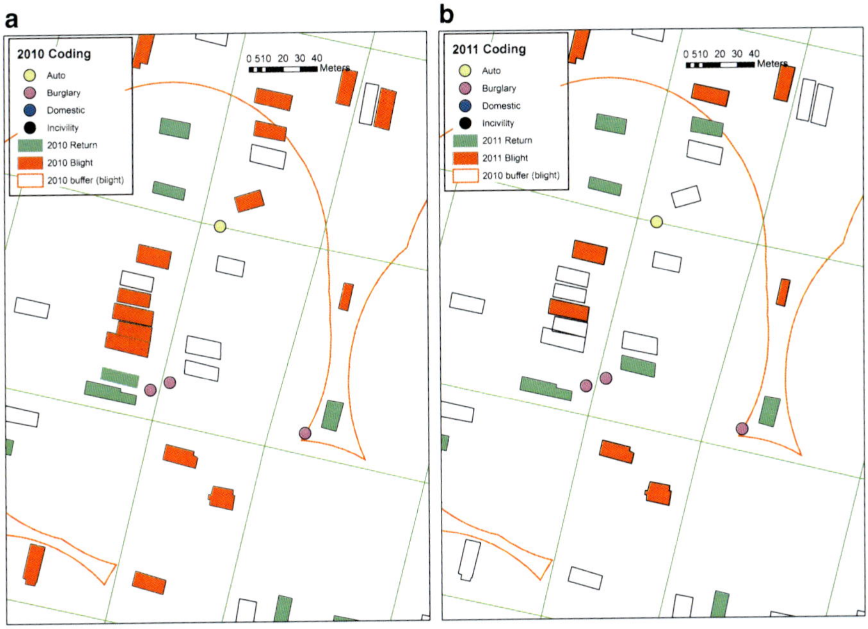

Fig. 9.7 (**a**) and (**b**) An example of the slow removal of disaster caused blight in the Lower 9th Ward

effects as the "… geographical range of humans is limited" (p 51), this might not hold for the Lower 9th Ward as it could be argued that the impediment of distance is lessened in post disaster landscapes. Obviously, this needs further investigation. On a similar note, one specific spillover effect to these crimes may occur along pathways, especially for the Lower 9th Ward where routes must be taken to crime attractors, both in terms of arriving at the neighborhood, and then to a final destination point such as that described in the newspaper quote regarding the dumping of a body. Again, linking spatial dependency to maps generated from the spatial video runs, Fig. 9.7a, b illustrate the vulnerability of occupied homes in a blighted and then cleared landscape. Three burglaries occur in blocks where only three homes are occupied. The rest of the immediate area is comprised of blighted buildings, some of which have been cleared by 2011. Is this an example of the attractiveness of the blighted properties increasing the risk for the occupied homes?

9.7.3 The Next Steps

In terms of future research, questions can be asked as to what constitutes an appropriate space around such potential crime attractors. For example Taniguchi et al. (2011) pose the question, how far from the street intersection does the influence of the gang set space extend? Similarly here, how far does the influence of a blighted building or an overgrown parcel extend? Spatial interdependency is widely acknowledged (Hipp 2007) but how does it influence crime in neighborhoods such as the Lower 9th Ward?

Finally, the spatial video is an archived data source that can be used for exciting new approaches to fine scale crime research. For example, it is possible to explore the type of neighborhood watchfulness view shed from a home's windows as suggested by Newman (described in Reynald 2011) by turning the types of maps displayed in this chapter into three dimensional street scenes. Similarly, aspects of dumping, and the transition from public to private space (or in this case, overgrown lot to private space) can also be mapped in three dimensions. As Pitcher and Johnson (2011) comment, GIS (and we extend this to geospatial technologies including field collection) now make fine scale, primary data driven, multiple time period research possible.

The rationale for conducting this type of work is double-edged in terms of reducing crime *and* improving neighborhood recovery. Paraphrasing Taniguchi et al. (2011:351), crimes concentrate in relatively few locations and so by understanding the place based characteristics of these spaces intervention strategies can be developed. Traditional examples of such location denial approaches include gang injunctions, though in the Lower 9th Ward it could simply be blight removal and grass cutting. To this end it will be interesting to see how crime patterns will change as a result of Mayor Mitch Landrieu's blight eradication strategy for New Orleans launched at the end of 2011.

9.8 Conclusion

A common thread for many recent place-based research papers is that fine scale environments are not only important theoretically in terms of the role these streets or buildings have on criminal action, but also in terms of accurate spaces for GIS based analysis. Broad area census aggregations do not capture these processes. Although crime pattern analysis can be conducted on events alone, linking these to other environmental variables has previously been difficult. Efforts that have attempted to capture detailed social and built environment information are expensive and time consuming, and therefore are either cross sectional or for a limited number of time periods. This chapter has produced further support for a geospatial approach that addresses these data issues. For example, a modified version of the methods used by Reynald (2011), such as a variant of the Block Environmental Inventory, and even counting pedestrians or other visible signs of activity in or around buildings and open areas to collect residential guardianship could be collected using the spatial video for multiple time periods. This allows for longitudinal fine-scale environmental data to be collected, and as such opens many new research avenues that had previously been relegated to discussion alone.

References

Accordino J, Johnson GT (2000) Addressing the vacant and abandoned property problem. J Urban Aff 22(3):301–315

Ackerman WV, Murray AT (2004) Assessing spatial patterns of crime in Lima, Ohio. Cities 21(5):423–437

Adams PR, Adams GR (1984) Mount Saint Helen's ashfall: evidence for a disaster stress reaction. Am Psychol 39(3):252–260

Bernasco W, Block R (2011) Robberies in Chicago: a block-level analysis of the Influence of crime generators, crime attractors, and offender anchor points. J Res Crime Delinq 48(1):33–57

Bess M (2008) Assessing the impact of home foreclosures in Charlotte neighborhoods. Geogr Pub Saf 1(3):2–5

Braga AA, Papachristos AV, Hureau DM (2010) The concentration and stability of gun violence at micro places in Boston, 1980–2008. J Quant Criminol 26(1):33–53

Braga AA, Hureau DM, Papachristos AV (2011) The relevance of micro places to citywide robbery trends: **a** longitudinal analysis of robbery incidents at street corners and block faces in Boston. J Res Crime Delinq 48(1):7–32

Brantingham PL, Brantingham PJ (1995) Criminality of place: crime generators and crime attractors. Eur J Crim Policy Res 3:5–26

Brantingham PL, Brantingham PJ, Vajihollahi M, Wuschke K (2009) Crime analysis at multiple scales of aggregation: a topological approach. In: Weisburd D, Bernasco W, Bruinsma GJN (eds) Putting crime in its place. Springer, New York

Brezina T, Kaufman JM (2008) What really happened in New Orleans? Estimating the threat of violence during the Hurricane Katrina disaster. Justice Q 25(4):701–722

Bursik RJ Jr, Grasmick HG (1993) Neighborhoods and crime. Lexington Books, New York

Cohen LE, Felson M (1979) Social change and crime rate trends: a routine activity approach. Am Sociol Rev 44:588–608

Cromwell P, Dunham R, Akers R, Lanza-Kaduce L (1995) Routine activities and social control in the aftermath of a natural catastrophe. Eur J Crim Policy Res 3(3):56–69

Curtis A, Mills JW (2011) Crime in urban post-disaster environments: a methodological framework from New Orleans. Urban Geogr 32(4):488–510

Curtis A, Mills JW, Kennedy B, Fotheringham S, McCarthy T (2007) Understanding the geography of post-traumatic stress: an academic justification for using a spatial video acquisition system in the response to Hurricane Katrina. J Conting Crisis Manag 15(4):208–219

Curtis A, Mills JW, McCarthy T, Fotheringham AS, Fagan W (2010a) Space and time changes in neighborhood recovery after a disaster using a Spatial Video Acquisition System. In: Showalter P, Lu Y (eds) Geospatial techniques in urban hazard and disaster analysis. Springer, Berlin

Curtis A, Duval-Diop D, Novak J (2010b) Identifying spatial patterns of recovery and abandonment in the Holy Cross neighborhood of New Orleans. Cartogr Geogr Inf Sci 37(1):45–56

Dalton E, Gradeck R, Mercaldo A (2008) Using maps of home foreclosures to understand national and local problems. Geogr Public Saf 1(3):5–8

Davila M, Marquart JW, Mullings JL (2005) Beyond mother nature: contractor fraud in the wake of natural disasters. Deviant Behav 26:271–293

Decker SH, Varano SP, Greene JR (2007) Routine crime in exceptional times: the impact of the 2002 Winter Olympics on citizen demand for police services. J Crim Justice 35:89–101

Enarson E (1999) Violence against women in disasters: a study of domestic violence programs in the United States and Canada. Violence Against Women 5(7):742–768

Felson M (2006) Crime and nature. Sage, Thousand Oaks

Foa EB, Stein DJ, McFarlane AC (2006) Symptomatology and psychopathology of mental health problems after disaster. J Clin Psychiatry 67(Suppl 2):15–25

Fothergill A (1996) Gender, risk, and disaster. Int J Mass Emerg Disaster 14:33–56

Fothergill A (1999) An exploratory study of woman battering in the Grand Forks flood disaster. Int J Mass Emerg Disaster 17:79–98

Frailing K, Harper DW (2007) Crime and hurricanes in New Orleans. In: Brunsma DL, Overfelt D, Picou S (eds) The sociology of Katrina. Rowman and Littlefield, Plymouth

Frailing K, Harper DW (2010) School kids and oil rigs: two more pieces of the post-Katrina puzzle in New Orleans. Am J Econ Sociol 69(2):717–735

Freudenburg WR (1986) The density of acquaintanceship: an overlooked variable in community research? Am J Sociol 92:27–63

Groff ER, Weisburd D, Yang SM (2010) Is it important to examine crime trends at a local "Micro" level?: a longitudinal analysis of street to street variability in crime trajectories. J Quant Criminol 26(1):7–32

Hipp JR (2007) Income inequality, race, and place: does the distribution of race and class within neighborhoods affect crime rates? Criminology 45:665–97

Immergluck D, Smith G (2006) The impact of single-family mortgage foreclosures on neighborhood crime. Hous Stud 21(6):851–866

Kinney JB, Brantingham PL, Wuschke K, Kirk MG, Brantingham PJ (2008) Crime attractors, generators and detractors: land use and urban crime opportunities. Built Environ 34(1):62–74

Kubrin CE (2003) Structural covariates of homicide rates: does type of homicide matter. J Res Crime Delinq 40:139–70

LaGrange RL, Ferraro KF, Supancic M (1992) Perceived risk and fear of crime: role of social and physical incivilities. J Res Crime Delinq 29(3):311–334

LeBeau JL (2002) The impact of a hurricane on routine activities and on calls for police service: Charlotte, North Carolina and Hurricane Hugo. Crime Prev Community Saf Int J 4(1):53–64

Loukaitou-Sideris A (1999) Hot spots of bus stop crime: the importance of environmental attributes. J Am Plan Assoc 65(4):395–411

Loukaitou-Sideris A, Liggett R, Iseki H (2002) The geography of transit crime. J Plan Educ Res 22(2):135–151

Mills JW, Curtis A, Pine J, Kennedy B, Jones F, Ramani R, Bausch D (2008) The clearinghouse concept: a model for geospatial data centralization and dissemination in a disaster. Disasters: J Stud Policy Manag 32(3):467–479

Mills JW, Curtis A, Kennedy B, Kennedy SW, Edwards J (2010) Geospatial video for field data collection. Appl Geogr 30(4):533–547

Morrow BH, Enarson E (1996) Hurricane Andrew through women's eyes: issues and recommendations. Int J Mass Emerg Disaster 14:5–22

Peed C, Wilson RE, Scalisi NJ (2008) Making smarter decisions: connecting crime analysis with city officials. The police chief (September). Available online: http://policechiefmagazine.org/magazine/index.cfm?fuseaction=display_arch&article_id=1603&issue_id=92008

Perkins DD, Taylor RB (1996) Ecological assessments of community disorder: their relationship to fear of crime and theoretical implications. Am J Community Psychol 24((1):63–107

Perkins DD, Meeks JW, Taylor RB (1992) The physical environment of street blocks and resident perceptions of crime and disorder: Implications for theory and measurement. J Environ Psychol 12(1):21–34

Perkins DD, Wandersman A, Rich RC, Taylor RB (1993) The physical environment of street crime. J Environ Psychol 13(1):29–49

Pitcher AB, Johnson SD (2011) Exploring theories of victimization using a mathematical model of burglary. J Res Crime Delinq 48(1):83–109

Reynald DM (2011) Factors associated with the guardianship of places: assessing the relative importance of the spatio-physical and sociodemographic contexts in generating opportunities for capable guardianship. J Res Crime Delinq 48(1):110–142

Reynald DM, Elffers H (2009) The future of Newman's defensible space theory: linking defensible space and the routine activities of place. Eur J Criminol 6:25–46

Robinson JB, Lawton BA, Taylor RB, Perkins DD (2003) Multilevel longitudinal impacts of incivilities: fear of crime, expected safety, and block satisfaction. J Quant Criminol 19(3):237–274

Rogers WH, Winter W (2009) The impact of foreclosures on neighboring housing sales. J Real Estate Res 31(4):455–479

Rushton G, Lolonis P (1996) Exploratory spatial analysis of birth defect rates in an urban population. Stat Med 15:717–726

Sampson RJ, Raudenbush SW (1999) Systematic social observation of public spaces: a new look at disorder in urban neighborhoods. Am J Sociol 105:603–651

Sampson RJ, Raudenbush SW, Earls F (1997) Neighborhoods and violent crime: a multi-level study of collective efficacy. Science 277(5328):918–924

Schuetz J, Been V, Ellen IG (2008) Neighborhood effects of concentrated mortgage foreclosures. J Hous Econ 17(4):306–319

Shaw CR, Zorbaugh F, McKay HD, Cottrell L (1929) Delinquency areas. University of Chicago Press, Chicago

Sherman L, Gartin PR, Buerger ME (1989) Hot spots of predatory crime: routine activities and the criminology of place. Criminology 27:27–55

Skogan W (1986) Fear of crime and neighborhood change. Crime Justice 8:203–229

Skogan W (1990) Disorder and decline: crime and the spiral of decay in American neighborhoods. University of California Press, Berkeley

Smith WR, Frazee SG, Davison EL (2000) Furthering the integration of routine activity and social disorganization theories: small units of analysis and the study of street robbery as a diffusion process. Criminology 38:489–524

Taniguchi TA, Ratcliffe JH, Taylor RB (2011) Gang set space, drug markets, and crime around drug corners in Camden. J Res Crime Delinq 48(3):327–363

Taylor RB (1997) Social order and disorder of street blocks and neighborhoods: ecology, microecology, and the systemic model of social disorganization. J Res Crime Delinq 34(1):113–155

Taylor RB (2001) Breaking away from broken windows: Baltimore neighborhoods and the nationwide fight against crime, grime, fear, and decline. Westview Press, Boulder

Taylor RB, Shumaker SA, Gottfredson SD (1985) Neighborhood-level links between physical features and local sentiments: deterioration, fear of crime, and confidence. J Archit Plan Res 2:261–275

Tierney K, Bevc C, Kuligowski E (2006) Metaphors matter: disaster myths, media frames, and their consequences in Hurricane Katrina. Ann Am Acad Polit Soc Sci 604:57–81

Tita GE, Cohen J, Engberg J (2005) An ecological study of the location of gang 'set space'. Soc Probl 52:272–99

Townsley M, Homel R, Chaseling J (2003) Infectious burglaries: a test of the near repeat hypothesis. Br J Criminol 43:615–33

Varano SP, Schafer JA, Cancino JM, Decker SH, Greene JR (2010) A tale of three cities: crime and displacement after Hurricane Katrina. J Crim Justice 38:42–50

Weisburd DL, Bushway S, Lum C, Yang SM (2004) Trajectories of crime at places: a longitudinal study of street segments in the city of Seattle. Criminology 42:283–321

Weisler RH, Barbee JG IV, Townsend MH (2006) Mental health and recovery in the Gulf Coast after Hurricanes Katrina and Rita. JAMA 296(5):585–588

Wilson JQ, Kelling GL (1982) Broken windows. Atl Mon 249(3):29–38

Wilson RE, Paulsen DJ (2008) Foreclosures and crime: a geographical perspective. Geogr Pub Saf 1(3):1–2

Xie M, McDowall D (2008) The effects of residential turnover on household victimization. Criminology 46:539–75

Chapter 10
Spatial Contagion of Male Juvenile Drug Offending Across Socioeconomically Homogeneous Neighborhoods

Jeremy Mennis and Philip Harris

Abstract This research investigates the influence of spatial contagion and neighborhood socioeconomic character on drug offense juvenile delinquency and recidivism among urban male youth. For this purpose we use a data set of all juvenile delinquents sentenced to court-ordered programs in Philadelphia, Pennsylvania from 1996 to 2003. U.S. Bureau of the Census data are used to capture various dimensions of socioeconomic character, including concentrated disadvantage, ethnicity, and residential mobility. A regionalization algorithm is used to generate neighborhood boundaries that maximize within-unit socioeconomic homogeneity while preserving large counts of juveniles so that reliable rates of drug offense delinquency may be calculated. Ordinary least squares (OLS) regression and spatial econometric modeling are employed to model the effects of neighborhood socioeconomic character and, where appropriate, spatial contagion, on delinquency, drug offense delinquency, and repeat drug offense delinquency outcomes. Results indicate that concentrated disadvantage enhances not only delinquency generally but also drug delinquency and repeat drug delinquency. In addition, modeled drug delinquency indicates spatial contagion among neighborhoods, while controlling for the influence of concentrated disadvantage, providing evidence for cross-neighborhood peer contagion.

Keywords Juvenile delinquency • Offense specialization • Peer contagion • Drug offense • Concentrated disadvantage • Regionalization

---

J. Mennis (✉)
Department of Geography and Urban Studies, Temple University,
1115 W. Berks St., 309 Gladfelter Hall, Philadelphia, PA 19122, USA
e-mail: jmennis@temple.edu

P. Harris
Department of Criminal Justice, Temple University, 1115 W. Berks St., 5th Floor,
Gladfelter Hall, Philadelphia, PA 19122, USA

M. Leitner (ed.), *Crime Modeling and Mapping Using Geospatial Technologies*, Geotechnologies and the Environment 8, DOI 10.1007/978-94-007-4997-9_10,

10.1 Introduction

Reducing juvenile delinquency and recidivism continues to be a major challenge to the juvenile justice system. Neighborhood level processes, including structural characteristics such as concentrated disadvantage, residential mobility, and segregation, are well-known mechanisms of juvenile crime (Shaw and McKay 1942). Cultural expressions of community trust and participation, as captured by concepts such as social disorganization and collective efficacy, have also been shown to be related to juvenile delinquency (Bursik 1988; Sampson et al. 1999). Juvenile delinquency may also be viewed through the lens of peer modeling of behavior, where criminal behavior is regarded as something that is learned from either an adult or other juveniles (Cloward and Ohlin 1960; Sutherland and Cressey 1992; Warr 1996).

In previous research we have argued that these neighborhood and social contextual mechanisms of juvenile delinquency do not occur independently but are rather intertwined (Mennis and Harris 2011). We consider that juvenile delinquency does not tend towards offense generalization, where juvenile crime is seen as a type of hooliganism, or youth 'running wild,' where youth may commit a variety offense types (e.g. violent crimes, property crimes, and so on). Rather, we view juvenile offending as a product of environmental and social contextual forces that produce offense specialization, where juvenile delinquents who do come in continued contact with the juvenile justice system tend to repeat the same type of offense. We reason that if criminal behavior among youth is indeed a learned behavior, then one would expect to see juveniles who interact socially behave in a similar manner. Likewise, given that juveniles' behaviors are, in part, a product of peer interaction, one would expect to see offense specialization at the neighborhood level as well, since juveniles living in the same neighborhood are likely to interact to a greater degree than juveniles living in different neighborhoods.

Of course, peer interaction among juveniles can also occur across neighborhood boundaries from one neighborhood to another. In such a case, one would expect that juvenile crime in a neighborhood would be the result not only of the structural and cultural characteristics of that neighborhood but also of the juvenile crime in nearby or adjacent neighborhoods. In other words, even for a particular neighborhood that had a relatively low measure of criminogenic neighborhood mechanisms (such as concentrated disadvantage), if it were located adjacent to a neighborhood with a high rate of juvenile delinquency, youth in the former neighborhood may be exposed to, and learn, criminal behavior from youth in the adjacent neighborhood. We consider such diffusion of youth criminal behavior a case of spatial contagion from one neighborhood to adjacent neighborhoods.

Critical to an investigation of neighborhood-level spatial contagion is the definition of 'neighborhood,' itself. We note that the vast majority of quantitative, neighborhood-level analyses of crime, as well as many other social and health outcomes, define neighborhood using a spatial unit of convenience, such as census enumeration units (e.g. U.S. Census tract), postal address units (e.g. U.S. zip codes), police districts, and so on. There are two major problems with this approach. First, one assumes that the neighborhood boundaries capture the process under investigation, in the case of the present research the neighborhood-level mechanisms of juvenile delinquency. This is problematic because spatial units of convenience such as those listed above do not necessarily conform to natural boundaries of regions of similar levels of concentrated disadvantage or collective efficacy. It is therefore difficult to recognize the relationship between such neighborhood-level mechanisms and juvenile crime outcomes when the neighborhoods themselves may be heterogeneous. Second, spatial units of convenience can vary widely in occurrences of juvenile crime. For example, certain impoverished urban Census tracts may be home to many juvenile offenders, while in some more affluent urban Census tracts there may be few juvenile delinquent residents. Low numbers of occurrences in certain spatial units can render the calculation of delinquency and recidivism rates unreliable and weaken the statistical power of the analysis.

In the present study we investigate neighborhood-level mechanisms and contagion of drug offense delinquency and recidivism in Philadelphia, Pennsylvania, U.S. We focus specifically on drug offense delinquency and recidivism as our own previous research has shown that juvenile drug offending, more than other types of offending, is prone to specialization and peer influence to a greater degree than other types of offenses (Mennis and Harris 2011). Drug offenses, as compared to offenses against persons (i.e. violent offenses) or property crimes, are less likely to be crimes of opportunity or passion, but are rather economically motivated. In addition, drug selling among youth implies involvement in a structured criminal organization where learning criminal behavior from peers and adults plays an important role (Little and Steinberg 2006; Nguyen and Bouchard 2012).

To address the issues associated with neighborhood level analyses discussed above, we employ a novel regionalization approach developed by Guo (2008) to generate socioeconomically homogeneous neighborhood units. Our motivation here is to use spatial units that (1) have spatial boundaries that capture the structural and cultural neighborhood-level mechanisms that may influence drug offense delinquency and recidivism, and (2) contain a sufficient number of juvenile delinquents to calculate reliable rates of drug offense delinquency.

10.2 Neighborhood Contagion and Juvenile Drug Offense Delinquency

The theoretical underpinning for this research rests on four principles. First, concentrated disadvantage enhances juvenile drug offense delinquency through both economic incentives for the offender as well as a lack of social control at the neighborhood level. Concentrated disadvantage, and its attendant characteristics of social disorganization, has long been known to catalyze juvenile delinquency (Cloward and Ohlin 1960; Shaw and McKay 1942). Concentrated disadvantage encompasses poverty, but also the idea that the effects of poverty are magnified by associated conditions of unemployment, low educational attainment, and racial and ethnic segregation that act to cut residents off from resources and opportunities for advancement out of poverty.

The economic motivation for engaging in criminal behavior that is profitable, such as drug selling, for poor, urban youth is clear when youth are confronted by a lack conventional job opportunities (Agnew 1994). These job opportunities may be constrained by structural characteristics, such as substandard education and a lack of access to available jobs, as well as cultural characteristics that may discourage employment within what is perceived as the mainstream culture while encouraging employment in high prestige criminal enterprises (Anderson 1999; Matsueda et al. 1992).

In addition, poverty is also associated with a lack of neighborhood collective efficacy that allows deviant behavior among youth to proliferate (Sampson et al.1999). Collective efficacy can be understood as a level of trust among neighbors, a willingness to help and assist others in the neighborhood, and a willingness to participate in activities with other neighbors to promote well-being in the neighborhood. In addition to poverty, residential mobility and ethnic heterogeneity have been suggested as suppressing collective efficacy, where such characteristics are seen as barriers to the development of social cohesion among neighbors (Shaw and McKay 1942; Twigg et al. 2010). Weak collective efficacy has been shown to be associated with drug offending among youth, as there is little social pressure in the community to limit illicit drug use and criminal activity (Grunwald et al. 2010).

The second theoretical principle claims that drug offending among youth is a learned behavior that is influenced by other adolescent (and adult) drug offenders with whom a youth interacts (Sutherland and Cressey 1992). Peer influence on adolescent substance use is well known (Valente 2003), and such influence has been tied explicitly to place, where certain neighborhood places hold particular value in their social characteristics relating to peer influence to use drugs. For instance, research suggests that adolescent substance use is influenced by the substance use of peers at certain locations that an adolescent visits routinely throughout their daily life (Mennis and Mason 2011). Likewise for delinquency, having delinquent friends is an important factor in delinquent behavior (Warr 2002), and social interaction among delinquents has been suggested as the mechanism for iatrogenic effects that have been observed for delinquency treatment programs (McCord 2003). Peer influence in adolescent deviant behavior, including drug use and delinquency, may be seen to operate through behavioral modeling, where youth may adopt the attitudes and behaviors of their peers, as well as the cultural norms within which innovative behaviors may be developed (Bandura 1996; Akers 1998).

The third theoretical principle is that the process of peer interaction among youth drug offenders produces specialization in drug offending that is extended to the neighborhood level, such that entire neighborhoods may be understood as specializing in drug criminal activity (Mennis and Harris 2011). Warr (1996) argues that offense specialization among youth is facilitated by group peer interaction, where groups of youth may focus on a particular type of criminal activity, and the persistence of a particular social group of juveniles is directly associated with the propensity for offense specialization. Nguyen and Bouchard (2012) discuss how such peer learning and socialization operates for juveniles involved in the drug trade, where the authors show how social capital enhances economic performance for youth within organizations focusing on marijuana cultivation.

We posit that such social groups of juvenile offenders are based on neighborhood residence, and that youth are likely to form social groups that facilitate offense specialization with delinquent youth who live in their own neighborhood. Casual within-neighborhood social contact may be the dominant form of social interaction for many youth, particularly in poor, urban neighborhoods where mobility may be limited due to a lack of resources for transportation, among other reasons. In our own previous research with the present data we have shown that juvenile delinquents do, indeed, specialize in particular types of offenses, and that offense specialization is strongest for drug offenses as compared to violent or property offenses (Mennis and Harris 2011).

The fourth, and final, principle underpinning this research is that drug offending among youth may be exported from high concentrated disadvantage neighborhoods to other nearby neighborhoods with relatively lower degrees of concentrated disadvantage through peer interaction across neighborhood boundaries. It stands to reason that if youth drug offending is influenced by youth living nearby, and that expressions of specialization of drug offending occur at the neighborhood level, that peer contagion from one neighborhood to another may be observed (Mennis et al. 2011). Importantly, we note that this may be the case even when one of the operative neighborhood mechanisms of juvenile delinquency, concentrated disadvantage, occurs only in the original neighborhood. Once specialization in drug offending among youth has been established in a disadvantaged neighborhood, it may be exported via social interaction to other nearby neighborhoods, even if those nearby neighborhoods are not necessarily disadvantaged.

Here, we address these theoretical principles by testing for neighborhood contagion in juvenile drug offense delinquency and recidivism while controlling for concentrated disadvantage, and other neighborhood characteristics that may be related to juvenile delinquency. For this purpose we use a spatial tessellation of neighborhood boundaries that optimizes within-neighborhood socioeconomic homogeneity. In this way, we aim to best capture, and distinguish between, mechanisms of drug offending that occur within each neighborhood versus those that occur as contagion between neighborhoods.

10.3 Data

Data on juvenile delinquency for Philadelphia, Pennsylvania for the period 1996–2003 were acquired from ProDES (Program Development and Evaluation System), a database of juvenile offenders assigned to court-order programs by the Family Court of Philadelphia, Pennsylvania. ProDES was developed by the Crime and

Justice Research Institute at Temple University under a contract with the City of Philadelphia to support program evaluation. ProDES collects data at the point of disposition, at program intake, at program discharge, and 6 months following program discharge to identify whether any new petitions have been filed (i.e. if the juvenile has committed a recidivating offense). This study focuses on the 7,323 male juveniles in ProDES who were in the system at least 6 months and therefore for whom we have data on recidivating offenses, if any. We removed females from the analysis because previous research, as well as our own preliminary analysis, indicates that the causes of delinquency differ substantially between girls and boys (Daigle et al. 2007).

Several outcome variables are derived from the ProDES database to capture rates of drug offense juvenile delinquency and recidivism, where recidivism is defined as the appearance of a new petition, i.e. offense, within 6 months of completion of the court-ordered program. To calculate these outcome variables, counts of youth age 10–20 years, delinquents with an instant drug offense (i.e. who entered the system initially due to a drug offense), and delinquents with a recidivating drug offense were summarized by neighborhood (the derivation of neighborhood boundaries are described below). We then calculated rates of delinquency, drug offense delinquency, and repeat drug offense. Table 10.1 provides definitions of the three outcome variables used in the study, and Table 10.2 provides summary rates for these variables that occur over the entire data set. Note that the delinquency rate is calculated as the proportion of all 10–20 year olds who are juvenile delinquents (of any offense type), whereas the drug offense delinquency rate is calculated as the proportion of juvenile delinquents who entered the Family Court system with an initial drug offense. The repeat drug delinquency rate is the proportion of juvenile delinquents with an initial drug offense who then re-offended with a second drug offense.

Neighborhoods were characterized using socioeconomic and crime variables collected from the U.S. Bureau of the Census (2000 Census) and the Philadelphia Police Department (for the period 2000–2002), respectively. Philadelphia contains 1,806 Census block groups, of which 32 were removed from the analysis because they were occupied almost exclusively by park land, industrial land, sports stadiums, or airports, (and which therefore have zero or very few people living within them), resulting in a data set of 1,774 block groups. The block groups range in total population from 16 to 3,988 with a mean of 855 and a standard deviation of 506. We chose a set of Census variables that capture neighborhood characteristics we hypothesize influence rates of drug juvenile delinquency and recidivism, as suggested by theories of social disorganization and concentrated disadvantage. These characteristics include race and racial segregation, residential mobility, isolation from 'mainstream' society due to linguistic or cultural barriers, unemployment and employment segregation with certain occupational categories as defined by the U.S. Bureau of the Census, educational attainment, and poverty. Table 10.1 provides definitions for all explanatory variables used in the analysis. Table 10.2 provides descriptive statistics for these variables.

Ethnic diversity within each block group was captured using a measure of local entropy calculated using the total number of Hispanic, non-Hispanic white, non-Hispanic African American, and non-Hispanic other minorities (i.e. non-Hispanics

Table 10.1 Variables used in the analysis

Category	Variable	Definition
Outcomes	Delinquency	Rate of juvenile delinquents for the population age 10–20
	Drug delinquency	Rate of juvenile delinquents entering the juvenile justice system with a drug offense among all juvenile delinquents
	Repeat drug delinquency	Rate of recidivating drug offenders among the juvenile delinquents who entered the Family Court system with a drug offense
Race	African American	Percent of population identifying as African American, non-Hispanic
	Other race	Percent of population identifying as other minority (non-African American), non-Hispanic
	Diversity	Entropy measure of within-unit racial diversity (see text for details of the calculation)
Residential mobility and cultural isolation	Renter	Percent of housing units occupied by renter
	Same house	Percent of population over the age of five living in the same house in 1995
	Linguistic isolation	Percent of households linguistically isolated (defined as when no person over the age of 5 years speaks English well)
	Foreign-born	Percent of population foreign born
Education	High school	Percent of population age 25 years or older with a high school diploma or equivalent
Employment	Employed	Percent of civilian population age 16 years or older employed
	Management	Percent of civilian population age 16 years or older employed in management, professional, and related occupations
	Construction	Percent of civilian population age 16 years or older employed in construction occupations (including trades)
	Services	Percent of civilian population age 16 years or older employed in service occupations
Poverty	Public assistance income	Percent of households receiving public assistance income
	Female-headed household	Percent of households that are female headed with children under 18
	Vacancy	Percent of housing units that are vacant
Crime	Violent crime	Per capita number of violent crimes (arrests for robberies, assaults, and homicides)
	Property crime	Per capita number of property crimes (arrests for burglaries, theft)

who are neither white nor African American). The value of diversity is calculated for each spatial unit j as

$$diversity_j = \sum_{i=1}^{n}\left[\left(\frac{P_{ij}}{P_j}\right)\ln\left(1/\frac{P_{ij}}{P_j}\right)\right]/\ln n \qquad (10.1)$$

Table 10.2 Descriptive statistics of block group-level and juvenile delinquency data

Block group-level variables (N=1,774)			
Category	Variable	Mean	SD
Race and ethnicity	Hispanic	9%	17%
	White	37%	37%
	African American	48%	39%
	Other race	6%	8%
	Diversity	0.41	0.26
Residential mobility and cultural isolation	Renter	34%	19%
	Same house	63%	16%
	Linguistic isolation	5%	8%
	Foreign-born	8%	10%
Education	High school	69%	16%
Employment	Employed	87%	12%
	Management	28%	18%
	Construction	7%	6%
	Services	22%	13%
Poverty	Public assistance income	11%	11%
	Female-headed household	14%	12%
	Vacancy	12%	9%
Crime	Violent crime	0.05 (per capita)	0.10 (per capita)
	Property crime	0.20 (per capita)	0.82 (per capita)
Overall Rates for Juvenile Delinquency Variables (N=7,323)			
		Rate	
Outcome	Delinquency	6%	
	Drug delinquency	24%	
	Repeat drug delinquency	25%	

Where P_{ij} is the proportion of population group i in spatial unit j (e.g. a block group) P_j is the population in spatial unit j, and ln n is the natural log of the number of population groups (here, n=4). Diversity varies from 0 to 1, with higher values indicating greater racial diversity (Apparicio et al. 2008). Philadelphia Police Department data yielded information on the street block locations of arrests for violent offenses (i.e. robbery, assault, or homicide) and property crimes (theft or burglary). Violent and property crime counts were summed for each block group and normalized by total population to yield block group-level per capita violent and property crime rate variables.

10.4 Defining Neighborhoods

10.4.1 Neighborhood Segmentation

As noted above, one of the major challenges in research addressing the influence of neighborhood characteristics on juvenile delinquency and other criminal behaviors is the definition of 'neighborhood' itself. Theoretically, the set of spatial units used in the analysis should reflect coherent neighborhoods within which the characteristics of that neighborhood are considered to exert some influence on the behavior under investigation. Practically, however, many studies simply rely on the boundaries of data enumeration for neighborhood definition, such as the common use of Census tracts, even though it is well established that such units often are a poor reflection of how neighborhood characteristics are spatially organized. We face a further challenge in defining neighborhoods in that we aim to calculate rates of drug offense juvenile delinquency and recidivism. Each neighborhood, therefore, ought to have enough juvenile delinquents present in order to calculate reliable rates.

To develop appropriate neighborhoods for analysis, we employ regionalization, the process by which a set of spatial objects are grouped together to form a set of spatially contiguous regions (Guo 2008). Typically, region-building is defined according to some objective function. Several types of approaches to regionalization have been proposed, including approaches that focus on clustering in attribute space followed by spatial aggregation (Haining et al. 1994), approaches that employ a heuristic algorithm based on iterative spatial partitioning (Openshaw 1977), and approaches that explicitly incorporate spatial constraints into the clustering algorithm itself (Assunção et al. 2006).

Here we employ a novel regionalization technique called REDCAP (Regionalization with Dynamically Constrained Agglomeration and Partitioning) developed by Guo (2008) that clusters a set of spatial units into a set of spatially contiguous neighborhoods composed of one or more of the original spatial units. This regionalization technique identifies neighborhoods by maximizing within-neighborhood homogeneity over a set of input spatial units with associated attribute values while maintaining spatially coherent (i.e. contiguous) neighborhoods. The technique also provides the option to partition neighborhoods in order to preserve a minimum within-neighborhood population.

REDCAP operates by first clustering the set of input spatial units based on their attribute values. A spatial constraint is then imposed on the spatial units to produce a hierarchical tree structure that encodes spatial adjacency relationships among units and thus facilitates the preservation of contiguity for hierarchically nested clusters. The second phase of the algorithm optimally partitions the tree into regions based on an iterative algorithm that minimizes the overall attribute heterogeneity that occurs over all the regions in the data set, where the heterogeneity of a single region is calculated as

$$H(R) = \sum_{j=1}^{d} \sum_{i=1}^{n_r} \left(x_{ij} - \overline{x}_j\right)^2 \tag{10.2}$$

Where $H(R)$ is the heterogeneity H of region R, d is the number of attributes, n_r is the number of spatial units in R, x_{ij} is the value for the jth attribute of the ith spatial unit, and $\overline{x}_j$ is the mean value for the jth attribute for all the spatial units in R (Guo 2008). The overall heterogeneity of a partition with k regions is simply the sum of all the regional heterogeneity values,

$$H_k = \sum_{j=1}^{k} H(R_j) \tag{10.3}$$

The technique provides alternatives for two parameterization settings for generating the initial tree structure. The first setting concerns the clustering method used, where the minimum distance in attribute space between regions is calculated using either the two nearest spatial units (single-linkage), the average distance of all spatial units (average-linkage), or the maximum distance between spatial units (complete-linkage). The second setting concerns the spatial constraint used to maintain spatially contiguous clusters, where single-order constraining considers the similarity only among adjacent spatial units between two clusters, while full-order constraining considers the similarity among all spatial units between each of two clusters.

We applied REDCAP to our block group-level explanatory variables (Table 10.1) using the complete-linkage clustering and full-order constraining, as these options were shown to provide the best overall quality in a comparative study (Guo 2008). In order to calculate reliable neighborhood-level rates of outcome variables in which the number of juvenile delinquents serves as the denominator, we enforced a requirement that all derived neighborhoods contain a minimum of 50 juvenile delinquents. We acknowledge that the threshold of 50 cases is somewhat arbitrary, however, given the overall rate of drug delinquency and repeat offending in the data set, having at least 50 delinquents within each neighborhood ensures substantial support for calculating these rates. Lower numbers of delinquents in a neighborhood may result in situations where the number of drug delinquents or repeat offenders is very low, problemetizing the derivation of reliable outcome variable rates. The regionalization yielded a set of 107 neighborhoods, which range in the count of member block groups from three to 116, with a median count of 11.5 block groups. The neighborhoods range in total population from 2,329 to 91,059 with a mean and standard deviation of 14,108 and 15,497, respectively. That the neighborhoods vary widely in the number of member block groups and total population indicates that the block groups themselves vary quite a bit in their number and rate of delinquents. This, of course, is precisely the problem we wish to solve by aggregating block groups together into neighborhoods, in order to ensure that each neighborhood has a sufficient number of delinquents for statistical analysis.

10.4.2 Creating Neighborhood-Level Variables

The 107 neighborhoods generated by the regionalization were used to recalculate the explanatory variables listed in Table 10.1 at the neighborhood-, rather than the block group-, level. Analogously, the outcome variables were also calculated at the neighborhood level by overlaying the post-program discharge address location of each juvenile delinquent with the newly created neighborhoods. Descriptive statistics for all neighborhood-level variables are presented in Table 10.3. Choropleth

Table 10.3 Descriptive statistics for neighborhood-level variables ($N = 107$)

Category	Variable	Mean	SD
Race and Ethnicity	Hispanic	14%	23%
	White	21%	27%
	African American	59%	36%
	Other race	5%	6%
	Diversity	0.45	0.27
Residential mobility and cultural isolation	Renter	35%	10%
	Same house	64%	9%
	Linguistic isolation	6%	7%
	Foreign-born	7%	7%
Education	High school	62%	12%
Employment	Employed	84%	7%
	Management	22%	9%
	Construction	7%	3%
	Services	25%	7%
Poverty	Public assistance income	15%	9%
	Female-headed household	18%	8%
	Vacancy	14%	7%
Crime	Violent crime	0.05 (per capita)	0.02 (per capita)
	Property crime	0.13 (per capita)	0.04 (per capita)
Outcome	Delinquency	8%	4%
	Drug delinquency	24%	10%
	Repeat drug delinquency	24%	12%

maps of the outcome variables are presented in Fig. 10.1. Clearly, the maps visually suggest that the highest rates of juvenile delinquency and drug offense delinquency tend to cluster in similar parts of Philadelphia. These areas can be identified as North Philadelphia (which, despite its name, is located approximately in the geometric center of the city extent) and West Philadelphia, which occupies the southwestern quadrant of the city.

Because several of the variables capture aspects of the same conceptual property (e.g. socioeconomic status, residential mobility), not surprisingly, many of our explanatory variables are highly correlated with one another. Such multicollinearity among explanatory variables can introduce substantial bias in regression results were they to be included together in a regression equation. We address this issue by employing a factor analysis with orthogonal rotation of all the explanatory neighborhood-level variables to yield a set of independent neighborhood-level factors. Such an approach also allows us to interpret the dimensions of variation within the context of conceptually defined causal mechanisms as considered through the lens of theories of social disorganization (Sampson et al. 1997).

From the 19 explanatory variables four factors were derived which explain 81% of the total variation. The factor loadings for each of the explanatory variables, the percent of total variation explained by each factor, and the interpretation of each factor are presented in Table 10.4. Choropleth maps of the factors are shown in Fig. 10.2. Generally, the factors appear to capture patterns of association between

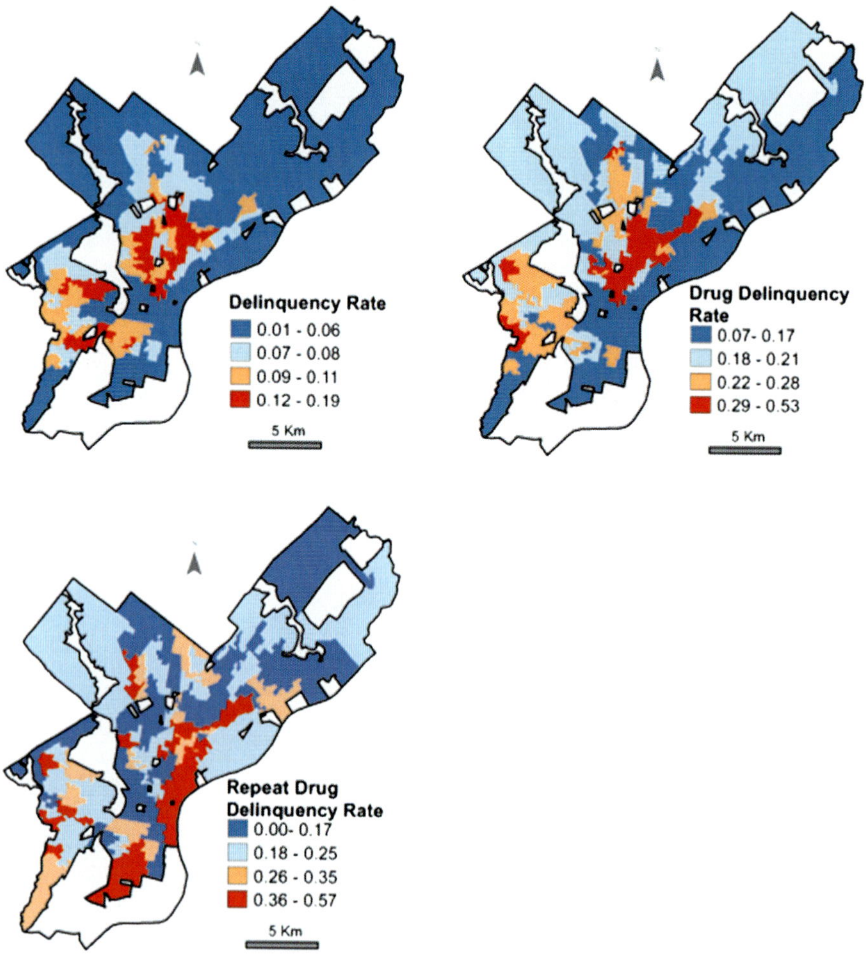

Fig. 10.1 Quantile-classified choropleth maps of outcome variables by neighborhood ($N=107$), where each variable mapped is expressed as a percentage (see text for definitions of variables): delinquency rate (*top left*), drug delinquency rate (*top right*), and repeat drug delinquency rate (*bottom left*)

race, socioeconomic status, and the transience of the population. Factor 1, which accounts for 34% of the variation, is clearly an indicator of concentrated disadvantage, reflecting characteristics of public assistance income, unemployment, female-headed households with children, vacant housing, low educational attainment, and violent crime. It is notable the highest values of factor 1 visually coincide with the highest rates of delinquency, as shown in Figs. 10.1 and 10.2.

Factor 2 is associated with *Hispanic* population, with a strong negative association with African American population. Employment is concentrated in construction occupations. Figure 10.2 shows that the neighborhoods with the highest values of factor 2 are clustered in Northeast Philadelphia, particularly in the southern portion

Table 10.4 Factor loadings for neighborhood-level variables ($N = 107$)

		Factors			
Category	Variable	1	2	3	4
Race and ethnicity	Hispanic	.48	.73	.15	.18
	White	−.69	.48	.12	.14
	African American	.25	−.83	-.36	-.24
	Other race	−.14	−.05	.94	.08
	Diversity	−.05	.46	.70	.30
Residential mobility and cultural isolation	Renter	.16	−.16	.04	.84
	Same house	.04	−.25	−.37	−.76
	Linguistic isolation	.38	.68	.46	.17
	Foreign-born	−.20	.16	.92	.03
Education	High school	−.87	−.40	.00	−.09
Employment	Employed	−.84	−.01	.19	−.27
	Management	−.79	−.24	.01	.38
	Construction	.02	.75	−.06	−.21
	Services	.68	−.46	.01	−.20
Poverty	Public assistance income	.89	.25	-.05	.19
	Female-headed household	.90	.05	−.04	.02
	Vacancy	.71	−.23	−.30	.34
Crime	Violent crime	.83	−.06	−.19	.37
	Property crime	.14	.16	.05	.74
Percentage of total variation explained		34%	18%	15%	14%
Interpretation		Disadvantage	Hispanic	Foreign	Mobility

of the northeast quadrant, which is well-known as the Hispanic hub of Philadelphia. Factor 3 reflects the *foreign-born* population, and is associated with racial diversity and linguistic isolation. Factor 4 is associated with *residential mobility*, as indicated by the high loadings of variables indicating staying in the same house for at least 5 years and proportion of renting households. Figure 10.2 shows the highest values occur in the downtown business district, and neighborhoods dominated by universities and students.

10.5 Analytical Strategy

The factors are entered as explanatory variables into Ordinary Least Squares (OLS) regressions of the outcome variables. We use the well known Moran's *I* (Moran 1948) to test for spatial dependency in the OLS model residuals (Lloyd 2007). In the current study, the spatial weights matrix is parameterized according to the commonly used rook's contiguity such that spatial units that share an edge boundary are considered neighbors.

If spatial dependency in the OLS model residuals is found, we hypothesize that the observed dependency is due to spatial contagion (diffusion) in the dependent variable, i.e. the delinquency outcome in one neighborhood affects the delinquency

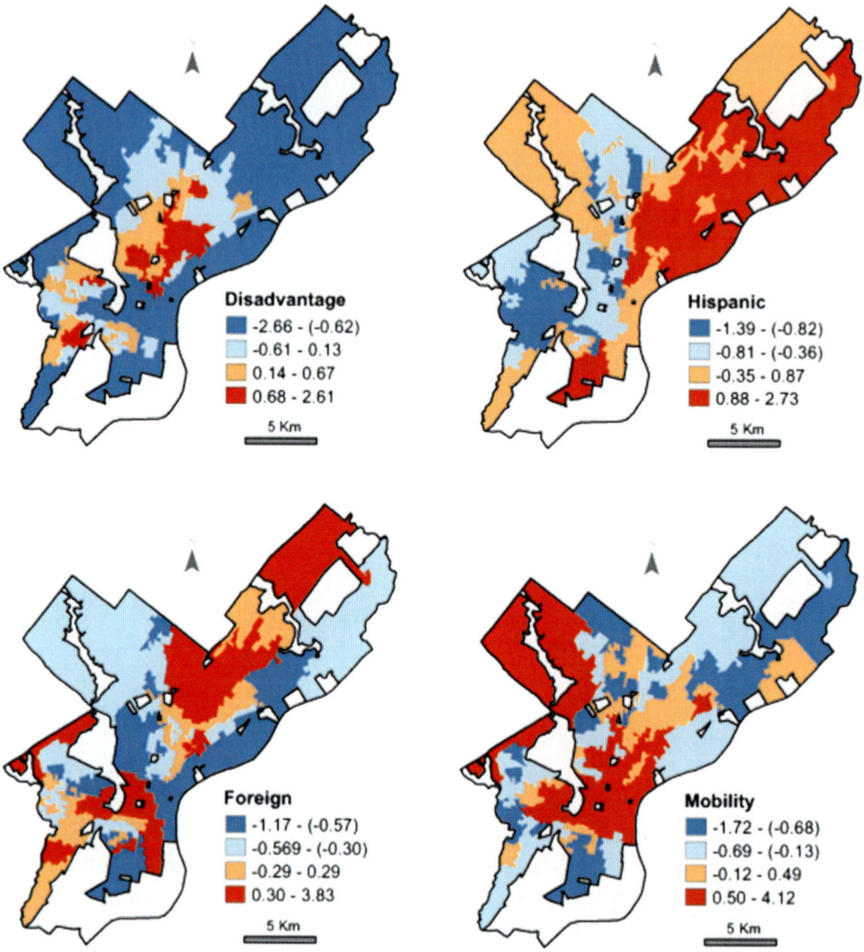

Fig. 10.2 Quantile-classified choropleth maps of explanatory variables, derived by factor analysis of Census variables listed in Table 10.4, by neighborhood ($N = 107$), where each variable mapped is a factor score: concentrated disadvantage (*top left*), Hispanic (*top right*), foreign-born (*bottom left*), and residential mobility (*bottom right*)

outcome in adjacent neighborhoods, even after accounting for the influence of other explanatory variables. We employ the Lagrange Multiplier (LM) statistics of the spatial error and spatial lag, as well as their robust forms, to investigate the nature of the spatial effects in the OLS models (Anselin et al. 1996). Significant values of the LM (lag) statistic, as compared to the LM (error) statistic, suggest that the nature of the spatial dependency in the model residuals are best accounted for using the spatial lag form of the spatial econometric model, expressed as

$$y = \rho Wy + X\beta + \varepsilon \tag{10.4}$$

where ρ is the coefficient of the spatial lag term of the dependent variable, and W is the spatial weights matrix. Maximum likelihood is employed to estimate the model (Anselin 1988). Again, we employ rook's contiguity to define the spatial weights, so that the spatial lag is calculated using adjacent spatial units. An alternative specification is the spatial error form of the spatial econometric model, where spatial dependency in the error term is accounted for. The equation for the spatial error model is similar to Equation 4, with the exception that the spatial lag term is derived not from the dependent variable but from the error term. The spatial error model is often used when the form of the spatial autocorrelation in the model residuals is considered to be a nuisance and the analyst is looking to account for spatial effects in order to better develop an unbiased model. The spatial lag model is more appropriate in cases where one hypothesizes spatial contagion as the mechanism for spatial dependency in the model residuals, where the process under observation is considered to 'spill over' from one spatial unit to another nearby spatial unit. Such is the case in the present study, where drug offending and delinquent behavior are considered to spread from adolescents in one neighborhood to adjacent neighborhoods through social contact.

10.6 Results

Table 10.5 presents the results of the OLS models for each of the outcomes. The model of delinquency indicates the strong positive influence of concentrated disadvantage on the juvenile delinquency rate, with the presence of foreign-born (non-Hispanic and non-African American) minorities having an additional suppressing effect. The explanatory power of this model is quite high, with these two variables explaining over 80% of the variation in delinquency rate. Drug offense delinquency is associated with concentrated disadvantage, the presence of Hispanics, and, to a lesser extent, residential mobility. The OLS model of repeat drug delinquency indicates the explanatory power is far less than for the other outcomes ($R^2=0.07$). Here, only concentrated disadvantage is significant.

An examination of the indicator for spatial autocorrelation indicates differences among the models. The Moran's I values for the OLS models of delinquency and repeat drug delinquency are not significant, indicating that the residuals are not spatially autocorrelated. This is not surprising for delinquency, given that most of the variation in this outcome is explained by concentrated disadvantage. On the other hand, the model for repeat drug delinquency offers relatively little explanatory power so the residuals are likely to be large and noisy.

The model of drug delinquency, however, does exhibit significant spatial dependency in the model residuals according to the Moran's I statistic. An assessment of the LM-error and LM-lag statistics reveals that the spatial lag model is more appropriate as compared to the spatial error model for modeling drug delinquency. Although both the lag and error forms of the LM statistic are significant,

Table 10.5 Results of OLS regression of outcomes ($N = 107$)

Explanatory variables	Delinquency	Drug delinquency	Repeat drug delinquency
Disadvantage	0.032***	0.061***	0.030**
Hispanic	−0.002	0.026***	0.015
Foreign	−0.010***	−0.009	0.015
Mobility	0.002	0.018*	0.010
Constant	0.083***	0.236***	0.242***
Adjusted R^2	0.83	0.46	0.07
AIC	−581.93	−251.19	−154.92
Moran's *I*	1.28	2.87***	0.25
LM (lag)		11.89***	
Robust LM (lag)		9.62***	
LM (error)		5.09*	
Robust LM (error)		2.82	

Reported values are standardized coefficients; $*p<0.05$, $**p<0.01$, $***p<0.005$

Table 10.6 Results of spatial lag regression of drug delinquency ($N = 107$)

Explanatory variables	Drug delinquency
Disadvantage	0.042***
Hispanic	0.015*
Foreign	−0.002
Mobility	0.009
Spatial lag	0.434***
Constant	0.135***
R^2	0.56
AIC	−261.94

Reported values are standardized coefficients (except for Spatial Lag, which is unstandardized); $*p<0.05$, $***p<0.005$

the robust forms of the LM statistic show that the LM (error) statistic becomes insignificant when the spatial lag is included in the model. This supports our hypothesis regarding the role of spatial diffusion of drug offending behavior across neighborhood boundaries as a mechanism for increasing drug offense delinquency.

We estimated a spatial lag model for the drug delinquency outcome. Results are presented in Table 10.6, which shows that the spatial lag term is highly significant. In addition, the model fit, as indicated by the *AIC*, shows an improvement over the analogous OLS model. The Moran's *I* statistic of the model residuals indicates the absence of spatial dependency in the spatial lag model. Interestingly, the presence of the spatial lag term causes the residential mobility variable to become non-significant, suggesting that the influence of this variable is an artifact of contagion of drug offense delinquency from nearby neighborhoods.

10.7 Discussion

These results clearly show the strong influence of concentrated disadvantage on neighborhood rates of juvenile delinquency, drug offense delinquency, and repeat drug offending. It is the only explanatory variable that is significant for all three outcomes. These findings support the influence of concentrated disadvantage on juvenile delinquency generally and drug offense delinquency specifically. Even repeat drug offending, i.e. the rate of juvenile drug offenders who recidivate with another drug offense, is enhanced by higher concentrated disadvantage. From the data available here we are unable to interpret the nature of this mechanism, whether it is due to economic motivation of youth due to poverty, a lack of collective efficacy at the neighborhood level, some combination of the two, or some other aspect of concentrated disadvantage, but these results provide strong evidence that concentrated disadvantage encourages the occurrence and persistence of drug offending among youth.

The high goodness of fit of the model predicting juvenile delinquency is also noteworthy. Delinquency generally is very closely related to concentrated disadvantage, so much so that nearly all the neighborhood-level variation in delinquency can be ascribed to variation in concentrated disadvantage. Compare this goodness of fit to models of drug delinquency and repeat drug delinquency. Though concentrated disadvantage is still significant for these two other outcomes the goodness of fit is substantially lower, suggesting that other mechanisms besides those associated with concentrated disadvantage are at work. The explanatory power of the model for repeat drug delinquency is particularly weak, and a look at the maps of the three outcome variables in Fig. 10.1 indicate that repeat drug delinquency bears no visually obvious relationship with either delinquency or drug delinquency. Given that individual juvenile drug offenders are more likely to offend with a drug offense than another type of offense (Mennis and Harris 2011), the lack of coherence between spatial patterns of drug offending and repeat drug offending is curious. It may be that that the low number of repeat drug offenders, as compared to the entire juvenile delinquent population, results in a noisy variable.

Besides concentrated disadvantage, the foreign-born explanatory variable is also significant in the model of juvenile delinquency, where communities with high levels of foreign-born tend to have lower rates of delinquency. We note that the foreign-born factor is likely capturing Asian communities (primarily Chinese and southeast Asian), which after whites and African Americans comprise the largest racial minority in Philadelphia, as the 'other race' variable (non-white, non-African American, and non-Hispanic) loads very highly and positively on this factor. At least in Philadelphia, juvenile delinquents are less likely to be found in such Asian communities. Neighborhoods with high Hispanic population, on the other hand, are associated with higher rates of drug delinquency. We note that the largest concentration of drug arrests (both adult and juvenile) in the city is located in the dominant Hispanic neighborhood (Mennis et al. 2011), and the association of the drug trade within this particular Hispanic community within Philadelphia is well-documented (ABC News 1995).

It is notable that spatial autocorrelation was observed in the model residuals for the regression of drug delinquency, but not for delinquency in general nor for repeat drug delinquency. This suggests that contagion of delinquent behavior from one neighborhood to other adjacent neighborhoods occurs for drug offending, but not for general delinquency. In addition, the tendency of delinquents in a neighborhood to repeat an initial drug offense with another drug offense does not appear to be influenced by such behavior in adjacent neighborhoods. The fact that such a high proportion of the variation in juvenile delinquency is explained by concentrated disadvantage (and foreign-born) indicates that there is little residual variation in delinquency left to explain by other mechanisms, such as spatial contagion. These results suggest that for rates of delinquency generally, within-neighborhood mechanisms are paramount. For drug delinquency, however, there is evidence that even after accounting for within-neighborhood characteristics, drug delinquency in one neighborhood influences the rate of drug delinquency in adjacent neighborhoods.

These findings support the idea that not only is drug offending in a neighborhood enhanced by concentrated disadvantage but also that such drug offending in a disadvantaged neighborhood can be exported to other youth in adjacent neighborhoods. As noted above, we attribute this process of neighborhood contagion to social contact among youth that promotes offense specialization. Such offense specialization is then expressed at the neighborhood level and spread to other neighborhoods through social contact among juveniles living nearby one another. Undoubtedly, the instrumental nature of drug selling contributes to its appeal (Baumer and Gustafson 2007).

Of course, drug selling differs in one significant way from other forms of delinquency: rather than involving a victim, it requires an exchange, in which both parties participate in an illegal act. If drug selling is viewed as a seller-buyer exchange, then the spread of drug selling to adjacent neighborhoods may be market driven. That is, if drug buyers reside in more affluent neighborhoods, then the business of drug selling may benefit from expanding its market to locations more attractive to buyers.

Interestingly, the spatial contagion term in the model of drug delinquency renders the residential mobility variable non-significant and reduces the influence and significance of the Hispanic variable. This implies that certain neighborhood characteristics that may be associated with drug offenses, such as ethnicity and instability in housing tenure, may be artifacts of spatial contagion. This is particularly interesting for interpreting the association between Hispanic and drug delinquency, as it suggests that the mechanism for the association between Hispanic ethnicity and drug offending is a proxy for a process of social interaction that spreads drug offending from neighborhood to neighborhood.

10.8 Conclusion

There are certain limitations to the present analysis that we acknowledge here. First, though we have used a novel regionalization method to develop optimal spatial units for analysis, the use of aggregated data limits our ability to make inferences about how neighborhood and social contextual mechanisms influence individual juveniles to engage in delinquent behavior, commit drug offenses in particular, or engage in repeat drug offending. By using rates of these outcomes within neighborhoods, we are able to model neighborhood-level mechanisms and contagion, but it is important to recognize that the causal processes of juvenile drug offending in which we are actually interested occur at the individual level.

We have made some headway in this regard by employing multilevel modeling to disentangle the influence of neighborhood disadvantage versus social capital on drug offense recidivism for individuals (Grunwald et al. 2010). Unfortunately, however, capturing the social interaction among juveniles is more challenging, as such social network data are typically collected using surveys specifically designed to extract information on the presence and strength of social contacts. However, the ProDES database does contain embedded information from which we can infer possible social interaction among juveniles. For instance, the database contains information on which juveniles co-offended or overlapped in the same treatment program at the same time. And, of course, we can calculate the distance between juveniles' residences to proxy likely neighborhood-based social interaction – for instance, where juveniles living on the same street block likely have at least a minimal level of social interaction as compared to juveniles living farther apart. In future research we plan to extract and use such information to develop more sophisticated models that incorporate social network effects into models of delinquency and recidivism.

Another limitation concerns the use of data relating to petitions in Family Court as a proxy for delinquent activity. It is certainly possible that juveniles engage in criminal activity for which they are not arrested. For example, a juvenile may be engaging in theft and vandalism but get caught for selling drugs. Thus, the offense for which a petition is filed may not be an accurate reflection of a juvenile's actual criminal profile. Again, however, a more detailed view on the criminal activity of such a large number of juveniles would necessitate a substantial prospective study that is beyond the scope of the present research. We note that the administrative data set used here provides a record of juvenile criminal activity that, while not perfect, is comprehensive in that it covers all juveniles committing crimes that warrant Family Court disposition to court-ordered programs over a period of time.

Methodological issues regarding the modifiable areal unit problem (MAUP) and the ecological fallacy also warrant consideration. In this study we sought to develop a tessellation of socioeconomically homogeneous neighborhoods under the constraint of a minimum sample of 50 cases within each neighborhood. However, it is well known that changing the tessellation of spatial units for spatially aggregated data, i.e. changing either the scale of aggregation or the partitioning scheme at a single scale, can alter the results of statistical analysis (Openshaw 1984; Fotheringham and Wong 1991). Theoretically, if the scale of analysis were made to be substantially coarser (i.e. aggregation to a lesser number of neighborhoods), the spatial contagion effect found in our study may disappear, as both advantaged and disadvantaged neighborhoods would be subsumed together within a single spatial unit. In a sense, we have sought to address this component of the MAUP by developing socioeconomically homogeneous spatial units, so that the spatial contagion effects we hypothesize are occurring can be detected. However, we acknowledge that the number of possible neighborhood tessellations is so great that the impact of the choice of specific neighborhood tessellation is unknown.

The ecological fallacy concerns the presumption that observations made for aggregated data apply to the individual level (Openshaw 1984). We interpret the process of spatial contagion of drug offending to occur from disadvantaged neighborhood to adjacent neighborhoods through the process of peer interaction. Such a mechanism assumes that disadvantaged juveniles in the disadvantaged neighborhood are influencing more advantaged juveniles who live in a different neighborhood nearby. However, each neighborhood is likely to have a mix of advantaged and disadvantaged juveniles. If, indeed, poverty itself is driving delinquency and drug offending, then it follows that the disadvantaged youth within more advantaged neighborhoods would purse drug offending, which could also produce the statistical spatial effects we found in this study. Unfortunately, without utilizing indicators of disadvantage at the individual level, we cannot distinguish the nature of spatial contagion, whether it is occurring among disadvantaged youth or between disadvantaged and more advantaged youth.

Despite these limitations, this research lends substantial support to theories of neighborhood and social contextual mechanisms of juvenile drug offending. Rates of delinquency, drug offense delinquency, and repeat drug offense delinquency are enhanced in neighborhoods with high concentrated disadvantage. Additionally, drug offense delinquency is subject to neighborhood contagion, where we speculate that drug offending is a learned behavior that juveniles in one neighborhood acquire from juveniles in adjacent neighborhoods though social interaction. We note that drug offending can be 'exported' in this manner from disadvantaged neighborhoods to adjacent, lesser-disadvantaged neighborhoods. Such mechanisms should inform interventions for reducing drug offending among youth that focus on neighborhood and social contextual mechanisms.

Drug prevention policies and programs focus almost exclusively on drug use; rarely is drug selling mentioned in the context of preventing drug involvement among young people. This exclusive focus on 'demand side' strategies rests on a large body of research that identifies motivations and risk factors found among drug-abusing youth and a policy history that pits images of criminal organizations and gangs against vulnerable and sometimes troubled children. Drug selling, however, can be an easy means of gaining income and status in disadvantaged neighborhoods. Additionally, youth with entrepreneurial skills and interests may provide a means for drug selling networks to spread to adjacent, more advantaged neighborhoods (Little and Steinberg 2006). Prevention strategies are needed that recognize and seek to counteract the appeal of drug selling, particularly in neighborhoods where legitimate sources of income and models of job-related success are scarce.

Acknowledgements This project was supported by a Summer Research Award by Temple University and Award No. 2006-IJ-CX-0022 awarded by the National Institute of Justice, Office of Justice Programs, U.S. Department of Justice. The opinions, findings and conclusions or recommendations expressed in this publication are those of the authors and do not necessarily reflect the views of the Department of Justice.

References

ABC News (1995) The badlands: death of an American neighborhood. http://abcnews.go.com/Nightline/video/badlands-death-american-neighborhood-10033816. Accessed 19 July 2011

Agnew R (1994) Delinquency and the desire for money. Justice Q 11(3):411–428

Akers RL (1998) Social learning and social structure: a general theory of crime and deviance. Northeastern University Press, Boston

Anderson E (1999) Code of the street: decency, violence and the moral life of the inner city. WW Norton, New York

Anselin L (1988) Lagrange multiplier test diagnostics for spatial dependence and spatial heterogeneity. Geogr Anal 20(1):1–17

Anselin L, Bera A, Florax RJ, Yoon M (1996) Simple diagnostic tests for spatial dependence. Reg Sci Urban Econ 26:77–104

Apparicio P, Petkevitch V, Charron M, (2008) Segregation analyzer: A C#.Net application for calculating residential segregation indices. Cybergeo Eur J Geogr 414:27

Assunção RM, Neves MC, Câmara G, Freitas CDC (2006) Efficient regionalization techniques for socio-economic geographical units using minimum spanning trees. Int J Geogr Inf Sci 20:797–811

Bandura A (1996) Social cognitive theory of human development. In: Husen T, Postlethwaite TN (eds) International encyclopedia of education, 2nd edn. Pergamon Press, Oxford

Baumer EP, Gustafson R (2007) Social organization and instrumental crime: assessing the empirical validity of classic and contemporary anomie theories. Criminology 45:617–664

Bursik RJ Jr (1988) Social disorganization and theories of crime and delinquency: problems and prospects. Criminology 26:519–552

Cloward RA, Ohlin LE (1960) Delinquency and opportunity. Free Press, Glencoe

Daigle LE, Cullen FT, Wright JP (2007) Gender differences in the predictors of juvenile delinquency. Youth Violence Juv Justice 5:254–286

Fotheringham SA, Wong DW (1991) The modifiable areal unit problem in multivariate statistical analysis. Environ Plan A 23:1025–1044

Grunwald H, Lockwood B, Harris P, Mennis J (2010) Influences of neighborhood context, individual prior history and parenting behavior on recidivism among juvenile offenders. J Youth Adolesc 39:1067–1079

Guo D (2008) Regionalization with dynamically constrained agglomerative clustering and partitioning (REDCAP). Int J Geogr Inf Sci 22(7):801–823

Haining RP, Wise SM, Blake M (1994) Constructing regions for small area analysis: material deprivation and colorectal cancer. J Public Health Med 16:429–438

Little M, Steinberg L (2006) Psychosocial correlates of adolescent drug dealing in the inner city – potential roles of opportunity, conventional commitments, and maturity. J Res Crime Delinq 43(4):357–386

Lloyd CD (2007) Local models for spatial analysis. CRC Press, New York

Matsueda RL, Gartner R, Piliavin I, Polakowski M (1992) The prestige of criminal and conventional occupations: a subcultural model of criminal activity. Am Sociol Rev 57(6):752–770

Mennis J, Harris P (2011) Contagion and repeat offending among juvenile delinquents. J Adolesc 34:951–963

McCord J (2003) Cures that harm: unanticipated outcomes of crime prevention programs. Ann Am Acad Pol Soc Sci 587:16–30

Mennis J, Mason MJ (2011) People, places, and adolescent substance use: integrating activity space and social network data for analyzing health behavior. Ann Assoc Am Geogr 10(2):272–291

Mennis J, Harris P, Obradovic Z, Izenman A, Grunwald H, Lockwood B (2011) The effect of neighborhood characteristics and spatial spillover on urban juvenile delinquency and recidivism. Prof Geogr 63(2):174–192

Moran PAP (1948) The interpretation of statistical maps. J R Stat Soc Ser B 10:243–251

Nguyen H, Bouchard M (2012) Need, connections, or competence? Criminal achievement among adolescent offenders. Justice Q DOI: 10.1080/07418825.2011.589398

Openshaw S (1977) A geographical solution to scale and aggregation problems in regionbuilding, partitioning, and spatial modelling. Trans Inst Br Geogr 2:459–472

Openshaw S (1984) The modifiable areal unit problem. Concepts and techniques in modern geography, vol 38. Geo Books, Norwich

Sampson RJ, Raudenbush SW, Earls F (1997) Neighborhoods and violent crime: a multilevel study of collective efficacy. Science 227(5328):918–924

Sampson RJ, Morenoff JD, Earls F (1999) Beyond social capital: spatial dynamics of collective efficacy for children. Am Sociol Rev 64:633–660

Shaw CR, McKay HD (1942) Juvenile delinquency and urban areas. University of Chicago Press, Chicago

Sutherland EH, Cressey D (1992) Principles of criminology, 11th edn. AltaMira Press, Lanham

Twigg L, Taylor J, Mohan J (2010) Diversity or disadvantage? Putnam, Goodhart, ethnic heterogeneity, and collective efficacy. Environ Plan A 42:1421–1438

Valente T (2003) Social networks influence on adolescent substance use: an introduction. Connections 25(2):11–16

Warr M (1996) Organization and instigation in delinquent groups. Criminology 34(1):11–37

Warr M (2002) Companions in crime: the social aspects of criminal conduct. Cambridge University Press, Cambridge

Chapter 11
Geospatial Modeling and Simulation of Property Crime in Urban Neighborhoods: An Example Model with Foreclosure

Jay Lee and Ronald E. Wilson

Abstract Based on neighborhood life cycles, this paper describes the development and the functions of an Urban Crime Simulator (UCS) that are based on a concept that neighborhood goes through cycles from newly established and energetic neighborhoods to matured and stabilized ones and then to deteriorated neighborhoods that await for new stimuli for revitalization. The UCS was developed to estimate changes in property crime rates as induced by changes in the socioeconomic characteristics of urban neighborhoods. UCS is fully integrated with geographically referenced data and is operational in GIS environment. It offers flexibility in the inclusion of neighborhood attributes that may best fit a specific localized context and knowledge of local neighborhoods and neighborhood attributes as suggested by criminological literature.

With UCS, urban neighborhoods are profiled by a selected set of attributes as defined by users. These neighborhoods are first classified into clusters by a hierarchical cluster analysis, which minimizes in-cluster differences and maximizes between-cluster differences. When attribute values of a target neighborhood are updated with projected or planned changes, UCS searches the entire area to find a reference neighborhood with an attribute profile that is the closest to that of the target neighborhood. Once the reference neighborhood is found, all neighborhoods in the reference neighborhood's cluster are statistically analyzed to yield an estimate for what a new crime rate may be for the target neighborhood with the projected changes.

J. Lee (✉)
Department of Geography, Kent State University, Kent, OH, USA
e-mail: jlee@kent.edu

R.E. Wilson
Department of Housing and Urban Development, Washington, DC, USA
e-mail: ronald.e.wilson@hud.gov

M. Leitner (ed.), *Crime Modeling and Mapping Using Geospatial Technologies*, Geotechnologies and the Environment 8, DOI 10.1007/978-94-007-4997-9_11,

UCS has a set of tools to assist its users. Correlation among included attributes can be easily calculated to detect if there is any issue of co-linearity. Global and localized spatial autocorrelation can be calculated to evaluate if any spatial dependency among their data would cause any concern in the simulations. Finally, global and localized regression models enable UCS users to assess how appropriate the selected attributes are with respect to explaining the variation in crime rates among the neighborhoods.

UCS is software designed for practical use by law-enforcement agencies that may not be able to take the necessary time to assemble a detailed comprehensive database as other modeling approaches require before carrying out such simulations.

Keywords Geospatial modeling • Simulation • Urban property crime • Place-based analysis • Neighborhood life cycles

11.1 Introduction

Expanding on the functionality of an Urban Growth Simulator (UGS, by Kent State University Applied Geography Laboratory with funding from the US Environmental Protection Agency in 1996) that was developed for simulating how urban neighborhoods grow under various growth management strategies, an extension was developed to explore how changes in urban settings affect neighborhood crime rates. Known as the Urban Crime Simulator (UCS), the extension is a set of software tools that uses GIS datasets of a local environment. The primary purpose for developing UCS is to allow city agencies a tool for estimating the degree of changes in crime rates based on their knowledge of upcoming changes in their urban neighborhoods.

Users of UCS are not required to assemble a specific form or combination of data layers to carry out simulations as many similar software packages require. Users of UCS can work with existing local datasets readily available to city agencies, such as specialized socioeconomic and demographic attributes at a micro-level of geographic detail. In addition, UCS users can select various combinations of neighborhood attributes that are more relevant to the local jurisdiction for tailored simulations to better reflect the user's own knowledge of their localities. Repeated simulations with different combinations of neighborhood attributes can assist in the validation of theories related to research that combines criminological and geographical theories. Since the combination of theories typically requires a custom set of variables at multiple geographical scales, a simulator can aid in the exploration of variable and scale combinations for many possible outcomes that may be used to develop policies for encouraging or containing urban growth.

UCS rests on a theoretical foundation that fuses a general neighborhood life cycle model together with theories of crime that have a geographic aspect. Geographical theories of crime emanated from the work of Patricia and Paul Brantingham on crime pattern theory that later became the basis for the sub-discipline of Environmental Criminology (Brantingham and Brantingham 1975). Brantinghams' seminal work created a framework that facilitates the examination of interactions between criminals and the geographic contexts in which they operate.

Geography enhances environmental criminology by specifically offering theories and principles about the changing urban landscape (spatial organization) and proximity effects between adjacent places (spatial dependency). Geography provides an understanding of why and how neighborhoods develop when disciplines that study individual behavior, such as criminology, do not. Combining criminology and geography theories for understanding neighborhood development provides an opportunity for crafting public policy that simultaneously addresses both individual and environmental factors. With UCS, local jurisdictions have the flexibility to whatever local data is available to examine urban change that affects crime and avoid having to conduct analysis with generalized data that may not accurately depict local conditions. Thus, the ability UCS has for exploiting local data to simulate the possible outcomes of urban change becomes immensely attractive to avoid making costly decisions that can emerge from unforeseen urban development consequences.

Growth and development has long been a source of contention amongst planners, policy makers, and government officials who are charged with solving problems related to urban change. Since UCS is designed to facilitate the exploration of crime problems associated with geographic change, this chapter will use a concrete example of simulating changes in property crime as a response to increased foreclosure rates within Census block groups.

The UCS has three main tools sets. First, the backbone of the simulation process in UCS is the automatic situating of each areal unit within a neighborhood life cycle stage from which to project change. Second, UCS extends this facet with a suite of spatial statistical tools to detect any degree of spatial dependency between adjacent areal units that might affect geographic change within any given neighborhood. Finally, UCS provides regression-based (classical and spatial) modeling tools for assessing the variation between a dependent variable that is explained by the variation in the user-defined set of independent variables for both global and localized trends.

The objective of this work is to provide a set of tools for city administrators, planners, and law enforcement agencies to examine the possible impacts from urban changes on current crime patterns and be more proactive toward preventing the emergence of new patterns. The analysis in the chapter will be limited to estimating property crimes as a starting point for two reasons. First, these crimes are highly structured geographically to land use related to where foreclosures occur. Second, property crimes are more prevalent with far higher rates than violent crime providing a larger sample for statistical projection.

11.2 Concepts and Theories of Neighborhood Change and Crime

Crime can be a change agent that can send neighborhoods into decline. Increased crime may also drive quality businesses away. Conversely, the reduction of crime can lead to economic investment and revitalization that bring positive changes.

According to crime pattern theory, opportunities for crime emerge, disappear or move as the geography changes across the urban landscape (Brantingham and Brantingham 1975). Crime is the result of underlying economic and demographic factors mixed with geographic environments that attract or repel crime (Wilson 2011), which makes geographical analysis an efficient way to analyze crime. Crime is far more predictable by place of occurrence than by individual offenders (Crow and Bull 1975; Roncek and Bell 1981; Sherman et al. 1989; Sherman and Weisburd 1995; Weisburd et al. 2004). This improved predictability is because once land is developed, buildings are erected, utility lines are installed, and transport systems are configured, a place becomes extraordinarily difficult to alter and does not change quickly. Any change that does occur is part of the general life cycle of the place, *e.g.*, a neighborhood.

The use of a neighborhood life cycle model presents a number of practical options toward simulating crime patterns based on changes in a place. Because many crimes have a spatial structure and form cohesive patterns (clusters) related to the geographic context they occur, place-based approaches can be efficient in analyzing change (Wilson 2007). For example, general property crimes and residential burglaries primarily happen in residential areas. Commercial robberies, likewise, mostly transpire in commercial districts. Auto thefts occur more frequently in places with large amounts of parking that are difficult to monitor. Homicide tends to occur across larger areas where poverty, inequality, physical deterioration, and economic decline have long been established across a number of neighborhoods.

Criminology has many mature geographic theories of crime that was borne out of Environmental Criminology with a rich set of empirical results about the interactions between criminal activity and place. With computing power and robust programming tools available today computational criminology has emerged to provide a framework for dynamically testing theories of crime activity across varying geographies that are otherwise impossible (Brantingham and Brantingham 1981, 2004; Bottoms and Wiles 2002). The primary objective of computational criminology is the development of quantitative approaches to analyzing and predicting crime incidences or distribution patterns by incorporating algorithms, data structures and software tools for geographic simulation. We take advantage of computational criminology to integrate the Deviant Places theory with the neighborhood life cycle concept as the basis for the design and development of a simulation procedure in the UCS.

11.3 Theory of Deviant Places

The theory of deviant places posits that high rates of crime and deviance persist in specific neighborhoods through a combination of population and ecological factors (Stark 1987). This theory weaves human and environmental interactions together in specific places that offer criminal opportunities. Stark set forth a series of 30 propositions that formed the basis of Deviant Places Theory that speculate a series of place-based changes that works to sustain crime. He identified five essential ecological factors that characterize deviant places — density, poverty, mixed land use, transiency and dilapidation. These five factors vary and change as a neighborhood changes along its life cycles.

Stark also cited a set of impacts and amplifiers of how people would respond to any negative ecological change that leads to moral and social decay, and inevitably increased crime. Increased crime, then, would lead to subsequent decreased levels of economic investment, quality of housing stock, quality businesses and of social services. This recursive decline in the neighborhood and increased crime ultimately leads to the establishment of a deviant place.

Deviant Places will persist because they are most likely populated by those in the "underclass" (Wilson 1987) who have little means or incentive to alter their lifestyle, intervene in incivilities, invest in their properties, or engage in the political process to improve their neighborhoods. Unless economic circumstances change, new investments are made, and the political will is exercised to revitalize the neighborhood, they will remain deviant places and crime will likely persist.

11.4 Foreclosures, Neighborhood Change and Crime

Over the past decade, many metropolitan areas have experienced extensive expansion of new or revitalized housing. Following this expansion was a massive amount of foreclosures from the collapse of the housing market caused by overzealous predatory lending and mortgage fraud (Immergluck and Smith 2005; Kaplan and Sommers 2009; Crossney 2010) and mortgage fraud (Fulmer 2009).

Foreclosure is a process which most often culminates in the removal of residents from their house from a mortgage default (Gilberto and Houston 1989). Foreclosures can have a direct, negative impact on neighborhoods. Furthermore, concentrated foreclosures can have a cumulative negative effect on housing values over an expanded geographic range if they remain vacant for an extended period of time (Simmons et al. 1998; Schuetz et al. 2008). With each foreclosure, the value of neighboring properties is estimated to decrease between 0.9 and 8.7% in value (Pennington-Cross 2004; Immergluck and Smith 2005; Lee 2008). The recognition of foreclosure as a process is critical to understanding the onset of crime from neighborhood change of this type. Although a small number of foreclosed properties might not completely initiate neighborhood change, their presence can still create opportunities for crime and disorder.

With concentrated foreclosures, there are several pathways to property degradation that can form potentially new geographies of opportunity for crime. Mass and clustered foreclosures can accelerate stage changes in neighborhoods where there is little or no population. This often leaves the properties vulnerable to degradation. Once a neighborhood downgrades a stage, crime begins to move in and acts as a reinforcing factor to hasten and deepen neighborhood decline. Houses that were once in good condition deteriorate from neglect, vandalism, weathering, or are not maintained. Theft, drugs, vandalism, vagrancy, prostitution, and arson can all move in behind a wake of foreclosures. These crimes are only the immediate impact and these neighborhoods can fall into disrepair. Properties in the neighborhoods become destitute, leaving them undesirable. Residents still remaining in these neighborhoods of abandoned properties face an increased risk of burglary, robbery or other violent crime. The long term impact could undo the significant progress that many metropolitan areas have made in the last few decades in improving the quality of life and making economic progress. Damage to these properties and the lack of reasons for attracting residents who will and can afford to repair or upgrade them lead to neighborhood decline toward becoming an impoverished neighborhood (Galster 2005).

Neighborhood change from foreclosures can be connected to the concept of the neighborhood life cycle (Metzger 2000). The neighborhood life cycle model posits a natural and predictable series of stages that neighborhoods go through as time transpires. All neighborhoods are settled within some stage of the life cycle and a complex interplay of changes in social, economic, ecological, and political characteristics that facilitates progression into another stage. This process is known as "filtering" and is the basic dynamic of the neighborhood life cycle concepts (Lowry 1960; Birch 1972; Yeates and Garner 1976; Winsberg 1989; Baxter and Lauria 2000). Exactly where a neighborhood is, and how fast it changes stages depends on numerous factors related, but not limited, to land use, housing stock, property conditions, population demographics, infrastructure maintenance, service provisions, school quality and economic conditions.

Many social science disciplines have used the neighborhood life cycle to explain demographic change (Hoover and Vernon 1939; Park 1952; Duncan and Duncan 1957; and Taeuber and Taeuber 1965; Baxter and Lauria 2000). Identifying what stage of the life cycle a neighborhood is in is important for establishing a base-line for comparison. From these stages changes in crime can be evaluated in relation to the impact of foreclosures through the life cycle model. As such, the neighborhood life cycle forms the foundation of the UCS by assuming that neighborhoods are moving into another stage of the model by matching attribute profiles to another neighborhood for prediction based on a shared range of characteristics.

11.5 A Hypothetical Example of Neighborhood Life Cycle Change and Crime

To help illustrate the combination of the neighborhood life cycle model and aspects of Deviant Places theory, we outline from Wilson and Paulsen (2010) a hypothetical set of changes in crime and neighborhood characteristics that may occur from a concentration of foreclosures.

Stage 1: *Healthy*. A neighborhood exhibits signs of well kept housing stock that is predominantly owner occupied but not crowded. These neighborhoods are usually new or revitalized. Income levels are similar, ranging between moderate to upper level. Properties are well maintained, damage is quickly repaired when it occurs, and upgrades to the housing structure are common. There is little noticeable crime except for minor incidents, such as juveniles acting out that range from simple vandalism to trespassing to theft from garages/yards. These crimes do not alarm the residents or prompt them to take much action.

Stage 2: *Incipient Decline*. Several houses in the neighborhood are abandoned from foreclosures. These houses begin to degrade and show variation in blight, leading to changing levels of attractiveness in the neighborhood. Real estate companies, banks, or private investors begin purchasing these properties to keep them occupied, maintained, and to make money on their holdings. Transiency becomes common from renters moving in and out. With renters, income levels of the neighborhood drop. Many renters are minorities with lower education levels. Longtime residents become concerned of the racial transition taking place. Vandalism, trespassing and thefts from garages, yards or automobiles become common with an escalation to instances of fights or underage drinking. Renters invest little in maintaining and upgrading the properties because they have no vested interest or funds for doing so. Landlords also invest little in hopes of maximizing their profits. Property values have declined to allow lower income buyers move in to the neighborhood. Burglary and/or thefts of vehicle happen for the first time. Concern about the neighborhood emerges and remaining residents consider the possibility of becoming victimized and think about selling their properties.

Stage 3: *Clearly Declining*. Houses that have remained vacant now require major repairs such as new roofs, siding, or windows. Routine maintenance has not been as common in the rented or newly purchased houses. Damages and neglect have left the neighborhood looking less attractive. Housing or property code violations occur for the first time or are mounting at particular addresses. A tipping point is being reached between the balance of owner and renter occupied housing units. Some houses are converted to multi-family dwellings and crowding grows. Residents who are concerned (often white) with the increasing racial heterogeneity grow more wary that minority will take over the neighborhood. With more minority children in schools, performance in education may decline and problems between juveniles emerge at or on the way home from school. Crime has settled into the neighborhood with vandalism having become common from the defacement of properties and other markings (graffiti). Burglary supplants theft as many residents no longer leave property out in the open. Aggravated assaults increase between neighbors who don't get along or are prompted from more frequent drug and alcohol use. Small scale drug dealing becomes present in still vacant houses. Police visit with some regularity. Fear of crime in the neighborhood has crystallized and housing prices have significantly declined. Nearby businesses begin to change to types that bear the characteristics of serving less affluent populations.

Stage 4: *Accelerating Decline*. The neighborhood looks terrible and many properties suffer from serious structural damage. Code violations have become regularity. Income levels have declined significantly with the influx of lower wage occupants. They have little in finances or incentive to upkeep properties. Unemployment becomes high from residents not having stable jobs and vagrancy becomes a model of daily activity. Schools have regular delinquency problems and there is a decline in the quality of teachers and other public services. Robberies become a regular occurrence. Drug markets have ample opportunities to form with little interference due to fear of crime and a physical environment that is conducive to these activities. Neighbors have little incentive to intervene. The neighborhood has gained a reputation across the metropolitan area as a place to stay away from, not move into, nor invest money into.

Stage 5: *Abandoned.* Housing for the most part is severely dilapidated with many left vacant and damaged. The neighborhood is now representative of the term "urban blight". With significant amounts of abandoned properties vagrancy, squatting, prostitution and arson are common, indicating that it can get no worse. Poverty has become concentrated and the reputation as a bad neighborhood has been firmly established. The neighborhood is rotationally designated as a "hot spot" from flare-up of criminal activity. It is a clear target for rehabilitation or for revitalization from a governmental perspective due to citizen outcry or investment opportunities. Success will depend on the ability of the structures to be renovated and proximity to amenities that would make would make investment worthwhile, if crime is controlled.

With this hypothetical framework of neighborhood change from foreclosures and increased crime, UCS can be used to explore how urban change might affect crime at neighborhood level. In the example outlined here, we use foreclosures, crime, and other socioeconomic variables at census block groups level to illustrate the feasibility of this approach. This analysis takes into account the current state of neighborhoods in the life cycle and uses only minor property crime as a measure for simulating commensurate changes in neighborhood characteristics toward decline and further increases from an increase in foreclosures.

Given the conceptual framework of neighborhood life cycle and the theory of deviant places, the appropriate unit of analysis will be a neighborhood or a geographic unit such as census blocks, blockgroups, tracts, or even counties. The level of data aggregation will no doubt have significant influence on results of simulating property crime in urbanized areas. The larger the geographic units used in the simulation processes, the less specific the simulation results will be.

Furthermore, it should be noted that both the neighborhood life cycle concept and the theory of deviant places would not work with individual households, streets, or similar entities in the simulation processes because neighborhoods are the conceptual unit. Unfortunately, this precludes the use of other excellent simulation approaches such as agent-based modeling and cellular automata.

11.6 Simulation Method

The simulation algorithm used in UCS is based on the outcome from the literature review. We developed the simulation methodology with the following features:

- That the integration of criminological and geographical theories would be better for explaining geographic patterns of crime. Blending deviant places and dependency theories with the concept of neighborhood life cycle, we have designed a simulation algorithm that categorizes neighborhoods so that they are most different between groups but most similar within each group.

- That not a fixed set of neighborhood attributes need to be applied to all geographic locations or in all situations as other simulation approaches often require. This led us to developing the algorithm so that analysts can define their own set of environmental, socio-economic, and demographic attributes (variables) of the neighborhoods being analyzed with whatever weights they see appropriate in assigning to the attributes. The selection and weighting of neighborhood attributes for the simulation can be based on users' expert experience or on their knowledge of local conditions, or from research outcome of existing literature.
- That a vector data structure is adopted in the developed algorithm to avoid limitations posed by a raster data structure. Raster data structure, such as that often used in cellular automata, does not work well at neighborhood level because the size of typical raster cells does not correspond to sizes of land parcels and/or neighborhoods well enough to have meaningful simulations.

The algorithm we developed for simulating changes in crime rates is based on changes in values of the selected attributes of the neighborhoods. It consists of four steps:

1. A hierarchical cluster analysis (George 2009; Long et al. 2010) that groups all area units (census block groups, or neighborhoods) into a hierarchy of clusters based on how similar those neighborhoods are between each other. This particular clustering method defines the cluster distance (difference) between two clusters to be the maximum distance, as measured by their attribute values, between their individual neighborhoods. At every stage of the clustering process, the two nearest clusters are merged into a new cluster. The process is repeated until all neighborhoods are agglomerated into one single cluster. Once the user defines the number of clusters desired, the cluster hierarchy at the corresponding level is used.
2. For an area (neighborhood) that is predicted or planned to have changes in the value of any included attribute(s) (as changes in the neighborhood's characteristics), UCS user would select the corresponding polygon that represents the neighborhood and update its attribute value(s) interactively within UCS. This neighborhood is the target.
3. UCS then searches among all neighborhoods to find one that has the closest attribute profile as that of the neighborhood being simulated, or the reference neighborhood.
4. UCS then statistically analyzes the trend of changes in crime rates among the neighborhoods in the same cluster as the reference neighborhood that was found to be the most similar to the updated neighborhood.
5. Output the statistical summaries of the crime rates among neighborhoods in the cluster that contains the newly updated neighborhood as the simulated result.

These steps were formulated using the concepts of neighborhood life cycle in that each neighborhood is said to have the potential to decline following the

Fig. 11.1 Simulation Algorithm: from selecting a target neighborhood, updating its profile, identifying a reference neighborhood and cluster membership, to outputting simulated results

hypothetical model above. Once a neighborhood experiences this cycle, it will likely need new stimulation, new investment, or renewal effort to revitalize it for another life cycle.

The deviant places theory provides the rationale and theoretical framework for selecting and including neighborhood attributes in the simulation processes. The Urban Crime Simulator provides users the flexibility of selecting any attributes with any desirable weight structure as input to the simulation process. In a graphic representation, the simulation methodology can be expresses as the steps in Fig. 11.1.

When assembling data sets for use in UCS, the selection of socio-economic, demographic and environmental attributes in the simulation is determined by the analysts based on their expertise, local experience, or knowledge learned from existing literature. Using the global and localized regression modeling tools and spatial statistical tools in UCS, users can fully explore the data set formed by their selection of neighborhood attributes so they can be aware of any spatial dependency or regional variation of the relationships between changes in crime rates and how the neighborhood attributes they selected change over space.

11.7 Software Development and Use

UCS was developed on the Windows platform with the C# programming language in Microsoft's Visual Studio 2005 with software components from ArcEngine 9.2 from ESRI (Redlands, California). The upside of this programming platform is that it enables speedy development by avoiding the need to re-program everything from ground zero. The downside of this decision is that the distribution of the finished software tool will need to assume that users have access to ArcEngine 9.2 or later.

11.7.1 Sample Model

As an example, a simple model is presented here to provide a snapshot for the functions and use of UCS. The data set used in this sample model is for Charlotte, North Carolina. While this sample model describes how UCS functions, it is not a standard model that UCS users should always follow. To have best simulated results for their own geographic areas, users should consider several factors:

1. *Data availability*. This is often the most critical factors to projects such as simulating urban crime. If users possess data that have different set of attributes from those used in this model, please carefully consider their relevance and select attributes to be included in the simulation model accordingly.
2. *Geographic resolution*. This is the level of geographic details needed for simulating urban crime. While UCS works with any level, the appropriate level should be determined by the users' own experience and expertise. Each locality has its unique characteristics. What works in one area may or may not work in other areas.
3. *Relevant attributes*. This is the set of selected neighborhood attributes that can best describe unique local characteristics. For the simulation of crime, this would depend on users' understanding of relevant criminological theories, own experience and expertise of local conditions.

11.7.2 Selection of Neighborhood Attributes

With the Charlotte, NC data set, all available attributes in our collection were first included in a stepwise regression model that was executed outside of UCS. This data set contains attributes that describe census block groups of Charlotte, NC in terms of demographic, socio-economic, and environmental attributes. The variables that are finally selected and used in the model here represent concentrated disadvantage, residential instability, and physical blight that are components of the deviant places theory. The regression model we used suggest the inclusion of the following attributes as the most relevant in explaining the variation of the dependent variable, PCRIMERATE00, or property crime rates in 2000.

Dependent variable	
PCRIMERATE00	*Crime rate, 2000*
Independent variable	
RWITE00	*Percent white population, 2000*
ROWNER00	*Percent owner-occupied housing units, 2000*
POP2000	*Total population, 2000*
RHH1F00	*Percent female-headed households, 2000*
RHISP00	*Percent Hispanic population, 2000*
FCLOSURE	*Percent of filed foreclosures*

RPOVERTY00	*Percent population under defined poverty, 2000*
R5NOMV00	*Percent population residing at the same address for the last 5 years, 2000*
MHVALE00	*Medium housing unit value, 2000*

Overall, this stepwise regression model has an adjusted-R^2 of 0.54, acceptable for use as an example with conventional statistical significance level of $p=0.05$. Following that, a quick run of calculating correlation coefficients between independent variables shows that no co-linearity issue exists. With all census block groups in Charlotte, NC, UCS and the selected independent variables, UCS suggests the optimal number of clusters as 10 for this simulation setup. Figure 11.2 shows the spatial distribution of the 10 clusters as the result of running the hierarchical cluster analysis using all census block groups in the city.

Because foreclosures and crime have both been shown to have negative spatial diffusion effects (spatial dependencies), we used UCS in the next step to learn which parts of the city the modeled relationship is stronger and which parts of the city this model may be weaker. This is achieved by calculating the global and local spatial dependency in the data set. In particular, localized regression models were calculated to account for known spatial dependencies of the sample model varies over space. With these results, simulations can be performed to estimate changes in property crime rates with projected/estimated changes in any selected attributes of these urban neighborhoods (*i.e.*, census block groups in this case).

11.7.3 User Interface

UCS allows users to load individual shapefiles (ESRI, Inc.) or open an existing ArcMap map document (.mxd). Once the data is loaded, the menus, tool buttons, are similar to those used in ArcMap.

Once data layers are loaded into UCS, users can select neighborhood attributes from the data layers for simulations. In addition, users can assign weights to each selected attribute. Figure 11.2 shows the dialog box in which users can select and weight neighborhood attributes to be included in the simulation. It also shows the interface for running functions for correlation, regression, and cluster analysis.

In this example, UCS suggested that 10 clusters are the best structure for minimizing the in-cluster variation and maximizing the between-cluster variation. Figure 11.3 shows the spatial distribution of the census block groups in 10 clusters. As can be seen in this display, neighborhoods of similar attribute profiles show a high degree of spatial clustering as well.

The output from running the Localized Regression Analysis provides an overview of the dependability of the simulation. Using the selected independent variables, Fig. 11.4 shows the spatial distribution of adjusted-R^2 values. Given the selected neighborhood attributes, the census block groups that are shown in darker colors provide better results from simulations in UCS. Extending from conventional

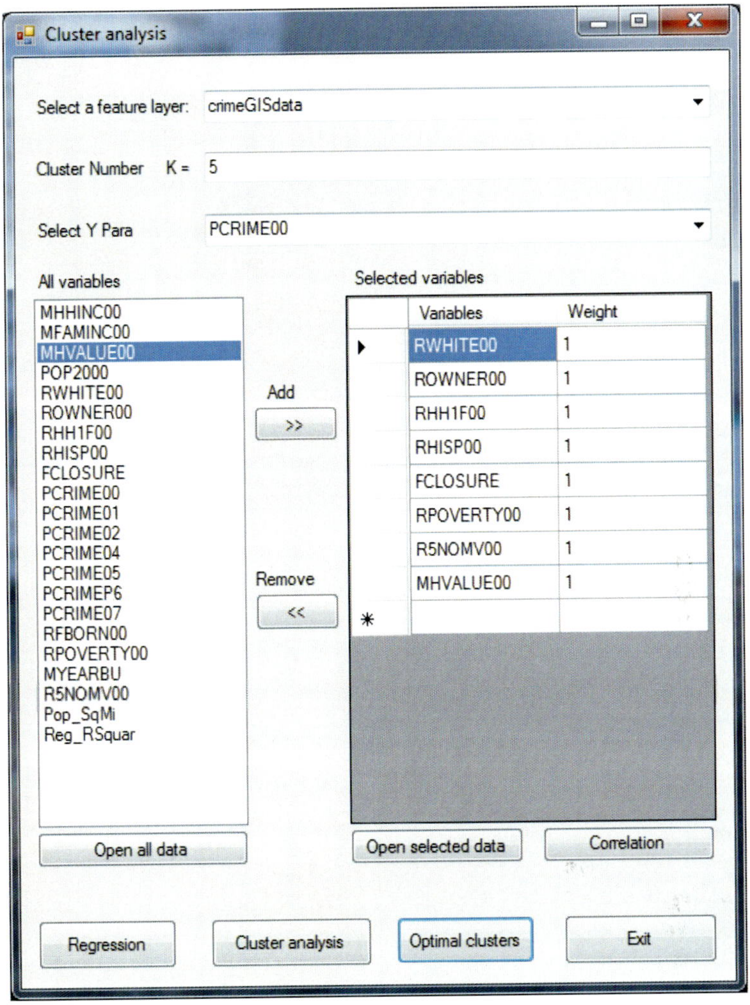

Fig. 11.2 Simulation setup and tools for exploratory analysis

regression models, the localized regression analysis incorporates a linear distance decay function in its calculation. For each neighborhood (census block group), its localized regression model only includes those neighborhoods that are within the distance threshold as defined by users. With an adjusted-R^2 calculated for each neighborhood, the map that shows how these localized adjusted-R^2 values distribute. It provides an indication of where in the city the simulations would perform better than other locations in the city.

Once the distribution of spatial dependency is assessed, the simulation starts by identifying a target neighborhood for which UCS would calculate its estimated changes in crime rates. The target neighborhood's attribute values are updated

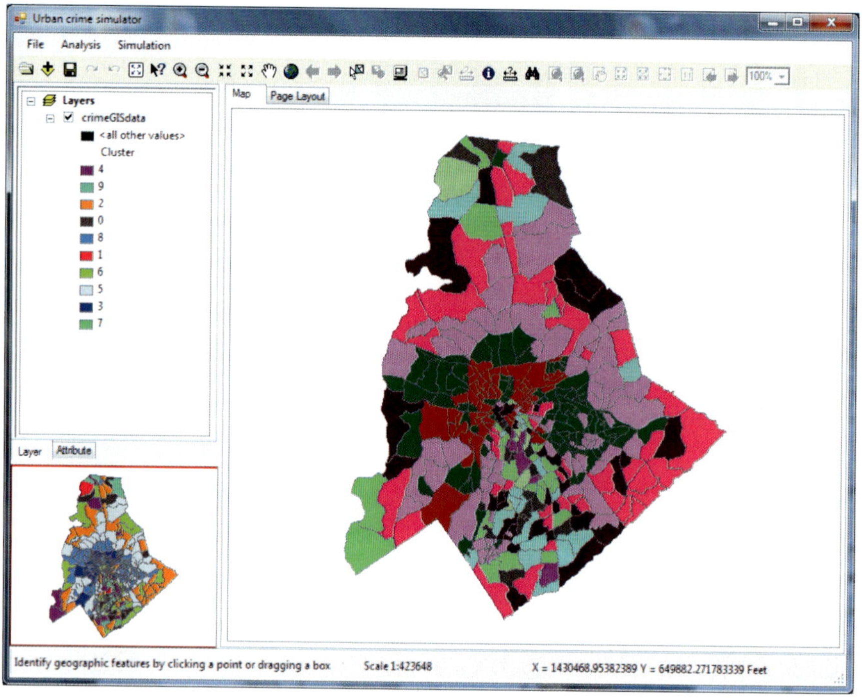

Fig. 11.3 Census Block groups in 10 clusters

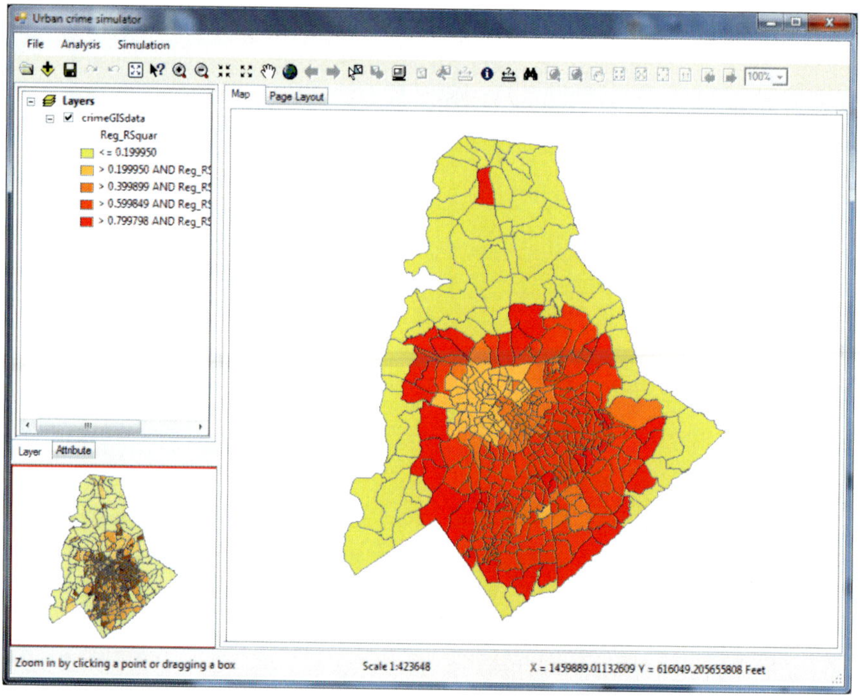

Fig. 11.4 Localized Regression Analysis

with known or planned changes to form a new attribute profile. Searching from all neighborhoods in the city, UCS selects a reference neighborhood whose attribute profile is the closest to that of the updated target neighborhood. UCS then statistically analyzes and summarizes the crime rates of neighborhoods in the same cluster as the reference neighborhood. The resulting crime rates statistics are the outcome of the simulation.

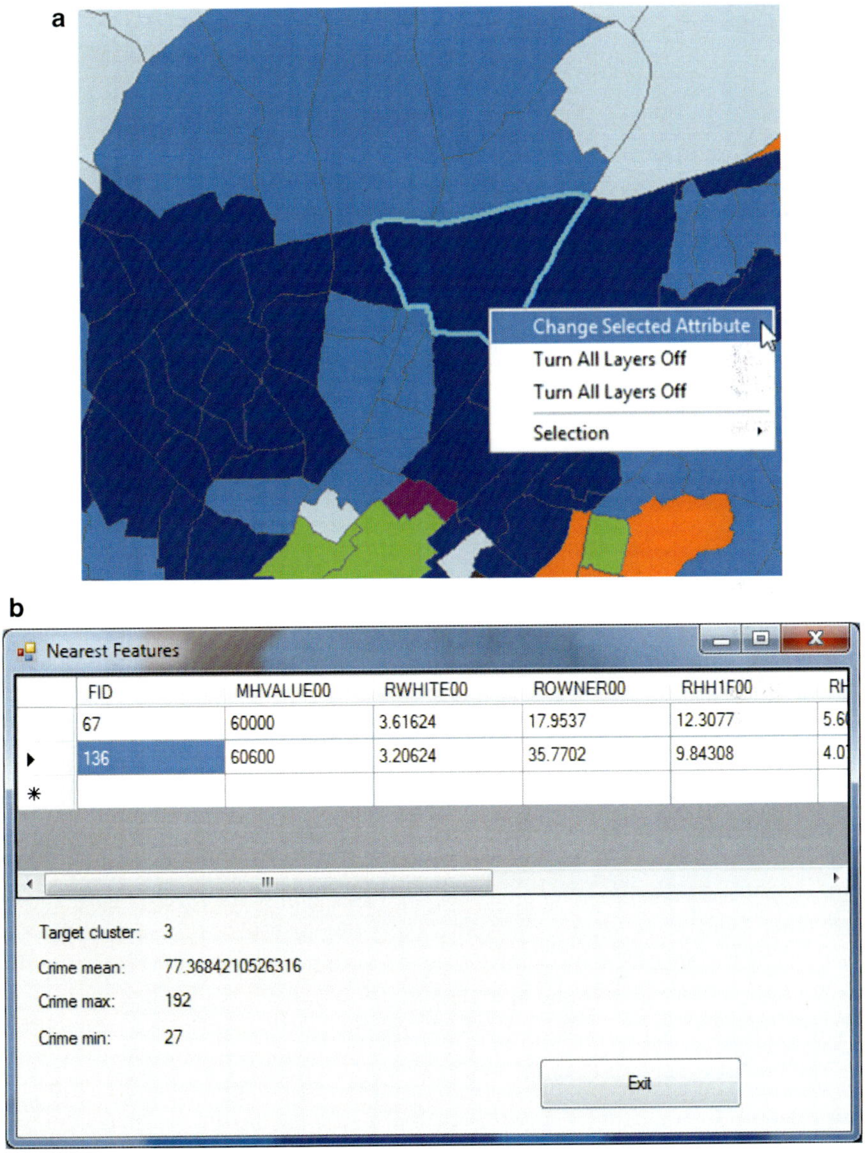

Fig. 11.5 (**a**) Target neighborhood to be updated/changed. (**b**) Output from simulation. (**c**) Locations of the target neighborhood and the reference neighborhood

c

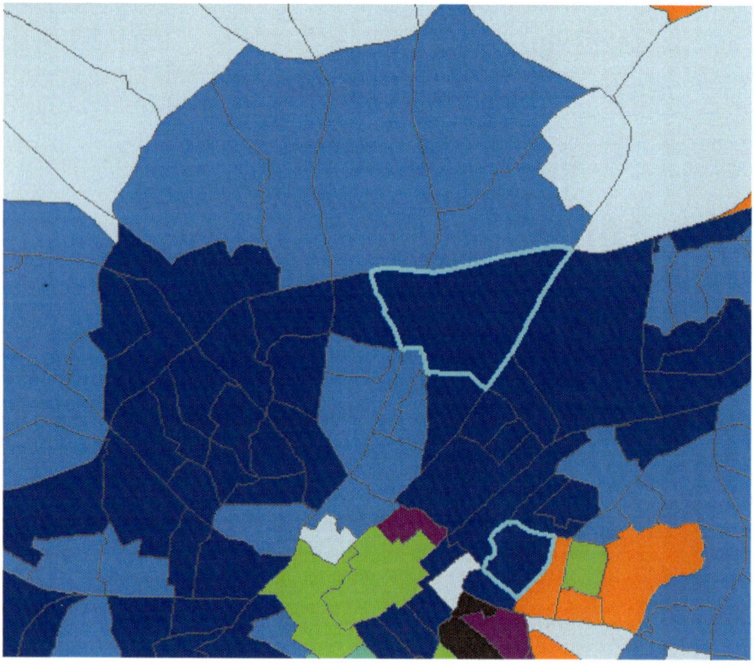

Fig. 11.5 (continued)

Below, Fig. 11.5a shows that a target neighborhood (polygon #67) is selected for changing/updating its attribute values. Figure 11.5b shows that a reference neighborhood (polygon #136) is identified by UCS and the simulated/estimated crime rate is likely around 77.368 within an interval between 192 and 27. Finally Fig. 11.5c shows the locations of the target neighborhood (upper) and the reference neighborhood (lower).

11.8 Concluding Remarks

Although several simulation approaches, such as cellular automata (Liang 2001; Lin 2008), agent-based modeling (Groff 2008; Malleson et al. 2010), among others, have been demonstrated to be useful tools in simulating various geographic phenomena, it is clear from the literature review we conducted for this study that they are limited to specific data structure and specific conditions. The lack of possible generalization from these approaches guided us to a different direction which resulted in the simulation algorithm we developed for Urban Crime Simulator.

We assume that not all local government agencies possess the needed technical expertise to conduct sophisticate simulations with such approaches such as cellular automata. We also assume that not all local government agencies possess the time and labor needed to develop a comprehensive set of rules for use in more sophisticate approaches such as agent-based modeling. Furthermore, it is apparent that not all cities would work with the same set of rules so they can take advantage of the benefits offered by cellular automata or by agent-based simulation modeling simulations. Associating attribute values to geographic units such as using census block groups as neighborhoods precludes the use of raster data structure. This is because all raster cells in the same neighborhood would need to be assigned the same attribute values when no data is available at finer geographic resolution. Developing a comprehensive rule base for each locality is often prohibitive to local government agencies given their other mandates.

The criminological literature suggests a wide spectrum of theories that may be useful in estimating changes in crime rates in urban neighborhoods. However, it is clear from this project that no one theory is comprehensive enough in dealing with all aspects of this process. From developing UCS, we have the following experience and thoughts for future development:

- That a synthesis of multiple theories works better for complex phenomena such as urban crime and its geographic distribution than any single theory does.
- That software tools such as UCS need to have the flexibility of allowing users to use data that they have, which may be at various levels of geographical aggregation/details.
- That local knowledge is the key to successful and meaningful simulations of urban crime simulation.
- That a better suite of data mining tools may be added to UCS to help users to explore spatial and temporal trends of their GIS data before defining their simulation models in UCS.

Finally, the UCS software, a user manual, and other information are available from the authors upon requests.

Acknowledgments This research was supported by funding from the South Carolina Research Authority (SCRA) and the National Institute of Justice (NIJ).

References

Baxter V, Lauria M (2000) Residential mortgage foreclosure and neighborhood change. Hous Policy Debate 11(3):675–699

Birch DL (1972) Toward a stage theory of urban growth. J Am Inst Plan 37:78–87

Bottoms AE, Wiles P (2002) Environmental criminology. In: Maguire M, Morgan R, Reiner R (eds) The Oxford handbook of criminology, 3rd edn. Oxford University Press, Oxford, pp 620–656

Brantingham P, Brantingham P (1975) Residential burglary and urban form. Urban Stud 12:273–284

Brantingham PJ, Brantingham PL (1981) Environmental criminology. Sage Publications, Thousand Oaks

Brantingham PL, Brantingham PJ (2004) Computer simulation as a tool for environmental criminologists. Secur J 17:21–30

Crossney KB (2010) Is predatory mortgage lending activity spatially clustered? Prof Geogr 62(2):153–170

Crow W, Bull J (1975) Robbery deterrence: an applied behavioral science demonstration – final report. Western Behavioral Science Institute. La Jolla, CA

Duncan O, Duncan B (1957) The Negro population of Chicago. University of Chicago Press, Chicago

Fulmer A (2009) Burning down the house: mortgage fraud and the destruction of residential neighborhoods. White paper submitted for a meeting on mortgage fraud, foreclosures and neighborhood decline. U.S. Department of Justice, National Institute of Justice, Washington, DC

Galster GC (2005) Consequences from the redistribution of urban poverty during the 1990s: a cautionary tale. Econ Dev Q 19(2):119–125

George D (2009) SPSS for windows step by step: a simple guide and reference, 16.0 Update. Allyn and Bacon, Boston

Gilberto SM, Houston AL (1989) Relocation opportunities and mortgage default. Am R Estate Urban Econ Assoc J 17(1):55–69

Groff E (2008) Simulating crime to inform theory and practice. In: Chainey S, Tompson L (eds) Crime mapping case studies: practice and research. Wiley, Hoboken

Hoover EM, Vernon R (1939) Anatomy of a metropolis. Harvard University Press, Cambridge

Immergluck D, Smith G (2005) There goes the neighborhood: the effect of single-family mortgage foreclosures on property values. Research report. Woodstock Institute, Chicago, IL

Kaplan DH, Sommers GG (2009) An analysis of the relationship between housing foreclosures, lending practices, and neighborhood ecology: evidence from a distressed county. Prof Geogr 61(1):101–120

Lee K (2008) Foreclosure's price-depressing spillover effects on local properties: a literature review. Community Affairs Discussion Paper, Boston, MA: Federal Reserve Bank of Boston. September.

Liang J (2001) Simulating crimes and crime patterns using cellular Automata and GIS. PhD disseration, University of Cincinnati

Lin L (2008) Artificial crime analysis system: using computer simulations and geographic information systems. IGI Global, Hershey

Long B, Zhang Z, Yu P (2010) Relational data clustering: models, algorithms, and applications. Chapman & Hall/CRC, Boca Raton

Lowry IS (1960) Filtering and housing standards: a conceptual analysis. Land Econ 36:362–379

Malleson N, Heppenstall A, See L (2010) Crime reduction through simulation: an agent-based model of burglary. Comput Environ Urban Syst 34(3):236–250

Metzger JT (2000) Planned abandonment: the neighborhood life-cycle theory and national urban policy. Hous policy debate 11:7–40

Park R (1952) Human communities. Free Press, Glencoe

Pennington-Cross A (2004) The value of foreclosed property. Journal of Real Estate Research 28:193–214.

Roncek DW, Bell R (1981) Bars, blocks, and crimes. J Environ Syst 11(1):35–47

Schuetz J, Been V, Ellen IG (2008) Neighborhood effects of concentrated mortgage foreclosures. NYU Law and Economics Research Paper No. 08–41

Sherman L, Weisburd DL (1995) General deterrent effects of police patrol in crime hot spots: a randomized controlled trial. Justice Q 12:625–48

Sherman LW, Gartin P, Buerger ME (1989) Hot spots of predatory crime: routine activities and the criminology of place. Criminology 27(1):27–55

Simmons RA, Quercia RG, Maric I (1998) The value impact of residential construction and neighborhood disinvestment on residential sales price. J R Estate Res 15(1/2):147–161

Stark R (1987) Deviant places: a theory of the ecology of crime. Criminology 25:893–909

Taeuber K, Taeuber A (1965) Negroes in cities. Aldine, Chicago

Weisburd D, Bushway S, Lum C, Yang SM (2004) Trajectories of crime at places: a longitudinal study of street segments in the city of Seattle. Criminology 42(2):283–321

Wilson WJ (1987) The truly disadvantaged: the inner-city, the underclass and public policy. University of Chicago Press, Chicago

Wilson RE (2007) The impact of software on crime mapping: an Introduction to a Special journal issue of *Social Science Computing Review* on crime mapping. Soc Sci Comput Rev 25:135–142

Wilson RE (2011) Guest editor's introduction: crime and urban form. Cityscape 13(3):1–6

Wilson RE, Paulsen DJ (2010) A theoretical underpinning of neighborhood deterioration and the onset of long-term crime problems from foreclosures. Working paper from a meeting on mortgage fraud, foreclosures and neighborhood decline, U.S. Department of Justice, National Institute of Justice, Washington, DC

Winsberg MD (1989) Life cycle neighborhood changes in Chicago suburbs, 1960–80. Growth Change 20:71–80

Yeates M, Garner BJ (1976) The North American city. Harper and Row, New York

Chapter 12
Measuring a Place's Exposure to Facilities Using Geoprocessing Models: An Illustration Using Drinking Places and Crime

Elizabeth Groff

Abstract The prominent role of facilities in influencing 'why crime happens where it does' has been widely recognized and vigorously researched. Criminological theories which focus on opportunity such as routine activity theory and crime pattern theory have provided the basis for such inquiries. Some of these investigations have targeted the role of facilities in fueling higher crime levels at places. They have usually developed facility-focused measures that quantify each facility's influence based on the crime experienced by the places located near it. Measures are calculated only at the locations with facilities present. However, improvements in data sources and software have allowed researchers to examine the population of small units of geography rather than focusing on only those with a facility present. Thus it is now possible to quantify the cumulative effect of nearby facilities on the crime rates of geographies of such street blocks and addresses. This chapter begins by discussing the traditional methods for exploring the relationship between facilities and crime. Next, the theoretical case for more sophisticated distance and activity-level based measures is made. The critical role of geoprocessing models in automating complex analysis processes is explained and a model developed to create three different exposure measures. Data describing the locations of drinking places and street block level crime are used to illustrate how measures produced by the model can be used to explore the relationships between exposure to facilities and an outcome such as crime. The output measures from the model are evaluated using descriptive statistics and then used as independent variables in an ordinary least squares regression. Local variation in the measures is examined using a bivariate LISA to highlight areas of negative and positive spatial autocorrelation between exposure to bars and crime. The chapter concludes with a discussion of the implications of the findings and probable next steps.

E. Groff (✉)
Department of Criminal Justice, Temple University,
550 Gladfelter Hall, 1115 Polett Walk, Philadelphia, PA 19122, USA
e-mail: groff@temple.edu

M. Leitner (ed.), *Crime Modeling and Mapping Using Geospatial Technologies*, Geotechnologies and the Environment 8, DOI 10.1007/978-94-007-4997-9_12,

Keywords Micro-spatial • Crime analysis • Facilities • Behavioral geography • Proximity • Activity space

12.1 Introduction

Since the 1970s, the characteristics of small areas and specific places in influencing crime patterns have received increasing attention. Some of this attention has been facilitated by theoretical advances that drew attention to the role of place in producing observed crime patterns. Theories under the rubric of environmental criminology hold that the context of crime events is critical to understanding crime patterns (Brantingham and Brantingham 1984, 1991a; Clarke and Cornish 1985; Cohen and Felson 1979). They focus on the convergence of individuals who are criminally inclined (i.e., motivated offenders) with people or things who are suitable targets, in situations where someone likely to intervene (i.e., capable guardian) is not present. How the three elements necessary for a crime to occur end up at the same place and time is primarily a function of the routine activities of each person. The characteristics of the place at which convergence occurs determine why the individuals who use the place are present and the behaviors they perceive to be norms for the place. In other words, different places facilitate differing levels of convergence involving differing types of individuals. Further, opportunity theories recognize that places do not exist in a vacuum but rather are situated in the context of their neighbors.

Empirical testing of these theories would not have been possible without the development of geographic information systems (GIS) software and the increased availability of data describing places. It was the combination of new technology and data describing places that facilitated empirical work relating the census characteristics of various census geographies, such as census tracts, census block groups, and census blocks to the types of crime which occurred there (see Weisburd et al. 2009 for a comprehensive history). In recent years, the focus of inquiry has shifted to smaller geographies and led to the discovery that significant variability exists in crime across smaller units of analysis such as addresses (Eck et al. 2000; Sherman et al. 1989) and street blocks (Groff et al. 2009; Taylor 1997; Weisburd et al. 2004). As Taylor (2010, p. 467) notes "different types of processes are likely to be involved at different spatial scales" and "more insight can probably be gained by examining impacts of smaller-scale contexts like streetblocks."

One aspect of places that is especially important is the number and type of facilities present since facilities draw people to the places (Brantingham and Brantingham 1995). Except for residents, people must travel to get to particular facilities they want to patronize. While there, they tend to also patronize other establishments nearby. This means places along frequently used routes of travel and near, but not necessarily at a facility, also have increased potential for convergence just by virtue of their proximity. Thus any an accurate description of the impact of facilities on the places near them must capture the probability of spatial interaction. The cumulative influence of surrounding facilities on a place can be thought of as that place's *exposure* to facilities. Exposure can be quantified for one type of facility, for example bars or schools, or multiple types of similar facilities, for example retail outlets or recreational facilities.

A major challenge to calculating an exposure measure lies in the considerable effort involved. Tools to automate the identification of polygon feature (i.e., area) neighbors are part of the functionality of GIS software and relatively easy to calculate using a three step process. This is not the case for measuring micro-level spatial interaction along street networks which requires almost twenty steps to compute. Manually completing each of those steps is both tedious and time-consuming. This chapter presents a geoprocessing model that automates the calculation of micro-level exposure measures. The chapter begins by exploring questions related to appropriately quantifying spatial interaction and establishing its geographic extent. The etiology of spatial interaction across places is explained. Three incrementally more complex measures of facility influence on nearby places are suggested and implemented in a geoprocessing model. An illustration of how the output of the model might be used to conduct further research is offered through an examination of drinking place influence on street blocks in Seattle, Washington.

12.2 Theoretical Background

Geographers have long known that nearby things tend to be related and the closer they are to one another the stronger the relationship (Tobler 1970). This observation suggests it is important to measure and analyze relations among near places (Miller 2004). Facilities shape human activity because they attract people to particular places. At the same time, individuals make choices about which facilities to patronize based on distance and preferences. Distance is essentially a proxy for the time/effort and cost of traveling to a particular facility and is weighed against the attractiveness of the facility. The important role of distance in human decision-making has been widely recognized. Zip's (1950) principle of least effort states people will try to minimize effort and cost, often equated with distance or time, when choosing destinations for shopping, recreation, seeing a movie etc. The distance decay function holds that the likelihood of interaction decreases with increasing distance (Brantingham and Brantingham 1984; Harries 1990, 1999; Katzman 1981; Rossmo and Rombouts 2008). Thus travel decisions have much to do with proximity and distance influences interaction through the spatial choices of individuals. Since the location of human activity is largely shaped by distance and opportunities in the built environment, it makes sense that near things tend to be related.

More broadly, individuals have programs of daily behavior that constitute their routine activity spaces (Carlstein et al. 1978; Hägerstrand 1970, 1973; Horton and Reynolds 1971; Pred 1967, 1969). These spaces encompass the places that are visited, termed nodes, and the routes taken among those places, referred to as paths. Often, activity spaces are expanded to incorporate newly discovered places of interest such as a restaurant. In addition, individuals tend to become familiar with the areas on the way to and from their nodes as well as places around their nodes. Because of this, the boundaries of activity spaces are fuzzy and tend to change over time.

Environmental criminologists recognize the applicability of spatial choice and human activity spaces to the study of crime patterns. Both crime pattern theory (Brantingham and Brantingham 1991a, b) and routine activity theory (Cohen and Felson 1979; Felson 1986; Felson and Clarke 1998) identify the convergence of motivated offenders and suitable targets at the same place and time as necessary elements for a crime to occur. The presence or absence of individuals who could potentially intervene is another necessary element (Cohen and Felson 1979; Eck 1995; Felson 1995). One type of capable guardian especially important in the context of facilities is place managers (Eck 1995; Felson 1995). Store employees, bartenders, parking lot attendants and other individuals who work at facilities are all place managers. In addition, crime pattern theory emphasizes the role of places characteristics and the urban backcloth in setting the stage for crime events.

In sum, the configuration of facilities within general areas of land use types shapes the routine activities of individuals which in turn influence the number of convergences which occur at places. This chapter demonstrates how geoprocessing models can be used to operationalize three different types of exposure measures. The output values can then be used in multivariate statistical models to represent exposure to facilities in a more nuanced fashion.

12.2.1 Quantifying Exposure

Exposure of areal units such as census blocks, block groups and tracts can be been easily measured using first-order (adjacent) or second-order (neighbors of adjacent units) neighbors. For example, first-order exposure is the total number of facilities in the set of adjacent areas and second-order exposure is the number of facilities in the neighbors of adjacent areas. Bernasco and Block (2011) recently undertook just such an analysis in Chicago using census blocks. They examined the effect of a variety of facilities on robbery using census blocks and their adjacent neighbors. However, using a finer spatial scale opens the door to new ways of more accurately quantifying potential spatial influence.

Past studies using micro-spatial scales have quantified the exposure of street blocks to facilities nearby using a simple count of the number of facilities within a distance threshold. Two of these studies used a questionnaire approach asking a sample of neighborhood residents whether certain facilities were located within three blocks of their home (Miethe and McDowall 1993; Wilcox et al. 2004).[1] The responses were aggregated to neighborhoods for the analysis. Another study

[1] Kurtz et al. (1998) examined the percent of the street block that was retail storefronts but did not consider adjacent areas.

examined every street block in a city and used a threshold distance of a quarter mile (just over three blocks) (Weisburd et al. 2011, 2012). This method had the advantages of: (1) being relative easy to compute using current GIS software and (2) producing an understandable outcome measure (e.g., two facilities within quarter-mile of a place).

However, there are three issues with the straight-line-distance, simple-count operationalization of place (in this case street blocks) exposure to facilities. Most importantly, it fails to account for the diminishing influence of facilities as distance increases. All facilities count equally regardless of whether they are on the same block or three blocks away. Second, it overestimates the geographic extent of a threshold distance because travel is limited by the transportation network. Third, it does not incorporate differences in facility influence. The influence of a facility may be tied to its physical size, the number of patrons, the amount of sales, the types of people who frequent it, and the times it is open for business just to name a few. These characteristics of facilities play a role in the amount and type of influence they have over crime at street blocks nearby. Recognizing these issues a recent study examined the places near bars using both Euclidean and street distance measures across a range of threshold distances and found street distance was a stronger measure of the relationship between facilities and crime (Groff 2011).

Building on the earlier work, this effort develops three new measures of facility influence that represent increasingly more sophisticated operationalizations of exposure to facilities. All three measure distance along a street network and thus address the concern that human activity patterns are not represented in a tradition count threshold measure of influence using Euclidean distance (i.e., 'as the crow flies') (Groff 2011).[2] The simplest measure, Count, is a count of the facilities within a threshold.[3] Each facility within the specified threshold distance counts as one, in this way the amount of influence attributed to each drinking place within a threshold is constant.

Measure 1:

$\text{Count}_i = \Sigma(\text{Number of facility locations})$
where the number of facilities within the threshold distance $_i$ from the place is counted.

To deal with the issue of distance from the place to the facility, inverse distance weighting (IDW) is used to 'discount' the potential influence of a place. A facility at a place counts as '1' while those farther away count less than one. Distances are once again measured along a street network. The outcome measure, IDW Count, represents the cumulative influence of facilities on a place discounted for distance, the larger the number, the greater the influence.

[2] Earlier work by Groff and colleagues has suggested that street distance buffers offer a more parsimonious representation of the spatial interaction occurring at specific distances than do Euclidean buffers (Groff 2011; Groff and Thomas 1998). Euclidean buffers often include events/facilities that cannot be reached using available travel routes.

[3] Threshold distances are measured from the midpoint of the street segment along the street network in all directions to the threshold distance specified. A street segment is included only if its midpoint falls within the threshold.

Measure 2:

IDW $\text{Count}_i = \Sigma(1/\text{Sqr}(d_{ij}))$
where d_{ij} = distance from the place *i* to each facility *j* within the threshold bandwidth

The final measure, Distance Weighted Activity (DWA), operationalizes 'exposure' as reduced by distance but increased by potential activity at a place. The DWA takes into account both the distance a facility is from a street block and the activity at that location and thus provides a more nuanced view of the relationship between exposure and crime. Measuring activity could be done in a variety of ways and will vary based on the type of facility. For example, total square footage, total sales, seating capacity, ticket sales, number of bar stools, or enrollment are a few possibilities. For drinking establishments, annual sales, number of bar stools, seating capacity, and square footage are all possibilities. Annual sales has the advantage of capturing actual dollars spent rather than the potential for people to patronize the facility as is the case with characteristics such as bar stools, seating capacity and square footage.

Measure 3:

Distance Weighted $\text{Activity}_i = \Sigma((1/\text{Sqr}(d_{ij})) * a_j)$
where d_{ij} = distance from the place *i* to each facility *j* within the bandwidth and a_j = total activity

12.2.2 Role of Geoprocessing Models

The three operationalizations of exposure measures can be created using the out-of-the-box software tools which exist in ArcGIS® but doing so requires a complicated series of roughly twenty individual actions on the part of the user. A way to automate the process is needed.

Geoprocessing models offer an attractive alternative to writing computer programs to achieve automation. Geoprocessing models are visual-based representations of processes. They allow nonprogrammers to automate tasks by dragging and dropping geoprocessing tools onto a canvas. Models can be easily passed from one user to another. Finally, they can be exported to a python script and then used as the basis for a more complex custom programming effort.

Automation of manual processes has many benefits including: (1) improved efficiency because tasks can be done more quickly and in less time (Kitchens 2006; Moudry 2012); (2) greater accuracy because there is reduced chance of human error (Fan 1998; Kitchens 2006; Moudry 2012); (3) greater consistency because the same steps are repeated without change from one analysis to the next (Fan 1998; Kitchens 2006; Moudry 2012); and (4) the creation of a record of steps taken which, in turn, enables exact replication of analyses (Moudry 2012; Thoma 1991). In addition, automation that involves computer programs or geoprocessing models can be easily shared and the analysis reproduced using different data.

Using geoprocessing models holds in common many of the same benefits familiar to users of syntax or log files. Models document what operations were conducted to produce the analysis and in this way they facilitate communication between analysts. This is extremely helpful when the same analysis has to be redone with new data or when someone else wants to conduct the same analysis using different data. In this way, models facilitate easy replication of research. Models also allow parameters to be easily and systematically changed to aid in sensitivity analysis. Finally, geoprocessing model provide flexibility because they run without human intervention or even presence.

12.3 Generating Measures of Facility Influence on Places

This chapter illustrates how a geoprocessing model can be used to operationalize and automate the collection of measures related to the exposure of places to facilities nearby. In keeping with recent research, the current study used street distance to measure spatial influence (Groff 2011) and examined very short threshold distances (Groff 2011; Ratcliffe 2011, 2012). Because the results are included for illustration purposes, this chapter discusses the results for only four threshold distances, one block (400 ft), two blocks (800 ft), three blocks (1,200 ft), and 2,800 ft (about 6 blocks or just over one half a mile).

A geographic information system (GIS) was used to generate the three measures quantifying place exposure to facilities. GIS are designed to facilitate spatial data creation, manipulation, storage, analysis, and visualization (Worboys and Duckham 2004). GIS have always had the ability to measure distances but recent versions have expanded the range of tools and made them more accessible. Most important to this research was the ability to measure distances along street networks from a set of origins to a set of destinations.

The Origin and Destination Matrix tool in Network Analyst (ArcGIS 9.3) was used for all the measurements. This tool is designed to measure the distance between an origin place or set of origins and a destination place or set of destinations. Here it was used to develop the three measures of exposure. Street block midpoints were used as the origins and the drinking places were the destinations. The software measured from each origin outward in all directions along the street network. If a destination was located within the threshold, a record was written to the matrix. The record contained the distance between the origin and the destination pair as well as information identifying the pair of locations involved.

At this point, the formulas for IDW Count and DWA were applied to the distance measures for each pair. In the case of IDW Count, each facility was inverse distance weighted by its distance from the street block (the closer the facility the greater the potential influence and thus the higher the score). The DWA measure multiplied the weighted score for each place-facility pair by total annual sales for that facility. The values for each measure were then summed for each street block to produce measures of the cumulative influence of facilities within a particular distance threshold.

The most challenging aspect of computing exposure measures is the large number of steps involved. As mentioned before, one reason gaps in our knowledge about the effect of facilities on the streets near them have remained is the time consuming nature of the inquiry. Computation of the final three measures from the output table required that fields be added, joins done, values calculated, and data summarized for each and then the entire process repeated for each distance threshold investigated. This is where the geoprocessing model was especially important to facilitating the multistep process required and ensuring consistent computation of measures (see Appendix for details of the model).

12.4 Using the Geoprocessing Model to Quantify Exposure to Drinking Places

To illustrate the value of facility exposure measures, an example analysis was conducted using drinking places in Seattle, WA and crime. Drinking places are ideally suited for a test case because there is a well-documented and positive relationship between bars and crime (Bernasco and Block 2011; Brantingham and Brantingham 1982; Frisbie et al. 1978; Groff 2011; Loukaitou-Sideris 1999; McCord and Ratcliffe 2007; Ratcliffe 2011, 2012; Rengert et al. 2005; Rice and Smith 2002; Roncek and Bell 1981). Accordingly, a positive relationship between bars and crime was hypothesized and the focus was on the ability of each measure of exposure to quantify the relationship.

12.4.1 Data

The street centerline file for Seattle, WA was provided by Seattle GIS. The locations of drinking places in 2004 (n = 157) were obtained from a private vendor.[4] Crime incident data for 2004 were provided by the Seattle Police Department and geocoded.[5] The number of crimes per street segment ranged from 0 to 515. The average street had 3.89 crimes (sd = 13.43). In Seattle, crime incident reports were generated

[4] Drinking places were identified using the NAICS code 7224 which defines them as follows: "This industry comprises establishments known as bars, taverns, nightclubs, or drinking places primarily engaged in preparing and serving alcoholic beverages for immediate consumption. These establishments may also provide limited food services." U.S. Census Bureau (2010) NAICS 7224: Drinking Places (Alcoholic Beverages) Retrieved 2/11/2010, from U.S. Census Bureau: http://www.census.gov/epcd/ec97/def/7224.HTM. InfoUSA provided the physical locations of the drinking places as well as their annual sales in thousands of dollars.

[5] All geocoding was done in ArcGIS 9.1 using a geocoding locator service with an alias file of common place names to improve the hit rate. The geocoding locater used the following parameters: spelling sensitivity = 80, minimum candidate score = 30, minimum match score = 85, side offset = 0, end offset = 3 percent, and Match if candidates tie = no. Manual geocoding was done on unmatched records in ArcGIS 9.1 and then in ArcView 3.2 using the 'MatchAddressToPoint' tool (which allowed the operator to click on the map to indicate where an address was located) to improve the overall match rate. Research has suggested hit rates above 85% are reliable Ratcliffe (2004). Geocoding Crime and a First Estimate of a Minimum Acceptable Hit Rate. *International Journal Geographical Information Science, 18*(1), 61–72. Our final geocoding percentage for crime incidents was 97.3 %.

Table 12.1 Descriptive statistics for exposure measures across street blocks (n=24031)

	Count				IDW count				Distance Weighted Activity (DWA)			
	Mean	SD	Min	Max	Mean	SD	Min	Max	Mean	SD	Min	Max[a]
800	.127	.535	0	9	.088	.374	0	6.919	46.67	355.58	0	12059.37
1200	.283	.917	0	12	.176	.576	0	8.246	95.82	513.00	0	12059.37
2800	1.416	2.89	0	21	.599	1.261	0	10.801	348.67	1056.55	0	12164.96

[a]The lack of change in the maximum values for DWA is due to the extremely high sales volume of a single drinking place. The sales of the single highest volume drinking place were over twice as high as the next and over three times as high as the rest of the drinking places. Consequently, its influence on the surrounding streets was higher than any other cumulative influence measured at any distance

by police officers or detectives after an initial response to a request for police service. Thus, they represent only those events which were both reported to the police and deemed to be worthy of a crime report by the responding officer. In this way, incident reports provided a measure of 'true' crime, at least the crime that was reported to the police. Specifically, we included all crime events for which a report was taken except those which: (1) occurred at an intersection,[6] (2) had an address of a police precinct or police headquarters; and (3) occurred on the University of Washington campus.[7]

12.4.2 Exposure Measures Produced

The values for the exposure measures vary based on two dimensions, the threshold distance and the operationalization (simple count vs. IDW vs. DWA). Looking first at the threshold distances, all the measures increase steadily as the distance threshold increases (see Table 12.1). This is expected since as the threshold size increases more drinking places are included in the measure. The finding provides additional confidence that the model is correctly specified.

The operationalization of the measures is consistent with the hypothesized relationships. As expected, the values for the Simple Count were consistently higher than the values for the inverse distance weighted count (IDW count) across all threshold distances (see Table 12.1). The average street in Seattle had .1 drinking

[6] Intersection crimes are excluded because incident reports at intersections differed dramatically from those at street segments. For example, traffic-related incidents accounted for only 3.77 % of reports at street segments, but for 45.3 % of reports at intersections. After excluding intersections, records that lacked a specific address, and records that could not be geocoded, there were 186,958 incident reports in 2004.

[7] Data on crime from the University of Washington campus were not provided to the Seattle Police Department after 2001. Efforts to obtain geocodable data directly from the University of Washington were unsuccessful.

Table 12.2 Descriptive statistics for exposure measures for only those street blocks within the threshold distance

		Count		IDW count		Distance Weighted Activity (DWA)	
	Streets	Mean	SD	Mean	SD	Mean	SD
800	1979	1.542	1.134	1.071	0.800	566.513	1113.667
1200	3804	1.787	1.618	1.112	1.028	605.088	1163.499
2800	11,256	3.022	3.604	1.278	1.589	744.142	1445.012

Note: Minimum and maximum are not included because the minimum is one for all measures and thresholds and the maximum is the same as in Table 12.1

places within about a block. The subset of only those street blocks with at least one drinking place within 800 ft averaged just over one drinking place (Table 12.2). The inverse distance weighted (IDW) counts were lower because they took into account the distance between the street and the drinking places within the threshold so they averaged .09 and 1.07 respectively. In contrast, the values of the DWA measure were much larger than either of the other two because it weights the sales of each drinking place (in thousands of dollars) by the distance between the street block and the facility. The average street block was exposed to approximately $46,670 distance weighted sales dollars within one block but among places with at least one drinking place within 800 ft the exposure was $566, 513.

Overall, the descriptive statistics reveal differences in the exposure values generated within measures as the threshold increases. There are also differences in the values from one operationalization of exposure to the next (i.e., between Count, IDW and DWA). All the observed differences are consistent with the way the measures were calculated. The next question is which combination of threshold size and operationalization most closely relates exposure to drinking places and crime at the street block.

12.4.3 Examining the Relationship Between Exposure to Drinking Places and Crime

To get an indication regarding which exposure measures were most strongly related to crime, a simple OLS regression was calculated using only the exposure measure and a spatial lag of total crime. The Corrected Akaike Information Criterion (AICc) was used to compare the different models and get preliminary feedback on model fit. Models with lower AICc scores fit the observed data better. The model adjusted R-squared was also used to explore whether different conceptualizations of exposure explained a greater percentage of crime controlling for the amount of crime on nearby street blocks.

$$\text{Total Crime} = B_0 + \text{Exposure} + \text{Crime Lag} \qquad (12.1)$$

Table 12.3 Regression results for exposure measures

Measure		Adjusted R^2	Beta	AICc	Rank
800					
	Count	0.1021	1.626**	190386.765	2
	IDW	0.1024	2.426**	190378.341	1
	DWA	0.1012	0.00351**	190412.227	3
1200					
	Count	0.1024	1.070**	190378.159	2
	IDW	0.1029	1.789**	190366.714	1
	DWA	0.1008	0.00013*	190423.359	3
2800					
	Count	0.0990	0.114	190470.709	
	IDW	0.0997	0.469*	190452.947	1
	DWA	0.0992	0.00364	190464.451	

Note: The entire population of Seattle street blocks is used, statistical significance is reported for comparison purposes only. Since the Koenker (BP) Statistic (Koenker's studentized Bruesch-Pagan statistic) was significant for all models, robust probabilities are reported
*p<.05; **p<.01; ***p<.001

There were only minor differences between exposure measures at specific thresholds (Table 12.3). The IDW measure produced better fitting models with slightly higher adjusted R-squared values at every distance threshold. Simple Count and DWA measures, respectively, had slightly lower explanatory power and poorer fit regardless of distance threshold. Consistent with previous research, the relationship between drinking places and crime was highest at shorter distances. At about six blocks (2800 ft), only the IDW exposure measure explained a significant amount of the variation in crime.

Overall, these measures more exactly quantify the local situation of each street block in Seattle and thus provide better inputs to multivariate models of exposure to drinking places. But the differences among the measures in explaining crime are relatively small, especially between Count and IDW. One explanation for the lack of large differences between the measures may be in the type and extent of local variation. If there is a good deal of local, micro-spatial variation it may be masked by using a global statistic such as OLS regression (even with a spatial lag).

To examine local rather than global relationships, a bivariate local indicator of spatial association (LISA) was used. The bivariate LISA allows investigation of whether the measure used to operationalize exposure to drinking places affects the relationship between drinking places and total crime (Anselin 1995, 2003). The statistic classifies the relationship between each place and its neighbors as one of four types and provides an assessment of statistical significance for each.

There are two general types of spatial autocorrelation, positive and negative. Positive spatial autocorrelation can take the form of places with a high exposure to drinking places being associated with high crime (High-High) or places with low exposure scores being associated low crime (Low-Low). Places exhibiting positive spatial autocorrelation are where the hypothesis regarding the relationship between bar influence and crime was supported. Negative spatial autocorrelation occurs

Table 12.4 Results of LISA by threshold distance and measure applied

	Simple count		IDW count		Distance Weighted Activity (DWA)	
	Frequency	Percent	Frequency	Percent	Frequency	Percent
800 ft						
Not significant	16,612	69.15	16,623	69.20	18,083	75.27
High-High	1,021	4.25	1,016	4.23	548	2.28
Low-Low	3,807	15.85	3,798	15.81	3,534	14.71
Low-High	2,090	8.70	2,087	8.69	1,771	7.37
High-Low	493	2.05	499	2.08	87	0.36
1200 ft						
Not significant	6,845	28.49	6,710	27.93	7,910	32.93
High-High	1,702	7.08	1,702	7.08	1,007	4.19
Low-Low	11,677	48.61	11,783	49.05	11686	48.65
Low-High	3,125	13.01	3,161	13.16	3258	13.56
High-Low	674	2.81	667	2.78	162	0.67
2800 ft						
Not significant	3006	12.51	3157	13.14	3,267.00	13.60
High-High	2,674	11.13	2,718	11.31	2,114.00	8.80
Low-Low	1,2872	53.58	12,555	52.26	13,783.00	57.37
Low-High	3,555	14.80	3,480	14.49	4,428.00	18.43
High-Low	1,916	7.98	2,113	8.80	431.00	1.79

where places with a high level of exposure were associated with low crime (High-Low) or places with low exposure scores were associated high crime (Low-High). Instances of the former relationship represent places that run counter to the general assumption that drinking places exposure is positively related to crime. These places are especially good candidates for further investigation. For this preliminary investigation, all four types of spatial autocorrelation are examined.

Comparing the three measures reveals a remarkable level of consistency between the Simple Count, IDW Count, and DWA measures of exposure (Table 12.4). The number of places that change classification from one measure to the other is relative small. The DWA is very different from the other two indicating the inclusion of sales in the calculation of exposure significantly changes the bivariate relationship between drinking places and crime. These differences are consistent as distance thresholds get larger. However, using larger thresholds spreads the influence of facilities over larger distances and increases the number of street blocks with exposure to drinking places. Not surprisingly, the number of significant relationships also increases. Of particular interest to understanding the influence of drinking places, the number of significant high exposure-high crime and high exposure-low crime places increase.

Bivariate LISA classifications for one part of Seattle are mapped to examine the patterns at the 800 ft and 1,200 ft thresholds for each of the exposure measures (Figs. 12.1 and 12.2). The color of the street indicates the type of spatial autocorrelation that exists between a street block and its neighbors. Only significant relationships are

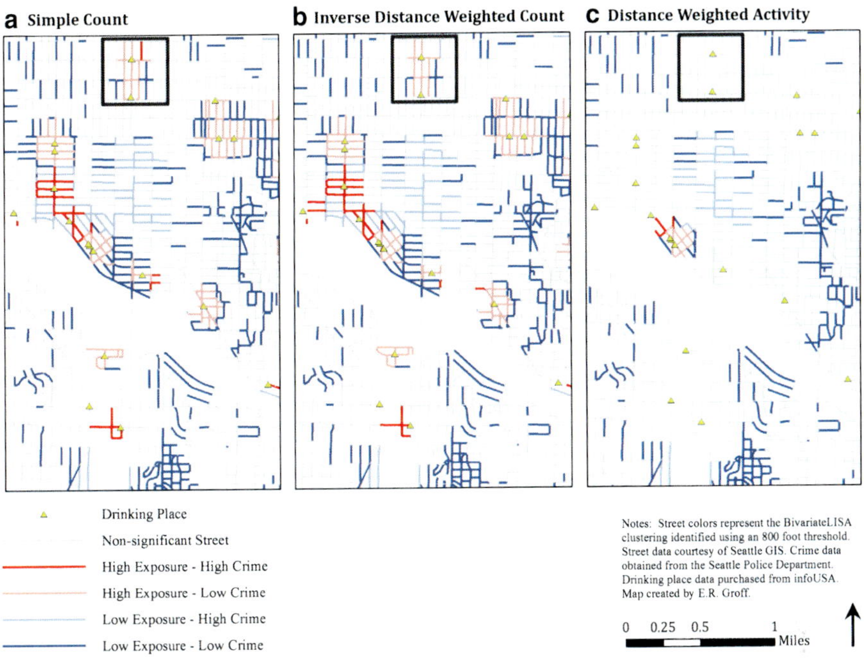

Fig. 12.1 (**a**) Simple count (**b**) Inverse distance weighted count (**c**) Distance weighted activity. Bivariate LISA Results for 800 Foot Threshold (Note: The *black square* near the top of all three maps highlights one area where change occurred across all three measures. There is no drinking place within 1200 ft to the north of the area shown)

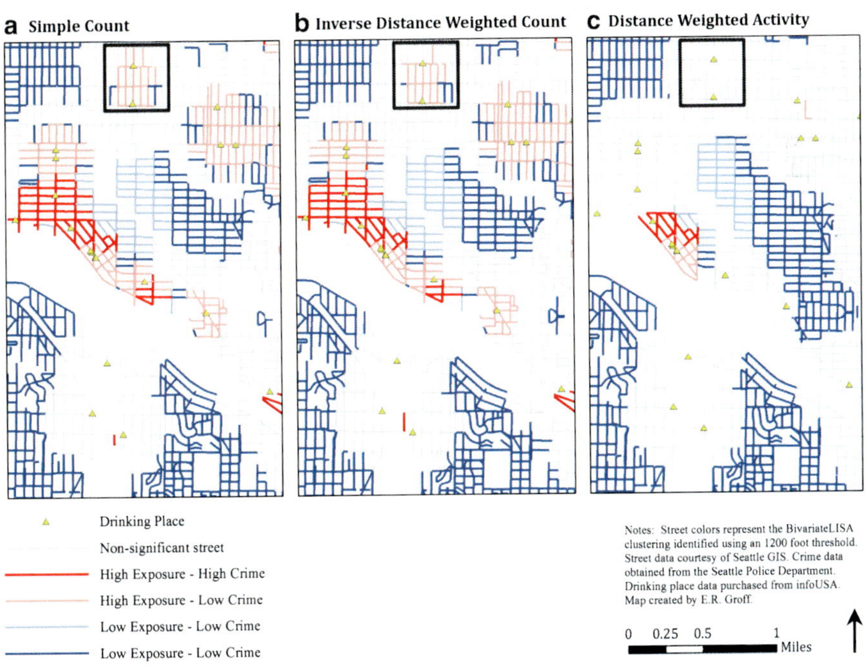

Fig 12.2 Bivariate LISA Results for 1200 Foot Threshold

shown ($p<.01$). Streets which have a high score on their exposure measure and are located near streets with high numbers of crime events are red. These are the streets where exposure is positively correlated with crime. Lighter colors on the map represent places with negative spatial autocorrelation. Streets scoring high on exposure to drinking places which are near streets with low crime appear pink. At these places, the presence of bars nearby is associated with low crime counts. Streets classified as low on exposure and surrounded by high crime places are light blue. These places are high crime but are not near a bar and thus are less interesting to the questions posed here. Finally, if a street is low on exposure and surrounded by low crime places, it is dark blue. These are the least interesting places in terms of exposure to facilities and crime because neither condition of interest is present.

The advantage to examining the results on a map is the ability to visualize where the relationships are located in Seattle. Consequently, locations where different measures classified the same place differently are evident. However, it is very hard to distinguish micro level patterns when mapping an entire city. Instead, one portion of the city is mapped as an illustration. Turning first to the relationships revealed at the 800 ft threshold by the simple count measure of exposure (Fig. 12.1a), there a few street blocks near drinking places (symbolized as yellow triangles) with high exposure and low crime (pink). In this area, streets adjacent to drinking places are most often classified as high exposure to drinking places associated with low crime. Given the relatively low number of drinking places, there are also many street blocks with low exposure and low crime (dark blue). Notice also the preponderance of thicker light grey streets across the city where there was low exposure to bars but still high crime. A few drinking places are surrounded by grey streets, in these places there was no significant relationship between the level of exposure and the amount of crime in their vicinity.

Figure 12.1b reveals the IDW measure of exposure has a pattern almost identical to the one produced by exposure as measured by a Simple Count. There was very little visual difference in the spatial autocorrelation of exposure measure when operationalized as a simple count or as a distance discounted count (IDW).

Incorporating the annual sales of the drinking places into a measure that represented the distance discounted presence of drinking places changed the spatial pattern quite a bit (Fig. 12.1c). Several areas that had significant associations between count-based measures of exposure and crime are no longer significant. At the same time, in some formerly non-significant (lightest) places near drinking places (areas not shown), there were now significant relationships to crime. A major difficulty in explaining these patterns lies in the complex nature of environment-crime relationships but some possible explanations for this finding are explored in the discussion section.[8]

As indicated by the tabular results, increasing the threshold to 1200 ft results in more street blocks with significant local relationships (between drinking place exposure and crime) (Fig. 12.2). Across the measures, once again Simple Count

[8] To save space only the results from the 800 and 1200 ft threshold are shown. Maps of additional thresholds are available from the author.

(Fig. 12.2a) and IDW (Fig. 12.2b) produce consistent patterns but DWA (Fig. 12.2c) is very different overall. The black square is again provided to highlight one area with differences visible across all three measures. In the area within the black square, Simple Count found more streets with significant High-Low relationships than IDW while DWA found none.

12.5 Discussion

This chapter demonstrates how geoprocessing models can aid research in two ways. One, by allowing the operationalization of new, geographically-based measures of exposure and two, by automating complex processes allowing the examination of measure performance across multiple thresholds. The ability to more precisely model spatial influence has increased due to greater availability of micro level data and new software functionality. Visual tools such as geoprocessing models provide the capability to automate processes within GIS. This automation is what makes it feasible to create more sophisticated distance-based measures of spatial interaction. Here a geoprocessing model was built to create three different measures of exposure to facilities. The model was then applied to the test case of quantifying the exposure of street blocks to drinking places in Seattle Washington.

The geoprocessing model provided the ability to easily create three different conceptualizations of exposure to drinking places and then test them under three different thresholds. Just computing the four different distance thresholds generated here would have involved over 240 manual steps and countless hours. Instead using the geoprocessing model running the model could be done in a few minutes and only involved setting the input data, threshold distance, and output data.

Analysis of model output found it was consistent with expectations. Exposure values were lowest for inverse-distance-weighted (IDW) because it weighted the presence of a facility by the distance from the street block. Exposure values were highest for distance-weighted-activity (DWA) because it represented inverse distance weighted by total sales volume. Simple-Count values fell in the middle.

Globally, there was little difference among the three measures at short distances. As distance increased, IDW emerged as producing the best fitting model for the illustrative example of drinking places and crime. The outcomes of a bivariate LISA analysis to examine local variation in the relationships of exposure and total crime revealed Simple Count and IDW were surprisingly similar but some local differences in the spatial patterns were evident in the maps. The results from the DWA measure were significantly different than the distance-only measures. Close inspection of the places that were classified as high DWA exposure and were significantly related to their neighbors reveals those street blocks were near either a single very high volume drinking place or near several low to moderate sales volume drinking places on adjacent streets.

In addition, the amount of activity at a facility is only one aspect of its potential influence. Perhaps equally important are the type of people and the place managers present at the facility. Madensen and Eck (2008) have proposed a model of bar-related violence that identifies bar 'theme', physical characteristics of the bar, and marketing strategies employed as critical factors to consider. Unfortunately, no data were available to represent of those facility attributes for all the drinking places in Seattle. Future research should include these aspects to get a more precise idea of what facility characteristics and place management characteristics interact.

One plausible explanation for the similarity in Simple Count and IDW has to do with the low number of drinking places in Seattle. The more facilities within a buffer the greater the difference will be between the Simple Count and IDW Count measures. With relatively few drinking places within each threshold there were fewer opportunities to weight the influence of facilities and thus the two measures were very similar. The short distance thresholds used, although empirically and theoretically justified, made it even more difficult to find multiple drinking places within a threshold. Greater differences among measures at short distances might be found in cities such as Philadelphia, PA, Washington DC, or Boston MA which have many more drinking places.

The measures created here tell only part of the story and do not control for how other features of the built and social environment might be interacting with the presence of a bar to determine the crime rate. Regardless of the type of facility being examined, in order to fully understand the differences observed, future research should test the measures in a multivariate model which could control for the differences observed at the local level. A multivariate model that incorporates the different measures of exposure as predictors would take into account those other factors and better evaluate the unique contribution of each measure to explaining variations in crime at street blocks. In addition, the differences in the pattern of bivariate correlations suggest the use of geographically weighted regression to examine the relationship in terms of multivariate predictors may be more illuminating than standard spatial regression models.

Finally, the model currently implements one of several possible weighting schemes such as exponential or inverse distance-weighted squared which would weight closer facilities much higher and the ones father away much lower. And of course, the current model needs to be applied to many different types of facilities in a variety of cities.

This chapter demonstrated the capability of geoprocessing models to provide a documented and replicable record of the analysis steps performed in GIS. Modeling exposure in different geographies or using different facilities in the same geography are straightforward and simply involve 'pointing' the model to look for different input data. The model can be easily shared with other researchers or practitioners interested in using the measures with their own data or those who want to build on the model to develop additional operationalizations of facility exposure. In sum, this research provides a theoretically sound spatial platform for achieving greater specificity in quantifying the influence of facilities.

Appendix: Documentation of Geoprocessing Model for Calculating Exposure to Facilities

A geoprocessing model was used to develop the cumulative measures of exposure to facilities. There are three main inputs to the model: (1) a network data set; (2) a feature class of point locations representing origins; and 3) a feature class of point locations representing destinations. The remainder of the Appendix provides a step-by-step description of the model. A snapshot of the section of the model appears first followed by a description of that step.

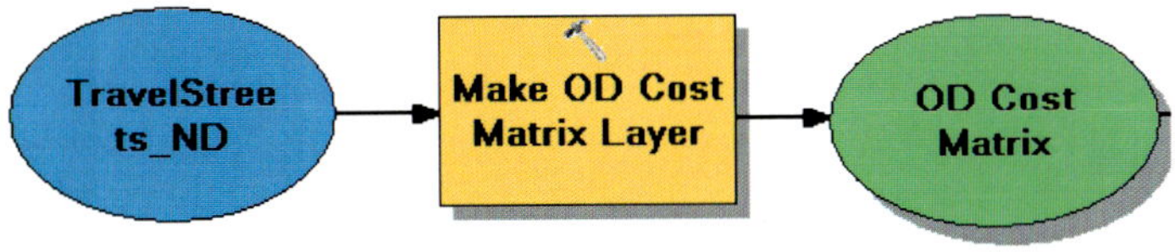

Step 1:
Make an OD Cost Matrix Layer – This tool makes an origin–destination (OD) matrix. The user sets the analysis properties for the matrix. This type of analysis is especially helpful when the goal is to specify the "costs of going from a set of origin locations to a set of destination locations" (Esri help 2011). This is where the user sets the impedance attribute and the threshold distance cutoff. The impedance attribute is the field is used to measure the distance or time between each origin and destination pair.

Output: OD Cost Matrix -

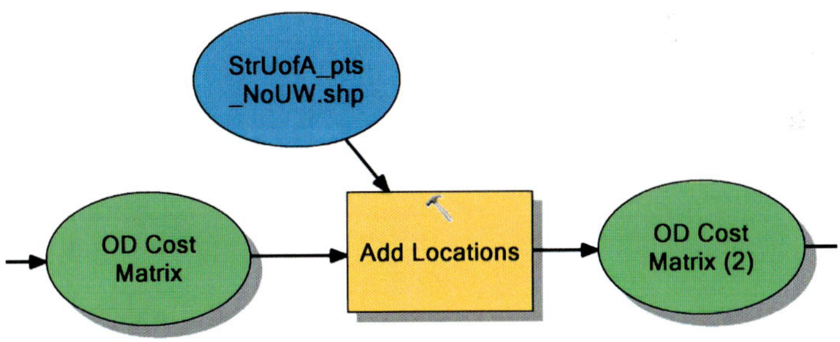

Step 2:
Add Locations – This tool adds locations which are network analysis objects to the network analysis layer. In this model, the first locations added are the **origin** locations which consist of the points representing the mid-points of street blocks (which by definition include both sides of the street between two intersections).

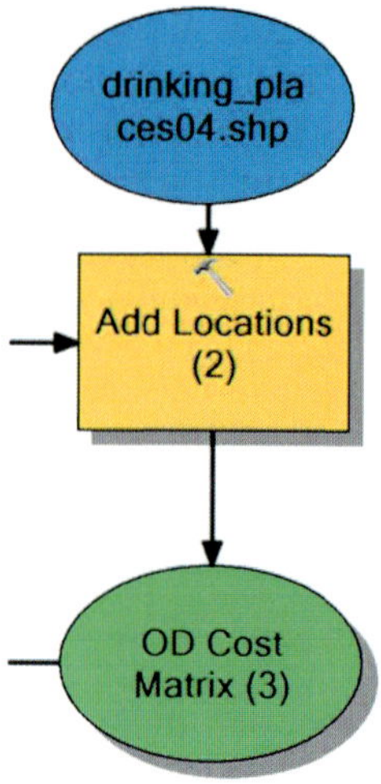

Step 3:
Add Locations – This tool adds locations which are network analysis objects to the network analysis layer. In this model, this second instance of Add Locations tool is adding the **destination** locations which consist of the points representing the type of facility (here it is drinking places).

Output: OD Cost Matrix (3)

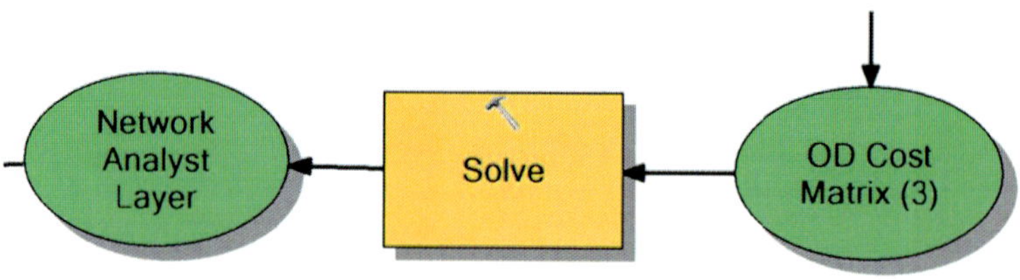

Step 4:
Solve tool – Uses OD Cost Matrix (3) and the properties identified earlier to measure the distances between origin and destination points.

Output: Network Analyst Layer – exists in virtual memory

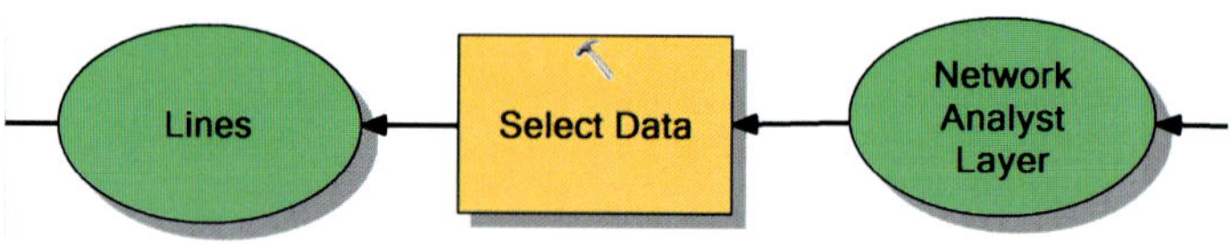

Step 5:
Select Data tool – Selects the Lines data element developed by the Solve tool which exists in the Network Analyst Layer stored in a geodatabase.

Output: Lines data

Step 6:
Copy Features (2) tool – Copies the features from the input feature class into a new layer in the geodatabase for further manipulation. Resides under work/scratch.gdb.

Output: outputodmatrix

The output matrix looks like this:

Table

outputodmatrix

OBJECTID *	Shape *	Name	OriginID	DestinationID	DestinationRank	Total_Length
1	Polyline	4162 - 34047	495	1	1	204.826254
2	Polyline	4662 - 34048	903	2	1	229.48744
3	Polyline	4663 - 34048	904	2	1	265.101566
4	Polyline	4665 - 34048	906	2	1	266.646389
5	Polyline	4668 - 34050	909	3	1	359.709663
6	Polyline	4669 - 34050	910	3	1	193.355371
7	Polyline	4670 - 34050	911	3	1	47.767859
8	Polyline	4672 - 34050	913	3	1	214.005087
9	Polyline	4706 - 34048	941	2	1	318.598799
10	Polyline	4708 - 34048	943	2	1	233.743104
11	Polyline	4712 - 34050	947	3	1	253.832426
12	Polyline	5636 - 25110	1774	4	1	255.337653
13	Polyline	5657 - 25109	1786	14	1	264.365483
14	Polyline	5719 - 7644	1810	5	1	135.778367
15	Polyline	5836 - 32024	1912	6	1	308.856527
16	Polyline	5837 - 32024	1913	6	1	103.178762
17	Polyline	5838 - 32024	1914	6	1	308.181503
18	Polyline	5859 - 32023	1934	7	1	279.4648
19	Polyline	5861 - 32023	1936	7	1	213.074911
20	Polyline	5865 - 32023	1939	7	1	223.026383

1 (0 out of 804 Selected)

Notice the 'Name' field has the origin node number followed by a dash and then the destination node number. In order to work with these, we need to get them into two separate fields. The field to contain the origin data is called 'UofA' and the field to contain the destination data is called 'DP'. The next several steps are to add the fields and then calculate their contents using the contents of the 'Name' field.

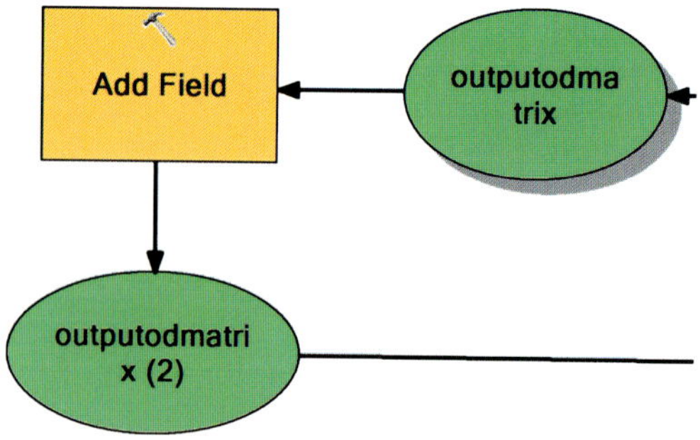

Step 7:
Add Field tool – Adds a field to the specified the input feature class (in this case outputodmatrix) in the geodatabase for further manipulation. The feature class resides under work/scratch.gdb/outputmatrix. The field added is called UofA (field type = double). The purpose of this field is to hold the five digit unique unit of analysis number.

Output: outputodmatrix (2)

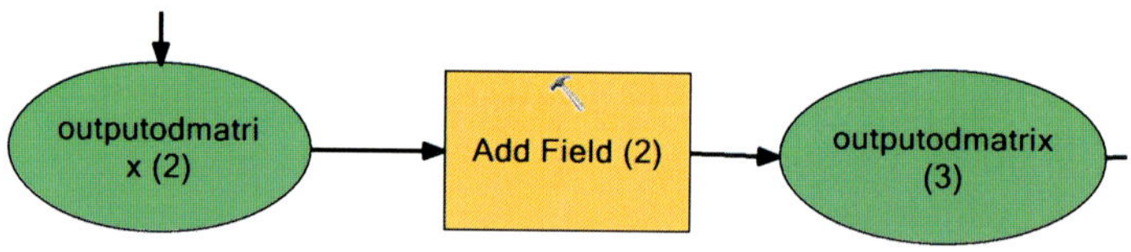

Step 8:
Add Field (2) tool – Adds a field to the specified the input feature class (in this case outputodmatrix (2)) in the geodatabase for further manipulation. The feature class resides under work/scratch.gdb/outputmatrix. The field added is called DP (field type = double). The purpose of this field is to hold the X digit unique unit of analysis number.

Output: outputodmatrix (3)

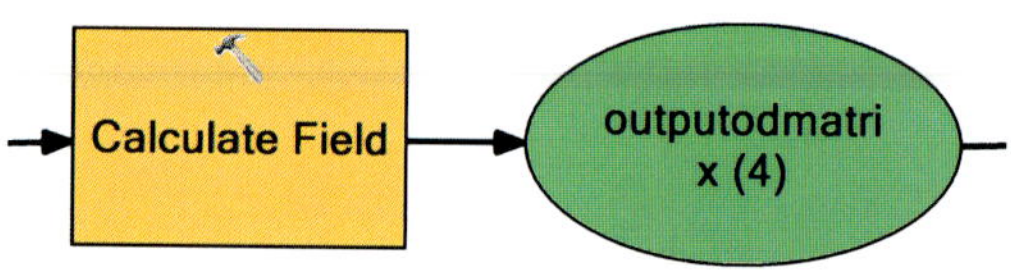

Step 9:
Calculate Field tool – Calculates the values in a field according to an expression supplied by the user. In this case, the expression (theValue) is created from a block of Visual Basic (VB) code which calculates the left side of the 'Name' equal to the 'UofA' field added in Step 7.

```
theVal
theName=[Name]
theLoc=Instr(theName,"-")
theValue=Left(theName,theLoc -2)
```

The first of code creates and sets a variable called 'theName' to be equal to the field called '[Name]'. The second line of code finds the position number of the dash in the contents of the 'theName' field. The third line subtracts 2 from the variable 'theLoc' (this was the position of the dash in the string). Output: outputodmatrix (4)

Output: OD Cost Matrix (2)

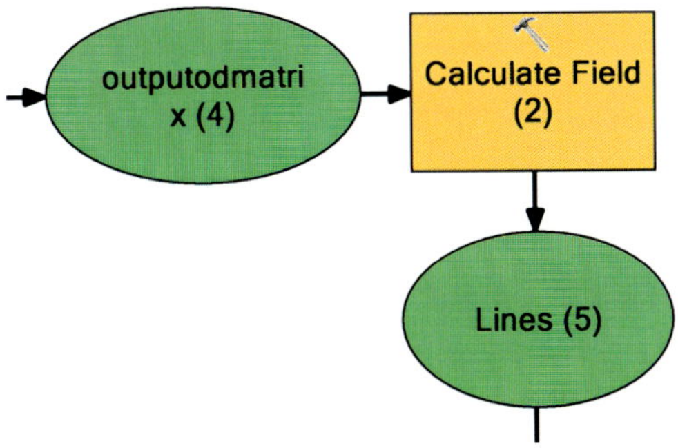

Step 10:
Calculate Field (2) tool – Calculates the values in a field according to an expression supplied by the user. In this case, the expression ('theValue') is created from a block of Visual Basic (VB) code which calculates the right side of the 'Name' equal to the 'DP' field added in Step 8.

```
theVal
theName=[Name]
theSize=Len(theName)
theLoc=Instr(theName,"-")
theValue=Right(theName,theSize – theLoc)
```

Similar to the code in Step 9, the purpose of this code is to extract the numbers representing the destination point's unique id. The first line of code creates and sets a variable called 'theName' to be equal to the field called '[Name]'. The second line of code creates and sets a variable called 'theSize' equal to the total length of 'theName' variable. For the first record in the sample below, 'theSize' =4. The third line creates and sets 'theLoc' variable to the position of the dash in the string (first line below 'theLoc' =6). The fourth line creates and sets a variable called 'theValue' to be the rightmost part of the string in 'theName' variable equal to the difference between 'theSize' and 'theLoc' (same example , 4–6=–2). The Expression box tells the computer to set the value of 'UofA' equal to 'theValue'.

Output: Lines (5)

After this step output matrix looks like this:

Table

outputodmatrix

OBJECTID *	Shape *	Name	OriginID	DestinationID	DestinationRank	Total_Length	Shape_Length	UofA	DP
1	Polyline	4162 - 34047	495	1	1	204.826254	0	4162	34047
2	Polyline	4662 - 34048	903	2	1	229.48744	0	4662	34048
3	Polyline	4663 - 34048	904	2	1	265.101566	0	4663	34048
4	Polyline	4665 - 34048	906	2	1	266.646389	0	4665	34048
5	Polyline	4668 - 34050	909	3	1	359.709663	0	4668	34050
6	Polyline	4669 - 34050	910	3	1	193.355371	0	4669	34050
7	Polyline	4670 - 34050	911	3	1	47.767859	0	4670	34050
8	Polyline	4672 - 34050	913	3	1	214.005087	0	4672	34050
9	Polyline	4706 - 34048	941	2	1	318.598799	0	4706	34048
10	Polyline	4708 - 34048	943	2	1	233.743104	0	4708	34048
11	Polyline	4712 - 34050	947	3	1	253.832426	0	4712	34050
12	Polyline	5636 - 25110	1774	4	1	255.337653	0	5636	25110
13	Polyline	5657 - 25109	1786	14	1	264.365483	0	5657	25109
14	Polyline	5719 - 7644	1810	5	1	135.778367	0	5719	7644
15	Polyline	5836 - 32024	1912	6	1	308.856527	0	5836	32024
16	Polyline	5837 - 32024	1913	6	1	103.178762	0	5837	32024
17	Polyline	5838 - 32024	1914	6	1	308.181503	0	5838	32024
18	Polyline	5859 - 32023	1934	7	1	279.4648	0	5859	32023
19	Polyline	5861 - 32023	1936	7	1	213.074911	0	5861	32023
20	Polyline	5865 - 32023	1939	7	1	223.026383	0	5865	32023

1 (0 out of 804 Selected)

outputodmatrix

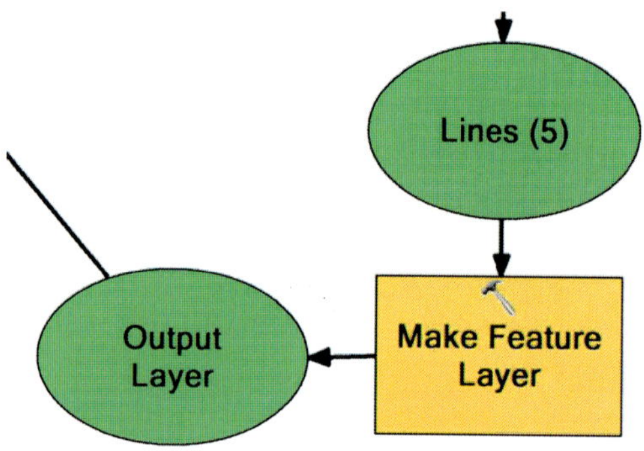

Step 11:

Make Feature Layer tool – Creates a feature layer from the lines representing the distances from each origin to each destination. This layer is temporary and will not persist after the session ends.

Output: Output Data Layer consisting of outputodmatrix_Layer

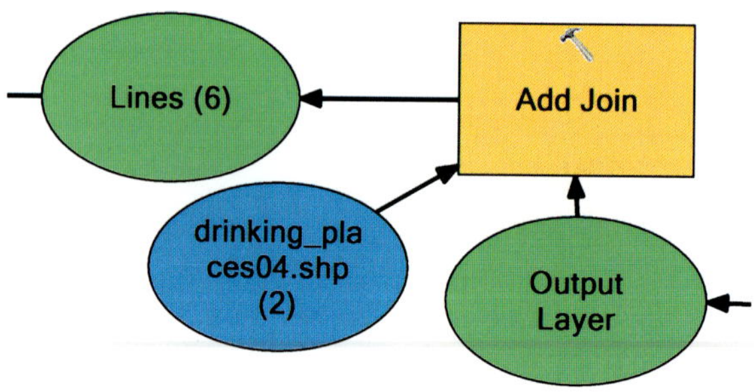

Step 12:
Add Join tool – Joins a layer or table view to another layer or table view based on a common field. In this case, the join is from the drinking places shape file to the Output Layer using the DP field. This allows the attributes attached to each drinking place to be used in the calculation of a measure.

Output: Lines (6) a composite of outputodmatrix_Layer

Step 13:
Copy Features tool – Copies the features of the input layer (in this case Lines (6)) to a new layer or feature class. The new feature class is stored in work\scratch.gdb\lines

Output: Lines (7) a composite of outputodmatrix_Layer

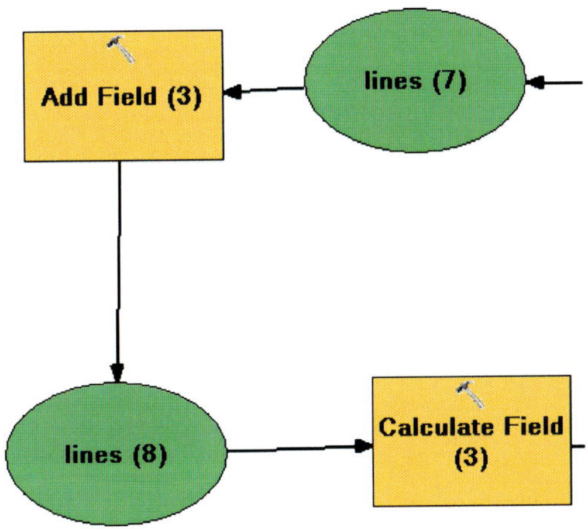

Step 14:
Add Field (3) tool – Adds a field to a table. In the model, this is adding the field 'i_Exp04' to hold the exposure values for inverse distance weighted count of drinking places in 2004. A value is calculated for each origin–destination pair within the threshold distance. In this case the influence of the drinking place is reduced from 1 based on the distance away from the origin. The farther the distance, the lower the resulting value attached to the drinking place.

Output: Lines (8) store in … work\scratch.gdb\lines

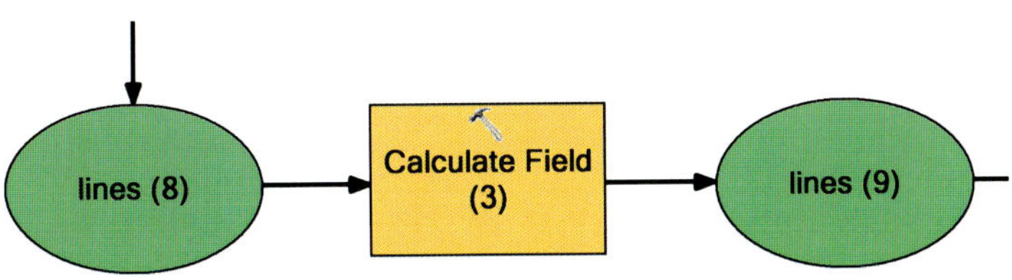

Step 15:

Calculate Field (3) tool – Calculates the field 'i_Exp04' equal to the formula in the Expression box. In the model, the expression is 1- Sqr (([Total_Length] /5280)) which is the formula for inverse distance weighting (IDW) of each drinking place.

Output: Lines (9) store in … work\scratch.gdb\lines

Step 16:

Add Field (4) tool – Adds a field to a table. In the model, this is adding the field 'i_ExpSal04' to hold the exposure values for inverse distance weighted annual sales a of drinking place in 2004. A value is calculated for each origin–destination pair within the threshold distance.

Output: Lines (10) store in … work\scratch.gdb\lines

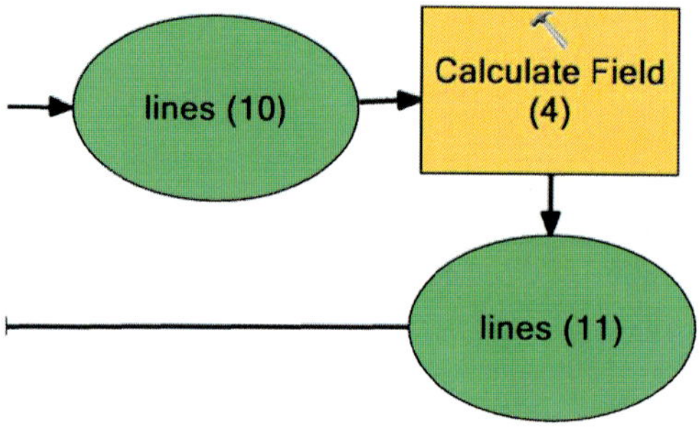

Step 17:

Calculate Field (4) tool – Calculates a value for the field 'i_ExpSal04' using the expression: (1- Sqr ([Total_Length] /5280))* [drinking_places04_num_sales]. This field holds the exposure values for inverse distance weighted annual sales of each drinking place within the threshold distance of each street block. A separate value is calculated for each origin–destination pair within the threshold distance.

Output: Lines (11) store in … work\scratch.gdb\lines

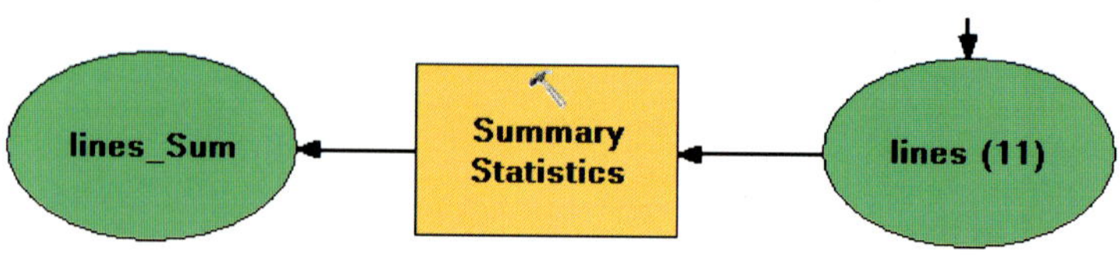

Step 18:
Summary Statistics tool – Calculates summary statistics for each field in a table. For each street block (UofA field is used as unique identifier), all the values of the 'i_Exp04' and the 'i_ExpSal04' are summed. This produces a table with one record for each street block and fields containing the cumulative values for length, 'i_Exp04' and 'i_ExpSal04'.

Output: Lines Sum – a new table that is written out and stored in … work\scratch.gdb\lines_Sum

References

Anselin L (1995) Local indicators of spatial association. Geogr Anal 27:93–115

Anselin L (2003) GeoDa 0.9 User's guide. Spatial analysis laboratory, University of Illinois, Urbana-Champaign

Bernasco W, Block R (2011) Robberies in Chicago: a block-level analysis of the influence of crime generators, crime attractors, and offender anchor points. J Res Crime Delinq 48(1):33–57

Brantingham PL, Brantingham PJ (1982) Mobility, notoriety, and crime: a study of crime patterns in urban nodal points. J Environ Syst 11(1):89–99

Brantingham PJ, Brantingham PL (1984) Patterns in crime. Macmillan, New York

Brantingham PJ, Brantingham PL (1991a) Introduction to the 1991 reissue: notes on environmental criminology. In: Brantingham P, Brantingham P (eds) Environmental criminology. Waveland Press Inc, Prospect Heights, pp 1–6

Brantingham PJ, Brantingham PL (1991b) Notes on the geometry of crime. In: Brantingham P, Brantingham P (eds) Environmental criminology. Waveland Press, Inc, Prospect Heights, pp 27–54

Brantingham PL, Brantingham PJ (1995) Criminality of place: crime generators and crime attractors. Eur J Crim Policy Res 3(3):5–26

Carlstein T, Parkes D, Thrift NJ (1978) Human activity and time geography. Halsted, New York

Clarke RV, Cornish DB (1985) Modeling offender's decisions: a framework for research and policy. In: Tonry M, Morris N (eds) Crime and justice: an annual review of research, vol 6. University of Chicago Press, Chicago, pp 23–42

Cohen LE, Felson M (1979) Social change and crime rate trends: a routine activity approach. Am Sociol Rev 44:588–608

Eck JE (1995) A general model of the geography of illicit retail marketplaces. In: Eck JE, Weisburd D (eds) Crime and place. Willow Tree Press, Monsey, pp 67–93

Eck JE, Gersh JS, Taylor C (2000) Finding crime hot spots through repeat address mapping. In: Goldsmith V, McGuire P, Mollenkopf GJH, Ross TA (eds) Analyzing crime patterns: frontiers of practice. Sage Publications, Thousand Oaks, pp 49–64)

Esri (2012) ArcGIS 10.0 help. Esri, Redlands, California.

Fan Y-H (1998) Effective defect detection and classification methodology based on integrated laser scanning inspection and automatic defect classification. Paper presented at the 1998 IEEE/Semi advanced semiconductor manufacturing conference and workshop, Boston, MA

Felson M (1986) Predicting crime potential at any point on the city map. In: Figlio RM, Hakim S, Rengert GF (eds) Metropolitan crime patterns. Criminal Justice Press, Monsey, pp 127–136

Felson M (1995) Those who discourage crime. In: Eck JE, Weisburd D (eds) Crime and place. Willow Tree Press, Monsey, pp 53–66

Frisbie DW, Fishbine G, Hintz R, Joelson M, Nutter JB (1978) Crime in minneapolis: proposals for prevention. Minnesota Crime Prevention Center, Minneapolis

Groff ER (2011) Exploring 'Near': characterizing the spatial extent of drinking place influence on crime. Aust N Z J Criminol 44:156–179

Groff, ER, and Thomas D (1998) Characterizing crime 'Within a distance of': a comparison of radius vs. street distance. In: ESRI Mid-Atlantic Users Group Meeting. Philadelphia, PA

Groff ER, Weisburd D, Morris N (2009) Where the action is at places: examining spatio-temporal patterns of juvenile crime at places using trajectory analysis and GIS. In: Weisburd D, Bernasco W, Bruinsma G (eds) Putting crime in its place: units of analysis in spatial crime research. Springer, New York, pp 60–86

Hägerstrand T (1970) What about people in regional science? Pap Reg Sci Assoc 24:7–21

Hägerstrand T (1973) The domain of human geography. In: Chorley RJ (ed) Directions in geography. Methuen, London, pp 67–87

Harries KD (1990) Geographic factors in policing. Police Executive Research Forum, Washington, DC

Harries KD (1999) Mapping crime: principle and practice. U.S. National Institute of Justice, Washington, DC

Horton FE, Reynolds DR (1971) Action space differentials in cities. In: McConnell H, Yaseen D (eds) Perspectives in geography: models of spatial interaction. Northern Illinois University Press, Dekab, pp 83–102

Katzman MT (1981) The supply of criminals: a geo-economic examination. In: Hakim S, Rengert GF (eds) Crime spillover. Sage, Beverly Hills, pp 119–134

Kitchens T (2006) Automating software development processes. Developer.* Retrieved from 05 Mar 2012 . http://www.developerdotstar.com/mag/articles/automate_software_process.html

Loukaitou-Sideris A (1999) Hot spots of bus stop crime: The importance of environmental attributes. J Am Plan Assoc 65(4):395–411

Madensen TD, Eck JE (2008) Violence in bars: exploring the impact of place manage decision-making. Crime Prev Commun Saf 10:111–125

McCord ES, Ratcliffe JH (2007) A micro-spatial analysis of the demographic and criminogenic environment of drug markets in Philadelphia. Aust N Z J Criminol 40(1):43–63

Miethe TD, McDowall D (1993) Contextual effects in models of criminal victimization. Social Force 71:741–759

Miller H (2004) Tobler's first law and spatial analysis. Ann Assoc Am Geogr 94(2):284–289

Moudry JA (2012) Real benefits of automated processes. Retrieved from 5 March 2012. http://www.nextgenpinews.com/files/Real%20Benefits%20of%20Automated%20Processes.pdf

Pred A (1967) Behavior and location: foundations for a geographic and dynamic location theory, part I. Gleerup, Lund

Pred A (1969) Behavior and location: foundations for a geographic and dynamic location theory, part II. Gleerup, Lund

Ratcliffe JH (2011) How near is near? Quantifying the spatial influence of crime attractors and generators. In: Andresen M, Kinney JB (eds) Patterns, prevention, and geometry of crime. Routledge, London

Ratcliffe JH (2012) The spatial extent of criminogenic places on the surrounding environment: a change-point regression of violence around bars. Geogr Anal 44(4):302–320

Rengert GF, Ratcliffe JH, Chakravorty S (2005) Policing illegal drug markets: geographic approaches to crime reduction. Criminal Justice Press, Monsey

Rice KJ, Smith WR (2002) Sociological models of automotive theft: integrating routine activity and social disorganization approaches. J Res Crime Delinq 39(3):304–336

Roncek DW, Bell R (1981) Bars, blocks and crimes. J Environ Syst 11:35–47

Rossmo DK, Rombouts S (2008) Geographic profiling. In: Wortley R, Mazerolle L (eds) Environmental criminology and crime analysis. Willan Publishing, Portland, pp 78–93

Sherman LW, Gartin P, Buerger ME (1989) Hot spots of predatory crime: routine activities and the criminology of place. Criminology 27(1):27–55

Taylor RB (1997) Social order and disorder of street blocks and neighborhoods: ecology, microecology, and the systemic model of social disorganization. J Res Crime Delinq 34(1):113–155

Taylor RB (2010) Communities, crime, and reactions to crime multilevel models: accomplishments and meta-challenges. J Quant Criminol 26(4):455–466

Thoma RS (1991) Automated phenylthiocarbamyl amino acid analysis of carboxypeptidase/aminopeptidase digests and acid hydrolysates. J Chromatogr 537:153

Tobler W (1970) A computer model simulation of urban growth in the detroit region. Econ Geogr 46(2):234–240

Weisburd D, Bushway S, Lum C, Yang S-M (2004) Trajectories of crime at places: a longitudinal study of street segments in the city of seattle. Criminology 42(2):283–321

Weisburd D, Bruinsma G, Bernasco W (2009) Units of analysis in geographic criminology: historical development, critical issues and open questions. In: Weisburd D, Bernasco W, Bruinsma G (eds) Putting crime in its place: units of analysis in spatial crime research. Springer, New York, pp 3–31

Weisburd D, Groff E, Yang S-M (2011) Understanding developmental crime trajectories at places: social disorganization and opportunity perspectives at micro units of geography. Office of Justice Programs, National Institute of Justice, Washington, DC

Weisburd D, Groff E, Yang S-M (2012) The criminology of place: street segments and our understanding of the crime problem. Oxford University Press, Oxford

Wilcox P, Quisenberry N, Cabrera DT, Jones S (2004) Busy places and broken windows? Toward defining the role of physical structure and process in community crime models. Sociol Q 45(2):185–207

Worboys M, Duckham M (2004) GIS: a computing perspective. CRC Press, Boca Raton

Zipf G (1950) Principle of least effort. Addison Wesley, Reading

Chapter 13
A Spatial Analysis of Methamphetamine Lab Seizures in the Midwest High-Intensity Drug Trafficking Area Before and After Federal Precursor Legislation

Aaron H. Gilbreath

Abstract This chapter uses spatial zero-inflated negative binomial regression to assess the relationship between methamphetamine lab seizures and county characteristics in the states of the Midwest High-Intensity Drug Trafficking Area for the years 2000–2010. I regressed meth lab seizure statistics from the El Paso Intelligence Center with county characteristics obtained from the 2000 and 2010 censuses. Two models were run to determine if the significant covariates for meth lab seizures changed as a result of the National Combat Methamphetamine Epidemic Act of 2005, which restricted precursor sales nationwide. The study does not find a significant difference in the covariates of the two models. In both cases, the most significant predictor of the presence of any meth lab in a county was their presence in neighboring counties, suggesting the agglomeration of methamphetamine production. In the count portion of the models, lab seizures were closely correlated with counties that were highly white but possessed the other characteristics associated with social disorganization.

Keywords Methamphetamine • Precursor legislation • Spatial regression • Zero-inflated models • Zero-inflated negative binomial regression

13.1 Introduction

This study, which assesses the relationship between domestic methamphetamine production and county characteristics in the Midwest High-Intensity Drug Trafficking Area (HIDTA) before and after the Combat Methamphetamine Epidemic Act of 2005, is positioned at the intersection of two burgeoning strands

A.H. Gilbreath (✉)
Department of Geography, University of Kansas, Lawrence, KS, USA
e-mail: ahg@ku.edu

M. Leitner (ed.), *Crime Modeling and Mapping Using Geospatial Technologies*, Geotechnologies and the Environment 8, DOI 10.1007/978-94-007-4997-9_13,

of literature regarding the illegal stimulant methamphetamine. The first deals with attempts to determine the covariates associated with domestic methamphetamine production in clandestine laboratories (Lu and Burnum 2008; Weisheit and Wells 2010; Gilbreath 2010). The second assesses the impact that various attempts to control domestic methamphetamine production have had on methamphetamine indicators (Cunningham and Liu 2003, 2005; Rueters and Caulkins 2003; McBride et al. 2008; Dobkin and Niciosa 2009).

The study fills holes in both literatures. First, it applies a spatial version of a zero-inflated negative binomial regression (ZINB) to seizure data to determine the significant covariates of methamphetamine production. Previous studies either did not use a zero-inflated model or were aspatial in their methodologies. Second, the study assesses the spatial impact of precursor laws. Other studies of this type have focused entirely on differences in the observation of methamphetamine indicators over *time* (before and after implementation), but not across *space*.

To conduct the analysis, I correlated seizure data at the county level from the El Paso Intelligence Center (EPIC) with demographic data from the United States Census using spatial zero-inflated negative binomial models. The goal is to determine what county characteristics were most closely associated with the domestic, clandestine production of methamphetamine before and after precursor legislation.

13.2 History of the Problem

13.2.1 Methamphetamine in the Midwest

The Midwest has a long and complicated history with methamphetamine. In the 1930s, when amphetamine products were first introduced as asthma medications, inhalers containing the drugs were abused by members of the Kansas City jazz scene (Rasmussen 2008). In the late 1960s high-dosage methamphetamine injection became a serious concern for the medical community nationwide, particularly in centers of the 1960s counter culture such as San Francisco and the East Village of New York. However, other locales not considered part of the hippie movement also experienced serious outbreaks of abuse. St Louis, for example, began to see intravenous methamphetamine use as early as the late 1950s (Rawlin 1968). When police crackdowns began to limit the supply of diverted, legally produced liquid methamphetamine, the country's first clandestine methamphetamine labs began to appear in 1962 and 1963. In 1971 the government made methamphetamine a Schedule II substance, greatly limiting its legal uses and forcibly reducing production.

Without legally produced meth available for diversion, clandestine production became the nation's primary source. Outlaw motorcycle gangs (OMGs) controlled most production and distribution between the drug's scheduling in 1971 and the late 1980s. These groups were responsible for bringing methamphetamine back to the Midwest from its enduring base in California. The drug supposedly gets one of its

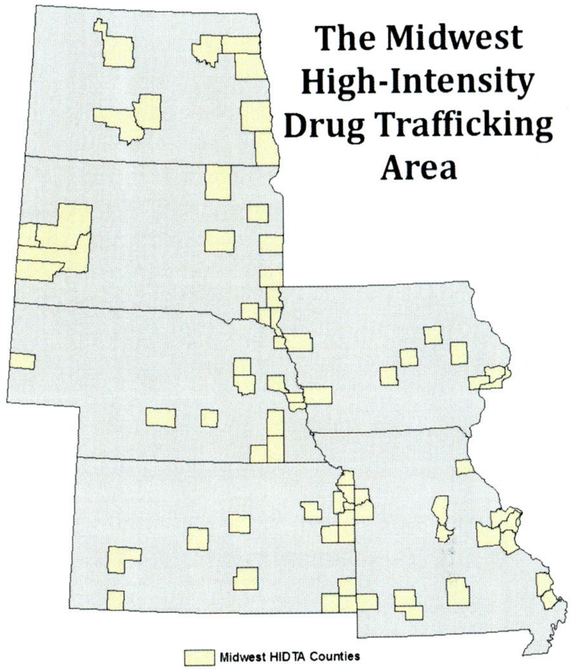

Fig. 13.1 The Midwest HIDTA

many nicknames, "crank," from the fact that OMGs frequently smuggled it in the crankcases of their bikes. Owen (2007) claimed that members of the Hell's Angels first brought meth production to southwestern Missouri in the 1970s, setting up labs in Mark Twain National Forest. The OMG meth presence in the region grew from there. In 1984 and 1985 members of a different biker gang, the Bandidos, were arrested throughout the Midwest for attempting to make and sell meth (*Los Angeles Times* 1985).

During the 1990s, new production methods that relied on easy-to-acquire precursors such as cold pills, rather than costly chemicals, spread among the drug-using populace, and caused a boom in methamphetamine labs and seizures. As one veteran member of the Independence, Missouri, Police Drug Taskforce put it, "in a matter of months [after the introduction of the new recipes], everyone was trying to cook dope" (Sweeny 2010). As availability of methamphetamine grew, so did demand. Between 1993 and 2003 admissions rates for people seeking treatment for methamphetamine abuse in the states of the Midwest HIDTA grew by an average of over 1,300%, and all but North Dakota exceeded the average national rate of 56 admissions for every 100,000 members of the population (DASIS 2006).

The federal government responded to the boom in methamphetamine use and production with the creation of the Midwest HIDTA in 1996. It is comprised of 73 counties in 6 states (Fig. 13.1). Cedar Rapids, Des Moines, Fargo, Kansas City, Omaha, Rapid City, Sioux City, St. Louis, and Wichita all fall within its scope. The HIDTA program had begun 8 years earlier as a way to fund and organize police efforts in key locations of drug trafficking and production. The program's goal is to disrupt drug smuggling and sales by coordinating the efforts of federal, state, and local police agencies. The program in the Midwest has produced mixed results. As mentioned above, rates for people seeking treatment have not been significantly reduced. The rate of labs seized per 100,000 people for the region (107.5 labs per 100,000 people) is also significantly higher than that for the nation as a whole (59.12 labs per 100,000) for the years 2000–2010, which are the focus of this study.

In 2005, the federal government passed the Combat Methamphetamine Epidemic Act, which removed all products containing ephedrine or pseudoephedrine from over-the-counter sales and required customers to show photo identification and to sign for their purchases. The law was an attempt by the government to remove the key ingredients of methamphetamine manufacture from the marketplace as a means to slow the continued growth of the drug's use and production across much of the country (Hunt et al. 2006).

The federal government has a long history of attacking the nation's methamphetamine problem through such supply-side interventions. The first attempt came with the rescheduling of phenyl-2-propanone in 1980. Further attempts occurred in 1988 with the Chemical Diversion and Trafficking Act, in 1993 with the Domestic Chemical Diversion and Control Act, and in 1996 with the Comprehensive Methamphetamine Control Act. In addition to federal action, in the early 2000s many states began to pass their own more stringent precursor legislation (McBride et al. 2008).

13.2.2 Previous Supply Side Intervention Analyses

Several studies have been undertaken to assess the efficacy of methamphetamine supply-side interventions. Cunningham and Liu (2003, 2005) found that the laws of 1988, 1993, and 1996 were effective in reducing both methamphetamine arrests in California and meth-related hospital admissions in California, Nevada, and Arizona. However, Rueter and Caulkins (2003) argued that the same laws had neither reduced methamphetamine use among the general population or arrestees nor significantly lowered the number of methamphetamine-related deaths. Dobkin and Niciosa (2009) observed that the Domestic Chemical Diversion Control Act had enabled the government to significantly disrupt the methamphetamine market in California, where prices soared, purity plummeted, and usage declined in the 18 months after the law was put into place. Weisheit and Wells (2010) found that the 2005 Combat Methamphetamine Epidemic Act significantly reduced the number of clandestine labs in operation across the country. McBride et al. (2008) argued that state precursor legislation generally resulted in significant reductions in the seizure of small toxic methamphetamine labs. To date, however, no study has explored the spatial ramifications of methamphetamine precursor laws by assessing whether the characteristics of the places that tended to have meth labs were changed by the Combat Methamphetamine Epidemic Act.

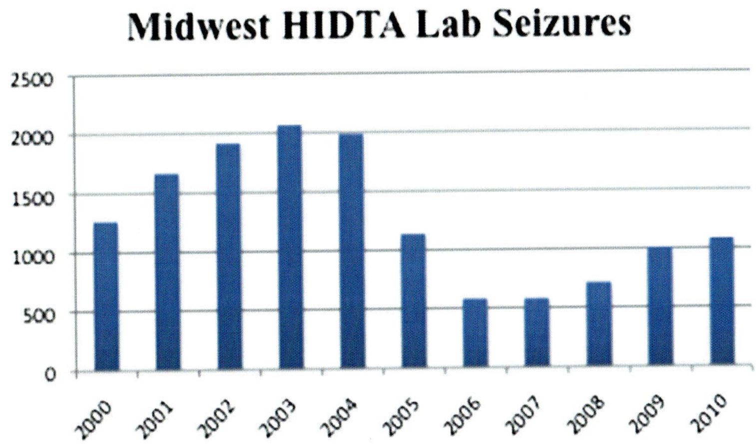

Fig. 13.2 Lab seizure totals by year for the Midwest HIDTA courtesy of the El Paso Intelligence Center

The 2005 Combat Methamphetamine Epidemic Act definitely reduced the number of labs seized nationwide (Weisheit and Wells 2010) and within the Midwest HIDTA.[1] However, the Midwest area has experienced significant growth in domestic methamphetamine production as producers sought to compensate for a dip in the availability of Mexican methamphetamine due to that country's restrictions on ephedrine- and pseudoephedrine-based products (NDIC 2009a, 2009b, 2010). That uptick in production is reflected in the total number of seizures in the region at the end of the decade (Fig. 13.2).

13.3 Methamphetamine Laboratories

13.3.1 Problems Created by Clandestine Laboratories

This study focuses on the domestic production of methamphetamine in clandestine labs in the Midwest HIDTA. Though a large percentage of the drug in the region comes from Mexico, domestic production is still a significant source and problem (NDIC 2010). Numerous ways exist to make methamphetamine. They have risen and declined in popularity over time as a result of police actions and varying demands for product. Some recipes for making the drug, such as the "Nazi Method," have distinctive odors associated with them and must occur far away from population centers to avoid detection. Other procedures have little tell-tale odor and can be

[1] The average annual number of labs seized per county between the years 2000 and the year 2005 was 3.14, and for 2007 through 2010 it was 1.59.

conducted virtually anywhere with enough space to shake a two-liter bottle full of precursors. Labs have been found in hotel rooms, homes, deserted outbuildings, trailers, and even the trunks of cars.

Labs vary in size and can produce quantities from a few grams to 50 pounds or more. Most labs found in the Midwest HIDTA are small-scale, designed to produce enough product for the cook's own use and some extra to sell to others. These operations are alternately referred to as STLs (small toxic labs) or "mom-and-pop" labs. "Superlabs," those producing quantities greater than ten pounds of the drug per batch, are generally associated with highly organized drug-trafficking organizations (DEA 2005). They are not common within the Midwest HIDTA.

Meth labs are incredibly dangerous places. Without any kind of intervention they often end in fires or explosions. Even when such calamity does not occur, a simple raid can be extremely hazardous. Taking apart an active lab can be like defusing a bomb because the production process involves numerous highly flammable solvents, explosive reactants like lithium and water, and noxious gasses.

The DEA estimates that, for every pound of methamphetamine created, five pounds of toxic waste are produced (DEA 2005). Not surprisingly, these byproducts are rarely disposed of properly. More often than not, they are dumped down a drain or left outside to leach into the ground, thus extending contamination well beyond the structure in which the meth was cooked. Though the police are responsible for removing lab equipment from a location, property owners must pay for "the cleanup of residual contamination after gross removal has occurred" (EPA 2009, 3). This can be a costly process. In a Rand Corporation study, Niciosa et al. (2009) estimated that meth labs alone (not including the social cost of meth use) cost the United States $61 million in 2005, not just in clean up and remediation, but also in injuries associated with their operation and seizure.

13.3.2 Understanding Lab Location

To better direct prevention efforts, and for other reasons, it is important to have a thorough understanding of where meth labs tend to locate. Much has been made of the fact that methamphetamine, unlike cocaine or heroin, is a synthetic drug. The nature of its method of production has caused many commentators to believe that it can be produced and used anywhere (Jenkins 1999). If that were the case, we would expect that traditional crime and drug market indicators would be insufficient for predicting the location of methamphetamine labs.

A number of different criminological perspectives exist from which one might select variables to explain lab location (Shaw and McKay 1942; Cohen and Felson 1979; Clarke 1980; Clarke and Felson 1993). But, given the scale at which we are operating (that of entire counties), variables associated with routine activity and rational-choice perspectives are not easily incorporated. We can, however, assess the efficacy of traditional social disorganization variables in predicting lab seizure locations.

Higher crime rates and drug markets tend to cluster in areas where a community has little ability to organize against them. Such lack of neighborhood efficacy is termed social disorganization. It frequently occurs in areas with high population turnover, a large number of renters, and high percentages of poverty, minorities, and single mothers (Shaw and McKay 1942; Sampson and Groves 1989; Kubrin and Weitzer 2003; Rengert et al. 2005, McCord and Ratcliffe 2007; Banerjee et al. 2008; Grattet 2009). If traditional indicators of social disorganization prove to be significant correlates to methamphetamine lab location, then we must reassess the way in which we discuss and analyze synthetic drugs.

Because the actual number and location of all methamphetamine labs is unknown and unknowable, we use lab seizures as a proxy. Obviously, no proxy is perfect, but seizures are the best measure available for domestic production and precedent exists in the literature for using such data to assess the spatial correlates of meth production. Lu and Burnum (2008), in an analysis of lab seizures around Colorado Springs, found lab location to be correlated with neighborhoods that had low median ages, predominantly white populations, and low levels of educational attainment. The authors used a Poisson regression model to assess the covariates, but do not appear to have accounted for spatial effects in their model. As will be made clear below, any analysis of spatial data that does not explicitly assess and account for spatial effects within its model is inherently flawed (see Sect. 13.4.1). Weisheit and Wells (2010) similarly failed to incorporate spatial effects in the regression portion of their analysis, which attempted to determine the covariates for lab seizures for the entire United States. Finally, in a study of lab seizures in Jefferson County, Missouri, Gilbreath (2010) found lab seizure locations correlated with census tracts having high unemployment, low population density, and longer distances from the center of the tract to the nearest police station.

13.4 Data

The seizure data for this project was acquired from the multiagency El Paso Intelligence Center (EPIC) via a Freedom of Information Act request. EPIC is the national clearinghouse for meth lab seizure data. They provided seizures by county for each year from 2000 to 2010. Rather than use just the 73 counties that officially make up the HIDTA, I included each county in the region's six states.[2] Although most of the officially designated counties of the Midwest HIDTA are found in the region's urban centers, methamphetamine has a reputation of being a rural drug, so I thought it was important to include these counties as well.

County-level data offers several advantages in this type of study (Messner et al. 1999). They are a resolution at which data from a number of other sources,

[2] Rather than sample all of Illinois, only Rock Island County was included, as it is the only Illinois county in the Midwest HIDTA.

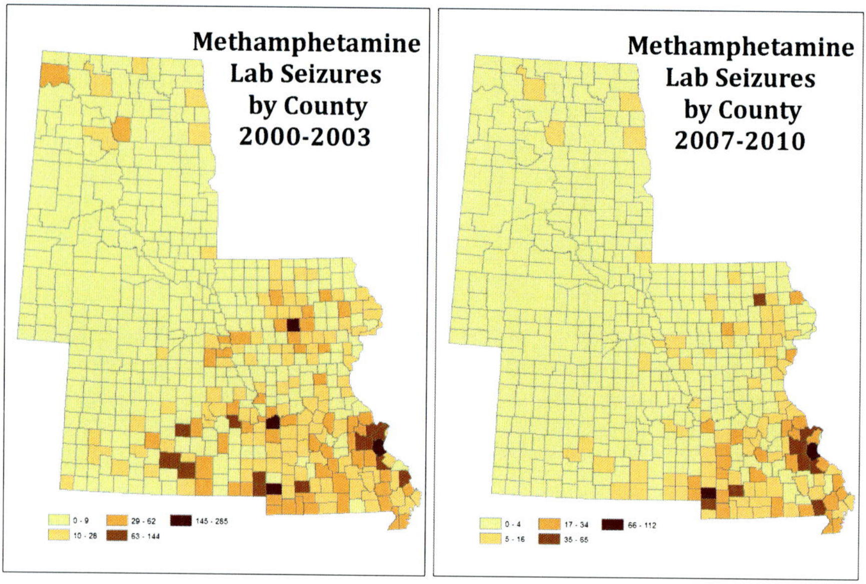

Fig. 13.3 The dependent variables of labs seized per county for the states of the Midwest HIDTA

particularly government agencies, are readily available. Counties also run the gamut from rural to urban, and from poor to rich, and the sheer number of counties in the study, 532, ensures adequate variation within the study area. Finally, this county-level data maintained by EPIC is the most detailed available for a study of this spatial scope.

Because the goal of this study is to assess the potential covariates for lab location both before and after the national precursor law went into effect, I created two dependent variables based on these distinct time periods (Fig. 13.3). The first is the total number of labs seized between 2000 and 2003, the peak years before state and federal precursor laws went into effect (McBride et al. 2008). The second is the total number of labs seized per county between 2007 and 2010.[3]

I selected potential explanatory variables based on a number of theoretical considerations. The goal was to select variables that combined the characteristics of users obtained from the Substance Abuse and Mental Health Services Administration (DASIS 2008; SAMHSA 2009; Hunt et al. 2006) with variables that are consistent with social disorganization theory and the existing geographic literature regarding methamphetamine lab location. [4]After testing a larger number of variables for collinearity, eleven were included in the model (Table 13.1).

[3] Because the Combat Methamphetamine Epidemic Act was not fully implemented until September 2006, my post-precursor law analysis begins with 2007. For the sake of covering the same time span between the two samples, peak years were cut at 2003.

[4] No study exists on the typical methamphetamine cook.

Table 13.1 Descriptive statistics for all variables

	2000–2003				2007–2010			
Variable	Minimum	Maximum	Mean	Standard deviation	Minimum	Maximum	Mean	Standard deviation
Count	0	285	12.99	26.79	0	112	4.35	10.51
Spat. Lag	0	119	13.52	18.26	0	59.75	4.54	8.32
Rent Occ.	12.8	56.8	26.03	6.54	12.8	58.2	26.64	6.65
Med. Age	20.6	51	38.88	4.36	23.5	53.4	41.90	5.55
Male Bach.	2.8	34.1	11.09	3.84	1.6	35.5	12.45	4.55
Poverty	3.5	42	11.45	4.55	4.2	62	14.29	5.97
Sing. Moth.	0.3	20.6	5.11	2.21	0.3	20.6	5.11	2.21
Vacant	3.5	52.9	13.20	7.04	4.8	53.7	15.30	7.37
Live Alone	13.2	40.3	27.44	3.48	14.8	42.6	29.01	3.77
Families	52.3	84.5	68.82	3.95	47.5	82.4	66.33	4.27
Mexican	0	35.5	1.96	4.03	0	49.5	3.39	5.74
White	4.5	99.7	93.25	12.00	2.9	99.2	91.69	12.58
PopDens	0.51	5624.12	56.49	281.47	0.47	5157.39	58.75	269.72
N=532								

Population density was included to test the connection between methamphetamine production and rural areas, as well as to control for differences in county sizes and populations (Herz 2000, United States Congress 2000).[5] I expected that lab seizures would have a positive correlation with population density, as anyone producing methamphetamine with the intent to sell would need to have a market. The percentage of a county whose population is white, the median age, the percentage of males over 25 with a bachelor's degree, and the percentage living in poverty were included based on the general characteristics of methamphetamine users, who tend to be white, young, poor, and under-educated (DASIS 2008; SAMHSA 2009; Hunt et al. 2006). These variables are also closely associated with social disorganization, the exception, of course, being percentage white. I expected the percentage of a county whose population is of Mexican origin to exhibit a negative correlation, assuming that such a county would have the potential of market penetration by Mexican drug-trafficking organizations and thereby eliminate the necessity for local production. I included the percentage of households with a single mother, percentage of homes occupied by renters, and percentage of vacant properties as additional indicators of social disorganization. I expected the percentage of households containing a family to have an opposite correlation to that of the social disorganization variables. The final variable in the models, spatial lag, is explained below (see Sect. 13.4.1). I collected two sets of independent variables: one from the 2000 census, and one from the 2010 census.[6]

[5] Unfortunately, the county level data for rural populations from the 2010 Census will not be available until October of 2012, so the percent of a county's population that is rural could not be used as our rural indicator.

[6] 2010 education attainment variables had to be obtained from the 5-year estimates of the *American Community Survey* after the long-form questionnaire was eliminated for the 2010 census. Economic data for 2010 variables are from the 2009 economic census.

13.5 Methods

13.5.1 Spatial Regression Models

To assess the covariates associated with lab seizures, this study uses a modified version of spatial regression models developed by Anselin (1988) and outlined in Ward and Gleditsche (2008). Such regression techniques are necessary because spatial data frequently exhibits what Getis (2007)) has called *the* fundamental concept of spatial analysis: spatial autocorrelation. Spatial autocorrelation is the clustering of similar values in space. It is frequently present in spatial data because collection units such as census tracts or neighborhoods have porous borders or exist only on maps. Human beings, biological vectors, economic forces, information and infrastructure all cross them at will. Actors in one area thus frequently have an impact on their neighbors. This impact is referred to in the literature as spatial dependence. The presence of spatial dependence, indicated by the significant clustering of similar values (significant spatial autocorrelation), is a sign of the violation of the independence assumptions inherent in most parametric inferential statistics. If significant spatial autocorrelation exists, and is not taken into account within a multivariate analysis, then "false indications of significance, biased parameter estimates, and misleading suggestions of fit" can result (Messner et al. 1999, 427).

Fortunately, several ways exist to account for spatial dependence within a model. If an investigator believes such spatial dependence is a result of actual interaction between observations, then he/she should consider using a *spatial lag* model. In such a model, a new independent variable is added to the regression equation to account for the existing spatial dependence. The lag variable, created using a spatial weights matrix, is usually some combination of the value of the dependent variable for all nearby units to each observation. Depending on the understanding one has of the process being modeled, the weights matrix can be based on some order of contiguous neighbors or on a distance-decay threshold.

It makes sense to use a lagged variable when one thinks of a dependent variable as continuous and potentially influenced by its neighbors. Baller et al. (2001) have associated a significant lag variable in the study of homicide with processes of diffusion, while Mennis et al. (2011) consider it evidence of spatial spillover in their study of juvenile delinquent recidivism. In the case of drug markets, Rengert et al. (2005) associated a significant lag variable with agglomeration.

On the other hand, if one assumed that the spatial effects in their model derives not from actual evidence of interaction between observations, but rather from model misspecification, missing independent variables, or some other statistical nuisance, then he/she might consider a *spatial error model*, in which the spatial dependence is accounted for in the error term.

In the case of the present study, spatial effects almost certainly result from interaction between counties as producers, suppliers, and information travel across borders. As such, a spatial lag model is most appropriate. To that end, I added a spatially lagged variable (based on first-order queen contiguity) to both regression

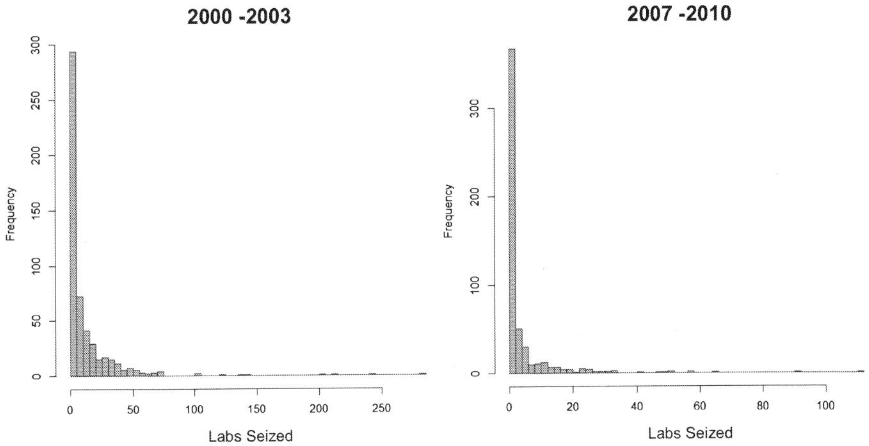

Fig. 13.4 Histograms of the two dependent variables

models. That is, for each county, the spatial lag variable is the average value of labs seized in all the neighboring counties it touches. This type of analysis has a long history in the study of crime, and was recommended by Anselin et al. (2000). The recent special issue of *The Professional Geographer* on the spatial analysis of crime also contains several good examples (e.g. Mennis et al. 2011, Andresen 2011). Baller et al. (2001) produced a spatial regression analysis of nationwide homicide rates that used county-level data much as this study does.

13.5.2 Zero-Inflated Regression Models

Most of the studies cited used ordinary least squares regression (OLS) for their analyses. However, count data (such as the number of labs seized) have several characteristics that make them ill suited for OLS techniques. They often contain a large number of zeros (areas with no observations of the dependent variable) and exhibit a severe positive skew. The two independent variables included here are no exception, since 24.4% of the counties in the 2000–2003 model had no labs seized within their borders and so did 47% in the 2007–2010 model (Fig. 13.4). When data have a disproportionate number of zeroes, a zero-inflated model should be substituted for the OLS one (McDonald and Lattimore 2010). Generally, either a zero-inflated Poisson regression, or a zero-inflated negative binomial regression is necessary.

One chooses between the two regression models based upon whether the distribution of the dependent variable is over-dispersed or not. In order to use a zero-inflated Poisson regression, the dependent variable's mean should be equal to its variance. If this is not the case, and the variance is significantly larger than the mean, then the distribution is overly dispersed, and a zero-inflated negative binomial model should be used (Atkins and Gallop 2007). In this study, the data were over-dispersed, making a zero-inflated negative binomial model the appropriate tool.

A zero-inflated regression model produces two different equations. For this reason it is often referred to as a mixed model. The first equation, sometimes referred to as the hurdle function, is essentially a logistic regression that determines the covariates associated with the probability of finding zero labs within a county. The second model, which is similar to a traditional OLS model (or a Poisson regression), determines which independent variables account for increasing lab seizures within those counties that have passed the hurdle of having no labs within them. It is entirely permissable to include different predictors in the two different models, although in this case I did not. A spatially lagged variable can be included in either side of the equation to account for spatial autocorrelation within the data. McCord and Ratcliffe (2007) and Rengert et al. (2005) used zero-inflated models with a spatial lag variable in their analyses of crime-count data.

For this study, I used *Open GeoDa* (Anselin et al. 2006) to create a spatial weights matrix based on first-order county contiguity, and then used this weights matrix to calculate a spatial lag variable for each county (mean labs seized in neighboring counties), which I then included in the zero-inflated regression models. The zero-inflated models were conducted using *R*. In addition, I assessed several characteristics of the spatial distribution of the dependent variables using the spatial statistics toolbox in *ArcGIS 10*.

13.6 Results

To begin the spatial regression process, I mapped the dependent variables and assessed them for spatial autocorrelation. Getis (2007), Baller et al. (2001), Messner et al. (1999), Anselin et al. (2000), and Eck et al. (2005) have all recommend such a test as the first step in exploratory spatial data analysis before any attempt at regression. An assessment of our dependent variable for 2000–2003 and 2007–2010 using Moran's I showed significant global spatial autocorrelation (2000–2003: $z=13.4$ $p<.001$ and 2007–2010: $z=20.7$ $p<.001$). An analysis of local autocorrelation, assessed using Anselin's Local Indicators of Spatial Autocorrelation (LISA) showed significant clustering of high values of lab seizures, indicating a violation of the independence assumption, and the necessity of spatial regression (Fig. 13.5).

Given the distribution of the data, the high number of zeroes, and the presence of significant spatial autocorrelation, I determined that a spatial zero-inflated model needed to be run. Analysis of the level of dispersion indicated that my data were overly dispersed, and that a zero-inflated negative binomial model was necessary ($=1.396$ (00–03) and 1.389 (07–10). In both cases the ZINB model compared favorably to a model containing only the intercept (likelihood ratio test chi square$=595.91$, $p<.001$ (00–03), and 477.54, $p<.001$(07–10)) and to a zero-inflated Poisson regression model using the same variables (Vuong test V statistic : -5.25, $p<.001$ (00–03) and −5.53 $p<.001$(07–10)).

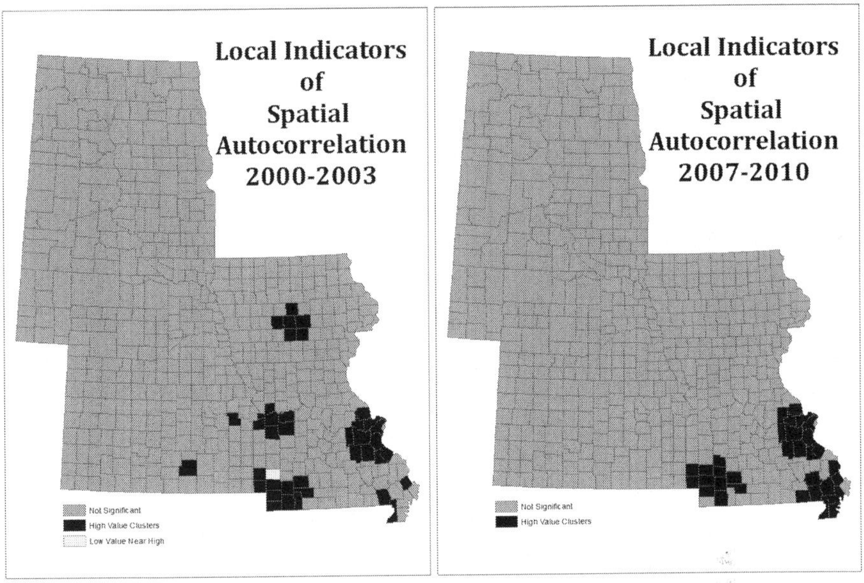

Fig. 13.5 Measures of local spatial autocorrelation in the dependent variables

For the 2000–2003 sample, the likelihood of no labs being seized within a county was negatively correlated with the spatially lagged count variable and with population density, and positively correlated with the percentage of homes that were vacant within the county. This means that an increase of one in the average number of labs seized in a county's neighbors reduced the likelihood of there being no labs seized within that county between 2000 and 2003 by 78.21%[7]. Similarly, an increase in a county's population density of one person per square mile further reduced the likelihood of no labs being found within the county over that time period by 12.95%. On the other hand, an increase of one in the percentage of housing units that were vacant in a county increased the probability of no labs being seized within its borders over the time period by approximately 25% (Table 13.2).

The count model of the ZINB regression determines coefficients for variables that contribute to increases in the number of labs seized. From 2000 to 2003, the spatial lag variable, the percentage of vacant properties, male educational attainment, the percentage of households with a single mother, and the percentage of the population that is white were all positively correlated with increasing lab seizures. In contrast, the percentage of households that contained families was the only negatively correlated variable. The poverty rate, population density, median age, percentage of properties that were renter-occupied, and percentage of the population that was Mexican were not significantly correlated with increasing lab seizures at $p \leq .05$ (Table 13.3).

[7] This value is calculated using the formula $(100^* (e^{-1.523904} - 1))$ where −1.523904 is the coefficient for the lagged variable (Atkins and Gallop 2007).

Table 13.2 Results for the logistic (zero predictor) portion of the ZINB model

	2000–2003				2007–2010			
	Estimate	S.E.	Z Score	Sig.	Estimate	S.E.	Z Score	Sig.
(Intercept)	−7.268903	17.207764	−0.422	0.673	1.744774	13.621925	0.128	0.898
Spat. Lag	**−1.523904**	**0.422713**	**−3.605**	**0.000**	**−1.544425**	**0.625519**	**−2.469**	**0.014**
Median Age	0.028997	0.191388	0.152	0.880	0.046823	0.086871	0.539	0.590
% Renter Occ.	0.047019	0.087023	0.54	0.589	0.007018	0.088431	0.079	0.937
% Vacant	**0.229066**	**0.079699**	**2.874**	**0.004**	0.012687	0.045984	0.276	0.783
% Sing. Moth.	−0.004149	0.402144	−0.01	0.992	−0.20066	0.294809	−0.681	0.496
% Poverty	−0.209626	0.165603	−1.266	0.206	0.152678	0.099806	1.53	0.126
% Male Bach.	0.030335	0.112135	0.271	0.787	0.090454	0.084447	1.071	0.284
Pop. Density	−0.138677	0.081793	−1.695	0.090	−0.012794	0.012149	−1.053	0.292
% White	−0.055387	0.059863	−0.925	0.355	−0.083181	0.067633	−1.23	0.219
% Mexican	−0.237321	0.209022	−1.135	0.256	−0.046148	0.056175	−0.821	0.411
% Families	0.178082	0.117213	1.519	0.129	0.03961	0.093412	0.424	0.672

Table 13.3 Results for the count portion of the ZINB model

	2000–2003				2007–2010			
	Estimate	S.E.	Z Score	Sig.	Estimate	S.E.	Z Score	Sig.
(Intercept)	–4.4920098	2.868234	−1.566	0.117	–2.7211056	3.7011464	–0.735	0.462
Spatial Lag	**0.0290827**	**0.0037927**	**7.668**	**0.000**	**0.0658063**	**0.0074477**	**8.836**	**0.000**
Median Age	–0.0470126	0.0249149	−1.887	0.059	**–0.0580102**	**0.0270718**	**–2.143**	**0.032**
% Renter Occ.	–0.0318169	0.0181003	−1.758	0.079	–0.0407613	0.0251276	–1.622	0.105
% Vacant	**0.0232354**	**0.0109036**	**2.131**	**0.033**	0.0021277	0.0121768	0.175	0.861
% Sing. Moth.	**0.5914383**	**0.0562411**	**10.516**	**0.000**	**0.5760556**	**0.0852761**	**6.755**	**0.000**
% Poverty	0.0399136	0.0211323	1.889	0.059	**0.0639268**	**0.0211418**	**3.024**	**0.002**
% Male Bach	**0.0756812**	**0.0199509**	**3.793**	**0.000**	0.0315928	0.0220197	1.435	0.151
Pop. Dens,	0.0003797	0.0002534	1.498	0.134	0.0003792	0.0003265	1.162	0.245
% White	**0.0726599**	**0.0129066**	**5.63**	**0.000**	**0.0738632**	**0.0207655**	**3.557**	**0.000**
% Mexican	0.0175243	0.0144087	1.216	0.224	–0.0262415	0.0184519	–1.422	0.155
% Families	**–0.038357**	**0.0200949**	**–1.909**	**0.056**	**–0.0632768**	**0.02635**	**–2.401**	**0.016**

For the sample of seizures after the precursor laws were in effect (2007–2010), the results were similar. The only significant covariate for the presence of lab seizures in the zero-inflation model was the spatially lagged count variable. Once again, it was negatively associated with the probability of no labs being seized within a county between 2007 and 2010. In this case, an increase of one in the lagged count variable decreased the probability of a zero count within the county by 78.66%. None of the other predictors were significantly associated with the zero-inflation model.

The covariates associated with increasing lab seizures were also similar to the 2000–2003 model. Once again, the spatial lag variable, the percentage of a population that was white, and the percentage of households with a single mother were all positively correlated with increasing meth lab seizures. Of these variables, the percentage of households with a single mother had the largest impact, raising the number of labs seized within the county by 3.32 for each increase of a percentage point.[8] In the 2007–2010 model the percentage of people living below the poverty level was also a significant positive covariant of increasing lab seizures, but the percentage of males over 25 with a bachelor's degree was no longer significant. The percentage of homes occupied by families was once again a significant negative covariate. In this model, the median age was also a negative covariate.

13.7 Discussion

As mentioned above, the 2005 Combat Methamphetamine Epidemic Act caused a dramatic dip in the number of labs seized both nationally and in the Midwest HIDTA. Seizures in the 2007–2010 period were only 35% of that for 2000–2003 (6,912 vs. 2,314). However, the their spatial distribution was much the same as before. The weighted mean center of lab seizures moved 66.4 miles southeast, but the size and orientation of the standard deviation ellipses for each time period were similar enough to argue that the general distribution of lab seizures was not significantly affected (Fig. 13.6).

The covariates of methamphetamine lab seizures also were not much affected by the Act. The most significant indicator of methamphetamine lab presence in a county (from the logistic portion of the ZINB) in both models was the presence of seizures in the counties that surrounded it (Table 13.1). In studies of drug markets, this is usually interpreted as the agglomeration of drug sales (Rengert and Robinson 2006, Rengert et al. 2005). That would also be an appropriate interpretation here.

Both of the count models similarly showed more similarities than differences. In the second model, more user characteristics proved to be significant in the manner that one would suspect, with median age being negatively correlated and the percentage of peo-

[8] This value is calculated by multiplying the percentage change (calculated in the same manner explained in the previous footnote) by the mean of the dependent variable for the data.

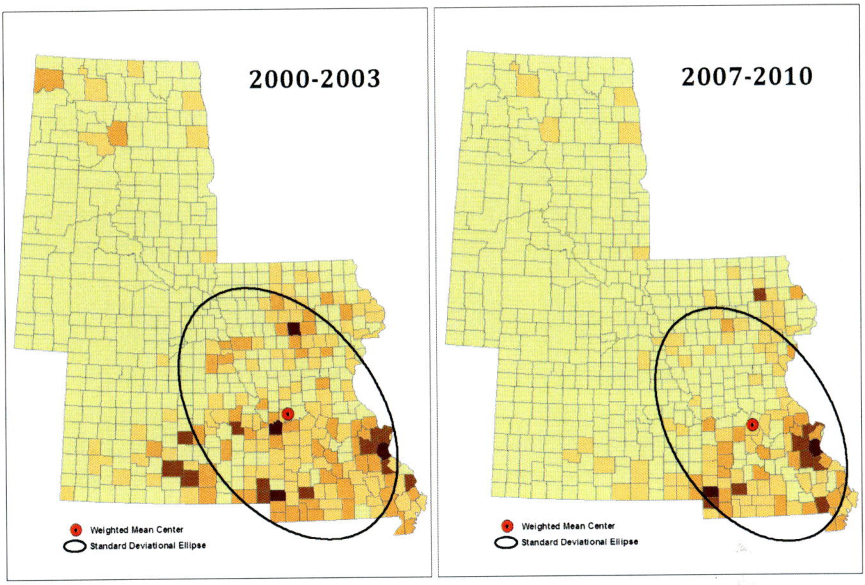

Fig. 13.6 Weighted mean center and standard deviational ellipse for lab seizures

ple living below the poverty level positively so. The percentage of white residents, even in a region deemed by many to be relatively homogenous, was consistently significant and was also the most impactful of the user characteristics in either model.

Percentage Mexican was clearly a poor proxy for methamphetamine market penetration by Mexican DTOs. Population density also was not significant in either count model, which might argue against the association between methamphetamine production and rural areas. When the percentage rural variable is made available for the 2010 census, this particular relationship will need to be reassessed.

Two of the social disorganization variables, percentage of households containing families and percentage of households with single mothers, were consistent across both models and had the expected correlation. The impact that the mothers variable had was surprising, as it surpassed many other variables that are more frequently associated with methamphetamine use or production, such as poverty or whiteness. Other social disorganization variables were less consistent. The percentage of homes occupied by renters was not significant in either model, and percentage of properties that were vacant was only significant in the first.

Given the overlap between the variables for user characteristics and social disorganization, it seems clear that methamphetamine production does indeed cluster in counties that exhibit social disorganization. The fact that the percentage of the county that is white (which is the opposite of a common indicator of social disorganization, the racial heterogeneity of a place, or percentage minority) was consistently significant does not alter this conclusion because whites make up the great majority of methamphetamine users (SAMHSA 2009).

Several limitations to this study exist that could be addressed in the future. The first is the temporal resolution of the data from EPIC, which was aggregated to yearly totals. The creation of the models' dependent variables was a negotiation between proximity to the date of the precursor law's implementation and the decennial census, which had the best available data. The best previous studies on precursor impacts used monthly data to assess their impact (Cunningham and Liu 2003, 2005; McBride et al. 2008; Dobkins and Nisocia 2009). However, none of them was dependent upon census data for independent variables. All of the seizure studies mentioned in the literature review relied on census data.

The spatial resolution of the data also limited this study. By operating at the county level, a number of variables associated with routine activities, rational choice, or a more nuanced ecological perspective could not be included in the model (Cohen and Felson 1979; Clarke 1980; Clarke and Felson 1993). For example, production might be highly correlated with the presence of pharmacies or big-box stores from which essential precursors and other materials might be purchased, or producers might situate labs near common commuter thoroughfares.

This study, the first of its kind to assess the covariates of methamphetamine production using a spatial zero-inflated negative binomial model, represents a significant step forward in our understanding of methamphetamine lab location. It proves that methamphetamine production at the county level, by concentrating in counties that exhibited most of the standard characteristics of social disorganization, behaves similarly to criminologists' basic understanding of other drug markets despite the unique nature of its synthesis. In addition, the study demonstrates that, although the Combat Methamphetamine Epidemic Act of 2005 did significantly reduce the number of lab seizures in the region, it did not alter the spatial characteristics of domestic methamphetamine production. In combination, these two conclusions argue that the strategy for policing methamphetamine does not need to be significantly different from that for other drugs and drug markets.

References

Andresen MA (2011) The ambient population and crime analysis. Prof Geogr 63(2):193–212

Anselin L (1988) Spatial econometrics. Kluwer, Boston

Anselin L, Cohen J, Cook D, Gorr W, Tita G (2000) Spatial analyses of crime. In: Duffee D (ed) Criminal Justice 2000: volume 4: measurement and analysis of crime and justice. National Institute of Justice, Washington, DC, pp 213–262

Anselin L, Syabri I, Kho Y (2006) GeoDa: an introduction to spatial data analysis. Geogr Anal 38(1):5–22

Atkins DC, Gallop RJ (2007) Rethinking how family researchers model infrequent outcomes: a tutorial on count regression and zero-inflated models. J Fam Psychol 21(4):726–735

Baller RD, Anselin L, Messner SF, Deane G, Hawkins DF (2001) Structural covariates of U.S. homicide rates: incorporating spatial effects. Criminology 39(3):561–590

Banerjee A, LaScala E, Gruenewald PJ, Freisthler B, Treno A, Remer LG (2008) Social disorganization, alcohol, and drug markets and violence: A space-time model of community structure. I. In: Thomas YF, Richardson D, Cheung I (eds) Geography and drug addiction. Springer, New York, pp 119–132

Clarke RV (1980) Situational crime prevention: theory and practice. Br J Criminol 20:136–147

Clarke RV, Felson M (eds) (1993) Routine activity and rational choice. Transaction, New Brunswick

Cohen LE, Felson M (1979) Social change and crime rate trends: a routine activity approach. Am Sociol Rev 44:588–608

Cunningham JK, Liu LM (2003) Impacts of federal ephedrine and pseudoephedrine regulations on methamphetamine-related hospital admissions. Addiction 98:1229–1237

Cunningham JK, Liu LM (2005) Impacts of federal precursor chemical regulations on methamphetamine arrests. Addiction 100:479–488

DASIS (2006) The dasis report: trends in methamphetamine/amphetamine treatment admissions. Substance Abuse and Mental Health Services Administration, Washington, DC

DASIS (2008) Primary methamphetamine/amphetamine admissions to substance abuse treatment. Substance Abuse and Mental Health Services Administration, Washington, DC

DEA (2005) Guidelines for law enforcement for the clean up of clandestine drug laboratories. Department of Justice, Washington, DC

Dobkin C, Nicosia N (2009) The war on drugs: methamphetamine, public health, and crime. Am Econ Rev 99(1):324–349

Eck JE, Chainey S, Cameron JG, Leitner M, Wilson RE (2005) Mapping crime: understanding hot spots. Department of Justice, Washington, DC

EPA (2009) Voluntary guidelines for methamphetamine laboratory clean up. Environmental Protection Agency, Washington, DC

Getis A (2007) A history of the concept of spatial autocorrelation: a geographer's perspective. Geogr Anal 40:297–309

Gilbreath A (2010) A spatial analysis of methamphetamine lab seizure data for Jefferson County, Missouri. Paper presented at the annual meeting of the association of American Geographers, 15 April 2010, Washington, DC

Grattet R (2009) The Urban ecology of bias crime: a study of disorganized and defended neighborhoods. Soc Probl 56(1):132–150

Herz DC (2000) Drugs in the heartland: methamphetamine use in rural Nebraska, research in brief. National Institute of Justice, Washington, DC

Hunt D, Kuck S, Truitt L (2006) Methamphetamine use: lessons learned. NIJ Grant report. Department of Justice, Washington, DC

Jenkins P (1999) Synthetic panics: the symbolic politics of designer drugs. New York University Press, New York

Kubrin CE, Weitzer R (2003) New directions in social disorganization theory. J Res Crime Delinq 40(4):374–402

Los Angeles Times (1985) 81 in cycle gang arrested. 22 February 1985, OC2

Lu M, Burnum J (2008) Spatial patterns of clandestine methamphetamine labs in Colorado Springs, Colorado. In: Thomas YF, Richardson D, Cheung I (eds) Geography and drug addiction. Springer, New York, pp 193–207

McBride DC, Terry-McElrath YM, Chiriqui JF, O'Connor JC, Vanderwaal CJ (2008) The relationship between state methamphetamine precursor laws and trends in small toxic lab (STL) seizures. Department of Justice, Washington, DC

McCord ES, Ratcliffe JH (2007) A micro-spatial analysis of the demographic and criminogenic environment of drug markets in Philadelphia. Aust N Z J Criminol 40(1):46–63

McDonald JM, Lattimore PK (2010) Count models in criminology. In: Piquero AR, Weisbrud D (eds) Handbook of quantitative criminology. Springer, New York, pp 683–698

Mennis J, Harris PW, Obradovic Z, Izenman AJ, Grunwald HF, Lockwood B (2011) The effect of neighborhood characteristics and spatial spillover on urban delinquency and recidivism. Prof Geogr 63(2):174–193

Messner SF, Anselin L, Baller RD, Hawkins DF, Deane G, Tolnay SE (1999) The spatial patterning of county homicide rates: an application of exploratory spatial data analysis. J Quant Criminol 15(4):423–450

NDIC (ed) (2009a) Midwest high-intensity drug trafficking area drug market analysis 2009. Department of Justice, Washington, DC

NDIC (2009b) Situation report: methamphetamine production in the United States rebounds to meet domestic demand. Department of Justice, Washington, DC

NDIC (2010) Midwest high-intensity drug trafficking area drug market analysis 2010. Department of Justice, Washington, DC

Nicosia N, Pacula RL, Kilmer B, Lindberg R, Chiesa J (2009) The economic cost of methamphetamine use in the United States, 2005. Rand, Santa Monica

Owen F (2007) No speed limit: the highs and lows of meth. St. Martin's, New York

Rasmussen N (2008) On speed: the many lives of amphetamine. New York University Press, New York

Rawlin JW (1968) Street level abuse of amphetamines. In: Russo JR (ed) Amphetamine abuse. Charles C. Thomas, Springfield, pp 39–50

Rengert GF, Ratcliffe JH, Chakravorty S (2005) Policing illegal drug markets: geographic approaches to crime reduction. Criminal Justice Press, Monsey

Rengert GF, Robinson JB (2006) Illegal drug markets: the geographic perspective and crime propensity. West Criminol Rev 7(1):20–32

Reuter P, Caulkins JP (2003) Does precursor regulation make a difference? Addiction 98:1177–1179

Sampson RJ, Groves WB (1989) Community structure and crime: testing social-disorganization theory. Am J Sociol 94(4):774–802

Shaw CR, McKay HD (1942) Juvenile delinquency and urban areas. University of Chicago Press, Chicago

Substance Abuse and Mental Health Services Administration (2009) Treatment episode Data Set (TEDS) Highlights—2007. Substance Abuse and Mental Health Services Administration, Rockville

Sweeny B. (2010) Personal interview. 31 March 2010, Independence, Mo

United States Congress, House Committee on the Judiciary Subcommittee on Crime (2000) Threat to rural communities from methamphetamine production, Trafficking, U.S. Government Printing Office, Washington, DC

Ward MD, Gleditsch KS (2008) Spatial regression models. Sage, Los Angeles

Weisheit RA, Wells LE (2010) Methamphetamine laboratories: the geography of drug production. West Criminol Rev 11(2):9–26

Part IV
Crime Mapping

Chapter 14
Comparing Fear of Crime and Crime Statistics on a University Campus

Sven Fuhrmann, Niem Tu Huynh, and Ruojing Scholz

Abstract Campus crime at colleges and universities has resulted in a call for more safety and preventive measures from policymakers, students, to citizens. While research highlights students' fear on campus crime, few studies have examined this topic from a spatial and cognitive perspective. In this chapter the authors report on a novel methodology to compare campus crime data with participants' self-reported cognitive fear of crime maps. In this study, 313 undergraduate students provided fear of crime maps at a middle-sized university in the southwestern United States. The students' perceptions were aggregated and compared to university crime statistics to produce five bivariate maps. These maps represent perceived fear of crime in relation to four broadly observed crime categories namely burglary, theft, harassment, and sexual assault. In this research effort, students' fear of crime is aligned with data for actual burglary and theft occurrences but their fear is exaggerated for harassment and sexual assault. The implications of this study are multifold, extending from potential safety improvements and better decision-making (e.g., aid law enforcement to target specific areas for crime monitoring) to developing educational workshops to dispel myths and present facts on campus safety. The introduced bivariate mapping technique provides another step towards safer university and college campuses.

S. Fuhrmann (✉) • R. Scholz
Department of Geography, Texas State University – San Marcos, San Marcos, TX, USA
e-mail: fuhrmann@txstate.edu; rw1283@txstate.edu

N.T. Huynh
Association of American Geographers, Washington, DC, USA
e-mail: nhuynh@aag.org

M. Leitner (ed.), *Crime Modeling and Mapping Using Geospatial Technologies*, Geotechnologies and the Environment 8, DOI 10.1007/978-94-007-4997-9_14,

Keywords Crime mapping • Campus crime • Cognitive maps • Fear of crime • Geographic information systems

14.1 Introduction

College campuses have been described as microcosms of society (Rund 2002). Although a generally safe place for students, faculty and staff, campuses are also subject to criminal activity like its surrounding communities. The majority of offences recorded on campuses are property crimes while violent crimes are rare (Robinson and Roh 2001). Nevertheless, the frequency and severity of events have made campus crime a priority for Congress and state-level policymakers with parents and students demanding more preventative measures (Fisher et al. 1997). College-related crime has received major public awareness with the murder of Jeanne Clery in 1986. Her case resulted in the Clery Act of 1990 (20 U.S.C § 1092(f)) which requires all institutions of higher education to disclose campus crime information publicly on an annual basis. The campus security report includes crime statistics, information about safety, prevention, and security measures on campus. Overall, the published reports suggest that campus crimes are relatively low compared to the larger community (Fox and Burstein 2010). However, there is continued fear of campus crime by students, administrators, faculty, and parents.

The current literature offers foundational research on student perception of campus crime but few studies have examined this topic from a spatial perspective. Some studies have accomplished this by producing maps marking the location of reported crimes (e.g., Robinson and Mullen 2001; Robinson and Roh 2007) while others have mapped participants' fear of crime (e.g., Brantingham et al. 1977; Nasar and Fisher 1993; Astor et al. 1999). Few studies have brought the two sets of data together as a way to compare cognitive maps of fear with reported crime incidents. This paper aims to bridge university crime statistics with fear of crime perceptions denoted on a map and analyzed using Geographic Information Systems (GIS).

Each human maintains a cognitive map – a subjective and dynamic spatial representation of the environment around them. These cognitive maps are complex and reflect personal assessments of the environment. Cognitive map content includes preferred spatial areas, landmarks, routes, experiences, and also areas humans want to avoid or are afraid of visiting. Cognitive maps include areas of fear, areas of comfort, and ways to move within these spaces (Downs and Stea 1973; Gould and White 1974). Using the cognitive map as data source, this research attempts to intersect two campus crime mapping efforts: (a) university police crime data with (b) the students' mental fear of crime maps. Both datasets enable the researchers to investigate two research questions: (a) What are the aggregated fear of crime patterns? and (b) Is there a relationship between students' fear of crime maps and reported crime activities?.

To answer the research questions, this research analyzes 313 undergraduates' fear of crime maps at a middle-sized university (29,105 full and part-time students enrolled during the fall 2009 semester) located in southwestern United States. The analysis provides a neoteric method that combines human perception and mapping of fear with geospatial data and technology to produce bivariate crime maps. These bivariate crime maps show similarities and discrepancies between reported crime and perceived areas of fear on campus. The practical use of the maps are to support university police and administrators to better assess and react to safety situations and concerns on campus.

14.2 Literature Review

14.2.1 Campus Crime Statistics and Policy Making

When parents send their children to colleges and universities they usually expect a safe living and learning environment. Unfortunately campuses are not excluded from violence and crime, which can range from theft and car break-ins to more violent crimes such as sexual assault to death. Up until 1990 institutions of higher education were not required to publish information about the safety of their campuses or to report crime statistics (McNeal 2007). It wasn't until the murder of Jeanne Clery, who lived in residence and studied at Lehigh University (Pennsylvania), that provided the major incident for developing and passing "The Student Right-to-Know and Campus Security Act" (Public Law 101–542) in 1990 (Janosik and Gregory 2003). Amendments to this act in 1998 renamed it to the "Jeanne Clery Disclosure of Campus Security Policy and Campus Crime Statistics Act" (20 U.S.C § 1092(f)). One purpose of the "Clery Act" was to provide students and parents with sufficient campus crime statistics to help them make informed college enrollment decisions. Another purpose of the law was to establish pressure on colleges and universities to take campus crimes seriously and to ensure a safe learning environment (Fisher et al. 2002). The Clery Act requires all institutions of higher education to disclose campus crime information to the public on an annual basis and to submit a crime report by October 1, reflecting three prior years, to the United States Department of Education (Lipka 2009). The report needs to include information about (a) the type of crime (e.g., criminal homicide, sexual offenses, robbery, aggravated assault, burglary, arson, motor vehicle theft, and arrests), (b) locations (e.g., on campus, on campus residential facilities, non-campus buildings, and public places), (c) a statement of current policies regarding procedures for reporting crime, (d) descriptions of security and crime prevention programs, (e) information about security and access to campus facilities, and (f) current policies concerning campus law enforcement (Fisher et al. 2002; Janosik and Gregory 2009; Lipka 2009).

The Clery Act provides better safety information and crime protection for college students, but two major drawbacks have been reported. First, few students and parents read the crime statistic reports and second, the numbers provided may not reflect true crime patterns and occurrences due to data entry errors and that not all crimes are reported (Gregory and Janosik 2002; Janosik and Gehring

2001; Lipka 2009). Furthermore, many students are completely unaware of the Clery Act (Janosik 2001; Gregory and Janosik 2002). Most students do not base their college decisions on it (Janosik 2002), and thus do not read the annual reports nor use them for on-campus safety decision making, e.g. which route to walk home at night. In addition to public awareness there have also been a number of implementation and compliance issues with the Clery Act (Carter 2001; Lipka 2009; McNeal 2007). Some of the issues are systematic while others are institution related. One criticism is the lack of guidance on compliance issues set by the U.S. Department of Education. Some institutions are reluctant to publish potentially negative information which is made worse by internal errors caused by data processing and reporting issues such as missing policies and personnel changes (Carter 2001; Lipka 2009; McNeal 2007). Sources interviewed by Lipka (2009) go as far as saying that crime statistics do not provide safety, only the integration of basic security reporting equipment (e.g., blue-light telephones, emergency signboard systems, and monitored security cameras) will lead to safer campuses.

14.2.2 Spatial Cognition and Fear of Crime

Downs and Stea (1973) define cognitive mapping as "a process composed of a series of psychological transformations by which an individual acquires, codes, stores, recalls and decodes information about the relative locations and attributes of phenomena in his everyday spatial environment. [...] The product of this process at any point in time can be considered as a cognitive map" (pp. 9–10). Cognitive maps are personalized memories about human spatial experiences. These spatial memories can consist of many components such as landmarks, paths, edges, images, stories, and areas of fear (Golledge 1999; Kitchin 1994; Kitchin and Freudschuh 2000).

The cognitive map research attracted attention from numerous fields including geography, planning and psychology. The theories, applications, methodologies and knowledge resulting from this research on cognitive maps collectively formed a significant contribution to the spatial cognition literature (e.g., Banai 1999; Kearney and Kaplan 1997; Kuipers et al. 2003; Singh 1996) and play a role in criminology research (Brantingham and Brantingham 1993). The extraction of information from people's cognitive maps has not been widely used in campus crime research although a thread of this research exists (e.g., Brantingham et al. 1977; Astor et al. 1999).

Brantingham and Brantingham (1995) describe fear of crime as a complex concept that includes fear of being attacked, suffering physical harm and/or loosing privacy and dignity. In the context of this manuscript we study 'fear of crime' as the fear of becoming a victim of crime in different parts on a university campus. Fear of crime is an emotional reaction to potential victimization and a natural response to an individual's perception of threat in their environment. It may influence peoples' actions to avoid becoming a victim (Balkin 1979; Baumer 1985). Fear of crime can be classified into three aspects: cognitive, affective, and behavioural fear of crime (Hale 1996; Jackson 2006). While behavioural fear of crime highlights people's actions to avoid becoming a victim; affective fear of crime tries to assess the fear towards a specific offense. This study investigates the cognitive aspect of students' fear of crime, in particular, participants' beliefs regarding the likelihood of becoming a victim on campus will be documented. Accuracy of student perception is of importance and the literature suggests peoples' cognitive perception of crime is generally reliable. Cognitive fear of crime generally matches occurrences and proximal locations of actual crime events within familiar areas such as neighbourhoods or districts (Brantingham et al. 1977; Lewis and Maxfield 1980).

Literature on campus crime generally focuses on three broad areas: (a) the role of legislature and university's role in crime prevention (Fisher and Sloan 1993; Gregory 2001; Hudge 2000; Janosik 2001), (b) the types and frequency of crime occurring on campus (Fisher and Nasar 1995; Fisher and Sloan 2003, 2007; Fisher et al. 1995; Robinson and Roh 2001), and (c) student perception of crime on campus (McConnell 1997; Nasar and Fisher 1993; Starkweather 2007; Tseng et al. 2004; Wilcox et al. 2007). An area of paucity in crime research is the inclusion of spatial research, and only a small number of publications can be found (Astor et al. 1999; Brantingham et al. 1977; Nasar and Fisher 1993; O'Kane et al. 1994; Rengert and Lowell 2005; Robinson and Mullen 2001; Robinson and Roh 2007). In the above listed studies, campus maps were used to visualize areas of criminal activity as clusters or hot spots while survey responses provided information about victimization fears.

One branch of criminology that applies maps to identify crime pattern analysis is known as environmental criminology. The concept, introduced by Brantingham and Brantingham (1993), provides a framework to analyze crimes in relation to the physical environment, paired with the activity patterns of both victims and offenders. The framework provides a geographic space to describe crimes, with observations that people commit offenses near central nodes (e.g., home, work), along paths that are connecting these nodes, and around edges that distinguish the landscape. Conclusions using this framework explain that victimization patterns are related to the *paths* and *nodes* of the victim (Brantingham and Brantingham 1993). The underpinning assumption is that offenders commit criminal activities close to their activity nodes and routine activity paths as these are the spaces they know best. Land use plays a critical role in environmental criminology, with crime occurring along physical and perceptual edges within a region (Brantingham and Brantingham 1993). Rivers, railroads, highways are physical edges while residential-commercial, or city-campus edges are perceptual borders (Brantingham and Brantingham 1993, 1995). While environmental criminology is an important theory within criminal justice research, this manuscript's focus is on comparing cognitive areas of fear with reported crime statistics.

14.2.3 Crime Mapping

Pin mapping can be described as the most elementary crime mapping technique in which dots represent the locations or concentration of crimes (Groff and La Vigne 2001; Vann and Garson 2001). Modern crime mapping has extended beyond this method. For example, Vann and Garson (2001) identified 21 different crime mapping and analysis functions while Groff (2007) describes agent-based modeling as a way to predict potential crime patterns. Criminology has been strongly influenced by crime mapping (Wilson 2007), which has gained prominence and acceptance. This method uses software to effectively combine crime theory with geographic analysis principles. New developments such as GIS and remote sensing have improved crime analysis in the past several years, supporting crime mapping and analysis through advanced spatial analytical methods.

A foundation for campus crime mapping was laid in the late 1970s. Seminal work include Brantingham et al. (1977) who analyzed and mapped residents' perceptions of crime sites in a student housing complex. The researchers encountered many residents in the housing complex who misperceived the location of the most dangerous crime sites and unconsciously increased their risk of victimization. Brantingham et al. (1977) concluded that general policing and management service improvements for the student housing complex are needed. Astor et al. (1999) investigated areas of violence on high school campuses. They asked students and teachers to mark the most violent events and dangerous areas on maps of their school campus. Results indicated that most crimes occurred in unsupervised spaces coined as "un-owned places" (Astor et al. 1999). The authors recommend students and staff to take ownership of these areas through increased security and interventions but also through getting to know students personally and extending behavioral interactions between students from the classrooms into the hallways, dining and parking areas. Robinson and Roh (2007) investigated crime statistics on a university campus in the southeast United States. They mapped out campus crime data for a 2 year period and found crime hotspots in campus dormitories (which the authors call "crime generators") and high-traffic areas between educational buildings and parking places. Hot spots are defined as places that have more than average occurrences of criminal activity (Sherman and Weisburd 1995; Fisher and Nasar 1995). Hot spots are classified into one of three categories: crime generators, crime attractors, and crime enablers (Clarke and Eck 2005).

Brower and Carroll (2007) reviewed alcohol-related aspects of a college town and used GIS to map the relationships among high-density alcohol outlets and different neighborhoods in the City of Madison, Wisconsin. The researchers mapped student address data and incident report data which they grouped in four low level crime categories: liquor law violations, assaults and batteries, vandalism, and noise complaints. Brower and Carroll (2007) found that noise complaints bordered along long-term resident and high-density student housing. Assaults and batteries peaked after bar closing time in the vicinity of student bars, and vandalism was mostly reported in the morning. The results of the study were applied to mitigate noise problems through a better university-community partnership, to change drinking policies in dormitories, and to modify practices for bar licensing and alcohol-related fines. Overall, map and GIS-based campus crime studies have shown huge potential in analyzing and mitigating crime on college and university campuses. This study borrows from and builds on the methods applied in these publications in order to add to the understanding of campus crime patterns.

14.3 Method

Two primary data sets are used in this study, (1) official crime data collected by the University Police Department and (2) students' self-reported fear of crime locations. A description of participants, measurement instrument, and the data collection procedure follows.

14.3.1 Participants

Three hundred and thirteen randomly participating undergraduate students across six different geography classes completed the survey in the fall 2009 semester. One hundred and seventy-eight undergraduate students were recruited from the world geography course (freshman level), 73 participants took the introduction to GIS course (sophomore level), 29 students volunteered in the junior level cartography course, and 33 subjects were recruited in two advanced level GIS courses (junior and senior level). One third of students surveyed were Geography majors while the remaining students specialized in a range of university disciplines (see Table 14.1).

Students participating in this study are representative of the larger student body. The gender was close to evenly split (48.2% females, 51.8% males) and reflected the general undergraduate enrollment ratio at the university well (53.4% females, 45.6% males, fall 2009 data). The average age of the total sample was 24.6 years (average undergraduate student age at university 21.5years; geography department 23.6 years) and included 91freshmen (mean age of 20.0 years old; university level 18.8 years; geography department 20.2 years), 57 sophomores (mean age of 24.6; university level 20.6 years; geography department 21 years), 69 juniors (mean age of 25.8; university level 22.3 years; geography department 23.8 years), 69 seniors (mean age of 26.9; university level 23.9 years; geography department 25.3 years). Twenty-seven students did not enter their study level information. Seven participant responses were excluded from the study because of incomplete answers. Overall the age and gender distribution of the study participants is representative of the undergraduate student population (see Table 14.2).

Since the university has a large commuter student enrollment the researchers also compared the areas of fear between commuter students and residents. Thirty percent of participants (n = 92) lived in student residences on campus, 33% (n = 103) of participants were living within the city limits, and 36% (n = 112) were commuters. In the fall 2009 semester 5,994 (30.3%) undergraduate students lived on campus, 4,760 (23.8%) undergraduate students lived within the city limits, while 9,193 (45.9%) of the university's undergraduate students commuted. Overall the residence distribution of the study participants is representative of the university's undergraduate student population.

Table 14.1 Participants' academic major

Major	#	Major	#	Major	#
Accounting	4	English	2	Math	4
Anthropology	2	Environmental studies	1	Music	1
Architecture	1	Exercise and sports science	7	Nutrition	1
Art	2	Fine arts	2	Political science	1
Biochemistry	1	Geography	99	Psychology	1
Biology	12	History	4	Recreation Admin.	1
Business	4	Interdisciplinary studies	4	Sociology	1
Communication design	2	Interior design	1		
Computer science	2	International business	7		
Criminal justice	1	International studies	16	Undeclared	19
Education	54	Marketing	2	Major not provided	54

Table 14.2 Characteristics of study participants compared to university undergraduate student body (data for fall 2009)

	Study participant characteristics	Geography department characteristics	University undergraduate student characteristics
Females	48.2%	30.3%	53.4%
Males	51.8%	69.7%	45.6%
Average age	24.6years	23.6 years	21.5 years
Average age freshmen	20 years	20.2 years	18.8 years
Average age sophomores	24.6 years	21 years	20.6 years
Average age juniors	25.8 years	23.8 years	20.6 years
Average age seniors	26.9 years	25.3 years	23.9 years

14.3.2 Measurement Instruments

A questionnaire and a campus map were used as primary survey tools. Participants were asked to provide three sets of information. First, demographic and residence information were collected, followed by marking and labeling campus areas the students considered unsafe to walk in. Finally, students were asked to identify areas that they were frequently visiting. In order to mark unsafe areas students received an official campus map that included all buildings, streets, parking lots, and other infrastructure information. The map also included a predefined reference grid that allowed the researchers to geo-reference and aggregate the collected data. In addition to identifying unsafe areas, participants were also asked to indicate the time of the day they considered these areas unsafe, what kind of crime they feared in these locations, and what improvements they would recommend. The questionnaire also assessed if the university escort service was used by the participants if they felt unsafe on campus.

14.3.3 Data Collection Procedures

Data were collected at the start of each class; the process taking approximately 15–20 min. The scope of the study was first explained followed by an invitation for voluntary participation. It was made clear that student identity would be anonymous and participants would not receive any extra course credit. Participants received a questionnaire and an official campus map on which they were asked to mark up to five areas that they would feel unsafe walking by themselves. Participants were asked to number these areas on the map and use the alphanumerical coordinate system (e.g., A1) to answer corresponding questionnaire questions. These questions were concerned with why and when students would feel unsafe at these areas. Once these tasks were completed, participants returned the questionnaire and the map to the study facilitator. Participants were thanked for their participation. No additional follow-ups or make-up opportunities were given to students who missed class.

14.4 Results and Discussion

14.4.1 Findings from University Police Department Data

The university police crime statistics database is freely accessible on the university website, providing data about reported crimes for the period 2007–2009. These crime incidents (539 cases) were categorized into five general offence types, plotted in their location of occurrence, and then aggregated into reference grids to create crime statistics maps (Fig. 14.1). The campus crime statistics revealed that the five major crimes were burglary (270 cases), theft (209 cases), assault (32 cases), sexual assault (16 cases), and harassment (10 cases). Thus, categories of the highest occurrence were selected to represent the five general offence types.

The majority of burglaries occurred in the parking lots whereas thefts mostly occurred in academic and administration buildings. Harassment, assault and sexual assault were recorded in and around the residence halls, whereas robberies (2 cases) and other criminal offences were not as widespread on campus. These findings are similar to but also depart from past studies. While other studies (e.g., Robinson and Mullen 2001; Robinson and Roh 2001) found drugs (legal and illegal) to be a leading criminal activity, this study did not find these results. Instead, non-violent crimes (burglary and thefts) followed by violent attacks (e.g., assault and sexual assault) were dominant on the investigated campus. In comparison, sexual assaults were higher on this campus than in Robinson and Roh's (2001) study. Hence, criminal activity numbers and types vary by campus. This study used the reported campus crime data for the spatial analysis and did not aim to include or model unreported campus crimes.

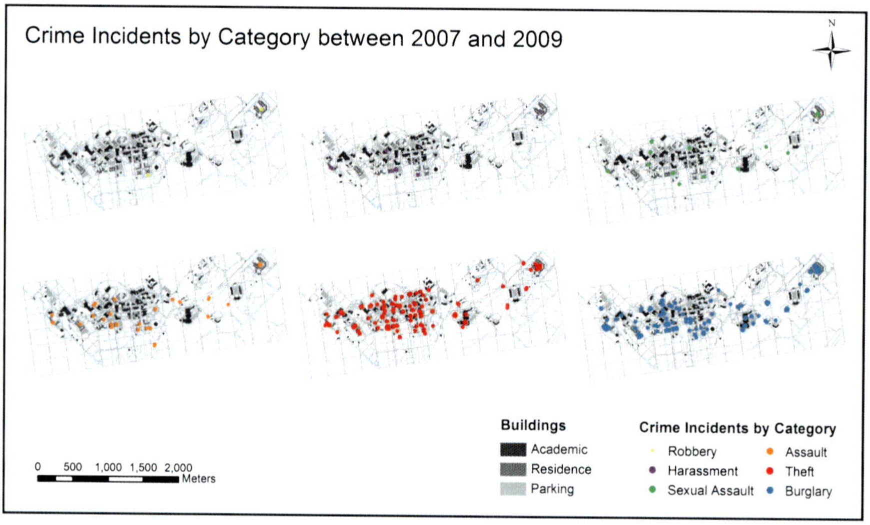

Fig. 14.1 Reported crime incidents between 2007 and 2009 (University Crime Statistics)

14.4.2 Generating and Generalizing Students' Fear of Crime Maps

The use of mapping to illustrate aggregated offenses is found in the crime literature (e.g., Groff and La Vigne 2001). Similar methods but different variables are used here. Students' fear of crime perceptions (mostly point locations) were aggregated into reference grids for each section of the campus and then used to create cognitive fear of crime maps (Fig. 14.2). Aggregating individual fear of crime perceptions into predefined grids allowed analysis using hot spot visualizations which helped with spatial pattern analysis. The grid size was based on the predefined reference grid of the original campus map and each grid represents 12.5 acres. A shortcoming of the aggregation method, formation of squares, is that there is less accurate spatial data compared to discrete (e.g. point) data. Openshaw (1984) and Ratcliffe and McCullagh (1999) describe this problem as the modifiable areal unit problem (MAUP) in which spatial data aggregation into artificial regions, e.g. administrative borders, grids, etc. might cause the generation of different spatial patterns. MAUP exists because of scale and zoning effects that change numerical representations depending on the scale and region categorization. These artificial spatial areas, i.e. the grid in this research, might change or mask spatial patterns. However for the purpose of this study the selected grid should work fine because cognitive maps are often distorted from reality. Mapping crime fears onto a reference grid will provide a reasonable area for fear of crime estimation.

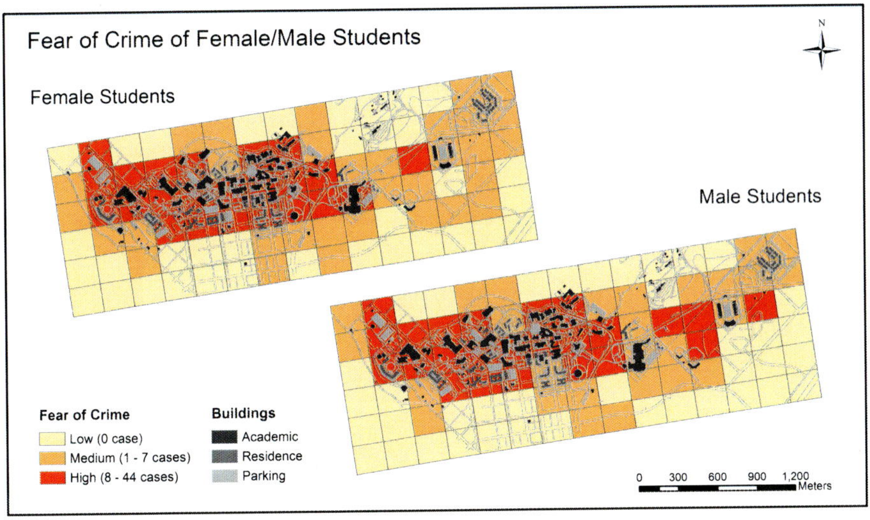

Fig. 14.2 Female and male students' fear of crime

14.4.3 Fear of Crime Patterns

The fear of crime maps of female and male participants show similar patterns (Fig. 14.2). Most fear is oriented towards the central campus and less towards the peripheral and remote areas of the campus. This is an interesting finding, since the central campus usually has a higher student/visitor concentration. However, the central campus has only a higher student/visitor concentration during the day. During night time there are fewer visitors in the central campus area which could lead to fear of crime, especially in the parking lots. In addition the central campus area contains mostly walk-only access with almost no car traffic during the evening/night, thus the mode of transportation might trigger a higher fear of crime perception in students. Current major construction work (large new buildings) and as a result smaller and dimmer pathways on the central campus cause several darker areas, thus a more fearful environment at night.

The participants' fear of crime explanations, provided on the questionnaire, highlight some of the reasons for fear. Most students describe that they fear crime in areas around academic buildings and in parking lots between the hours of 8:00 PM and 7:00 AM. Areas outside academic buildings, such as walkways, have been described as "not well lit", "wooded", having "no security presence", being "unmonitored" and providing "hiding places". Students described parking lots as "isolated", "dark", "vacant" and also as being "not monitored". One student wrote: "Nobody will see if you get attacked at night."

Residency status (students living on campus and off campus) provides a change in spatial crime perception patterns. These different patterns are related to parking spaces that are in remote areas off campus. It is noticeable that commuter students (who park on the outer parking spaces) have a higher fear of crime in these areas than students who live on campus and park their car close to their residence.

Campus residents do not feel safe at night around the areas outside their residential halls since these are perceived to be "very dark", "not well lit" and have only "few people" present. One student described it as being "a very scary walk" when passing through those areas. However, students continue with their activities at night and go out for dinner, visit a friend living in another part of campus, or go shopping. Some students noted that they do not feel safe inside the residential halls either, because they received police warnings about recent home invasions and heard stories about break-ins and assaults in the students' residences. Overall these findings will need a follow-up study to investigate in greater detail the extent of fear of crime on this specific campus.

14.4.4 Bivariate Crime Mapping

In a final data analysis and mapping step, bivariate maps comparing crime statistics and fear of crime were created. The observed crime and fear of crime rates (residents and commuters) for four categories namely burglaries, thefts, harassments, and sexual assaults were classified into three classes using the quantile classification method. The resulting classes were named low, medium, and high crime rate as well as low, medium, and high fear of crime. The individual values for each of these three classes are plotted in Figs. 14.3, 14.4, 14.5, 14.6, and 14.7. A set of four bivariate maps was created. These maps represent fear of crime in relation to observed burglaries, thefts, harassments, and sexual assaults. A final bivariate map compares the total number of observed crimes against fear of crime values. All bivariate maps contain a special bivariate color scheme to visually represent low-low to high-high fear-reality relations, wherein light violet represents low-low and dark violet visualizes high-high fear-reality relations (Figs. 14.3, 14.4, 14.5, 14.6, and 14.7). Low-low fear-reality relations indicate that observed and fear of crime are both low whereas high-high fear-reality relations visualize that observed and fear of crime are both high. In both cases the fear-reality relations indicate that students' fear of crime maps match observed crime statistics. All low-high (red) or high-low (blue) fear-reality relations indicate a mismatch between students' level of fear and actually observed crime. Thus red areas indicate regions of major concern in the created bivariate maps. Data combinations that result in medium results (e.g. low-medium, medium-high) were categorized and represented in light yellow. In so doing, the maps represent major agreement and differences between observed and perceived crime rates.

None of the bivariate maps revealed an inverse relationship between fear of crime and actual crime rate (i.e., low fear of crime and high crime rate). Generally, participants overestimated crime incidences. These so called "fear spots" are areas that do not have frequent offenses but are feared to be dangerous (Fisher and Nasar 1995).

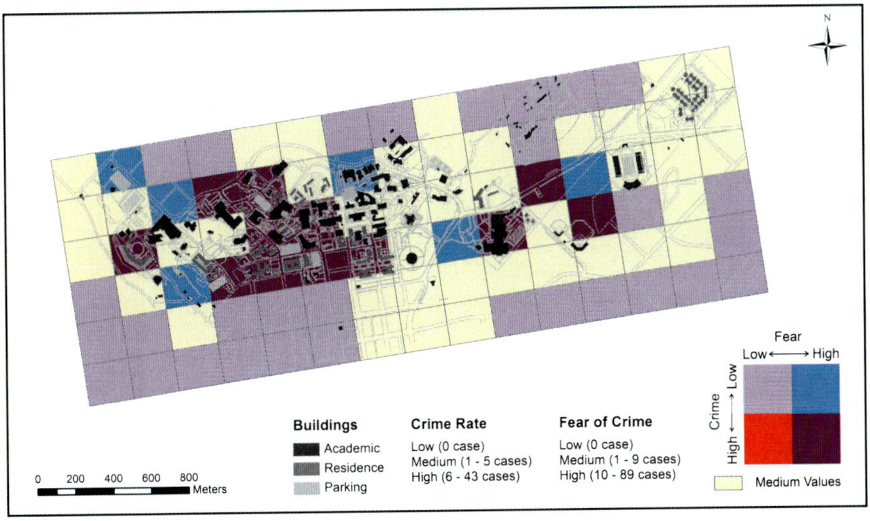

Fig. 14.3 Crime rate vs. Fear of crime: burglary

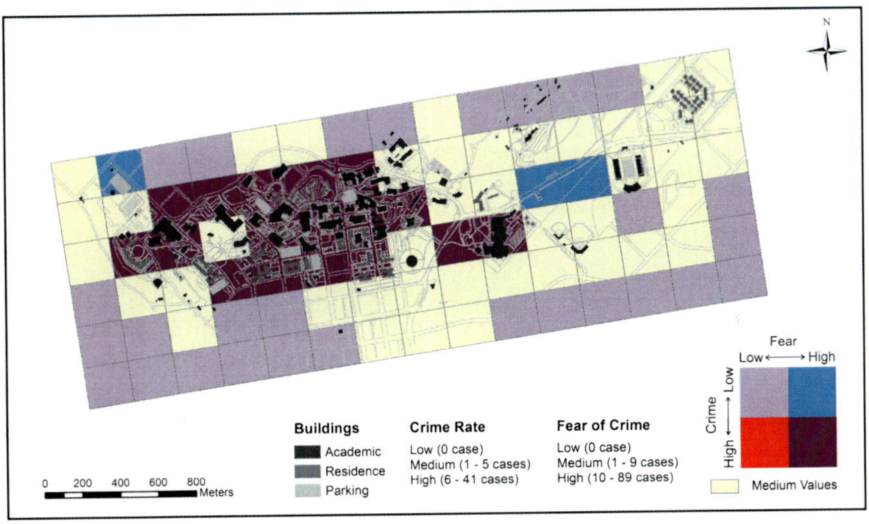

Fig. 14.4 Crime rate vs. Fear of crime: theft

Additional research needs to be conducted to better understand why participants overestimate crimes in certain areas.

The bivariate burglary and theft maps show that student fears match the actual crime rate for the central campus in most areas. Remote areas of campus generally have higher crime fears than recorded statistics (Figs. 14.3 and 14.4). This may be attributed to student observations (residents and commuters) of dim lighting, minimal

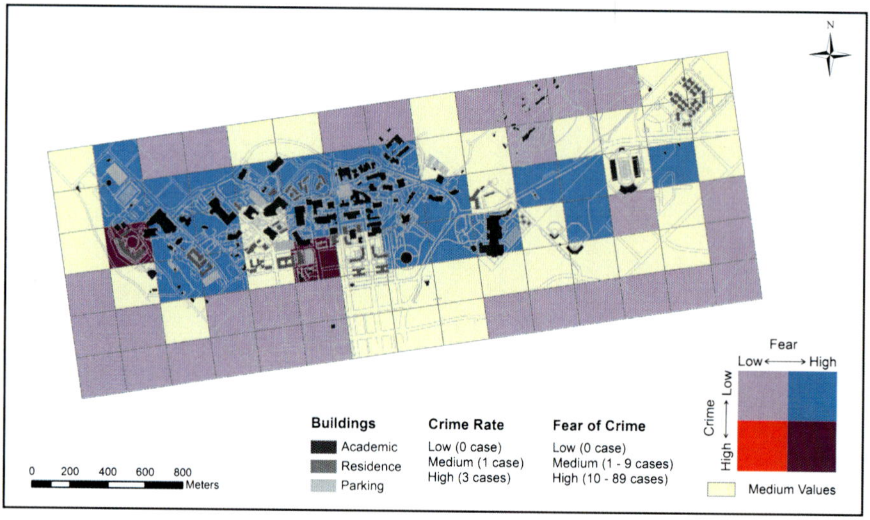

Fig. 14.5 Crime rate vs. Fear of crime: harassment

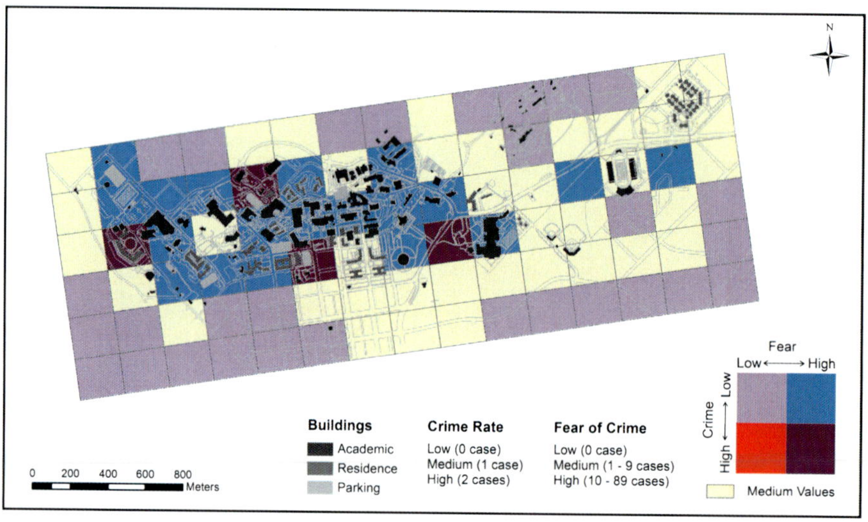

Fig. 14.6 Crime rate vs. Fear of crime: sexual assault

patrol/surveillance in these areas or the notion that the chance of theft is directly proportional to the distance object is away from its owner. In addition participants reported hearing stories about attempted rape and assaults in these remote and isolated campus areas.

The harassment and sexual assault bivariate maps highlight different patterns. Both maps show high fears of crime for the central campus but an actual low observed crime rate. These patterns are true for most of the central campus except

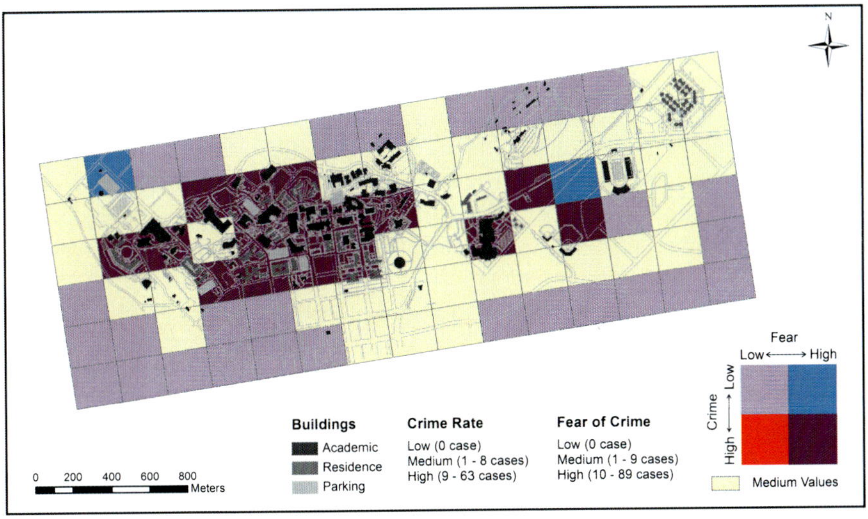

Fig. 14.7 Crime rate vs. Fear of crime: all crimes reported

for a few grids (mostly in student housing areas) in which a higher crime rate matches a higher fear of crime indicator (Figs. 14.5 and 14.6). Thus, the areas of fear might not only to be in public campus areas but in student residences itself which some researchers also describe as "crime generators" (e.g., Brantingham et al. 1977, 1995; Robinson and Mullen 2001; Robinson and Roh 2001). Overall the bivariate maps in this study suggest that students might over estimate cases of sex related offenses on campus, or that some cases might not have been reported to law enforcement. However, the high perception of sex crimes matches with the data of this campus and also nationwide. The rate of forcible rate per 100,000 people was 12.38 in 2006 and dropped slightly in 2007 to 12.18. This is the highest value amongst listed violent crimes of robbery and aggravated assault (Fox and Burstein 2010).

Figure 14.7 compares all crime observations for 2007–2009 and the reported fear of crime. The data reveal a close match between areas with higher crime occurrences and higher fears of crime. Two areas in Fig. 14.7 report higher fears of crime and lower crime rates. These areas represent remote parking facilities in the football stadium area and the recreation center. Both locations are about a 20–30 min walk away from campus center. University police and administrators might consider reviewing these areas to assess if additional patrolling or safety features are needed in these regions.

14.5 Conclusions

Campus crimes at universities and colleges are an unfortunate reality. Some immediate solutions to address campus crime are a stronger campus police force and campus monitoring/patrolling, quick responses to crime events through university police and other law enforcement agencies, and mitigating crime potentials through student education. Learning about students' fear of crime could help law enforcement target specific areas for crime monitoring and also provide information for educating students about taking special precautions in some areas of a university campus.

The research presented here attempts to address this issue by introducing a non-invasive data collection method that combines cognitive fear of crime representations and observed crime rates into an effective bivariate mapping method. The resulting graphic representations identify (a) potential student misconceptions of safe versus unsafe areas on campus, and (b) potential underreported crime areas. In this study, student fear of crime is aligned with data for burglary and theft but is over predicted, compared to the reported cases, for harassment and sexual assault. It may be that students are more fearful than statistics suggest or in the case of sexual crimes students learn of them but these cases are less likely to be reported than loss of personal property. This outcome needs to be further investigated.

This study has shown several benefits of the bivariate mapping approach for improved campus safety. First, it is a proof of concept of the bivariate mapping approach and thus provides, despite limited, answers towards interpreting campus crime pattern. Since the participant demographics are similar to those of the whole student body, the current findings serve as initial indicators and should be generalizable for the university. Second, the introduced bivariate mapping approach can be replicated for any other campus in the U.S. or internationally. The results will provide researchers with an essential view on campus safety and necessary tools for campus safety improvements. Follow-up studies should investigate spatial analysis at multiple scales, usability aspects for decision making, and public accessibility issues.

This study has some limitations that need to be addressed in future research. In order to create representative results about broader fear of campus crime graduate students, faculty and staff should also be included into the research study. Here the researchers compare only a representative group of the undergraduate population to campus crime data; although the number of undergraduate participants is similar to other studies, e.g., Fisher and Nasar (1995). While focusing on one population may be a short coming, it should also be noted that the undergraduate population is probably the most vulnerable demographic on campus and deserves the research focus. Next, a larger participant group with a higher number of fear indicators might also allow analyzing the cognitive and observed data with a finer reference grid. Individual reference grids might be necessary for each campus, if this study gets replicated by other researchers in other regions. Finally, the data may not accurately reflect the actual crime occurrences due to non-reporting of all crimes and data entry errors (Groff and La Vigne 2001; Robinson and Roh 2001).

This research contributes to applied campus safety in several ways. First, general student fear of crime assessments may help law enforcement to quickly target specific areas for additional crime monitoring. In addition, law enforcement could inform the public to take special precautions in some areas of a university campus until safety measures are in place. Secondly, the bivariate analysis addresses campus safety by combining alternative spatial mappings in the form of cognitive fear of crime maps with observed crime rates. This novel representation of two datasets allows identification of students' fear of crimes in areas where actual crime rates are low, and also can show regions where fear of crime is low, but observed criminal activity is high. The latter case would allow university administrators and police to educate students about potential threats in these areas, increase patrolling, and consider other crime mitigation measures. The bivariate crime maps are effective representations for university administrators and university police in decision and policy making tasks, for crime mitigation, and public awareness purposes. Overall this research provides a promising data collection method, analytic approach, and visualization technique for building safer university and college campuses.

References

Astor R, Meyer H, Behre W (1999) Unowned places and times: maps and interviews about violence in high schools. Am Educ Res J 36(1):3–42

Balkin S (1979) Victimization rates, safety and fear of crime. Soc Probl 26(3):343–358

Banai R (1999) A methodology for the image of the city. Environ Behav B Plan Des 26(1):133–144

Baumer TL (1985) Testing a general model of fear of crime: data from a National sample. J Res Crime Delinq 22(3):239–255

Brantingham PL, Brantingham PJ (1993) Nodes, paths, and edges: considerations of the complexity of crime and the physical environment. J Environ Psychol 13:3–28

Brantingham PL, Brantingham PJ (1995) Criminality of place: crime generators and crime attractors. Eur J Crim Policy Res 3(3):5–26

Brantingham PJ, Brantingham PL, Molumby T (1977) Perceptions of crime in a dreadful enclosure. Ohio J Sci 77(6):256–261

Brantingham PJ, Brantingham PL, Seagrave J (1995) Crime and fear at a Canadian university. In: Fisher BS, Sloan JJ (eds) Campus crime: legal social, and policy perspectives. Charles C. Thomas Publisher, Springfield, pp 123–155

Brower AM, Carroll L (2007) Spatial and temporal aspects of alcohol-related crime in a College Town. J Am Coll Heal 55(5):267–275

Clarke RV, Eck JE (2005) Crime analysis for problem solvers in 60 small steps. U.S. Department of Justice, Center for Problem Oriented Policing. Retrieved 24 Sept, 2012, from http://www.popcenter.org/library/reading/PDFs/60steps.pdf

Carter SD (2001) Campus violence symposium – introduction. Stetson Law Rev 32(1):1–6

Downs R, Stea D (1973) Cognitive maps and spatial behavior: process and products. In: Downs R, Stea D (eds) Image and environment: cognitive mapping and spatial behavior. Aldine Publishing Company, Chicago, pp 8–26

Fisher BS, Nasar JL (1995) Fear spots in relation to microlevel physical clues: exploring the overlooked. J Res Crime Delinq 32(2):214–239

Fisher BS, Sloan JJ (1993) University responses to the University Campus Security Act of 1990: evaluating programs designed to reduce campus crime. J Secur Adm 16(1):67–80

Fisher BS, Sloan JJ (2003) Unraveling the fear of victimization among college women: is the 'shadow of sexual assault hypothesis' supported? Justice Q 20(3):633–659

Fisher BS, Sloan JJ (eds) (2007) Campus crime, 2nd edn. Charles C. Thomas, Springfield

Fisher BS, Sloan JJ, Wilkins DL (1995) Fear and perceived risk of victimization in an urban university setting. In: Fisher BS, Sloan JJ (eds) Campus crime: legal, social, and policy perspectives. Charles C Thomas, Springfield

Fisher BS, Sloan J, Cullen F, Lu C (1997) The on-campus victimization patterns of students: implications for crime prevention by students and post-secondary institutions. In: Lab SP (ed) Crime prevention at a crossroads. Anderson Publishing, Cincinnati, pp 101–126

Fisher BS, Hartman JL, Cullen FT, Turner MG (2002) Making campuses safer for students: the clery act as a symbolic legal reform. Stetson Law Rev 32(1):61–89

Fox JA, Burstein H (2010) Violence and security on campus: from preschool through college. Praeger, Santa Barbara

Golledge RG (1999) Wayfinding behavior: cognitive mapping and other spatial processes. The John Hopkins Press, Baltimore

Gould P, White R (1974) Mental maps. Penguin, Harmondsworth

Gregory DE (2001) Crime on campus. Compliance, liability, and safety. Campus Law Enforc J 31(4): 27–29, 31–32

Gregory DE, Janosik SM (2002) The Clery Act: how effective is it? Perceptions from the field – the current state of the research and recommendations for improvement. Stetson Law Rev 32(1):7–59

Groff ER (2007) 'Situating' simulation to model human spatio-temporal interactions: an example using crime events. Trans GIS 11(4):507–530

Groff ER, La Vigne NG (2001) Mapping an opportunity surface of residential burglary. J Res Crime Delinq 38:257–278

Hale C (1996) Fear of crime: a review of the literature. Int Rev Victimol 4(1):79–150

Hudge MA (2000) The campus security act. J Secur Adm 23(2):25–27

Jackson J (2006) Introducing fear of crime to risk research. Risk Anal 26(1):253–264

Janosik SM (2001) The impact of campus crime Awareness Act of 1998 on student decision making. NASPA J 38(3):348–360

Janosik SM (2002) Parent's views on the Jeanne Clery Campus Crime Act and campus safety. Educational Policy Institute of Virginia Tech, Blacksburg

Janosik SM, Gehring DD (2001) The impact of the Jeanne Clery disclosure of campus crime Statistics Act on college student behavior. Educational Policy Institute of Virginia Tech, Blacksburg

Janosik SM, Gregory DE (2003) The Clery Act and its influence on Campus Law enforcement practices. NASPA J 41(1):182–199

Janosik SM, Gregory DE (2009) The Clery Act, Campus Safety, and the perceptions of senior student affairs officers. NASPA J 46(2):208–227

Kearney A, Kaplan S (1997) Toward a methodology for the measurement of knowledge structures of ordinary people: The conceptual content cognitive map (3CM). Environ Behav 29(5):579–617

Kitchin RM (1994) Cognitive maps: what are they and why study them? J Environ Psychol 14(1):1–19

Kitchin RM, Freudschuh S (2000) Cognitive mapping: past, present and future. Routledge, London

Kuipers B, Tecuci D, Stankiewicz B (2003) The skeleton in the cognitive map: a computational and empirical exploration. Environ Behav 35(1):81–106

Lewis DA, Maxfield MG (1980) Fear in the neighborhoods: an investigation of the impact of crime. J Res Crime Delinq 17(2):160–189

Lipka S (2009) In campus-crime reports, there's little safety in the numbers. Chron High Educ 55(21):A1

McConnell EH (1997) Fear of crime on campus: a study of a Southern University. J Secur Adm 20:22–46

McNeal LR (2007) Clery Act: road to compliance. J Pers Eval Educ 19(3–4):105–113

Nasar JL, Fisher BS (1993) Hot spot's of fear and crime: a multitude-method investigation. J Environ Psychol 13:187–206

O'Kane J, Fisher R, Green L (1994) Mapping campus crime. Secur J 5(3):172–179

Openshaw S (1984) The modifiable areal unit problem. Geo Books, Norwich

Ratcliffe JH, McCullagh MJ (1999) Hotbeds of crime and the search for spatial accuracy. Geogr Syst 1(4):385–398

Rengert G, Lowell R (2005) Combating campus crime with mapping and analysis. Crime Mapp News 7(1):1–5

Robinson MB, Mullen KL (2001) Crime on campus: a survey of space users. http://www.pscj.appstate.edu/campuscrime.html. Last viewed on16 Feb 2012

Robinson MB, Roh S (2001) Crime on campus: a survey of space users. Crime Prev Community Saf Int J 3(4):33–46

Robinson M, Roh S (2007) Spatial aspects of calls for service at a Regional Comprehensive University. In: Fisher BS, Sloan JJ (eds) Campus crime: legal, social, and policy perspectives. Charles C. Thomas, Springfield

Rund JA (2002) The changing context of campus safety. New Dir Stud Serv 99:3–10

Sherman L, Weisburd D (1995) General deterrent effects of police patrol in crime "hot spots": a randomized, controlled trial. Justice Q 12(4):625–648

Singh RR (1996) Adapting geographic information systems to sketch planning needs. Unpublished Master thesis, Massachusetts Institute of Technology, Cambridge

Starkweather S (2007) Gender, perceptions of safety and strategic responses among Ohio University students. Gend Place Cult 14(3):355–370

Tseng CH, Duane J, Hadipriono F (2004) Performance of campus parking garages in preventing crime. J Perform Constr Facil 18(1):21–28

Vann IB, Garson GD (2001) Crime mapping and its extension to social science analysis. Soc Sci Comput Rev 19(4):471–479

Wilcox P, Jordan PE, Pritchard AJ (2007) A multidimensional examination of campus safety victimization, perceptions of danger, worry about crime, and precautionary behavior among college women in the Post-Clery Era. Crime Delinq 53(2):219–254

Wilson RE (2007) The impact of software on crime mapping – an introduction to a special journal issue of social science computing review on crime mapping. Soc Sci Comput Rev 25(2):135–142

Chapter 15
Testing the Usability of Time-Geographic Maps for Crime Mapping

John D. Morgan and Philip E. Steinberg

Abstract Time geography offers a rich framework for representing movement across space and time. An extension of time geography to crime mapping, as proposed by the models discussed in this chapter, requires an accounting for victim and offender mobility under event-related constraints (e.g. accessibility to a crime scene). This chapter discusses results from a study that evaluates the usability of 3D space-time cube maps for representing crime patterns. Also considered is the utility of the time-geographic framework for exploring crime events that occur at unknown points in space and time. To this end, this chapter discusses the problem of crime activities that are not amenable to point-based mapping, potential alternative visualization methods using time-geographic techniques, and the procedures and results of usability tests wherein participants were asked to interpret maps that incorporated various time-geographic attributes. The overall purpose of the study was to assess the practicality of using time geography within a crime mapping context.

Keywords Crime mapping • Time geography • Map usability • Mobility • Accessibility

J.D. Morgan (✉)
National Environmental Modeling and Analysis Center, CPO 2345, One University Heights, University of North Carolina at Asheville, Asheville, NC 28804, USA
e-mail: jdmorgan@unca.edu

P.E. Steinberg
Department of Geography, Florida State University, Room 323 Bellamy Building, 113 Collegiate Loop, PO Box 3062190, Tallahassee, FL 32306-2190, USA
e-mail: psteinberg@fsu.edu

M. Leitner (ed.), *Crime Modeling and Mapping Using Geospatial Technologies*, Geotechnologies and the Environment 8, DOI 10.1007/978-94-007-4997-9_15,

15.1 Introduction

The use of mapping for exploring and analyzing the spaces of crime has grown considerably in recent years. Uncovering hotspots, criminal networks, flows, and investigative leads have become common goals in the application of crime mapping by both researchers and practitioners. Further, particular focus on the influence of the urban physical form for revealing geographic patterns of crime, known as environmental criminology (Brantingham and Brantingham 1993), has become the basis for modern crime pattern theory. This perspective, whose focus on nodes and pathways draws heavily on the work of architectural theorist Kevin Lynch (1960), pays particularly close attention to the ways in which structural forms influence behavioral patterns, of both perception and movement. Crime pattern theory, in turn, has come to inform many state-of-the-art policing practices, particularly concerning the allocation of scarce police resources.

Crime pattern theory is typically applied in crime mapping through routine activities theory (Felson and Clarke 1998), which holds that a criminal event is necessitated by three elements coming together in space and time: a victim (or target), an offender, and an opportunity. The investigative application of routine activities theory, popularly termed geographic profiling (Rossmo 2000), is utilized to identify patterns of offense and develop search strategies (LaVigne et al. 2000). Building on Newton's (1988) pioneering work in geoforensics, geographic profiling centers on using available data to locate an offender's base of operations, or haven, so that police resources can then be concentrated around that base and the offender can be apprehended (Leitner et al. 2007). Although it receives little direct attention in the literature, cartographic representations, or maps, play a key role in geographic profiling, both for identifying the offender's haven and for communicating that location to on-the-ground police personnel.

A particular challenge for geographic profiling, and crime mapping in general, is that some crimes, such as identity theft, defy traditional cartographic representation because of the fragmented nature in which such incidents occur across space and time (Hubers et al. 2008). Other crimes, such as thefts on crowded public vehicles in transit, do occur at singular locations in space and time, however they too challenge cartographers because the precise time-space locations of these crime-events are rarely if ever known. It is to this latter type of crime that we turn in this chapter. In these types of crime, the mobility of both the perpetrator and the victim confound conventional forms of investigative mapping. These types of crime events require a novel approach to mapping crime spaces.

Techniques such as travel demand modeling and (route) kernel density estimation (KDE) are found in the existing crime mapping literature and do address the topic of mobility and crime. For instance, in an effort to identify incident routes between known offender addresses and robbery locations, Levine (2010) utilized a street network grid of Chicago applying various impedance function calculations to develop a gravity model based on data from robberies in 1997 and 1998. KDE is an increasingly popular technique favored for its flexibility when setting parameters such as the grid cell size and bandwidth (Chainey et al. 2008). Further, KDE provides a visually pleasing smooth surface which can be utilized to identify crime hot spots. However, applications of KDE are limited to cases where the location of the crime event is *known*. The research discussed in this chapter addresses problems encountered in mapping crime events that, although still profoundly geographic, are not reducible to single points in space and time.

The consideration of mobility and corresponding space-time factors as part of a spatial analysis of events (e.g. incidents of disease or crime) is hardly novel. Indeed, one of the most widely adopted space-time approaches to spatial analysis of events is the Knox test (Knox 1964). The Knox test is a statistical technique used to determine whether events are clustered in both space and time. And while the Knox test is commonly cited in the epidemiology literature (Schmertmann et al. 2010), it has also found use in crime mapping (Grubesic and Mack 2008). Brunsdon et al. (2005) provide an excellent review of a variety of methods utilized for visualizing space and time crime patterns. However, notably absent from their review are any applications of Hägerstrand's time geography.

Time geography (Hägerstrand 1970) offers a rich framework for representing movement across space and time (Kwan and Lee 2004; Miller 2005; Pred 1977). Discrete activities, viewed from the perspective of an individual's mobility, are recognized as being bounded by defined constraints. Given these constraints, time geography utilizes visual semantic tools to explain individual movements in space and time. An extension of time geography to crime mapping, as proposed by the models discussed in this chapter, requires an accounting for victim and offender mobility under event-related constraints (e.g. accessibility to a crime scene).

To explore the potential for time geography-based visualizations to depict the spatial dynamics of mobile crimes, and to assess the practicality of these visualizations for interpreting crime patterns and allocating police resources, this chapter reports on results from a study conducted in 2009–2010 in which nine scholars and ten practicing crime investigators participated in a series of usability tests in which they were asked to manipulate and interpret visualizations that, to increasing degrees, incorporated time-geographic techniques. Quantitative analysis of participants' success in interpreting the maps revealed a high level of usability, even for the most sophisticated applications of time geography. However, qualitative results from the interviews suggest that while many participants in both groups recognized the *potential* utility of time-geographic techniques for crime mapping, usability issues constrained many of them from fully realizing these benefits (usefulness), and this was especially so for members of the practitioner group. These findings suggest that time geography has much to contribute to crime mapping and analysis, but that first advances must be made to improve the technique's usability.

15.2 A Brief Explanation of Time Geography

Although a full literature review of time geography is beyond the scope of this chapter it is worthwhile to explain the fundamental constructs that make up the approach since it has seen little application within the context of crime mapping. The explanatory tools behind time geography rest on the fundamental tenet that human activities have both spatial and temporal dimensions and that these cannot be meaningfully separated. Individual mobility in these dimensions is recognized as being bounded by certain space and time constraints. Hägerstrand (1970) categorizes the concept of constraints as fitting within one of three types: capability, authority, and coupling. These three categories of constraints are seen as being interrelated rather than additive, and they manifest themselves by dictating the space-time control points between which activities are undertaken to achieve predetermined goals (i.e., projects) (Carlstein 1978; Miller 1991; Neutens et al. 2007; Zillinger 2005).

Capability constraints address the physical limitation of individuals, such as those imposed by the need to eat or sleep. Suppose that a single offender were carrying out a succession of pickpocket acts across a given shopping district area. Then a key assumption would be that the offender has the capability, on foot or otherwise, to reach the different criminal opportunity points. *Authority constraints* reflect the influence of organizations external to the individual that control access to different places at different times (Ratcliffe 2006). Authority constraints refer specifically to levels of access at an individual level. For the rationalizing criminal, such as described by Clarke and Felson (1993), authority constraints play a prominent role in the decision making process leading up to an act of offense. *Coupling constraints* are recognized as the necessity of certain activities to form production, consumption, social, and miscellaneous activity bundles (Pred 1977). Practically, coupling constraints are defined by socially accepted modes of behavior such as shop operating hours or bus departure times. In order for a certain exchange to take place between two people, they must come together in time and space (e.g. a waiter taking an order from a customer). The concept of coupling constraint fits well with Cohen and Felson's (1979) routine activities theory. To commit a given crime implies the ability to take advantage of a given opportunity for that crime. With certain exceptions (such as mail fraud or internet-based crime), criminal activities are bounded by the coupling constraint of the perpetrator and victim meeting in space and time. For instance, in order to steal a victim's wallet, a pickpocket must be proximate, in both space and time, to the victim (e.g. on the same elevator or on the same street).

Cartographically, an implementation of the time-geographic framework takes place within a 3D (three-dimensional) map environment, where the third, vertical, dimension represents time (Fig. 15.1). Although the framework has developed a rich array of possible visual semantic symbologies, we will focus on the four fundamental ones: cube, point, path and prism.

The space-time cube (Fig. 15.1) is used as a tool to represent the spatial two-dimensional (2D) axes, x and y, along with a third temporal axis, z. Empirical evidence suggests that the space-time cube representation is advantageous in conveying complex spatiotemporal data to users (Kristensson et al. 2008). Further, the space-time cube is capable of representing, simultaneously, the whole space-time continuum and the position of events in this continuum (Gatalsky et al. 2004).

There is very little mention of the *space-time point* as a construct in the time-geographic literature. Instead the space-time point is often seen as a

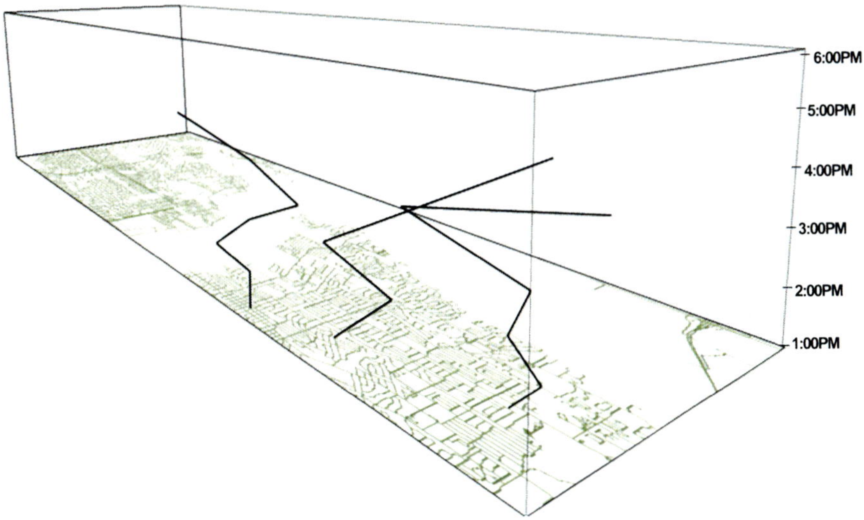

Fig. 15.1 Space-time cube with space-time paths

necessary element in the construction of the space-time path. Miller (2004) refers to space-time points as control points and describes the space-time path (Fig. 15.1) as the linkage of a series of space-time points. However, in a case where we have no need, or are not interested in, particular path information, the 0-dimensional space-time point is all that is required (Hendricks et al. 2003). An individual's *space-time path* is constructed by drawing straight lines connecting known space-time points such as those provided by travel-diary survey data (see Kwan 2000). The visual aspect of the space-time path concept represents an individual's known trajectory illustrated on a two-dimensional plane. The *space-time prism* (Fig. 15.2) leads to the idea of the time budget, in which a person can move away from the start location, limited only by the maximum travel velocity and the next known point (Tessmann 2006). A space-time prism gathers all space-time paths an individual might have drawn during a specific time budget and delimits the feasible set of opportunities within a person's reach (Dijst and Vidakovic 2000; Forer 1998). In theory, the prism is the intersection of two cones, called beads by Hornsby and Egenhofer (2002) and (Hariharan 1999). The lower cone represents the possible paths, referred to as *potential paths*, of travel from a given starting control point, while the upper cone represents the same path possibilities, in space-time, approaching the destination control point (Fig. 15.2).

Conceptually building on the space-time path constructed from two space-time points (x, y, t), Hariharan (1999) defines the apexes of a bead as either being collocated in space, but shifted in time (x_0, y_0, t_0 and x_0, y_0, t_1), or shifted by both space and time. As a further explanation, Hariharan (1999) explains that if the two

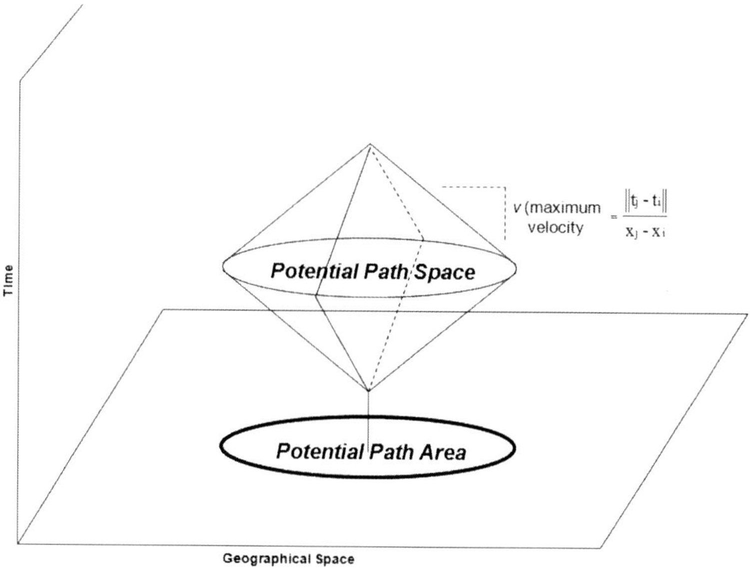

Fig. 15.2 Space-time prism (right bead)

space-time points are separated only by time then a *right bead* is formed. In this case the two half cones form a circle located in the plane

$$t = \frac{(t_0 + t_1)}{2} \tag{15.1}$$

around the center

$$\left(x_0, y_0, \frac{(t_0 + t_1)}{2} \right) \tag{15.2}$$

with a radius of

$$r = \frac{(t_0 + t_1)}{2} \tan\phi \tag{15.3}$$

The projection of the right bead (prism) onto the 2D plane forms a circle. The other possibility, called an oblique bead (prism), is when the two space-time points are shifted by both space and time, or are not spatially aligned.

The area of the prism is termed the *potential path space* (PPS), and the projection of the oblique bead (prism) onto the 2D plane forms an ellipse, or *potential path area* (PPA) (Fig. 15.3).

The slope of the cone shows a given possible maximum velocity for the represented individual from a known point in space, while the space-time path indicates an individual's activities in both space and time. The mathematical formulation for velocity is given as:

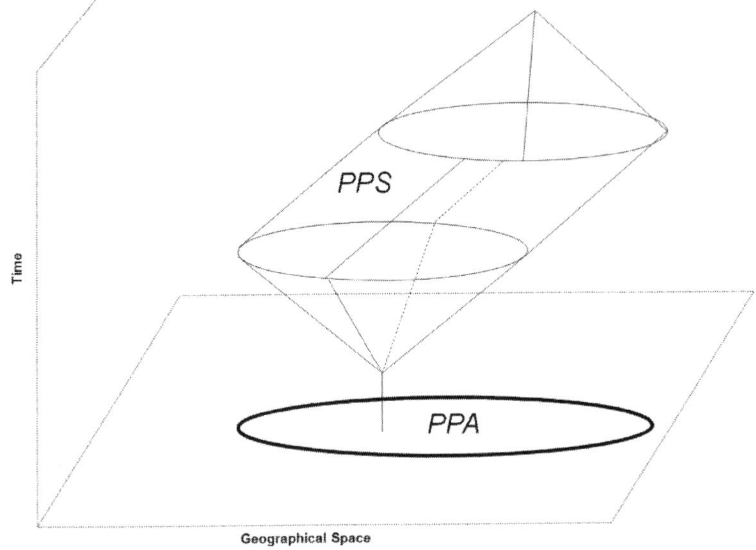

Fig. 15.3 Space-time prism (oblique bead)

$$vi_j = \frac{\|x_j - x_i\|}{t_j - t_i} \tag{15.4}$$

where || || is the vector norm or distance between the locations (x_i, x_j). Placing Eq. (15.4) within the context of the prism represented in Fig. 15.2 we can see that this equation allows us to scroll through locations in the space–time path using time as an index (Miller 2005). Using the maximum velocity assumption (Wu and Miller 2001), we can then conceptualize how an offender's potential path space, represented by the interior of the cone, intersects with a crime incident site x_j, by showing all of the locations in space and time that the offender could have occupied during the time budget interval (t_i, t_j). The range of offender travel capability then is constrained only by a defined maximum velocity, signified as v. This velocity is, mathematically, the subtraction of a known time segment, $t_j - t_i$, divided by the distance between known control points, $x_j - x_i$. The concept of velocity, expressed as PPA, is integral to a time-geographic approach. Therefore a potential path representation is used to show the points in space and time that the person could occupy during this travel episode (Miller 2005).

15.3 Usability Issues in Time-Geographic Visualizations

While there seems little doubt that a time-geographic visualization can contain much more information than a static, two-dimensional map, a map functions only if it *communicates* information. And frequently the barrier to a map communicating information is not that it contains too little information but that it contains too *much* (Gahegan 1999).

Therefore, before any recommendations can be made for applying time geography to crime mapping, time-geographic crime maps must be tested for their usability. In particular, crime maps are typically used for two functions: strategic and operational/tactical (Walsh and Ratcliffe 2005). A map that is so rich in information that it escapes comprehension by a practicing crime analyst is not useful for either function and, as Harrower et al. (2000) note, it is not uncommon for a map to be over engineered to a degree that it becomes difficult or impossible for use by a practitioner who has not undergone specialized training.

To investigate the applicability of time geography to crime mapping, this study relied on the history of subjecting both time-geographic and crime mapping methods to usability testing. Snook et al. (2007) and Paulsen (2006) have carried out map usability testing with police officers, asking them to mark an 'X' on the map where they thought that a serial burglar lived based on a 2D crime map pattern. Kristensson et al. (2008) have carried out usability testing comparing baseline 2D maps with 3D space-time cube maps. The Kristensson et al. (2008) test was conducted with an audience of novice users with the purpose of verifying the visual utility of the space-time cube.

The research reported in this chapter sought to combine the objectives of these two groups of tests, assessing the usability of 3D space-time cube maps for representing crime patterns and thereby testing the utility of the time-geographic framework for exploring crime events that occur at *unknown* points in space and time. The overall purpose of the study was to assess the practicality of using time geography within a crime mapping context.

Experts – both researchers and practitioners – were recruited based on their involvement in the crime mapping community, and testing was done in a manner consistent with methods that have been previously applied in usability studies evaluating the effectiveness of geovisualizaiton methods (Olson and Brewer 1997). Separate interviews were conducted with experts individually, thereby avoiding the influence of group opinion as might occur, for instance, in a focus group (Stewart and Shamdasani 1990). Participants were selected from two groups: practitioners (practicing crime analysts) and researchers (scholars with expertise in crime analysis, mapping, or both). Individual law enforcement agencies were contacted to set up interviews with practitioners, and additional practitioner interviews (as well as the interviews with researchers) were conducted at professional and academic conferences.

At each interview session participants were presented with a crime scenario (pickpocketing in a crowded shopping district) that assumes a single offender acting on two victims within a given range of time. While victim space and time paths were known and provided (as likely would be the case for investigators who were attempting to solve an actual crime), no offender data was given other than the constraint of velocity (which also mimics the circumstances of an actual crime investigation). The concepts of time geography were explained at the start and again as they were introduced during the exercise. The semi-structured interview consisted of cognitive walkthroughs through five map iterations of the same crime scenario.

With each iteration the tools of time geography were further incorporated, with the final map incorporating many of the visualization techniques fundamental to time geography. At each iteration, participants were asked questions to test their understanding of the tools of time geography based on the visual representation of the map. Each interview lasted approximately 45 min.

15.4 The Usability Interview Process

The usability interview process consisted of a pretest briefing and the actual interview. At each session, the participant was presented with the crime in which incidents of pickpocketing were occurring while victims were moving through a crowded shopping district. The primary maps used in the interviews were developed within a hybrid-GIS environment combining a base map from desktop GIS (MapWindow) and a 3D modeling software (Google SketchUp). This map interface allowed for the interviewees to be able to interact with the 3D time-geographic maps through interactive tools such as pan, orbit and zoom.

The pretest briefing was meant to set up the actual interview. At the start of the interview the participant was categorized as a practitioner or researcher. Also, the participant was asked if he or she was familiar with the concepts of geographic profiling or crime pattern recognition. Next the participant was presented with some basic assumptions of the scenario or key facts of the case:

- The maximum velocity for each map is set at 88 ft/min based on a leisurely pace in a crowded shopping district.
- This area is approximately one half of a mile by one half of a mile wide.
- Though this velocity is applied to the victim space-time path, it is also assumed to apply to the offender.
- Therefore, a key assumption for these maps is that all three individuals – *the two victims and the single offender – are traveling the same maximum velocity*.

After the crime scenario and key facts were explained, the participant went through a semi-structured interview process conducted as a cognitive walkthrough, a usability testing method commonly used in engineering wherein a research subject has set goals, performs actions, and evaluates feedback (Polson et al. 1992). Each user was given a table of two known victim space-time points and a brief explanation of the visual tools of time geography to be utilized. Harrower et al. (2000), in conducting usability studies of cartographic interface tools, note that novel interfaces may not result in improved performance unless sufficient training is provided in how to use them. For some of the participants, particularly in the practitioner category, the concept of 3D time-geographic maps would be a new concept. Therefore, care was taken in explaining the time-geographic concepts, e.g. space-time cube, space-time path and space-time prism.

Once the pretest briefing was complete each interview commenced with an introduction to the first map. Set in the crowded and heavily touristed La Rambla shopping district of Barcelona, known as the pickpocketing capital of the world (Adams 2009), participants were presented with a scenario in which they were to assume that a single offender had picked the pockets of two victims within a given range of time. And while victim space and time paths were known and provided, no offender data was given other than the constraint of velocity. The concepts of time geography were explained at the start and again as they were introduced during the exercise.

With each map iteration the tools of time geography were further incorporated. Participants were asked questions throughout to test their understanding based on the visual representation of the map. During each map iteration participants were given both a paper color map and access to the computer-based map (laptop) with Google SketchUp. Also, the participant was given basic instruction on how to navigate the map interface (e.g. pan and orbit) by use of the mouse-computer interface. To structure the process the participants were asked to answer a set of questions and complete a specific task regarding each map as it was introduced.

The first map iteration (Fig. 15.4) utilized the 2D *Flow Map* containing point symbology at the known point locations for each victim along with directional flow arrow symbology indicators between known points. Participants were asked the following questions and instructed to complete the following task:

Question 1: Can you tell me what is going on in this map?

Question 2: At what time do you think the two victims were at their closest?

Task 1: Please circle the area on the map where you think the pickpocket operated out of based on the visual information provided in this map.

Follow up question to Task 1: How or why did you select this area?

Question 1 in this iteration was designed to acclimatize the participant to the map interface and cognitively connect the content of the pretest briefing to the map interface. *Question 2* in this iteration was designed to engage the participant in the concept of space and time (or lack of) at the map interface. Essentially this question was not answerable by the map alone because time had not yet been visually incorporated into the map as is revealed in later iterations. Therefore, *Task 1* in this iteration was meant to test the *perceived* usefulness and usability.

Next the participant was provided with the *Potential Path Area Map* containing the potential paths for the victims between known points based on the assumptions of the scenario (Fig. 15.5). At this point the concept of PPA was again briefly explained. The participant was then asked to answer the following questions and complete the following task regarding the map:

Question 1: What does this map add to better identify a search strategy for the offender?

Question 2: At what time do you think the two victims were at their closest?

Task 1: Please circle the area on the map where you think the pickpocket operated out of based on the visual information provided in this map.

Follow up question to Task 1: How or why did you select this area?

As was the case in the first iteration, the second question in this iteration was not actually answerable from this map, since time was not incorporated into the visualization. Thus, once again, the second question, as well as the task, were designed to

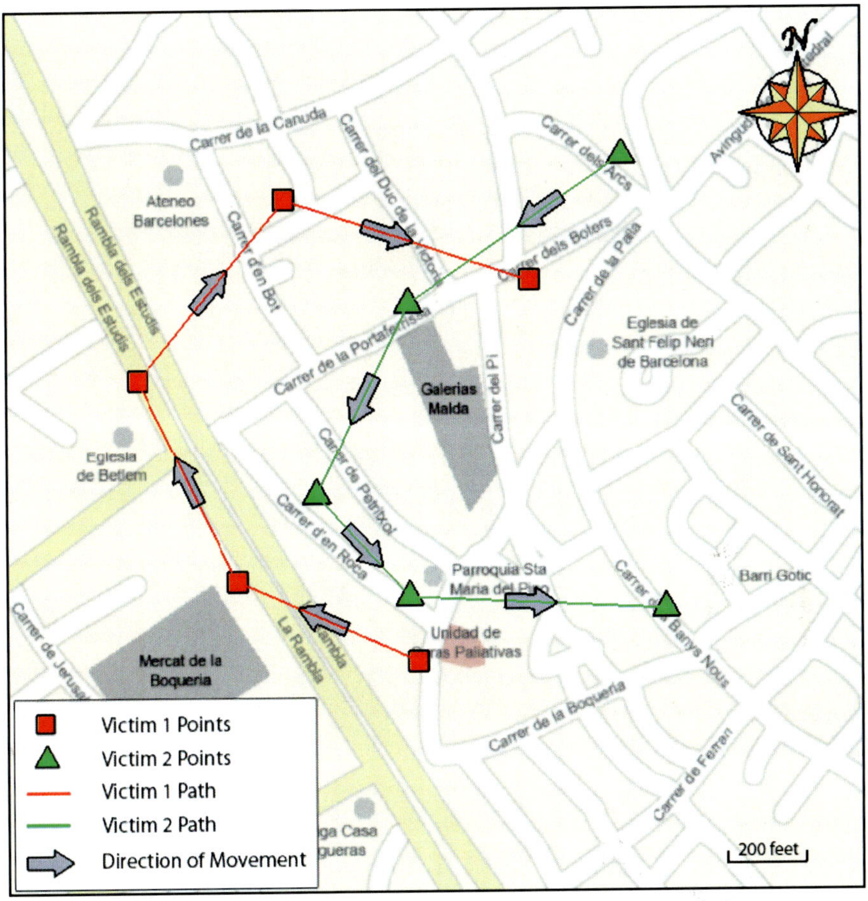

Fig. 15.4 Flow map

test *perceived* usefulness and usability rather than the actual capacity of the map to communicate information.

Next the participant was provided the same *Potential Path Area Map* with only the intersections of the two victims (Fig. 15.6). This concept was briefly explained as a filtering, or a further illumination, of a certain aspect of the *Potential Path Area Map*. The participant was then asked to answer the following questions and complete the following task regarding the map:

*Task 1:*Please circle the area on the map where you think the pickpocket operated out of based on the visual information provided in this map.

Follow up question to Task 1: How or why did you select this area?

During the next iteration the participant was provided a *Space-time Path Map* incorporating the two victims' paths in both space and time (Fig. 15.7). The concepts of space-time paths and space-time cubes were reiterated. Also, the participant was directed to utilize the mouse and interactive features of the 3D map within the

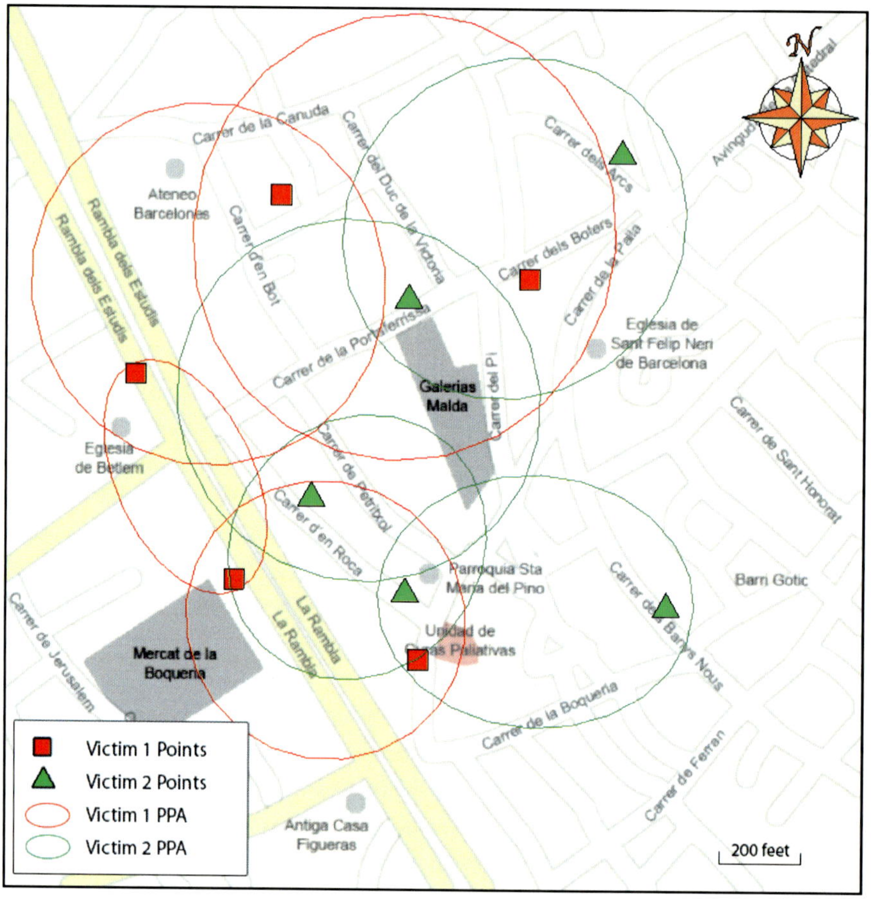

Fig. 15.5 Potential path area map

computer-based map. The participant was then asked to answer the following questions and complete the following task regarding the map:

*Question 1:*What does this map add to better identify a search strategy for the offender?

*Question 2:*At what time do you think the two victims were at their closest?

*Task 1:*Please circle the area on the map where you think the pickpocket operated out of based on the visual information provided in this map.

Follow up question to Task 1: How or why did you select this area?

Question 1 in this iteration was again a question meant to illicit the participant's perceptions of usefulness. However, *question 2* in this iteration was designed as a direct test of usability, and was answerable, with some certainty, by interacting with the map interface since it now contained temporal information.

In the final iteration, the participant was provided with a map that replaced space-time paths with space-time prisms (Fig. 15.8). The concept of the space-time prism was reiterated. And again the participant was directed to utilize the mouse and

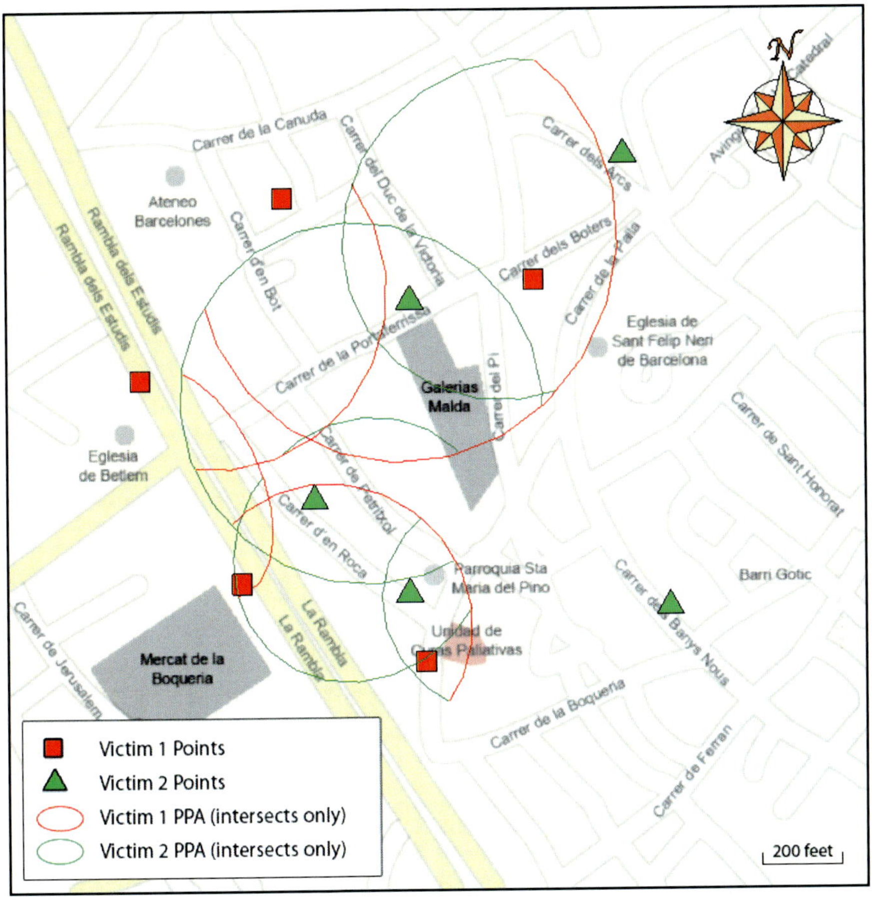

Fig. 15.6 Potential path area map intersections only

interactive features of the 3D map within the computer-based map. The participant was then asked to answer the following questions and complete the following task regarding the map:

Question 1: What does this map add to better identify a search strategy for the offender?

Question 2: At what time do you think the two victims were at their closest?

Task 1: Please circle the area on the map where you think the pickpocket operated out of based on the visual information provided in this map.

Follow up question to Task 1: How or why did you select this area?

Finally, the participant was asked an open ended question meant to asses satisfaction and potential utility (usefulness) of the tools of time geography:

Final Question: Are these types of tools practical in your job or research?

During this interview there were five map iterations (Figs. 15.4, 15.5, 15.6, 15.7, and 15.8), and with each map iteration the participant was asked to develop a search strategy to reveal the location of the offender. And though it is impossible

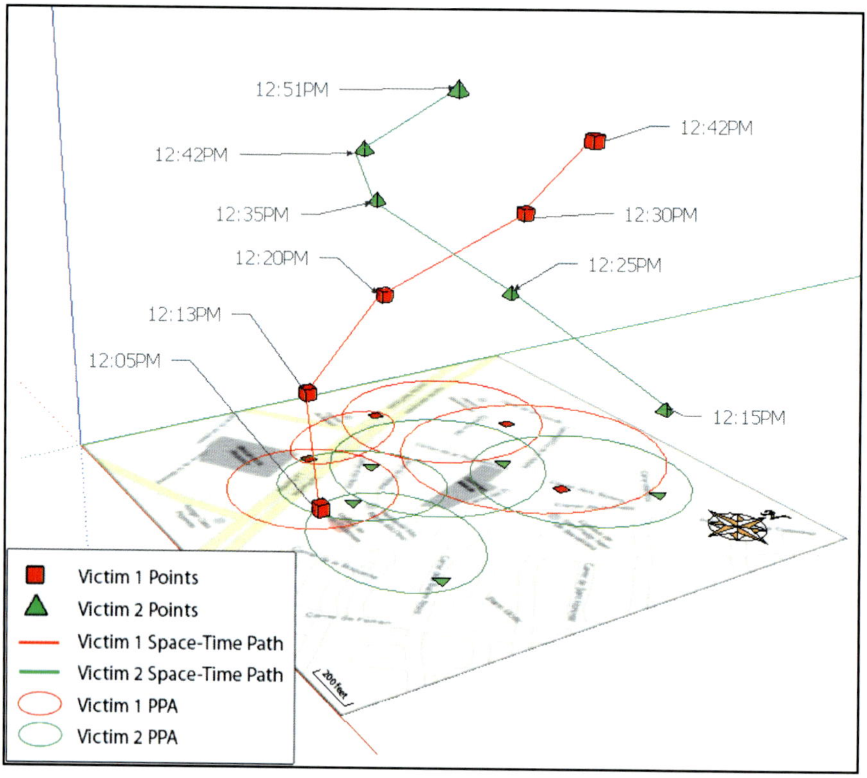

Fig. 15.7 Space-time path map

to know for sure where the offender carried out the crime, a search strategy, it was explained, in this case represented a hypothesis of where the offender's base of operations was. To assist the participants when developing their search strategies, verbal cues were given throughout that would assist the participant in connecting the exercise with practical decision making that occurs when one is attempting to thwart or catch offenders, such as selecting a location for police patrols or installation of security cameras. Thus each participant generated five different search strategies by circling areas on the map, with each subsequent search strategy being supplemented by increased utilization of the tools of time geography.

15.5 The Usability Interview Results

The time-geographic tools provided an ability to represent individual contextual factors such as victim speed constraints, while the street map itself represented environmental and place-based context such as street layout and building locations. What could not be represented by the maps was the unknown, which was in this case the actual location

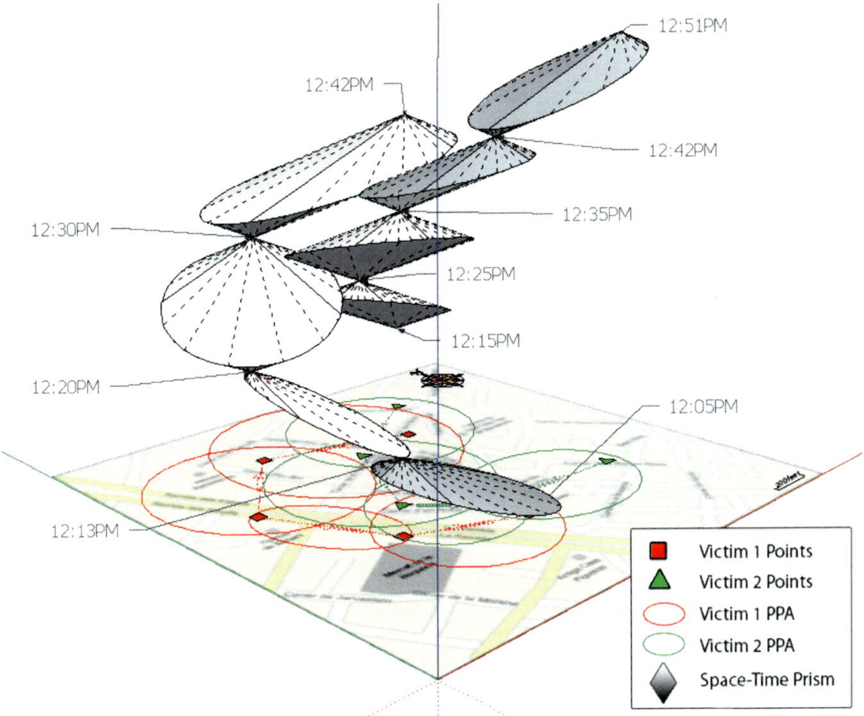

Fig. 15.8 Space-time prism map

of the crimes and the offender. It was left to the map user (the interviewee) to draw conclusions about the unknown (offender data) from the known (victim data and the time-space constraints of the geographic environment).

15.5.1 Locating Space on a Time Geography Map

At each iteration in the interview, participants were asked to develop a search strategy for a single offender by circling an area on the map. Often the participants wanted to select multiple areas but they were encouraged to try to limit their selection to a single best search strategy area. Equally relevant were the reasons cited for making their search strategy selections. It is interesting to note that most participants modified their search strategy areas as each new map, and hence an additional time-geographic tool, was introduced. However some participants felt strongly about their previous choices and tried to stick as closely as possible to those choices, even when confronted with new data.

In the series of maps that follow (Figs. 15.9, 15.10, 15.11, 15.12, and 15.13), the areas that participants selected have been generalized to areas indicated by capital

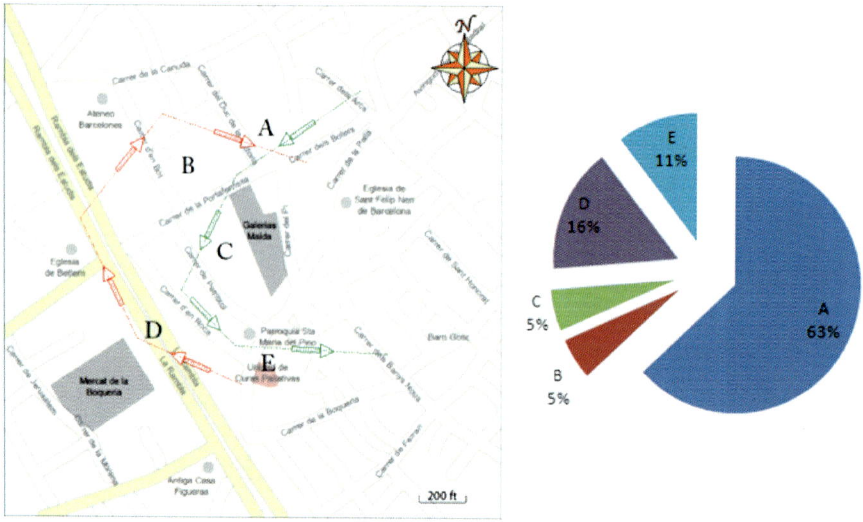

Fig. 15.9 The search strategy areas selected by participants from the flow map

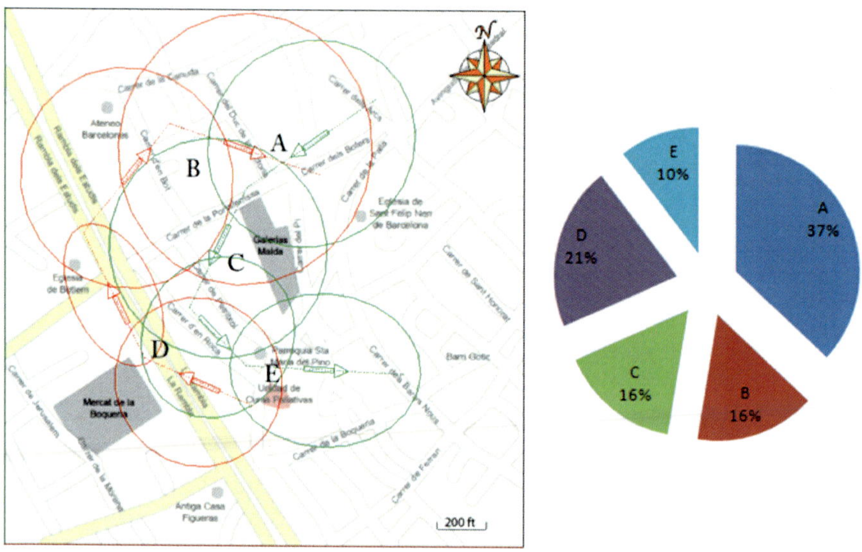

Fig. 15.10 The search strategy areas selected by participants from the potential path area map

letters. On each map, the associated pie chart shows the percentage of participants who selected each area. While most participants when presented with this simple flow map (Fig. 15.4) relied on map symbology (the directional arrows) in reaching their recommendations for a search strategy, some of the participants focused more

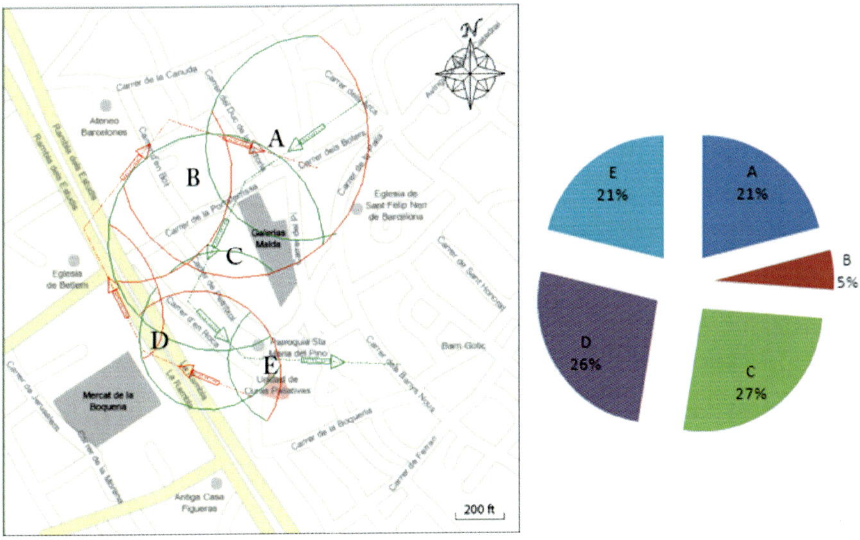

Fig. 15.11 The search strategy areas selected by participants from the potential path area intersections map

on the environmental and situational context to develop their search strategies. For instance, four of the practitioner participants justified their search strategies by noting that a pickpocket would likely frequent a university or market area (areas D, C, or E on Fig. 15.9). One of these participants stated, "I am looking strongly at the built areas and the influence on the likelihood of the crime. For example the market area is likely to have a lot of opportunities for distraction." However the majority of participants (63%) identified the location where the two victims' paths crossed (area A) as the most likely candidate for the offender's base of operations.

With the introduction of the PPAs to the map, participants became less certain that the area where the two victims' paths crossed (area A) was the best place for a search strategy, although it remained the predominant choice (Fig. 15.10). Again, several of the participants, in particular those who were practitioners, cited environmental factors in selecting their search strategies. One participant, calling the PPA intersections "convergences," commented, "Given the area of convergence I am thinking that [around] the Galleria area that is open and the offender could have had more opportunity. The offender could see more about where the victims were coming and going." This trend might best be explained by specific knowledge gained from experience of seeing how crime is actually carried out within context. This same participant went on to comment about the limited way in which time was represented on this map: "Time and space are important in geographic profiling to locate the most likely area at which the victim and offender will meet with an opportunity. But, I don't see time here as of yet." The participant is correct to note that time had not yet been explicitly represented in the map. However, the PPAs

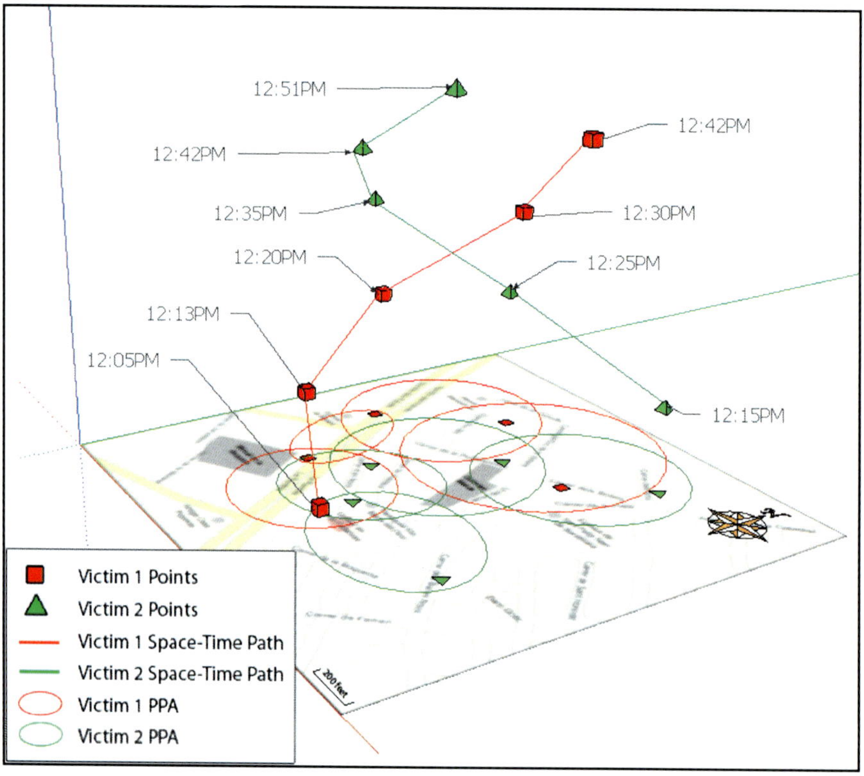

Fig. 15.12 The search strategy areas selected by participants from the space-time path map

themselves were explained as being based on the speed of travel and amount of time between two known points. The introduction of PPAs definitely gave the participants more to consider in developing their search strategies. Another participant noted that the PPAs told him more about where the victims might have spent time in a common area, or near each other. And, another participant commented, "… the narrower the PPA there is likely to have been a choke point forcing them into a likely crowded area and increasing vulnerability to pick pocketing."

One of the challenges with visually analyzing the PPA map is in discerning the different PPA intersections. One participant replied that he reached his search strategy choice "… by looking at the intersecting circles." But he followed up this answer with, "Actually there are bunches of intersecting PPAs so this does get a little confusing. I was influenced by the original intersection [when I was viewing the Flow Map] but now I am reconsidering. Along the edges the offender would be able to isolate the victims. If I was going to allocate foot patrol officers I would consider that they get bored very easily patrolling in a small area. I would choose a larger area of intersection." And another participant remarked, "I am still most certain about the crossing paths (area A) but with the intersecting PPAs. Also, I am considering this road [La Rambla]

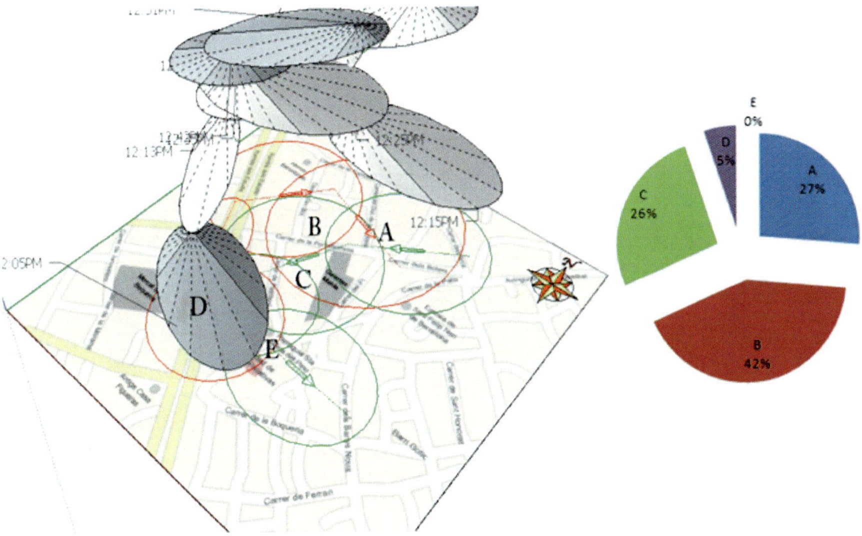

Fig. 15.13 The search strategy areas selected by participants from the space-time prism map

and am assuming it is a busy road with a lot of traffic." These quotations reveal a desire to supplement incomplete (or, perhaps, confusingly presented) time-geographic information with contextual data.

Anticipating the challenge of discerning the different PPA intersections, the next map iteration (Fig. 15.11) removed all of the PPA information except for the intersections. This map drew the participants' attention more toward the areas of densest PPA intersection such as areas A and E. Area E definitely seems to have the tightest intersections and some participants noted this. One participant remarked, "The areas with the tighter intersections (smaller) also become[s] an area of interest." But still other participants felt strongly persuaded by their original line of reasoning, with one saying "… with this one it is still close to the market but the intersections move me a little bit."

The next iteration (Fig. 15.12), along with the addition of 3D, introduced explicit time data that had been noticeably missing from the previous maps (as one participant noted during the previous iteration, "If I knew a range of times then I could rule out places where they were too far apart"), and it was anticipated that this new data might lead to a radical shift in participants' search strategies. Because of the addition of 3D, at this point participants were encouraged to interact with the map via the computer mouse. Now the participants were using the interactive tools of orbit, pan and zoom to visually inspect the space-time paths.

The addition of time onto the map interface was a new concept to some participants. One practitioner participant noted, "This is interesting! I am not used to seeing time in a map in this way. What I am used to is time as a bar chart or histogram that accompanies the map. This is quite different!" The resulting search strategies coming out of the introduction of the space-time paths drew some

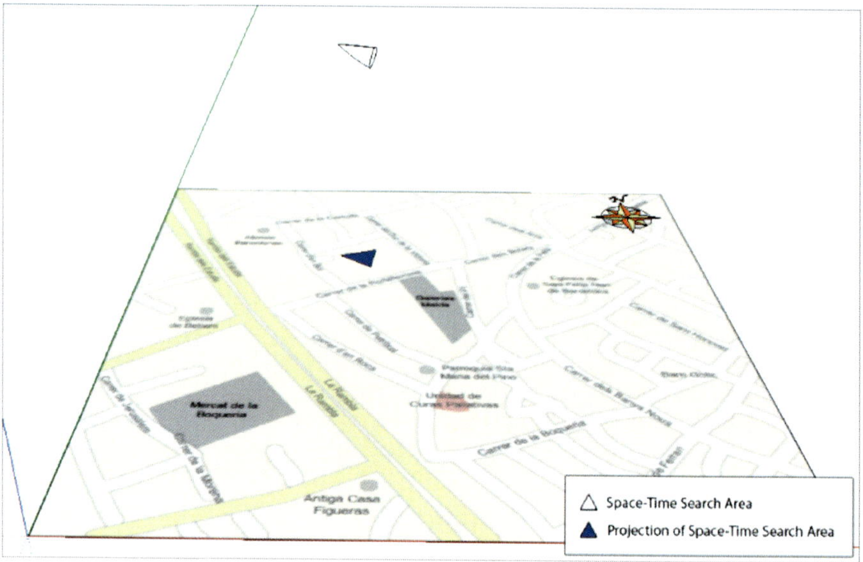

Fig. 15.14 Area where two victims were at their closest in space and time

participants more towards area C. Likely more influential at this area than the Gallerias is the fact that participants were attempting to see the location on the ground (the 2D map) where the space-time paths appeared to be at their closest. Most participants did this by orbiting to a top-view of the map and attempting to see where the two space-time paths were at their closest. This was in fact area C, and indicates positive usability of the space-time paths.

With the final map iteration, the *Space-Time Prism Map* (Fig. 15.13), participants were offered a chance to select a new search strategy. And while some participants were not swayed from their previous selections the plurality of participants (42%) selected Area B, which had rarely been selected on any of the previous four maps. Though some expressed a challenge in doing so, participants who selected this area did so by orbiting the map to realize the area where the prisms intersected on the ground (the 2D map).

As it turns out, Area B is indeed the location at which the two victims were at their closest in space and time (Fig. 15.14). And so, based solely on the metric of distance and known information, the correct area was found by 42% of the participants (in contrast with the original, more conventional flow map which led to a correct choice by only 5% of the participants). One practitioner participant in selecting Area B noted, "It looks like this is where the two cones are closest together," demonstrating an awareness of the concepts embodied in the space-time prism map as well as an ability to read it as a visualization.

However some participants expressed frustration with this map's usability and, therefore, its usefulness. One participant noted, "This is difficult for a layperson to utilize these tools." Additionally, some participants noted a particular challenge in cognitively connecting the prisms to the 2D map. One practitioner participant noted

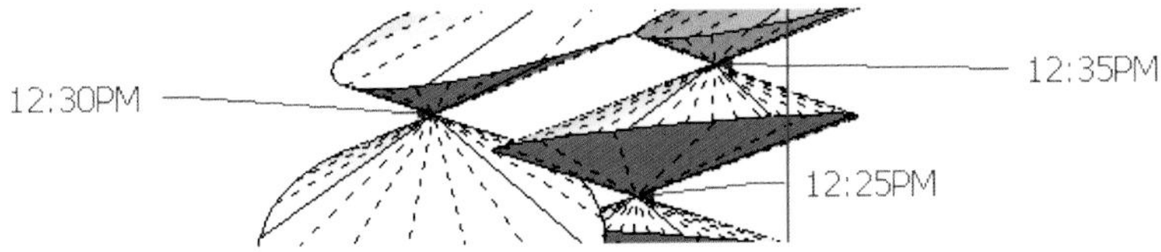

Fig. 15.15 The space-time prism intersection that can be found only by map interaction

while panning and orbiting that he was interested particularly in the area of intersection of the prisms "... but I don't know how to show it on the map. It is difficult to get from the space-time prism to the map. The victimized area should be in the vicinity of the intersection of the prism." Still another researcher participant, with a professional background in emergency management, clearly stated that he felt the 3D environment was too complex for mapping crime events saying, "I feel like a 2D environment would be more efficient. When looking at a crime event you have to act rapidly and this is too complex for that. The 2D could depict the same if you labeled the time at the points."

Finally, some participants, particularly those focusing on environmental context, continued with a search strategy that focused on the market areas or what they perceived to be busy thoroughfares. These participants seemed to be drawing on their experience or knowledge about crime. As one participant noted regarding his search strategy selection, "An offender operates in a place where he can victimize someone as they are leaving the shopping area. The offender can then move on the next potential victim. You don't want all of the victims coming together at once." The reality is that even with the constraint of similar velocity to the victims the offender's potential paths could have overlapped the victims' in many different areas. Therefore, a single correct answer to the best search strategy was not really possible. And it was very informative from both a crime mapping and time-geographic perspective to hear the researcher and practitioner participants' feedback regarding what they thought were the best search strategies.

15.5.2 Locating Time on a Time Geography Map

During the usability interviews participants were shown first a space-time path map and then a space-time prism map. Participants were then tasked with locating the time at which two victims, represented in the maps, were at their closest. With the addition of the space-time prism, participants should have been able to conclude an answer in the range of 12:25 and 12:30. During this iteration participants were encouraged to interact with the map through the available tools (pan, orbit and zoom) which were required to answer this question with certainty through visual inspection. This certainty would come only from visual inspection of the only intersection of the two space-time paths by the two represented victims (Fig. 15.15).

And while some participants answered by selecting a certain narrow time range, others selected a wider range signifying the uncertainty of their answers. The results

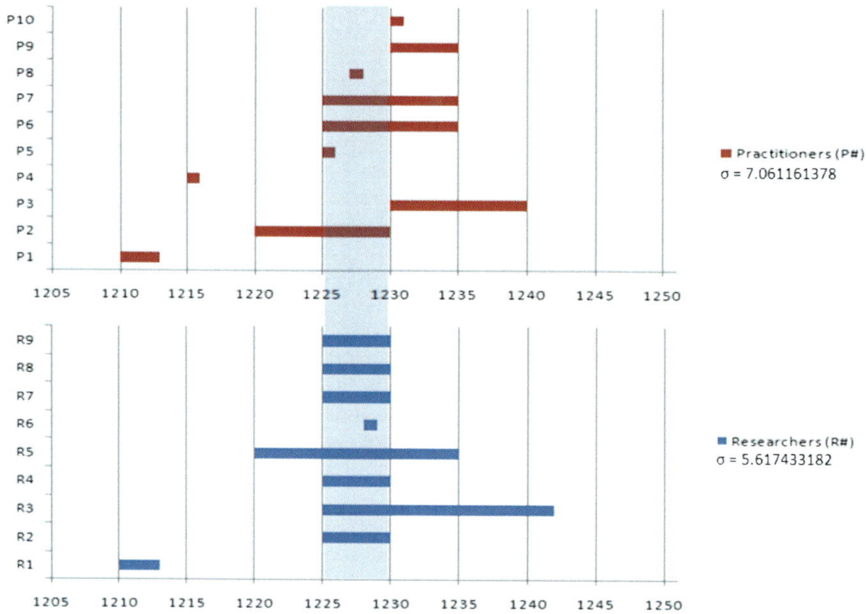

Fig. 15.16 The participant space-time prism selections

in Fig. 15.16 illustrate a notable difference between the researchers and the practitioners with regard to perspective usability of the space-time prisms to tell the time at which the two victims were at their closest.

Any selected time between the 12:25 and 12:30 range indicates a correct reading, and positive usability, of the space-time prisms. Any selected time outside of the 12:25 and 12:30 range indicates a lack of usability. This usability was dependent, of course, on the user's ability to utilize the space-time prism as a tool. Six of the nine researchers (66%) were able to use the space-time prism to locate the time at which the two victims were at their closest. However, only two of the ten practitioners (20%) demonstrated this ability. The standard deviation for the practitioners was 7.061 compared to the researchers' 5.617 (denoted by lower case Greek letter sigma, σ). This difference in results may be accounted for by the fact that researchers were more likely to have been exposed previously to alternative methods of mapping (e.g. 3D maps). Practitioners were more likely to have only been exposed to mainstream commercial GIS products and processes. One practitioner interviewed at a large metropolitan police department remarked when introduced to space-time paths and prisms that "this is something quite different than we are used to … we are point people." On further clarification the practitioner explained that it is the standard on her police department's crime mapping team to represent crimes primarily as dots on a map and that time-geographic crime maps were a novel idea.

During each map iteration the participants were asked the question, "At what time do you think the two victims were at their closest?" For the first two maps, the

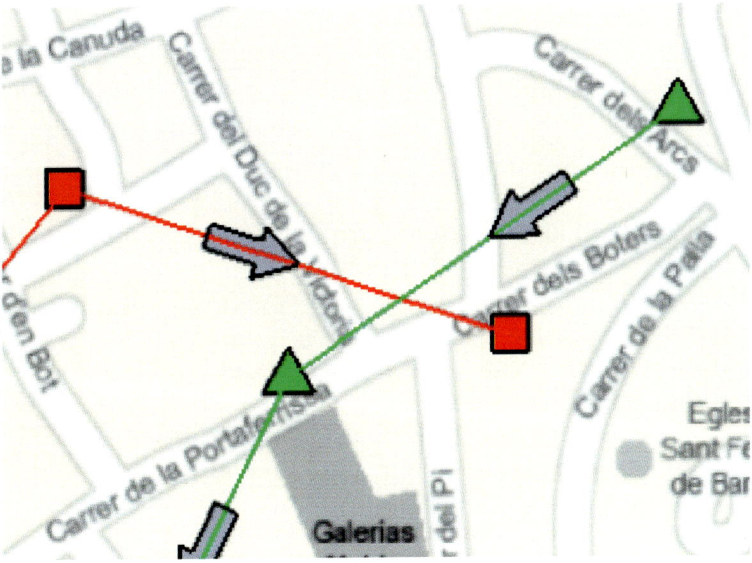

Fig. 15.17 Victim paths that cross in space, but not necessarily time

2D *Flow Map* (Fig. 15.4) and the *Potential Path Area Map* (Fig. 15.5), the question was not answerable based solely on the visual information represented on the map. However, this question was asked in order to gauge the participants' understanding of the map area and to see their thought process as it related to incorporating temporal questions into the map. During the asking of this question the participants were given a hint that the question was not answerable by emphasizing that their assessment should be made using "the visual information presented in this map." Absent from the *Flow Map* were any time labels. In retrospect the addition of time labels into the 2D maps might have been useful. However, they were left out because the temporal axis had not yet been introduced. Nonetheless, an interesting result of this question was that most of the participants were drawn early on to the intersection of the 2D paths from the *Flow Map* as their choice of the area where the two victims were at their closest regardless of the fact that they had no data to support this conclusion (Fig. 15.17). One participant on introduction to the *Flow Map* remarked with certainty about where the victims were at their closest, "I would say where they crossed paths!".

Upon further consideration, and as the interviews progressed, most participants realized that they could not conclude with certainty, from the 2D maps, that the victims were at the closest at the 2D intersection. One practitioner participant noted, "At a glance you can't tell what time they were at their closest. Intuitively I want to say they were moving from different start points moving at the same speed, but you can't really tell for sure." Though the questions were asked with the map as the focus the participants still had access to the tabular data from the pretest briefing which was the known victim space-time point data. And some participants referred back to this data but noted that it would take some time to correlate the tabular data with that presented on the 2D flow map to determine the point at which the two victims were at their closest.

15.6 Conclusion and Research Implications

A key factor in testing the usability of the maps within this chapter is the determination of whether the 3D time-geographic approach can reveal patterns where traditional 2D GIS methods usually cannot. To this end it seems natural to correlate the goals of geographic profiling with those of geographic visualization. MacEachren (2001) describes geographic visualization as the use of visual geospatial displays to explore data and through that exploration to generate hypotheses, develop problem solutions, and construct knowledge. This description correlates well with the objective of geographic profiling as a criminal investigative technique that attempts to provide information on the likely "base of operations," or offender residence, of offenders thought to be committing serial crimes (Harries 1999; Rich and Shively 2004; Rossmo 2000).

A primary assumption of a geographic profile is that the offender's base of operations lies within the distribution of crime incident sites (Rossmo 2000). Combining the concept of anchor points with the crime triangle of routine activities theory, a crime event occurs when both the victim and offender are within proximity to the offender's anchor point. Further placing these concepts within a time geography framework, an anchor point will lie within a given victim's space-time prism (potential paths) along his or her space-time path (known paths). Therefore, time geography can lend itself to helping to incorporate time explicitly (and visually) into developing geographic profile search strategies. Snook et al. (2007) have found that with appropriate training police investigators could be as accurate in their predictions as actuarial driven computer predictions of crime areas (see also Paulsen 2006). Potentially, the human deductive element in crime analysis could be further enhanced through engagement with advanced visualization techniques. However, at some point, as visualization systems become ever more complicated, the system becomes unusable to all but the most highly trained professional. Through an investigation of the usability of time-geographic crime visualizations, this chapter has suggested that limits do exist, but they are surmountable. Certainly many of the respondents – especially those who were practicing crime analysts rather than academics who were more accustomed to time-geographic visualizations – were uncomfortable with some of the more sophisticated maps. However, the fact that so many of them were able to use even the most complex time-geographic map correctly suggests that, with training, these visualization techniques could achieve large-scale adoption by the crime-mapping community.

The potential application of time-geographic and associated space-time analysis methods to crime mapping should be bolstered by advances in technologies such as satellite navigation systems and land-based navigation systems. For instance, in a related vein of research, recent strides have been made in developing methods that utilize global positioning systems (GPSs), along with time-geographic methods, to study pedestrian movement patterns in urban spaces such as tourist centers (Pettersson and Zillinger 2011; Shoval and Isaacson 2006; van der Spek et al. 2009). The findings from this research on pedestrian movement, along with the usability findings presented in this chapter, suggest that the potential contributions of time geography extend beyond crime mapping to a range of human mobility activities.

By including the variable of time explicitly within the map space, time geography adds certain contextual factors as it maps elements of human activity spaces that typically are absent in mapping. And the construction of the time-geographic tools is informed by context through the application of constraints. Additionally, recent efforts to account for such factors as varying velocities between known space-time points are adding to the potential ways in which time-geographic tools can be applied (Miller and Bridwell 2009). Still, despite the innovate approach of time geography to the process of map design, the technique remains vulnerable to the pitfalls of generalization that are faced by all forms of cartographic representation. This generalization is required because all maps are smaller (and necessarily less complex) than the realities they represent (Monmonier 1996).

While many participants in the usability study conducted in this chapter found the tools of time geography (e.g. the space-time prisms) to be usable in map reading, others (primarily in the practitioner group) found them less useful and preferred to focus on environmental context factors such as nearness to market areas. These participants felt that context was an equally or more important consideration in selecting their search strategies than the metrical distances that could be calculated between two victims' space-time paths. For instance, one practitioner participant noted that it would be helpful to see how many other people (non-victims) were present in each area of the study space. The importance of the difference between seeing movement as an abstract mathematical phenomenon and seeing it as an embodied practice performed by environment-interpreting agents (both offenders and victims) was not lost on these participants.

The enduring importance of spatial context leads us back to Lynch's (1960) focus on the influence of (city) form and cognition on human activity. Arguably the participants in the map usability study who focused on context (e.g. the market as a site conducive to pickpocketing activity) over metrical space (e.g. the point in time and space at which the victims were at their closest) were influenced more by their perceptions of urban form. People's perceptions of space, and in turn, their behavior therein, are strongly influenced by their mental images of what different localities mean to them (Lynch 1960; Buttimer 1980; Cosgrove 1984). Some participants expressed a strong interest in what they perceived as particular spatial contexts that were relevant to a pattern of crime. And the influence of urban form on patterns of crime is clearly communicated in Brantingham and Brantingham (1991) theory of environmental criminology. Indeed, regardless of time geography's potential utility in crime mapping and investigation, this study also reinforces the findings of critical cartographic theorists (e.g. Del Casino and Hanna 2005; Kitchin and Dodge 2007) that a map, when considered as an *object*, can never fully capture the cognitive feedback loop between perception and practice. As long as this limit remains (which we believe will be the case for the foreseeable future), no map will ever achieve complete reliability in modeling the experience (and hence the geography) of the crime event, as practiced by the opportunity-seeking offender. Nonetheless, by incorporating time into a form of visualization that can be used by the crime analyst, time geography provides an innovative and potentially practicable tool for those who wish to understand – and intervene in – the spatial patterns and processes of criminal activities.

Acknowledgements This research was funded by National Science Foundation Doctoral Dissertation Research Improvement grant no. 0927850.

References

Adams B (2009) Barcelona "worst city for pickpockets". http://www.telegraph.co.uk/travel/travelnews/6229558/Barcelona-worst-city-for-pickpockets.html. Last updated on 25 Sep 2009

Brantingham PJ, Brantingham PL (1991) Environmental criminology. Waveland Press, Prospect Heights

Brantingham PL, Brantingham PJ (1993) Nodes, paths and edges: considerations on the complexity of crime and the physical environment. J Environ Psychol 13:3–28

Brunsdon C, Corcoran J, Higgs G (2005) Visualising space and time in crime patterns: a comparison of methods. Comput Environ Urban Syst 31:52–75

Buttimer A (1980) Home, reach and the sense of place. In: Buttimer A, Seamon D (eds) The human experience of space and place. St. Martin's Press, London

Carlstein T (1978) Innovation, time allocation and space-time packing. In: Carlstein T, Parkes D, Thrift N (eds) Timing space and spacing time, vol. 2, human activity and time geography. E. Arnold, New York

Chainey S, Tompson L, Uhlig S (2008) The utility of hotspot mapping for predicting spatial patterns of crime. Secur J 21:4–28

Clarke RV, Felson M (1993) Routine activity and rational choice. Transaction, New Brunswick

Cohen L, Felson M (1979) Social change and crime rate trends: a routine activities approach. Am Sociol Rev 44:588–608

Cosgrove D (1984) Social formation and symbolic landscape. Croom Helm, London

Del Casino VJ, Hanna SP (2005) Beyond the "binaries": a methodological intervention for interrogating maps as representational practices. ACME An Int E-J Crit Geogr 4:34–5

Dijst M, Vidakovic V (2000) Travel time ratio: the key factor of spatial reach. Transportation 27:179–199

Felson M, Clarke R (1998) Opportunity makes a thief: practical theory for crime prevention, police research series paper 98. Home Office, Policing and Reducing Crime Unit, Research, Development and Statistics Directorate, London

Forer P (1998) Geometric approaches to the nexus of time, space, and microprocess: implementing a practical model for mundane socio-spatial systems. In: Egenhofer MJ, Golledge RG (eds) Spatial and temporal reasoning in geographic information systems. Oxford University Press, New York

Gahegan M (1999) Four barriers to the development of effective exploratory visualization tools for the geosciences. Int J Geogr Inf Sci 13:289–309

Gatalsky P, Andrienko N, Andrienko G (2004) Interaction analysis of event data using space-time cube. In: Banissi E et al (eds) IV 2004. 8th international conference on information visualization, proceedings, London, UK, 14–16 July 2004. IEEE Computer Society, Los Alamitos, pp 145–152. ISBN 0-7695-2177-0, ISSN 1093-9547

Grubesic TH, Mack EA (2008) Spatio-temporal interaction of urban crime. J Quant Criminol 24:285–30

Hägerstrand T (1970) What about people in regional science? Papers of the Regional Science Association. Sci Assoc 24:7–21

Hariharan R (1999) Modeling intersections of geospatial lifelines. Master's thesis, Pennsylvania State University College of Information Sciences and Technology

Harries K (1999) Mapping crime: principles and practice. National Institute of Justice, Washington, DC

Harrower M, MacEachren MA, Griffin AL (2000) Design and assessment of geographic visualization tool to support earth science learning. Cartogr Geogr Inf Sci 27:279–293

Hendricks M, Egenhofer MJ, Hornsby K (2003) Structuring a wayfinder's dynamic space-time environment. Lect Notes Comput Sci 2855:75–92

Hornsby K, Egenhofer M (2002) Modeling moving objects over multiple granularities. Ann Math Artif Intell 36(1–2):177–194

Hubers C, Schwanen T, Dijst M (2008) ICT and temporal fragmentation of activities: an analytical framework and initial empirical findings. Tijdschr Econ Soc Geogr 99:528–54

Kitchin R, Dodge M (2007) Rethinking maps. Prog Hum Geogr 31:331–44

Knox G (1964) The detection of space-time interactions. Appl Stat 13:25–29

Kristensson PO, Dahlback N, Anundi D, Bjornstad M, Gillberg H, Haraldsson J, Martensson I, Nordvall M, Stahl J (2008) The trade-offs with space time cube representation of spatiotemporal patterns. Department of Computer and Information Science, Linkoping University. http://arxiv.org

Kwan M-P (2000) Interactive geovisualisation of activity-travel patterns using three-dimensional geographical information systems: a methodological exploration with a large data set. Transp Res 8:185–203

Kwan M-P, Lee J (2004) Geovisualization of human activity patterns using 3D GIS: a time-geographic approach. In: Goodchild MF, Janelle DG (eds) Spatially integrated social science. Oxford University Press, New York

LaVigne NG, Fleury JK, Szakas J (2000) Auto theft and detecting chop shop locations. In: Turnbull LS, Hendrix EH, Dent BD (eds) Atlas of crime: mapping the criminal landscape. Oryx Press, Phoenix

Leitner M, Kent J, Oldfield I, Swoope E (2007) Geoforensic analysis revisited: the application of Newton's geographic profiling method to serial burglaries in London, UK. Police Pract Res Int J 8:359–370

Levine N (2010) CrimeStat: a spatial statistics program for the analysis of crime incident locations (v 3.3). Ned Levine & Associates/National Institute of Justice, Houston/Washington, DC

Lynch K (1960) The image of the city. MIT Press, Cambridge

MacEachren AM (2001) Cartography and GIS: extending collaborative tools to support virtual teams. Prog Hum Geogr 25:431–444

Miller HJ (1991) Modeling accessibility using space-time prism concepts within geographical information systems. Int J Geogr Inf Syst 5:287–301

Miller HJ (2004) Activities in space and time. In: Stopher P, Button K, Haynes K, Hensher D (eds) Handbook of transport 5: transport geography and spatial systems. Elsevier, Amsterdam/London

Miller HJ (2005) A measurement theory for time geography. Geogr Anal 37:17–45

Miller HJ, Bridwell SA (2009) A field-based theory for time geography. Ann Assoc Am Geogr 99:49–75

Monmonier MS (1996) How to lie with maps. University of Chicago Press, Chicago

Neutens T, Witlox F, Van de Weghe N, De Maeyer P (2007) Human interaction under uncertainty. Transp Res 7:335–352

Newton MB (1988) Geographical discovery of the residence of an unknown dispersing localized serial murder. Unpublished Manuscript, Louisiana State University

Olson JM, Brewer CA (1997) An evaluation of color selections to accommodate map users with color-vision impairments. Ann Assoc Am Geogr 87:103–134

Paulsen D (2006) Human versus machine: a comparison of the accuracy of geographic profiling methods. J Investig Psychol Offender Profil 3:77–89

Pettersson R, Zillinger M (2011) Time and space in event behaviour: tracking visitors by GPS. Tourism Geogr Int J Tour Space Place Environ 13:1–20

Polson P, Rieman J, Wharton C, Olson J, Kitajima M (1992) Usability inspection methods: rationale and examples. In: The 8th human interface symposium (HIS92). Kawasaki, Japan, pp 377–384

Pred A (1977) The choreography of existence: comments on Hägerstrand's time-geography and its usefulness. Econ Geogr 53:207–221

Ratcliffe JH (2006) A temporal constraint theory to explain opportunity-based spatial offending patterns. J Res Crime Delinq 43:261–291

Rich T, Shively M (2004) A methodology for evaluating geographic profiling software. Abt Associates, Cambridge

Rossmo DK (2000) Geographic profiling. CRC Press, Boca Raton

Schmertmann CP, Assunção RM, Potter JE (2010) Knox Meets Cox: adapting epidemiological space-time statistics to demographic studies. Demography 47:629–650

Shoval N, Isaacson M (2006) The application of tracking technologies to the study of pedestrian spatial behaviour. Prof Geogr 58:172–183

Snook B, Haines A, Taylor PJ, Bennell C (2007) Criminal profiling belief and use: a study of Canadian police officer opinion. Can J Police Secur Ser 5:169–179

Stewart DW, Shamdasani PN (1990) Focus groups: theory and practice. Sage, London

Tessmann S (2006) Time geography for efficient shared-ride trip planning in transportation networks. Diploma thesis, Institute for Geoinformatics, University of Münster, Germany. Retrieved from http://www.psu.edu/

van der Spek S, van Schaick J, de Bois P, de Haan R (2009) Sensing human activity: GPS tracking. Sensors 9:3033–3055

Walsh P, Ratcliffe JH (2005) Strategic criminal intelligence education: a collaborative approach. J Int Assoc Law Enforc Intell Anal 16:152–166

Wu Y, Miller HJ (2001) Computational tools for measuring space-time accessibility within transportation networks with dynamic flow. J Transp Stat 4:1–14

Zillinger M (2005) A spatial approach on tourists' travel routes in Sweden. Etour, Östersund

Chapter 16
Understanding Spatiotemporal Patterns of Multiple Crime Types with a Geovisual Analytics Approach

Diansheng Guo and Jiang Wu

Abstract Comprehensive crime data sets have been collected over time, which contain the location and time of different crime types such as aggravated assault or burglary. To understand the patterns and trends in such data, existing mapping and analysis methods often focus on one selected perspective (e.g., temporal trend or spatial distribution). It is a more challenging task to discover and understand complex crime patterns that involve multiple perspectives such as spatio-temporal trends of different crime types. In this Chapter we used a data mining and visual analytics approach to analyze the crime data of Philadelphia, PA, which has all the crimes reported from January 2007 to June 2011. Specifically, the adopted approach is a space-time and multivariate visualization system (VIS-STAMP) and the analysis examines the spatial and temporal patterns across six crime types, including aggravated assault, robbery, burglary, stolen-vehicles, rape and homicide. The geovisual analytic tool provides the capability to visualize multiple dimensions simultaneously and be able to discover interesting information through a variety of combined perspectives.

Keywords Spatial-temporal analysis • Crime analysis • Visual analytics • Data mining • Multivariate mapping

D. Guo (✉) • J. Wu
Department of Geography, University of South Carolina,
709 Bull Street, Room 127, Columbia, SC, 29208, USA
e-mail: guod@sc.edu

M. Leitner (ed.), *Crime Modeling and Mapping Using Geospatial Technologies*, Geotechnologies and the Environment 8, DOI 10.1007/978-94-007-4997-9_16,

16.1 Introduction

With the increasing academic interests in place-based crime theories since late twentieth century (Anselin et al. 2000), a large body of literature has discussed the relationship between spatial locations and crimes. Crime analyses span across a wide range of topics, such as identifying the crime concentration in a study area (Chainey and Ratcliffe 2005; Craglia et al. 2000; Eck et al. 2005; Murray et al. 2001; Ratcliffe and McCullagh 1998; Wu and Grubesic 2010), discovering the underlying social/physical factors or built environment that may account for spatial patterns of crime activities (Gorman et al. 2001), investigating theoretical roots of how space exerts influences on the crime pattern (Messner and Anselin 2004), establishing effective models used for law enforcement and legitimate prevention programs (Hunt et al. 2008; Ratcliffe 2004), and developing methodologies for spatial and statistical analyses of crime incident data (Anselin et al. 2000; Bernasco and Elffers 2010; Levine 2006).

Crime data may be divided into two major categories based on spatial representation: point data (with point locations of crime incidents) and areal data (crime incident counts aggregated to predefined boundaries). Point crime data can be converted (aggregated) to areal data when needed. To visualize and analyze point patterns, commonly used methods include quadrat count, distance statistics, and kernel density estimation (KDE). Some regard KDE as the most suitable spatial analysis technique for visualizing point data (Chainey et al. 2008; Eck et al. 2005). For visualizing areal crime patterns, choropleth map, scatterplot and/or variogram coupled with classic statistics including Moran's I and Geary's C (Cliff and Ord 1970), distance-based statistics and LISA statistics (Anselin 1995) are frequently applied.

In addition to spatial distribution patterns, the temporal trend of crimes at various temporal scales is also of critical interest to both researchers and law enforcement (Felson and Poulsen 2003; Henry and Bryan 2000; Rengert 1997; Townsley 2008; Townsley et al. 2000; Weisburd et al. 2004; Weisburd et al. 2009). For example, Ratcliffe and McCullagh (1998) propose a framework called 'aoristic crime analysis' to detect spatio-temporal crime patterns where the exact offense time is unknown (Ratcliffe 2000, 2002). To visualize spatio-temporal patterns of crime activities, Brunsdon et al. (2007) compare and evaluate three major techniques, including map animation, the COMAP (Brunsdon 2001) and isosurfaces. In addition, hotspot plot (Townsley 2008), space-time cube (Nakaya and Yano 2010) and CrimeViz (Roth et al. 2010) provide alternative approaches for space-time analysis of crimes.

It has also been recognized that it is important to include additional factors, such as crime types or socioeconomic environment of the crime neighborhood, to understand the context and underlying process that influence the spatial and temporal variation of crime activities (Hagenauer et al. 2011). However, there are not many methods that can simultaneously visualize and analyze the spatial, temporal, and multivariate dimensions related to crime activities. Hangenauer et al. (2011) propose a framework to identify the spatial and temporal characteristics of crime patterns by incorporating the socioeconomic and environmental attributes of the neighborhoods, which consists of three steps. First, a spatial scan statistic is applied to identify significant spatio-temporal crime hotspots. Second, a self-organizing map (SOM) is used to cluster neighborhoods based on their social-economic and environmental attributes. Finally, the hotspots are mapped in the SOM visualization so that one may see the correlation between crime hot spots and contextual factors. The main limitation of this approach is that the visualization (i.e., U-matrix and Component Planes) in the framework cannot display or link to actual spatial locations and therefore it is difficult to perceive spatial distribution patterns or correlations. Additionally, as noted by Hagenauer et al. (2011), different crime types often exhibit different patterns in relation to space, time, and context, and therefore should be separated and compared in analysis, which is not addressed in the framework.

In the research field of geovisualization and visual analytics, it is an active and challenging research topic to develop new approaches for visualizing complex datasets that contain geographic locations, time series, and multiple variables (Andrienko et al. 2010; Guo et al. 2006). To map and visualize patterns across multiple variables and dimensions, dimension reduction techniques are often used. Guo et al. (2005) developed a multivariate mapping approach (named SOMVIS) that integrates self-organizing map (a dimension reduction and clustering method), color encoding, and multidimensional visualization to map multiple variables in a single map. Guo et al. (2006) extended the SOMVIS approach to accommodate the temporal dimension and visualize spatio-temporal and multivariate information simultaneously. This new approach is called VIS-STAMP, an acronym for Visualization for Space Time and Multivariate Patterns. VIS-STAMP first constructs an *overview* of the data, from which the analyst can easily perceive complex patterns across all dimensions and then explore specific patterns through user interactions such as selection and linking different views. In this chapter, we apply VIS-STAMP to the analysis of crime data in Philadelphia (2007–2011) that involves spatial, temporal, and multivariate information. We will briefly introduce the VIS-STAMP approach in Sect. 16.3.

This chapter is organized as follows. Section 16.2 introduces the study area and the crime data. Section 16.3 briefly explains the methodology and how the data were processed and prepared for the analysis. The analysis results are presented in Sect. 16.4, with subsections focusing on different types of patterns being discovered from the data. We provide discussions on the methodology, analysis, and future directions in Sect. 16.5.

16.2 Study Area and Data

Philadelphia County, PA covers approximately 143 square miles with a population of 1,526,006 according to U.S. Census 2010. Crime incident reports are available from the online service provided by the Philadelphia Police Department (http://citymaps.phila.gov/CrimeMap/StepByStep.aspx). There were in total 169,829 crime incidents for six crime types (aggravated assault, robbery, burglary, stolen vehicle, rape and

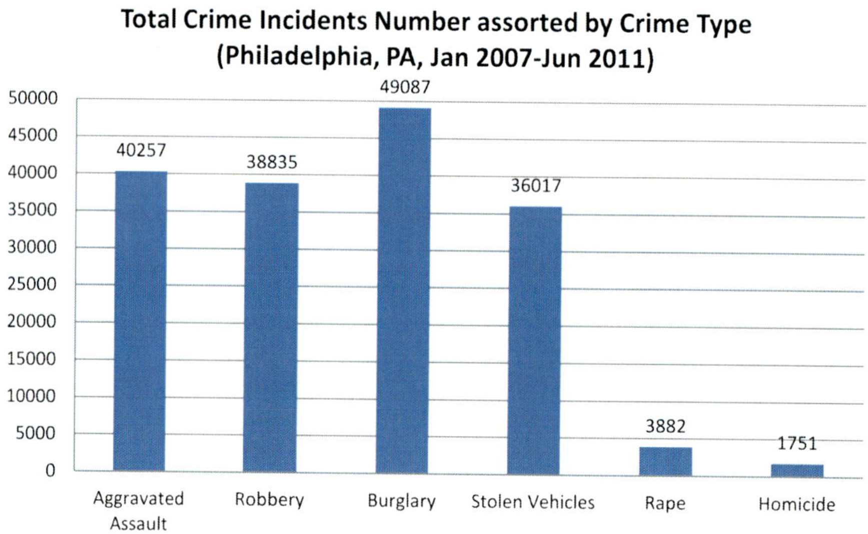

Fig. 16.1 Total crime incidents by crime type (Philadelphia, PA, Jan 2007–June 2011)

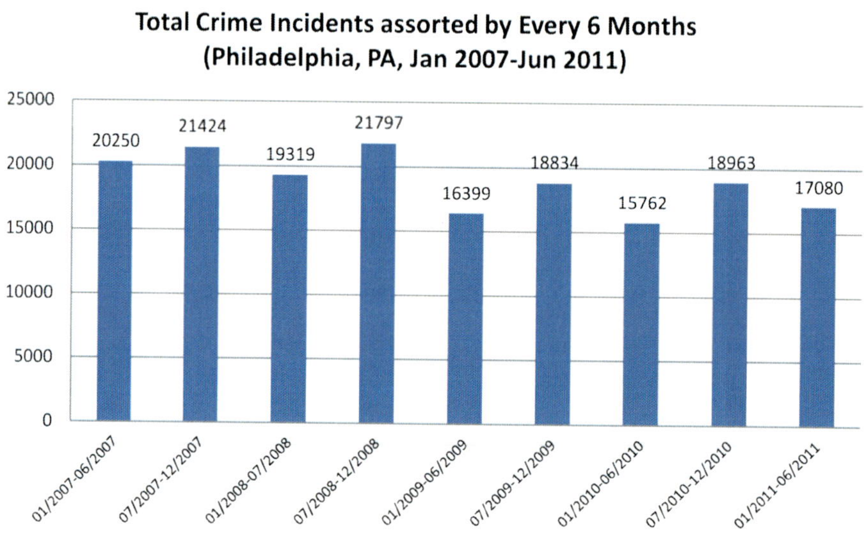

Fig. 16.2 Total crime incidents for every 6 months, since Jan. 2007

homicide) from January 2007 to June 2011. Each crime incident record has a type, date, police dispatch time, and its geographic location (x, y coordinates).

Figure 16.1 shows the total number of crime incidents for each crime type. Aggravated assault, robbery, burglary, and stolen vehicles are the four most frequent types, and rape and homicide have much fewer incidents. Figure 16.2 shows the distribution of crimes (of all six types) across time, aggregated for every six months.

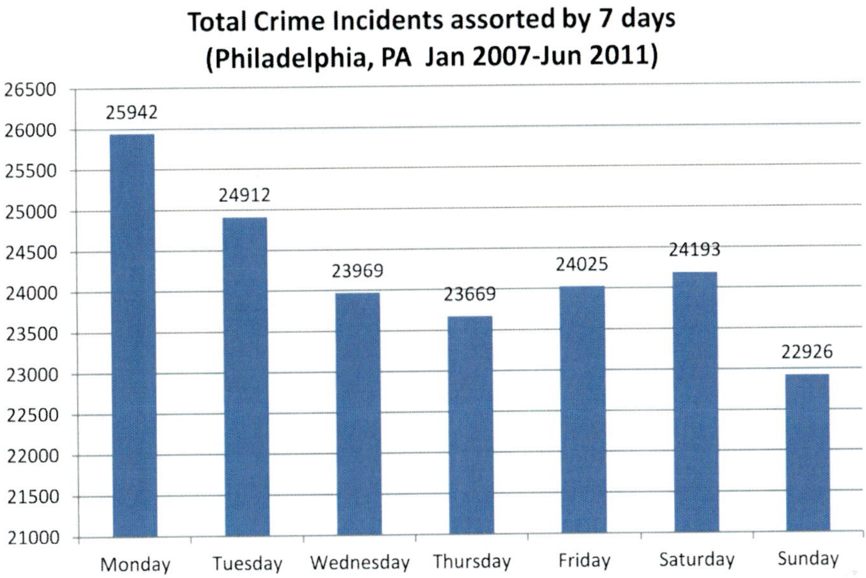

Fig. 16.3 Total crime incidents number assorted by 7 days per week

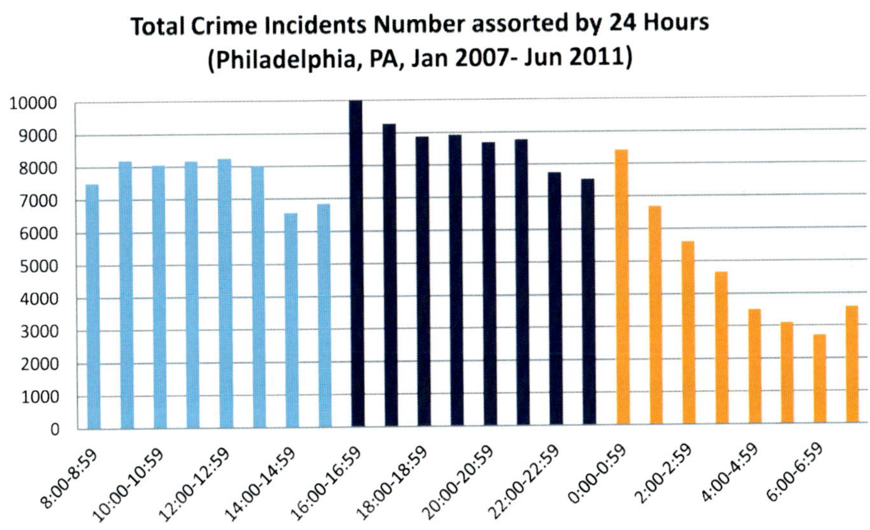

Fig. 16.4 Total crime incidents number assorted by 24 h per day

The annual crime totals dropped a little bit after year 2008. For each year, there appeared to be more crimes in the second half (Jul–Dec) than in the first half (Jan–Jun). In addition to these general trends over years, we also explored the trends over 7 days of a week (Fig. 16.3) and 24 h in a day (Fig. 16.4). Figure 16.3 shows an interesting trend over the 7 ways in a week, with Monday and Tuesday having the

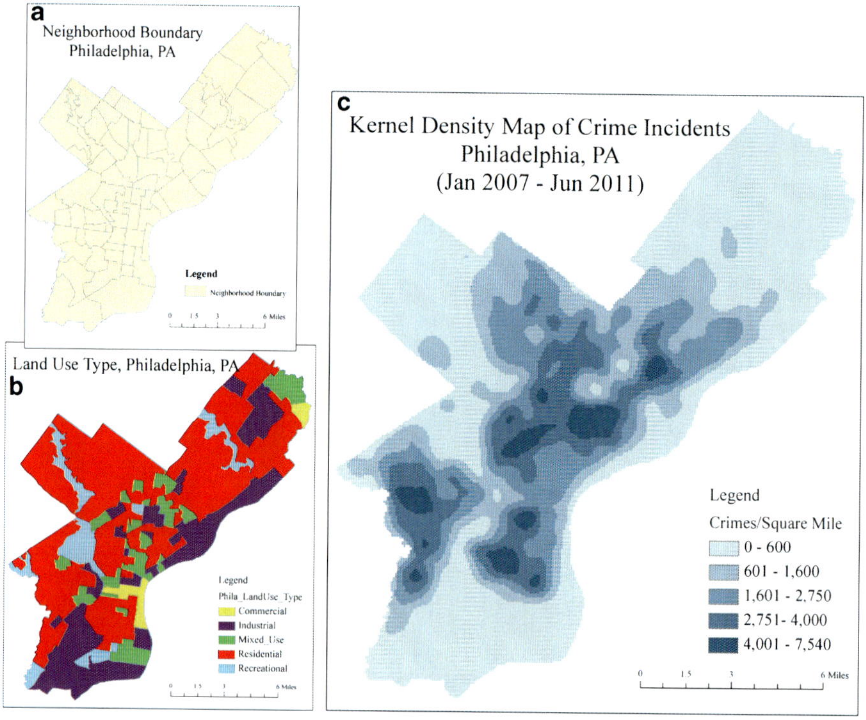

Fig. 16.5 (**a**) NIS neighborhood boundaries (Neighborhood Information System, http://cml.upenn.edu/nis/), Philadelphia, Pennsylvania. (**b**) Land use types aggregated from census tracts. (**c**) Crime density, including 169,829 crime incidents from Jan. 2007 to June 2011

most crimes while much less on Sunday. Figure 16.4 shows the aggregated crime counts for each hour, regardless of day and type. Based on three abrupt surges in crimes as shown in Fig. 16.4, in our analysis we divided a day into three 8-h time periods: Early Morning (0:00–7:59), Day Time (8:00–15:59) and Evening/Night (16:00–23:59).

The overall spatial distribution of the 169,829 crimes is presented by a kernel density map (Fig. 16.5c). To help understand the context of the crime locations, five major land use types (commercial, industrial, mixed use, residential, recreational) are mapped for the study area (Fig. 16.5b), which are extracted and generalized based on the Philadelphia Zoning Code. The land use type data is aggregated with census tracts and parcel data. If the area of a specific land use type exceeds 50% of the total properties within a tract, this particular land use type is assigned to the entire tract. If none of four types covers more than 50% of a tract, the tract is deemed 'Mixed Use'. The land use map (Fig. 16.5b) shows that the residential area covers most of the western and northern Philadelphia. Recreational areas mainly include public waterfront parks and sport facilities in the county. The water front belt in the northeast and the lower south are occupied by industrial area.

The research question is that, although Figs. 16.1, 16.2, 16.3, 16.4, and 16.5 can help understand crime patterns or trends from a specific perspective (e.g., crime type, time, or space), it is difficult to examine the data across multiple perspectives, such as the patterns of different crime types and their distribution and trends over space and time. For example, Fig. 16.5c may show the spatial concentration of crimes, but it cannot reveal the characteristics of crimes (such as composition of crime types) or temporal trends at different places. In this research, we use the VIS-STAMP method to gain insights on crime patterns that involve multiple perspectives. We use 69 neighborhoods (see Fig. 16.5a for a map of neighborhoods and detailed definition at http://cml.upenn.edu/nbase/nbAbout.asp) as the spatial units, along time and crime types, to aggregate crimes for further analysis (see Sect. 16.3 for details on different data aggregations). We use neighborhoods as the base units for two reasons. First, each neighborhood has a meaningful community definition that directly supports policy analysis and planning efforts in the city. Second, neighborhoods are sufficiently large to examine its internal crime characteristics across several dimensions such as time or crime type.

16.3 Data Preprocessing and Methodology Overview

16.3.1 Data Aggregation and Preprocessing

The input data is a set of 169,829 geocoded crime incidents, which are aggregated into a *data cube* and transformed in different ways depending on the analysis task. Figure 16.6 shows an illustrative view of the data cube. The three dimensions in the data cube include: *spatial dimension* (69 neighborhoods), *temporal dimension* (which can be of three different temporal scales), and *multivariate dimension* (e.g., crime types). The three temporal scales are semi-annual (i.e., half-year periods as shown in Fig. 16.2), day of the week (regardless of month and year, as shown in

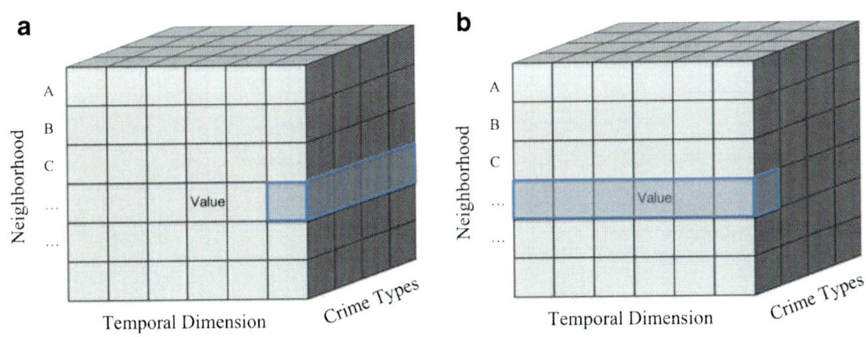

Fig. 16.6 The data cube, which is a space-time-attribute aggregation of crime data: (**a**) A composition of crime types for the same place and time is highlighted. (**b**) A time series for a place and crime type is highlighted

Fig. 16.3), and hour of the day (including three 8-h periods, regardless of day, month, and year, as shown in Fig. 16.4). Other temporal scales can be easily accommodated as well. For the multivariate dimension, additional crime-related variables (such as offender's age, offender's modus operandi, victim's information, etc.) can be used, which unfortunately are not available in the Philadelphia data. Each *cell* in the data cube is a unique combination of a spatial unit (e.g., neighborhood A), a time period (e.g., Monday), and a crime type (e.g., aggravated assault). The value for a cell is the total number of crimes in the cell, e.g., the total number of aggravated assaults occurred in neighborhood A on Monday.

Once the data cube is constructed as described above, it should be normalized or standardized, which depends on the analysis tasks. In this chapter, we primarily focus on two different analysis tasks:

1. Task 1: How crime characteristics vary across space and change over time. Figure 16.6a highlights a sequence of crime counts (i.e., a composition of different crime types) for a specific neighborhood at a specific time. We want to understand how such crime compositions change from place to place and from time to time. For this type of analysis, we normalize the data cube by dividing the crime count of each cell in the sequence by the total crimes of the sequence. In other words, after the normalization each cell value becomes a percentage value representing the proportion of each crime type in the sequence (for a specific neighborhood and at a specific time). There two reasons for choosing this standardization procedure instead of dividing the crime composition by area size or population. First, crime activities are not necessarily related to population density or area of predefined boundaries. Second, we are more interested in the crime composition (or characteristics) of a place and time.
2. Task 2: How temporal trends differ in different places and for different crime types. This is to look at the data cube from a different perspective by focusing on the time series, one for each neighborhood/crime type combination. Figure 16.6b highlights a time series for a specific neighborhood and a specific crime type. Here we want to understand how the temporal trend varies across space and crime types. For example, one type of crime may have been increasing while another type(s) of crime(s) may have been declining for some part of the city. For this purpose, we divide the crime count in each cell in a time series by the total crime count of the time series. In other words, each cell value now becomes a percentage value representing the proportion of crimes in each time period for a specific neighborhood and crime type.

Given a normalized data cube, the VIS-STAMP system will view it differently depending on the analysis task. For the first task described above, VIS-STAMP treats the cube as a set of multivariate vectors arranged in a space-time matrix (see Fig. 16.6a), where a multivariate vector is a sequence of percentage values representing proportions of crime types for a place and time. For the second analysis task, VIS-STAMP treats the cube as a set of time series arranged in a space-crime matrix (see Fig. 16.6b), where each time series is a set of percentage values representing the temporal trend of crimes for a place and crime type. VIS-STAMP performs clustering with the set of vectors (which are either multivariate vectors or time series) and visualizes them across two other dimensions.

16.3.2 Multivariate Mapping and Space-Time-Attribute Visualization

The VIS-STAMP approach extends the Self-Organizing Map (SOM) (Kohonen 2001) to extract clusters from the set of multivariate vectors, project the clusters onto a two-dimensional space, and use a 2D color scheme (Brewer 1994) to color each cluster so that similar clusters have similar colors. Clusters are visualized in map(s) and other visual representations, such a re-orderable space-time matrix. Each cluster and the data items in the cluster are of the same color (assigned by the SOM) in all visual components. In other words, similar colors in a map represent clusters of similar multivariate vectors. A parallel coordinate plot (PCP) is used as the 'legend' to show the multivariate meaning that each color represents. Details on the methodology can be found in (Guo et al. 2005, 2006). Below we use an example analysis to help explain the methodology.

Figure. 16.7 presents a multivariate map of the crime data described in the previous section, *without considering temporal variations*. In other words, this is a special case of the cube in Fig. 16.6a, where there is only one time period (entire 54 months, January 2007– June 2011). Therefore, each neighborhood has a multivariate vector, representing the composition of different crime types in the neighborhood. A weight can be assigned for each variable. The analysis involves five variables, including four major crime types (i.e., aggravated assault, robbery, burglary, and stolen vehicles) and the Median Household Income for each neighborhood. The goal is to see the crime compositions in different neighborhoods and their possible relation with income level. We included the Median Household Income in the PCP visualization but did not use it in the SOM clustering so that clusters are constructed independent of income levels. Due the limitation of the software, which does not allow a zero weight, we assigned an extremely small weight to Median Household Income to effectively exclude it in the clustering step. All other variables, i.e., crime types, are assigned the same weight (see Fig. 16.7a).

SOM groups the 69 neighborhoods into 38 clusters based on their multivariate vectors (i.e., crime compositions). SOM arranges the clusters with a 2D U-matrix so that similar clusters are close to each other (Fig. 16.7b) without considering their locations. Based on a 2-D color scheme, each cluster is assigned a color so that similar clusters have similar colors (Guo et al. 2005). The parallel coordinate plot (PCP, see Fig. 16.7c) shows the mean vector of each cluster. Each axis in the PCP represents a variable and uses a nested-means scaling (Guo et al. 2005), which puts the mean value of that variable at the midpoint of its axis. The PCP also provides several other linear scaling options. Each cluster is a string in the PCP, with the same color as it has in the SOM. The width (thickness) of the string represents the number of neighborhoods in the cluster, i.e., larger clusters (having more neighborhoods) is represented with a wider string.

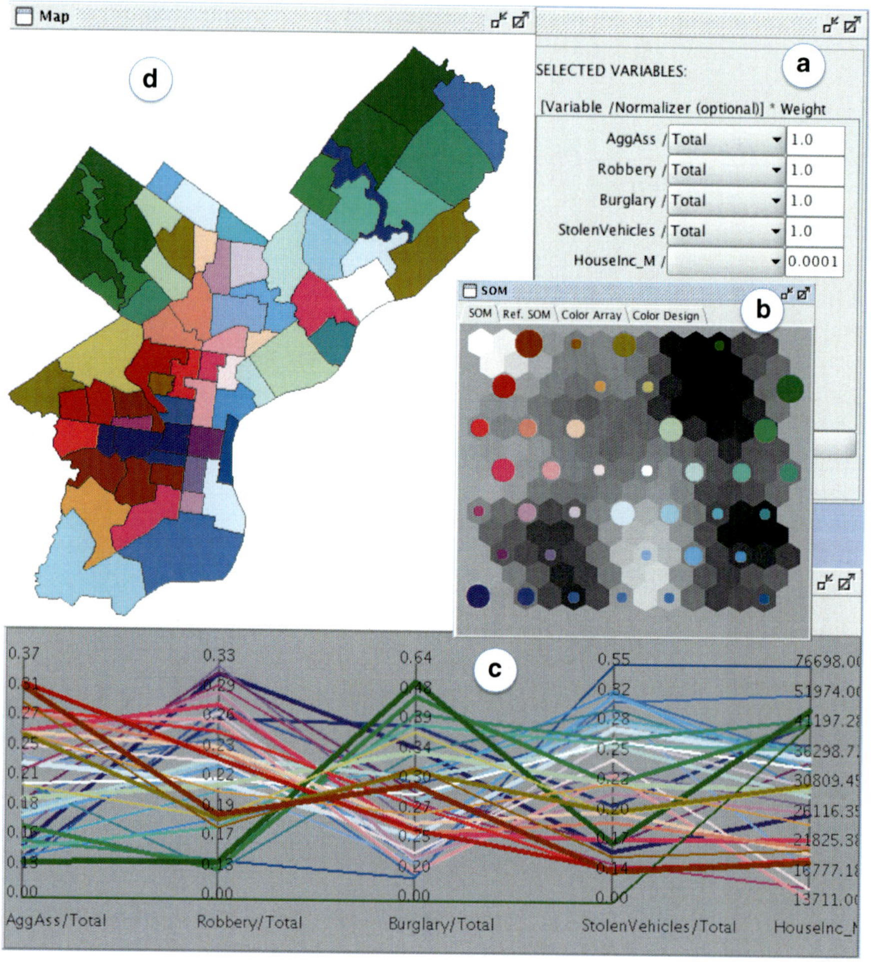

Fig. 16.7 A multivariate map of four major crime types. (**a**) Data processing and configuration. (**b**) Self-Organizing Map, where the each circle is a cluster. (**c**) PCP. (**d**) Multivariate map

From the PCP in Fig. 16.7c, for example, we can tell that the reddish clusters represent neighborhoods with a relatively high proportion of crimes being aggravated assault (about 30%) and a low median household income (less than 20 k/year). The map (Fig. 16.7d) shows each neighborhood in the same color as that of the cluster that the neighborhood belongs to. With the PCP and the map, we understand not only the crime compositions of the reddish clusters but also where they are in the geographic space (i.e., concentrating in West Philadelphia, such as Strawberry Mansion, Mill Creek and Haddington). The greenish clusters represent neighborhoods with primarily burglary crime threat (which accounts for nearly 50% of the crimes in those neighborhoods) and such neighborhoods mainly locate in the northwestern and northern parts of Philadelphia, which have a higher median household income (more than 40 k/year). Similarly, one can interpret the meaning of other clusters, such as the purple and blue clusters, in both the geographic and multivariate spaces.

To add the time dimension to the analysis, each neighborhood will have a unique multivariate vector for each time period (for the first analysis task) or a unique time series for each crime type (for the second analysis task). VIS-STAMP will then group similar crime compositions or time series into clusters. Essentially, the set of multivariate vectors or time series are reduced to a set of clusters and encoded in colors, which are visualized in other components such as maps (with one map for each time period or for each crime type) and a re-orderable matrix, where the vertical dimension is ordered by spatial units and the horizontal dimension is either the set of time periods or the set of crime types. We will explain this in detail in the next section with a variety of analysis results.

16.4 Analysis Results

16.4.1 Spatio-Temporal Patterns of Crime Compositions (Task 1)

We first examine how crime compositions change across space (neighborhoods) and time (every six months). Figure 16.8 shows the result with VIS-STAMP, which includes a re-orderable matrix (and in this case it can be called a space-time matrix), a map matrix (each map represents a crime composition map for a time period), and a PCP. In the re-orderable matrix, the rows represent the 69 neighborhoods and the columns represent nine 6-month periods from January 2007 to June 2011. The rows are ordered so that similar neighborhoods in terms of crime compositions over time are next to each other. Columns are in the naturally temporal order. Each column in the space-time matrix corresponds to a map in the map matrix. In other words, the re-orderable matrix and the map matrix show the same data from two different perspectives, with the former focusing on revealing temporal patterns while the latter focusing on spatial and spatio-temporal patterns. Essentially, with colors representing multivariate information (i.e., crime compositions in this case), the re-orderable matrix or the map matrix is an *overview* of the data cube. To ensure meaningful analysis, if a cell in the data cube has less than 20 crime incidents, the cell (i.e., the multivariate vector) will be excluded from the analysis and its corresponding elements in the map matrix or re-orderable matrix will be colored in gray.

In Fig. 16.8, the analysis again uses the four major crime types, same as in Fig. 16.7. The difference is that, in Fig. 16.7, there are 69 multivariate vectors (i.e., crime compositions) while here there are $69 \times 9 = 621$ vectors. From the PCP, one can understand the meaning of each cluster (and thus the meaning of each color). For example, reddish clusters represent a composition with a high proportion of aggravated assault, average on robbery and low on stolen vehicle. In the

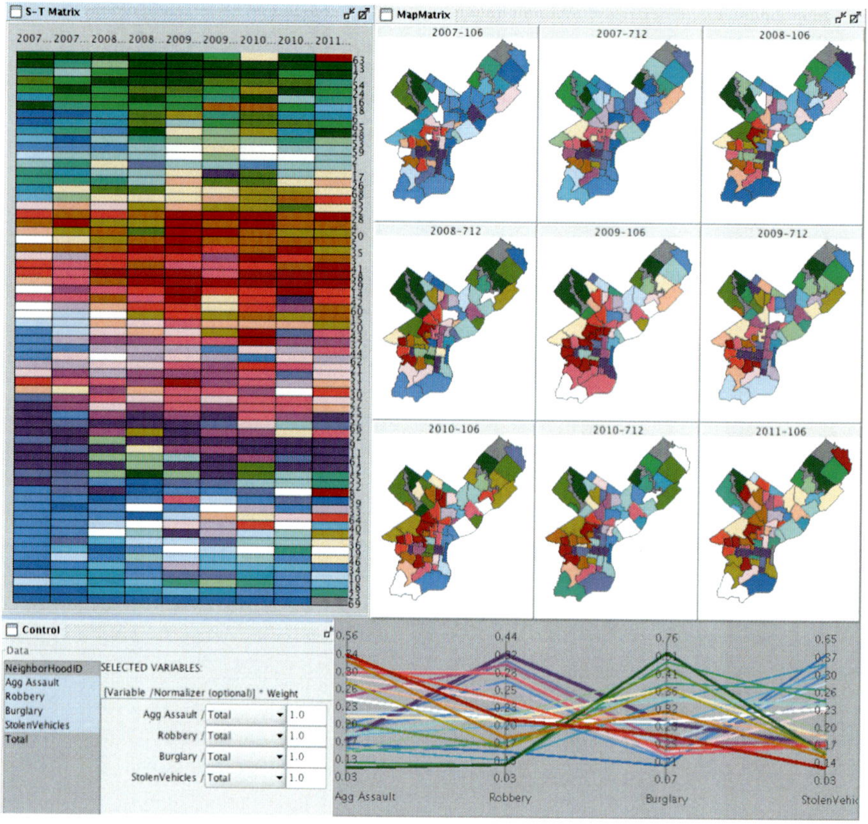

Fig. 16.8 Multivariate crime patterns across space and time (every 6 months from Jan 2007 till June 2011). The view includes a re-orderable matrix (*top-left*), map matrix (*top-right*), self-organizing map (not shown, see Fig. 16.7b from an example), and a parallel coordinate plot (*bottom-right*)

re-orderable matrix, it is clear that there are more reddish cells after 2008. In the map matrix, it is also evident that reddish neighborhoods are expanding. On the other hand, bluish (from light blue to navy blue) clusters represent high proportion of stolen-vehicle crimes while relatively less of other crimes. From both the re-orderable matrix and the map matrix, it is evident that the number of bluish neighborhoods had decreased over time.

To take a closer look at these two opposite trends, one can select the two groups of clusters in the PCP (see Fig. 16.9). In other words, both the reddish clusters and the bluish clusters are highlighted. Now it is very easy to perceive these two types of patterns described above, involving space, time, and crime types. From Jan 2007 to June 2011, the neighborhoods with high percentage of stolen vehicle crimes are shrinking dramatically, especially since Jan 2009. On the opposite, the threat of aggravated assault surged since January 2009. Comparing Figs. 16.9 and 16.5b, we may explore the possible relation between crime types and land uses. For example,

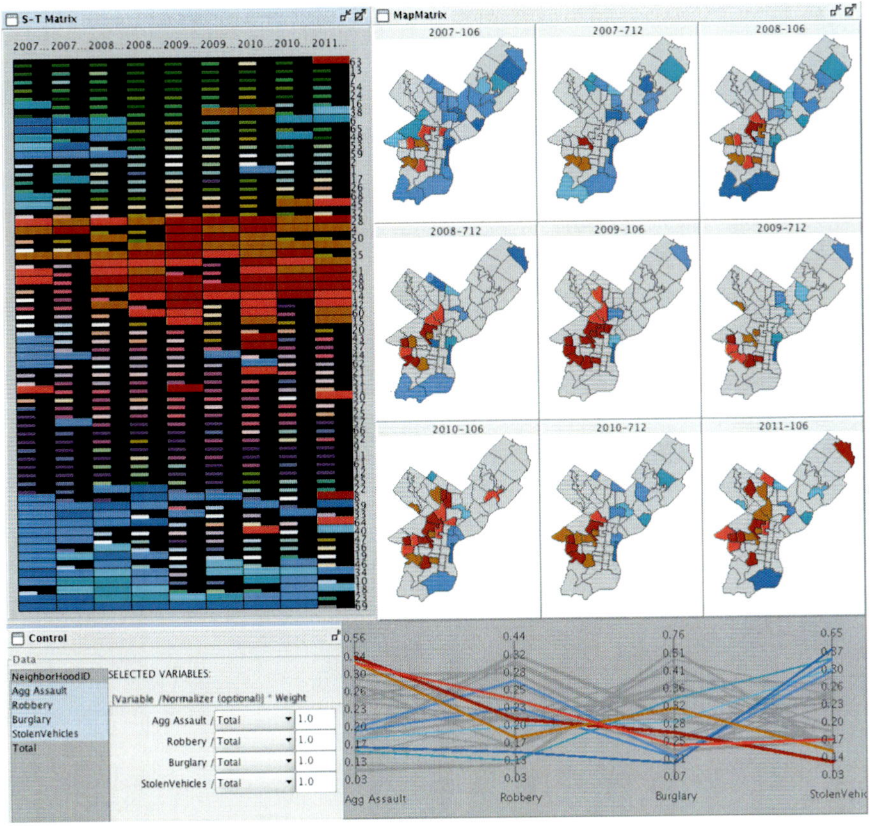

Fig. 16.9 Aggravated assault and stolen vehicles exhibit opposite spatio-temporal trends. Bluish clusters on average have at least 30% of their crimes being stolen vehicle, while reddish clusters are those with at least 30% being aggravated assault. While bluish clusters have been diminishing over time and space, reddish clusters are expanding and growing

neighborhoods with high percentage of stolen vehicles are primarily of residential and industrial land use types. Through interactive exploration, one may also examine many other patterns that are present in the overview (Fig. 16.8).

By changing the temporal scale to seven days of the week (regardless of month and year), we can discover different patterns from the data. Figure 16.10 shows the result of the same data except that it uses seven weekdays on the temporal dimension. The clusters and colors for this analysis are very similar to those in the previous analysis, e.g., reddish clusters are dominated by aggravated assault and bluish clusters mainly represent stolen vehicles activities. It is interesting to see that reddish clusters grow/expand steadily from Monday to Sunday, with weekend days threatened most by such violent crimes. This specific pattern becomes even more evident if we select the reddish clusters with 25% or more crimes being aggravated assault (Fig. 16.11). Both the re-orderable matrix and the map matrix show that

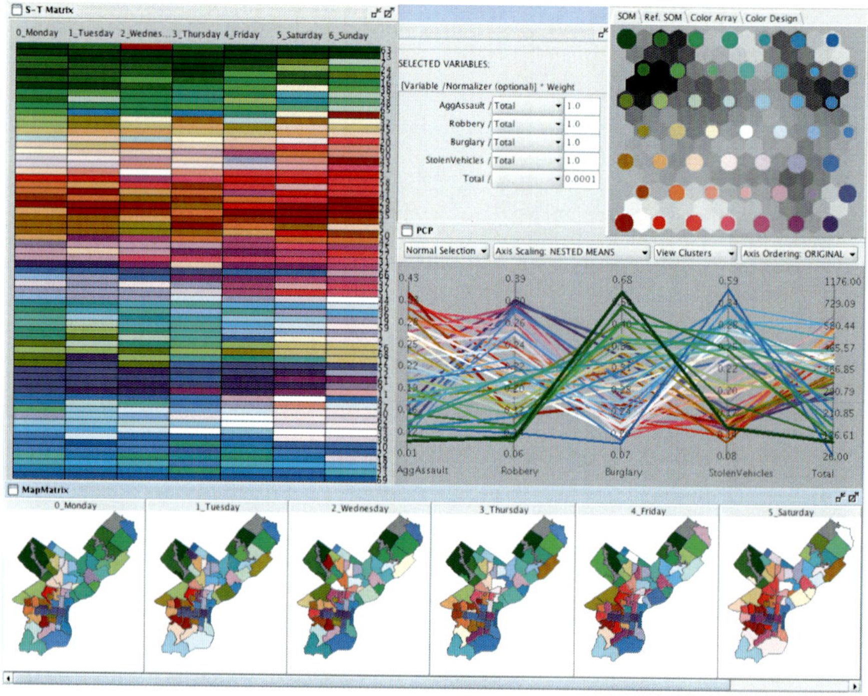

Fig. 16.10 Crime composition patterns across space and 7 days a week

spatial concentration and escalating temporal trend of the crime. Note that, in Fig. 16.11, the PCP shows each individual vector instead of the clusters, which is an option that the user can choose.

One may also focus on the greenish clusters, which represent a high percentage of burglary crimes, which exhibit a different trend, i.e., more incidents during the week days (Monday-Friday) than weekends. As for robbery crimes (in purple), it shows a persistent spatial concentration at the downtown area of Philadelphia but no noticeable temporal variation. One may also use 24 h or several time periods of a day to analyze daily patterns of crimes. Due to limited space, we do not include these analyses here.

16.4.2 Spatial-Crime Differences of Temporal Trends (Task 2)

VIS-STAMP can also treat time series as "multivariate vectors" and support the analysis of temporal trends and their differences in space and for different crime types. As introduced in Sect. 16.3.2, this is a different view of the same data cube as used in the previous analyses, with a slightly different normalization procedure. Each time series is normalized so that each value represents the percentage of total crimes (for a specific spatial unit and a crime type) in each time period. For example,

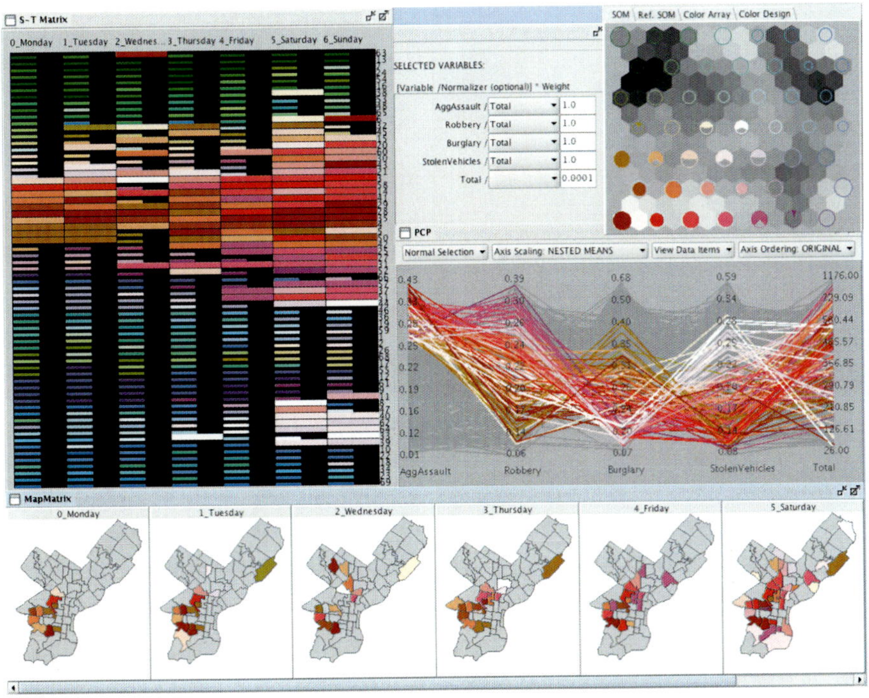

Fig. 16.11 Space-time trend of aggravated assaults (>25 %) over seven weekdays

neighborhood A had 100 robbery crimes, out of which 10 crimes occurred for time period T1, then the value for T1 is 10% for this time series for neighborhood A and robbery crime.

VIS-STAMP extract clusters with all time series and assign a color to each cluster. As such, similar colors now represent similar temporal trends. Figure 16.12 shows the analysis result with Philadelphia data, with four major crime types and nine 6-month periods. The PCP shows the clusters of time series, with each axis representing one time period. For example, a dark green cluster represents a declining trend, with more crimes in earlier times than in later time periods. The reddish clusters, on the other hand, represent a surging trend with more crimes in recent time periods. Each map in the map matrix shows the overall spatial distribution of temporal trends for a specific crime type. For example, it is obvious that the stolen vehicle crime has dropped significantly over time, as evident in its map that is dominated by greenish colors. Conversely, aggravated assault and burglary crimes have been rising lately, as shown in the "reddish" maps for both crime types. If we take a closer look at the two maps, we can also notice the spatial differences in temporal trends, with some neighborhoods getting better while others getting worse. For example, burglary surged in recent months for some neighborhoods in the central and the northeast portion of Philadelphia, while the northwest in purples and blues has improved lately (relative to their worse time around 2008 and 2009). Blueish and purplish clusters represent trends with a peak in the middle (in 2008

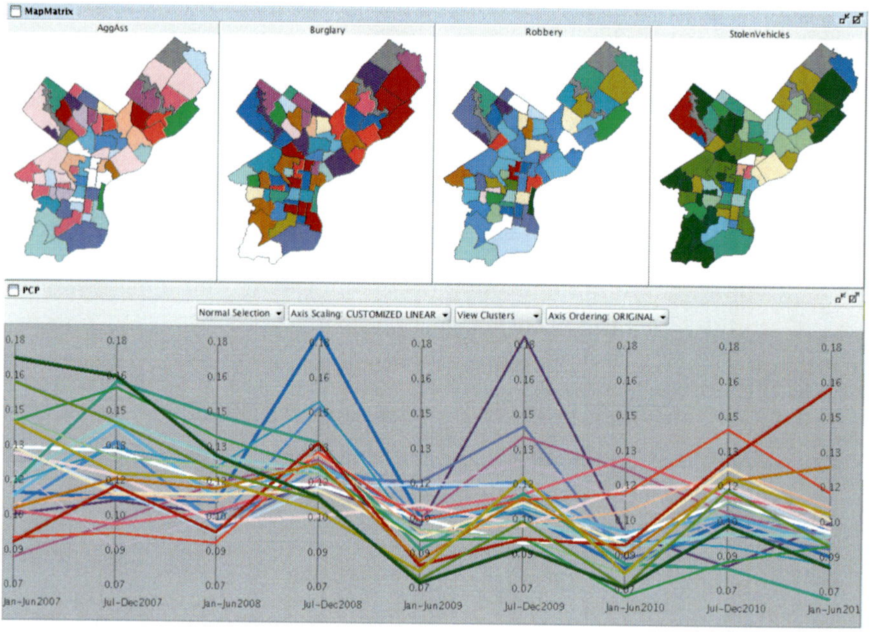

Fig. 16.12 Temporal trends vary across space and crime types. Different from the PCP in other figures, the PCP here uses a linear scaling with fix minimum and maximum values on each axis so that the height of a curve can be compared

and 2009) but relatively low crimes recently. Therefore, the robbery map indicates that robbery crimes have dropped as well, relative to the peak 2 or 3 years ago.

There is also an obvious temporal oscillation pattern shown in Fig. 16.12, where the second half (Jul–Dec)of a year often has more crimes than in the first half (Jan–Jun). Similar to the diagram presented in Fig. 16.12, we can also change the temporal scale to weekdays or hours of the day to analyze difference temproal patterns and their variation over space and crime types. One can also examine the patterns interactively through selection and linking. Due to space limitation, we will not present more and different analysis results.

16.5 Discussions and Conclusion

This Chapter presents an exploratory approach to discover and understand complex crime patterns that involve multiple perspectives such as spatio-temporal trends across different crime types. The VIS-STAMP approach is adopted to analyze the crime data in Philadelphia reported from Jan. 2007 to June 2011. The analyses focus on four major crime types: aggravated assault, robbery, burglary, and stolen-vehicles. All crimes are aggregated into a data cube with space (69 neighborhoods), time, and crime types as the three dimensions. VIS-STAMP can effectively construct an overview of the major patterns in the data cube, allowing the analysis and understanding of complex patterns across all dimensions and supporting interactive exploration of specific patterns through highlight selection and multiple-view linking. A variety of interesting patterns have been found in the Philadelphia crime data, including spatio-temporal variations of different crimes and the shifting temporal trends across space and crime types. Compared to conventional methods such as density mapping or temporal analysis alone, the VIS-STAMP approach provides an alternative way to discover more complex patterns across multiple perspectives.

The crime data is originally a point data set, which is converted and aggregated to an areal data set based on neighborhoods and other dimensions. This may be a limitation since it reduces the data resolution by using a predefined set of boundaries. An alternative solution, which we do not include in the analysis, may be to create a raster density surface for each crime type and for each time period, and then treat each raster pixel as a "spatial unit" in subsequent analysis. However, this approach has its own limitations, such as the excessive spatial autocorrelation introduced by the kernel density estimation and the uncertainty in the "interpolated" data especially when there are no sufficient data points for certain locations and time periods. Although crimes may be influenced by different factors at different scales, neighborhood is a reasonable choice in examining crime patterns in Philadelphia for two reasons, as we explained earlier. First, neighborhoods are meaningful communities that are directly related to policy analysis and planning efforts in the city. Second, neighborhoods are sufficiently large to examine its internal crime pattern across several dimensions such as time or crime type. Choosing a suitable spatial scale is important in using the VIS_STAMP approach in order to avoid statistically unstable crime measures (such as percentages), which requires that each areal unit contains a sufficient number of crime incidents.

As an exploratory approach, VIS-STAMP currently lacks rigorous statistical testing procedures to evaluate the significance level of discovered patterns such as clusters and trends. Future work may integrate statistical testing through Monte Carlo simulation to assess patterns, in addition to visual exploration. Findings discovered through VIS-STAMP and confirmed with subsequent testing may help in crime modeling and prediction and related policy making for crime control. Given the complexity, unknown factors, and time-varying characteristics of crimes, data driven and exploratory approaches are indispensible for understanding crime data and providing timely information for response.

Acknowledgments This work was supported in part by the National Science Foundation under Grant No. 0748813. We greatly appreciate the constructive comments and suggestions from anonymous reviewers.

References

Andrienko G, Andrienko N, Bremm S, Schreck T, Von Landesberger T, Bak P, Keim D (2010) Space-in-time and time-in-space self-organizing maps for exploring spatiotemporal patterns. Comput Graph Forum 29:913–922. doi: 10.1111/j.1467-8659.2009.01664.x

Anselin L (1995) Local indicators of spatial association—LISA. Geogr Anal 27:93–115

Anselin L, Cohen J, Cook D, Gorr W, Tita G (2000) Spatial analyses of crime. Crim Justice 4:213–262

Bernasco W, Elffers H (2010) Statistical analysis of spatial crime data. In: Piquero AR, Weisburd D (eds) Handbook of quantitative criminology. Springer, New York, pp 699–724

Brewer CA (1994) Color use guidelines for mapping and visualization. Visual Mod Cartogr 2:123–148

Brunsdon C (2001) The comap: exploring spatial pattern via conditional distributions. Comput Environ Urban Syst 25:53–68

Brunsdon C, Corcoran J, Higgs G (2007) Visualising space and time in crime patterns: a comparison of methods. Comput Environ Urban Syst 31:52–75

Chainey S, Ratcliffe J (2005) GIS and crime mapping. Wiley, Chichester

Chainey S, Tompson L, Uhlig S (2008) The utility of hotspot mapping for predicting spatial patterns of crime. Secur J 21:4–28

Cliff AD, Ord K (1970) Spatial autocorrelation: a review of existing and new measures with applications. Econ Geogr 46:269–292

Craglia M, Haining R, Wiles P (2000) A comparative evaluation of approaches to urban crime pattern analysis. Urban Stud 37:711–729

Eck JE, Chainey S, Cameron JG, Leitner M, Wilson RE (2005) Mapping crime: understanding hot spots. In: NIJ special report. https://www.ncjrs.gov/pdffiles1/nij/209393.pdf

Felson M, Poulsen E (2003) Simple indicators of crime by time of day. Int J Forecast 19:595–601

Gorman DM, Speer PW, Gruenewald PJ, Labouvie EW (2001) Spatial dynamics of alcohol availability, neighborhood structure and violent crime. J Stud Alcohol 62:628

Guo D, Gahegan M, MacEachren AM, Zhou B (2005) Multivariate analysis and geovisualization with an integrated geographic knowledge discovery approach. Cartogr Geogr Inf Sci 32:113–132

Guo D, Chen J, MacEachren AM, Liao K (2006) A visualization system for space-time and multivariate patterns (VIS-STAMP). IEEE Trans Vis Comput Graph 12:1461–1474

Hagenauer J, Helbich M, Leitner M (2011) Visualization of crime trajectories with self-organizing maps: a case study on evaluating the impact of hurricanes on spatio-temporal crime hotspots. In: Proceedings of the 25th conference of the International Cartographic Association, Paris

Henry LM, Bryan BA (2000) Visualising the spatio-temporal distribution of motor vehicle theft in Adelaide, South Australia. In: National Centre for Social Applications of GIS (GISCA). http://pandora.nla.gov.au/nphwb/20010320130000/http://www.aic.gov.au/conferences/mapping/henry.pdf

Hunt ED, Sumner M, Scholten TJ, Frabutt JM (2008) Using GIS to identify drug markets and reduce drug-related violence. In: Thomas YF, Richardson D, Cheung I (eds) Geography and drug addiction. Springer, Dordrecht

Kohonen T (2001) Self-organizing maps. Springer, Berlin/New York

Levine N (2006) The CrimeStat program: characteristics, use and audience. Geogr Anal 38:41–56

Messner SF, Anselin L (2004) Spatial analyses of homicide with areal data. In: Janelle DG, Goodchild MF (eds) Spatially integrated social science. Oxford University Press, Oxford

Murray AT, McGuffog I, Western JS, Mullins P (2001) Exploratory spatial data analysis techniques for examining urban crime. Br J Criminol 41:309–329

Nakaya T, Yano K (2010) Visualising crime clusters in a space time cube: an exploratory data analysis approach using space time kernel density estimation and scan statistics. Trans GIS 14:223–239

Ratcliffe JH (2000) Aoristic analysis: the spatial interpretation of unspecific temporal events. Int J Geogr Inf Sci 14:669–679

Ratcliffe JH (2002) Aoristic signatures and the spatio-temporal analysis of high volume crime patterns. J Quant Criminol 18:23–43

Ratcliffe JH (2004) Crime mapping and the training needs of law enforcement. Eur J Crim Policy Res 10:65–83

Ratcliffe JH, McCullagh MJ (1998) Aoristic crime analysis. Int J Geogr Inf Sci 12:751–764

Rengert GF (1997) Auto theft in central Philadelphia. In: Homel R (ed) Policing for prevention: reducing crime, public intoxication and injury. Criminal Justice Press, Monsey, p 7

Roth R, Ross K, Finch B, Luo W, MacEachren A (2010) A user-centered approach for designing and developing spatiotemporal crime analysis tools. In: Proceedings of GIScience, Zurich, Switzerland

Townsley M (2008) Visualising space time patterns in crime: the hotspot plot. Crime Pattern Anal 1:61–74

Townsley M, Homel R, Chaseling J (2000) Repeat burglary victimisation: spatial and temporal patterns. Aust N Z J Criminol 33:37–63

Weisburd D, Bushway S, Lum C, Yang SM (2004) Trajectories of crime at places: a longitudinal study of street segments in the city of Seattle. Criminology 42:283–322

Weisburd D, Morris N, Groff E (2009) Hot spots of juvenile crime: a longitudinal study of arrest incidents at street segments in Seattle, Washington. J Quant Criminol 25:443–467

Wu X, Grubesic T (2010) Identifying irregularly shaped crime hot-spots using a multiobjective evolutionary algorithm. J Geogr Syst 12:409–433

Part V
Applications and Implementations

Chapter 17
The Use of Geospatial Information Technology to Advance Safer College Campuses and Communities

Gregory Elmes and George Roedl

Abstract In this collaboration between university researchers and practitioners from two adjacent law enforcement jurisdictions, crime incidents are examined for spatio-temporal trends stretching across jurisdictional borders. The goal of the partnership is to increase the safety of students both on and off campus by identifying crime clusters which enables proactive policing efforts specifically targeted to high crime areas. The applied research confirms spatial and temporal crime clusters across a shared boundary. The implications for this partnership suggest that researchers and multiple law enforcement jurisdictions can work together to identify and solve community problems. In this chapter, background information is presented on research-practitioner partnerships, campus crime, student victimization and the Clery Act, and the goals and objectives of the collaboration. After presenting three analyses on data gathered over the first year, policing and research implications are discussed.

Keywords Cross-jurisdiction • Campus crime • Spatio-temporal analysis • Hotspot mapping • Researcher-practitioner partnership • Community-oriented policing • Problem-oriented policing

17.1 Introduction

Although the effective and efficient reduction of crime is increasingly dependent on spatial analysis, the availability and accessibility of such methods varies widely from one law enforcement jurisdiction to another (LeBeau and Leitner 2011; McCarthy and Ratcliffe 2005). A growing volume of literature is dedicated to the reduction of

G. Elmes (✉) • G. Roedl
Department of Geology and Geography, West Virginia University,
330 Brooks Hall, Morgantown, WV 26506-6300, USA
e-mail: Greg.elmes@mail.wvu.edu; Groedl@gmail.com

M. Leitner (ed.), *Crime Modeling and Mapping Using Geospatial Technologies*,
Geotechnologies and the Environment 8, DOI 10.1007/978-94-007-4997-9_17,

crime through the use of spatial analysis by law enforcement agencies. However, the representation of campus communities and cross-jurisdictional analyses in this literature is limited. With some notable exceptions, the collaboration between law enforcement agencies, researchers, and the community is under-reported in available literature (National Institute of Justice 2009). To add to the available literature, this chapter documents and discusses the objectives and results of a collaborative crime mapping and analysis program implemented between campus police, municipal police, and academic researchers.

The Increasing Student and Community Safety (ISaCS) partnership, an applied research project funded by the National Institute of Justice, Office of Justice Programs, brings together university researchers, campus police, and municipal police to accomplish four mutually beneficial objectives. The first establishes the capacity to perform cross-jurisdictional crime mapping and analysis to assist in the promotion of a safer campus and municipality. The second objective utilizes spatial analytical technologies to provide spatial and non-spatial information needed for problem-oriented, intelligence-led decision making and resource allocation. Objective three enhances community-oriented policing methods and reduce crime through greater public awareness and participation and the fourth objective establishes partnerships with additional law enforcement agencies and researchers. The current project resulted from the convergence of several initiatives, including the interest of the West Virginia University (WVU) and Morgantown Chiefs of Police in the integration of information and the increased use of crime analysis. This interest was supplemented by the establishment of a course in crime mapping and analysis in the geography program at WVU and by support of a Researcher-Practitioner Partnership grant from the National Institute of Justice (2009). This chapter can aid criminal justice practitioners and academic researchers interested in applied studies that have the potential for lasting effects on a community.

Campus and municipal police share many similar objectives and methods, yet there are significant differences in their missions. Although municipal police departments are charged with preventing and solving crime in the general public arena, campus police deal primarily with a specific segment of the population — students, and to a lesser extent with faculty, staff, and visitors to campus. This, in effect, often leads campus police to be subject to greater scrutiny from parents, campus administrators and public officials at all levels of government (Rengert et al. 2001). Campus police are required to conform to the Jeanne Clery Disclosure of Campus Security Policy and Crime Statistics Act (Clery Act) and accordingly report Clery statistics, whereas municipal police departments can be regulated by state code, the Uniform Crime Report (UCR) handbook and the more comprehensive National Incident Based Reporting System (NIBRS). These differences in established priorities and reporting mechanisms pose a challenge in creating a cross-jurisdictional crime mapping and analysis program.

The chapter begins by introducing the researcher-practitioner partnership concept as the basis for collaboration, and then proceeds by situating the context of crime on college campuses which, in turn, serves as a prelude to a discussion of the Clery Act and its assessment as a tool for crime prevention and reduction.

A review of law enforcement organization and jurisdictional relationships is provided, along with a description of the ISaCS partnership. Example analyses of crime incidents across jurisdictions after the first year of the partnership follow and the implications of the analyses and partnership are discussed. This chapter contributes to crime mapping literature in several ways. First, highlighting a collaborative applied project adds to the available collection of case studies. Researchers and practitioners regularly engage a variety of community stakeholders, public officials, and decision makers to gather data, assess local conditions, and make decisions which ensure maximum effect of limited resources. We demonstrate cross-jurisdictional crime mapping is feasible, yields additional insight into various analyses of crime, and is relevant to proactive problem-solving strategies. Second, we examine a segment of the population that is vulnerable to victimization, in addition to being contributors to crime. Although our focus is on college students, we believe this work provides a template for building collaborations that examine other groups of interest. Third, the analyses and results of this collaboration are appropriate for a broad audience including campus administrators, law enforcement agencies, policy makers, students, and community stakeholders. Lastly, our work establishes a lasting capacity to perform crime analysis in a campus police department and culminates with results of interest to anyone conducting crime mapping or considering the possibilities of implementing their own crime mapping and analysis program.

17.2 Background

17.2.1 Research-Practitioner Partnerships

In recent years the National Institute of Justice has funded many projects that support researcher-practitioner partnerships (Backes 2009). In a typical partnership, a researcher works within a law enforcement agency (LEA) to develop, conduct and evaluate needed criminal justice research (see also McEwen 1999, 2003). Encouraging the researcher and practitioner to work closely with one another over a prolonged period is anticipated to lead to a better understanding of the roles and relations of research, practice, and policy within a particular law enforcement agency (NIJ 2009). The importance of the practitioner's involvement is acknowledged throughout the research process — from formulating the initial objective, research questions and accessing data, to advising a study as it progresses and helping to ensure practical perspectives in the analysis of data and report writing. In one documented collaborative effort between law enforcement agencies, researchers, community representatives, victim advocate groups, city government, and social service agencies in the Memphis Strategic Approaches to Community Safety Initiative (SACSI), participants viewed the researchers as most effective in identifying the problem and assessing the impact of sexual assault victimization among teenage girls (Coldren and Forde 2010).

Policing research has more than a 30 year tradition, but the relationships between practitioners and researchers have not always been harmonious, stemming from or leading to misunderstandings on both sides. Beal and Kerlikowske (2010) indicated researchers generally have projects in mind that seldom match practitioner's needs. Research needs to be meaningful to the LEA (Buerger 2010) and not simply a determining factor of police activity (Scott 2010) It is evident that the two widely different cultures, academic researchers and police practitioners, have different goals and objectives with respect to reward frameworks and measures of success leading to a 'Dialogue of the Deaf' (Bradley and Nixon 2009). However, the attainment of more effective policing requires researchers and LEAs to work together (Scott 2010). Despite the differences in reward structures, there is a continuing and underlying desire for research which improves evidence-based and intelligence-led policing. Cordner and White (2010) suggested that LEAs believe researchers can offer them something useful. Boba (2010) attributed successful applied collaborations to physical proximity, grant funding, practical knowledge by the researcher, research knowledge by the practitioners and trust. Davis (2010) argued that the gap between research and practice exists in every field and the tendency to put research to use in policing was much better than other areas. Citing successes from community policing to hot spot policing, Rosenbaum (2010) demonstrated that researchers have made valuable contributions to police policy. Similarly, Ikerd (2010) and Scott (2010) cited successful partnerships that have helped shape effective problem-oriented policing efforts. However, Gilling (2005) stated the problem with partnerships focused on problems such as crime, is that approaches are "means to ends" (see Goldstein 1979) and therefore attract less interest than merited, inevitably resulting in fewer publications.

Criminological police studies provide neutral feedback and scientific rational for policy (Das 2010). While the institutionalization of police studies may be challenging, Wood and Bradley (2009) adamantly proclaimed that such partnerships matter and Rosenbaum (2010) viewed partnerships as critical to mutual learning. Recognizing the existing separation between policing and science, Weisburd and Neyrou (2011) provided a comprehensive discussion on the merits of partnerships and the implications for future disconnects. According to Engel and Whalen (2010) research may address one or more of four areas of policing refinement: operational effectiveness and efficiency, external validity, cooperative transparency, and the information technology revolution. The present chapter focuses on the last area: partnership research in information technology.

17.2.2 Campus Crime

Crimes are prevalent across the U.S. and have a large impact on university campuses, commonly involving violence, protests, drug and alcohol use, sexual assault, identity theft, fraud, and theft of expensive equipment (Stafford and Rittereiser 2007). According to the National Center for Victims of Crime (2009), over 88,000 crimes were reported on college campuses in 2008; of these, 97% were property crimes. Baum and Klaus (2005) reported that, of the crimes against college students between 1995 and 2002, violent crime rates declined 54% and that 93% of all incidents occurred off- campus, with an annual rate of 60.7 per 1,000 students. Additionally, the Tennessee Bureau of Investigation (2007) identified an increase in college student victimization off- campus. Although there was little statistical difference in the crime victimization rates of students living on- or off-campus, Baum and Klaus (2005) further identified that college students aged 18–25 were less frequently victimized than non-students within the same age cohort.

In general, college campuses are relatively safe communities which experience lower crime rates than the surrounding communities (Bromley 1995; Henson and Stone 1999; National Center for Victims of Crime 2009; Volkwein et al. 1995). Fox and Hellman (1985) compared campuses located in urban, suburban, and non-urban locations and found that urban campuses generally experience the lowest crime rates, but that in general, location had little impact on campus crime rates. Subsequently, Sloan (1994) conducted a similar comparison of rural, small town, and large urban campuses. That research showed violent crimes were correlated with surrounding community crime rates, but property crimes were not. A positive correlation between on-campus crime and the crime rates of the surrounding cities was found by Bromley (1995). Although he recommended additional research to further understand campus crime and prevention strategies, the consensus suggests that crime rates on college campuses are generally lower than the surrounding community and have been declining.

Research findings examining campus crime and its relationship to internal and external factors vary from study to study. McPheters (1978) provided a catalyst for campus crime research when he associated higher crime rates at 38 college campuses with higher proportions of on-campus student residents, and proximity to urban areas with high unemployment rates. His research linked campus crime to both on- and off-campus variables. Fox and Hellman (1985) identified inverse relationships between the crime rates of 222 campuses and tuition costs. They further revealed that crime rates were significantly lower in locations with a higher ratio of campus police staff to students. Volkwein et al. (1995) correlated campus crime rates for 416 institutions with 23 various community, organizational and student measures. They identified an increased likelihood of both violent and property crimes at medical schools and health science centers. Bromely (1995) linked increased crime levels for 265 institutions to total student population and total male student population. Other significant factors included the number of buildings, acreage size, amount of fraternity activity, and total female population. In an examination of 546 universities, Sloan (1992a) found campus crime rates strongly correlated with seven variables, including tuition costs, total student population, resident population, fraternity/sorority presence, and student/security ratio. Sloan (1994) continued his work and further associated violent crime with minority enrollment. Overall, previous research indicates higher campus crime rates are associated with higher levels of on-campus student residency, student economic status (as implied by higher tuition costs), and student/police ratio.

The majority of crimes against college students occur off-campus (Baum and Klaus 2005; National Center for Victims of Crime 2009) and these observations suggest that any measure of total potential crime risk to students attending college should consider crime rates for both on-campus and the adjacent off-campus community. They also suggest the necessity for campus officials and police to work closely with municipal officials and police to protect students and provide a safer community. Furthermore, although prior research correlates crime rates with different associated variables, it does little to explain criminal behavior or provide paths to reduce it. For example, although dorms may be considered "hotspots" of crime on campus (Bromley 1995), large universities have many dorms, which may experience different rates and types of crime. Additional research should examine why rates and types of crime vary between similar places across campus. Also, as Volkwein et al. (1995) pointed out, campus medical schools and health science centers have an unexplained higher likelihood of crime incidents.

Relatively few publications examine campus crime, with most research coinciding with the federal campus crime disclosure acts in the 1990s, which received significant national attention resulting from legislative action and high profile court rulings. More recent interest in campus crime has been spurred by media reports of high profile campus incidents, such as the Virginia Tech shooting rampage in 2007 which killed 32 people and wounded 25 others (Virginia Tech Review Panel 2007). Published research has limitations stemming from data and methods, as well as a failure to keep up with changing methods in police strategies and prevention techniques, criminology theory, campus infrastructure transitions and growth trends, and spatial science tools, to identify a few examples relevant to the current partnership.

17.2.3 Student Victimization and the Clery Act

The following section provides a brief overview of the Clery Act and its reporting requirements. Although never designed as an instrument to prevent crime, the Clery Act sets crime reporting standards among campuses receiving federal financial aid money while recognizing campus differences and permitting flexibility in compliance. Since the Clery Act was first passed into law in 1990, it has been amended four times to address some of its shortcomings. Institutions failing to comply with Clery Act standards are subject to civil penalties, including fines and loss of federal financial aid funds. Despite the strengths of the Clery Act, it has been criticized for not doing enough to protect students. We will conclude with a review of our findings that emphasize the need not only for crime awareness programs, but for geo-visualization tools able to represent complex data.

The Jeanne Clery Disclosure of Campus Security Policy and Campus Crime Statistics Act, (US Code 20 USC 1092 (f)) part of the Higher Education Act of 1965, requires colleges and universities participating in federal student aid programs to disclose specific information about campus crime and security policies. Under the act, universities and colleges are required to provide an annual campus security report of mandatory crime statistics for a 3 year period. The institution's police or security department is also required to maintain a public log of all reported crimes, containing the nature, date, time and general location of each crime and the disposition of the complaint. Compliance standards require the institution to disclose crimes statistics which are covered by the Clery Act. Crime statistics are classified as criminal offenses, hate crimes, and arrests and referrals for disciplinary action. Criminal offenses are reported in seven major categories, with sub-categories: (1) Criminal Homicide (a) Murder and Non-negligent Manslaughter and (b) Negligent manslaughter, (2) Sex Offenses (a) Forcible and (b) Nonforcible, (3) Robbery, (4) Aggravated Assault, (5) Burglary, (6) Motor Vehicle Theft, and (7) Arson. Crimes classified as hate crimes include those in the previous seven categories, plus theft, simple assault, and vandalism/destruction of property, if motivated by prejudice. Additionally schools are also required to report the following three types of incidents if they result in either an arrest or disciplinary referral: liquor law violations, drug law violations, and illegal weapons possession.

The Clery Act has received mixed reviews in the research literature. Declines in campus crime since 1990 suggest the legislation has prompted institutions to take preventative measures (Volkwein et al. 1995) and has increased confidence in campus policing (Fisher 1995; Janosik 2001). However, as Janosik (2001) stated, reports by themselves do little to protect students or change behavior. Volkwein et al. (1995) suggested the reporting requirements actually overestimate campus crime directed toward students, and Bruno (2009) highlighted inconsistencies in enforcement and reporting of various criminal activities. For example, the distinction between theft (a non-reportable statistic) and burglary (a reportable statistic) is often difficult to make and there may be a propensity to classify burglaries as theft in an effort to achieve lower reportable crime rates.

In a survey conducted by Janosik (2001), only 29% of college students were aware of the Clery Act, which was designed to allow students to consider crime risk information when making decisions about college selection. Of the students who remembered receiving crime statistics from their institution, 79% admitted to not reading the material. Furthermore, fewer than 4% of respondents considered campus crime statistics relevant to college selection. Our own investigation of the ten universities with largest student populations indicated no relationship between published Clery statistics and student enrollment (Roedl and Elmes 2010). In contrast, the Janosik (2001) survey reported that campus crime awareness and prevention programs reached over half the respondents with most of them reporting changes in their behavior as a direct result. Although many different conclusions could be drawn, campus crime research in general indicates that crime awareness and prevention programs are more effective than simply providing statistical summaries.

In terms of the current research and analysis, Clery statistics (adapted from the FBI's Uniform Crime Reporting handbook) are not directly comparable to state-defined legal codes defining crimes and subsequent report systems or even the more comprehensive National Incident Based Reporting System/NIBRS. Exact direct comparison between campus crime and off-campus crime therefore become difficult when comparing a state-defined crime or NIBRS-defined crime to the federally-defined UCR or Clery crime. Another impediment concerning Clery Act statistics is their limited spatial content. As stated in the Clery Act, incidents are classified as occurring on-campus, within residence halls, or in public spaces such as streets and sidewalks. Until the most recent revision, non-campus controlled buildings, such as fraternities and research facilities were considered off-campus for Clery Act reporting purposes. According to the Clery Act, campus police are required to request crime statistics from local LEAs responding to incidents within the campus geography. However, local LEAs are not obligated to satisfy the request. In instances that local LEAs do provide crime statistics, there may be a propensity to provide either too much or too little data because of the differences between UCR and Clery definitions.

As a majority of student victimization occurs off-campus, both on-campus incidents and those of the surrounding community should be examined together to produce a more holistic assessment of student safety. The locations of the incidents should be more detailed and a single reporting standard should be established for all involved LEAs. Also, considering that many students are in new and unfamiliar locations, visual representations of crime locations, such as physical and web-based maps, should become a standard. Although students may not know the names of all the nearby streets, they can easily recognize a point on a map in relation to their daily route. The effectiveness of crime maps as visual aids for campus crime awareness and crime prevention and their potential for interaction represents an important direction for geospatial awareness research.

17.2.4 Jurisdictions: Campus and Municipal Police

Stimulated by new ideas and new innovations, police research on crime control has advanced rapidly and benefited immensely from powerful research methods and databases which have provided new insight (Sherman 1992). Campus and municipal police departments alike have employed various policing philosophies, practices, and strategies over time in pursuit of crime reduction (see Sloan 1992b and Uchida 1997 for campus and municipal police histories respectively). The dominant policing philosophy for police departments, including campus police departments, throughout the U.S. (including those involved in this project) is community policing (Stafford and Rittereiser 2007). Not surprisingly, community policing is a major theme of police research as demonstrated by its dominance in publications (Bartholomew et al. 2009). Despite the prevalence of community policing by LEAs and within the policing literature, there is no single definition, no single correct way of implementation, and approaches are as varied as the communities and LEAs (Chaiken 2001; Roberg et al. 2008). Regardless of policing philosophy, LEAs (municipal and a majority of campus LEAs) utilize at least three common tactics: patrols (random and directed), rapid response to service calls, and retrospective crime investigations by detectives (Moore 1992).

The reporting procedures of the two law enforcement agencies involved in this project have different audiences. Those of the West Virginia University PD are driven by regulations established under the Clery Act as described in the previous section. On the other hand, the Morgantown municipal PD is directly responsible for enforcing state and federal laws, and responds to the imperatives of municipal authorities and citizens. Despite these considerations, the two LEAs have established a long-time working relationship. Campus LEAs depend upon municipal support in the event of large scale incidents, such as the aforementioned Virginia Tech shootings. A good working relationship ensures that quality data conforming to Clery Act requirements is obtained from municipal LEAs when requested. Having a working relationship with designated liaisons is important for establishing primary contact personnel to make and fulfill such requests, and for generating emergency notifications of threats as well as responding mutually. Recognizing the importance of municipal and campus police working together to protect college students, the Major Cities Chiefs Association and Bureau of Justice (Major Cities Chiefs Association 2009) developed the *Campus Security Guidelines* as a means to improve ongoing relations between campus and municipal LEAs. This comprehensive document resulting from surveys of campus and municipal LEAs provides recommendations for developing policies, preventing and preparing for incidents, responding to incidents, and actions to follow after responding.

17.2.5 The ISaCS Partnership

The WVU-led partnership provides local law enforcement agencies in university and municipal jurisdictions with the capacity to perform spatial analysis for problem-oriented, intelligence-led policing, and response to crime incidents. The partnership between practitioners and researchers, furthermore, enhances community-oriented policing efforts by providing current crime incident maps utilizing an internet map server. The overall aim is to identify emerging crime trends, enabling law enforcement agencies to inform the public and solicit community participation, while developing crime reduction strategies tailored to providing safer campuses and communities. The four specific goals (listed in the introduction of this chapter) required to accomplish this overall objective were developed mutually by the researchers and practitioners. The current partnership has resulted in local law enforcement agencies gaining an understanding of the operational benefits of crime mapping and analysis while providing researchers with the ability to apply their skills and knowledge to an applied research problem capable of creating safer communities and informing policy.

The ISaCS partnership takes place in Morgantown, WV. West Virginia University consists of three connected campuses: the Downtown campus, the Evansdale campus, and the Health Sciences campus. The campus population for 2010 was approximately 29,000 students, 2,300 faculty, and 3,300 staff employees. The resident population of the city of Morgantown had grown over the last decade at a rate of

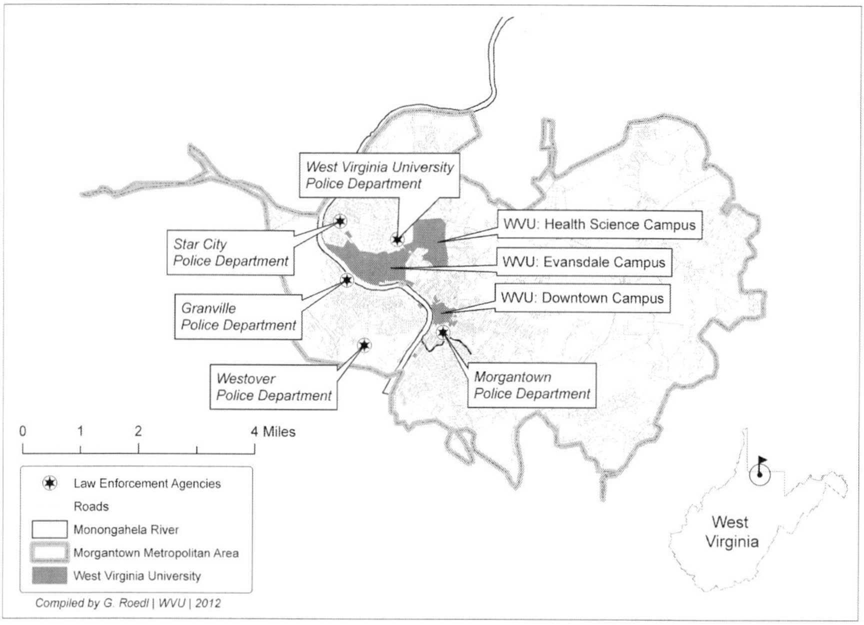

Fig. 17.1 Morgantown Metropolitan Area with locations of local LEAs and WVU campuses

17% to a 2010 total of 30,000 with an additional 70,000 residents in smaller towns throughout the county. The Morgantown metropolitan area has five police jurisdictions: Morgantown, Star City, Westover, Granville, and West Virginia University police departments (Fig. 17.1). All three WVU campuses are bounded by the Morgantown police jurisdiction.

Responsibilities for objectives and goals of the partnership are divided among the two LEAs and two researchers (a research faculty member and a graduate student) and were outlined in a Memorandum of Understanding (MOU) prior to project implementation to ensure compliance and specify expectations. Daily crime incidents (exported from each LEA in spreadsheet form) are passed from the LEAs to the researchers to be stored in a secure geospatial database containing supporting ancillary data such as campus and municipal building footprints, aerial photos, and street networks. Over the course of the project, attributes for the ancillary data have undergone considerable revisions to provide more useful information to support operational use and research objectives by practitioners and researchers. This on-going revision has further enabled the creation of a composite geo-coder capable of locating features by address, proper name (e.g., Store XYZ), campus feature (e.g., tennis court), and all known aliases or derivative names. Although time consuming and tedious to create and maintain, the use of such ancillary data and alias information in the geo-coding process increased positional accuracy substantially. For internal use, the incident data is kept intact, while data intended for the public is abridged to only include date/time, cleansed-address, incident category, and incident report number. The process is automated with a customized script to enable practitioners to create maps from a basic template without extensive GIS training; a situation accentuated by limited numbers of staff. To overcome disparities in reporting between LEAs, the script also automates the classification of state statute codes into the ten categories examined by this project. The geospatial database stores uploaded incident data, populates the map templates used for operational analysis across jurisdictions, and serves spatial data to two dynamic internet maps; one designed to be accessed by the public and the other by LEAs. To enable rapid deployment of the initial maps and underlying data, the ESRI ArcGIS Viewer for Flex with pre-compiled sample codes (http://help.arcgis.com/en/webapps/flexviewer/) was utilized. User-submitted enhancements were incorporated into the viewer to improve functionality.

17.3 Analyses and Discussion

The use of crime reports from two adjacent LEAs allows for the development of research questions that compare and contrast incidents, examine jurisdictional interactions, and enable cross-jurisdictional analyses of criminal activity useful for establishing relationships with socio-demographic variables and geographic features. Although researchers are able to advance criminology knowledge in general, such analyses are also relevant and useful to law enforcement practitioners. As the partnership progresses, we will present specific research results that are practical, as well as contribute to the researcher-practitioner, cross-jurisdictional policing, and campus crime literature. The following discussion will provide examples of temporal, spatial, spatio-temporal and spatial regression analyses performed on data collected during the first year of the partnership. As results are presented, implications for researchers and practitioners will be discussed.

Based on 12 months of data for 2010, crime rates at WVU were found to be lower than corresponding rates of Morgantown: 0.66/10,000 vs. 4.21/10,000 for property crimes and 0.19/10,000 vs. 1.11/10,000 for violent crimes. These crime rates are unofficial estimates and likely to be higher than the true values since our data is based on original daily incident reports and includes unfounded or non-cleared incidents. Having observed that WVU crime incidents corresponded to published research findings of lower rates on campus than in the surrounding community, we focused on some more specific comparisons that characterized local crime using simple crime mapping and analysis techniques.

17.3.1 Temporal Data Analysis

The first analysis presented is an exploratory spatial data analysis (ESDA) of temporal trends. ESDA is a critical step for visualizing and characterizing patterns, clusters,

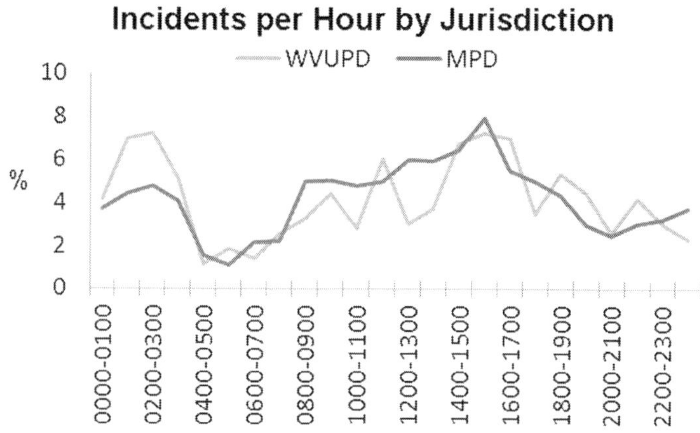

Fig. 17.2 Percentage of MPD an WVUPD crime incidents by hour of day

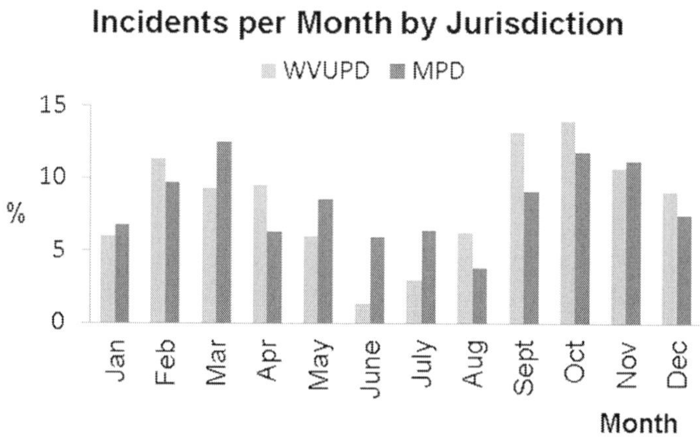

Fig. 17.3 Percentage of MPD and WVUPD crime incidents by month

and outliers in data (Baller et al. 2001). Using the capabilities of GIS as an ESDA tool provides a flexible way of examining data, generating new hypothesis, and identifying unexpected spatial patterns (Eck et al. 2005). For example, a comparison of all 2010 incidents from Morgantown and WVU by hour of day reveals a very similar overall pattern in both jurisdictions (Fig. 17.2). Although there are slight variations, crime generally increases in the early morning hours followed by a mid-morning drop then a gradual increase that peaks in the mid-afternoon. Although not shown here, similar line graphs of crime categories also present remarkable similarity for both WVU and Morgantown incidents. Bar graphs comparing volume of crime incidents by month also follow similar trends (Fig. 17.3). With some subtle variation, there are spring and fall peaks with the number of summer incidents much lower than the rest of the year. Incidents graphed by day

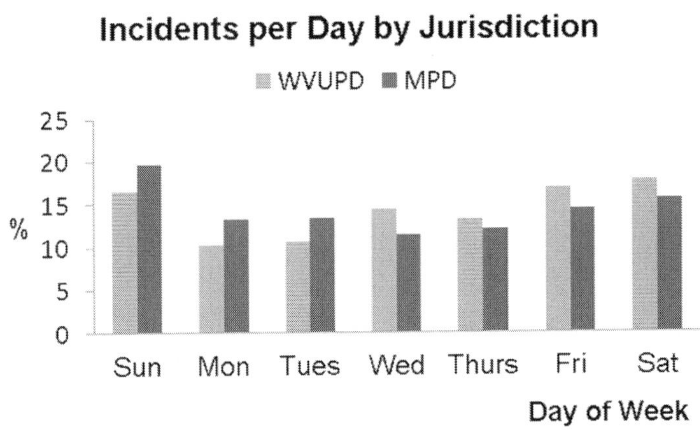

Fig. 17.4 Percentage of MPD and WVUPD crime incidents by day of week

of the week reveal variations during the weekday with significantly more crime on weekends (Fig. 17.4). Again, although not shown here, comparisons of crime categories reveal more similarity than difference between Morgantown and WVU. Theft, destruction/vandalism, and assault/battery account for the largest portion of incidents within each jurisdiction (Fig. 17.5).

From the researcher's perspective, ESDA is a powerful tool for exploring potential hypotheses. Again, for example, we initially thought the high number of WVUPD incidents on Wednesdays were due to several Wednesday basketball games. However, further analyses revealed that only two offenses occurred within the vicinity of the Coliseum (the basketball arena) and most of the Wednesday incidents occurred in the off-season, resulting in rejection of this hypothesis. From the practitioner's perspective, ESDA permits LEAs to allocate needed resources to the identified days, times and locations more likely to experience crime events and ideally prevent or at least quickly respond to a call for service. Temporal characterization of incidents served as an important starting point for understanding trends between and across jurisdictions. The logical subsequent step was to characterize spatial trends.

17.3.2 *Spatial Data Analysis*

Although crime may be random and unpredictable, crime mapping and analysis is a tool researchers and practitioners can use to discern spatial and temporal trends. Under certain conditions, times or locations, crime exceeds the average or expected rates to form distinct hot spots (Anselin et al. 2000; Clarke and Eck 2003; Eck et al. 2005; Rivero and Pepper 2010). These hot spots are represented as significant clustering of crime events within small geographic areas or time spans (Braga 2001, 2005). Sherman et al. (1989) observed crime clustered into hot spots in only a relatively

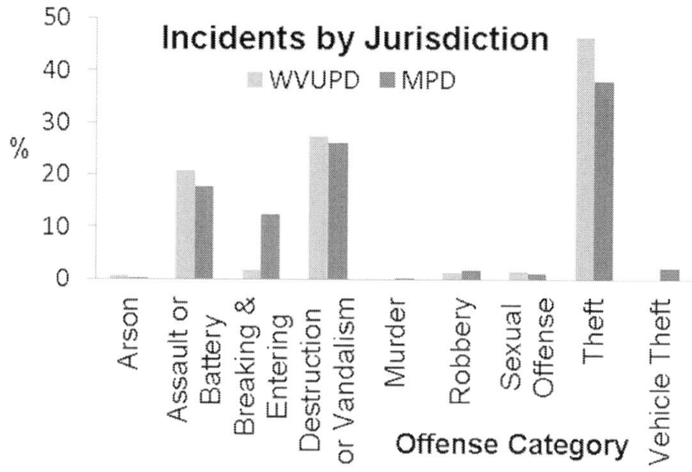

Fig. 17.5 Percentage of MPD and WVUPD crime incidents by type

few discrete areas, even in the most crime ridden neighborhoods. Although, WVU and Morgantown are relatively small geographic areas with experienced officers employing knowledge-based policing strategies, hot spot maps show changing patterns which provides additional information to law enforcement officers. Hot spot locations are mapped routinely to provide geo-visualization of weekly problem areas. A point density analysis is applied to determine if crime events are random, dispersed, or clustered. Point density analysis aids in determining if patterns are true clusters that deviate significantly from spatial randomness or if they are only visual perceptions and not statistically significant (Anselin et al. 2000).

LEA officers may be familiar with the hot spot locations within their own beat or shift, but they may be less familiar with hot spots in other beats or shifts. Furthermore, they are even less likely to be familiar with hot spots in neighboring jurisdictions and to be unaware of crime clusters across jurisdictions. Hot spot mapping across police jurisdictions provides LEAs with a more informed and complete picture of crime. Absent edge-effects inherent in single jurisdictional analyses, hot spot clusters are often identified stretching across Morgantown and WVU jurisdictions (Fig. 17.6) creating a shared boundary problem between the two jurisdictions (see Eck 2002). Raising awareness of clustering across jurisdictional boundaries enables LEA officers and supervisors to work together using intelligence-led strategies as a valuable addition to knowledge-led practices.

17.3.3 Spatio-Temporal Data Analysis

Despite the recognition of both space and time as relevant hot spot dimensions, most crime analyses treat space and time as separate entities (Assunção et al. 2007; Bernasco and Block 2009; Bernasco and Elffers 2010; Grubesic and Mack 2008;

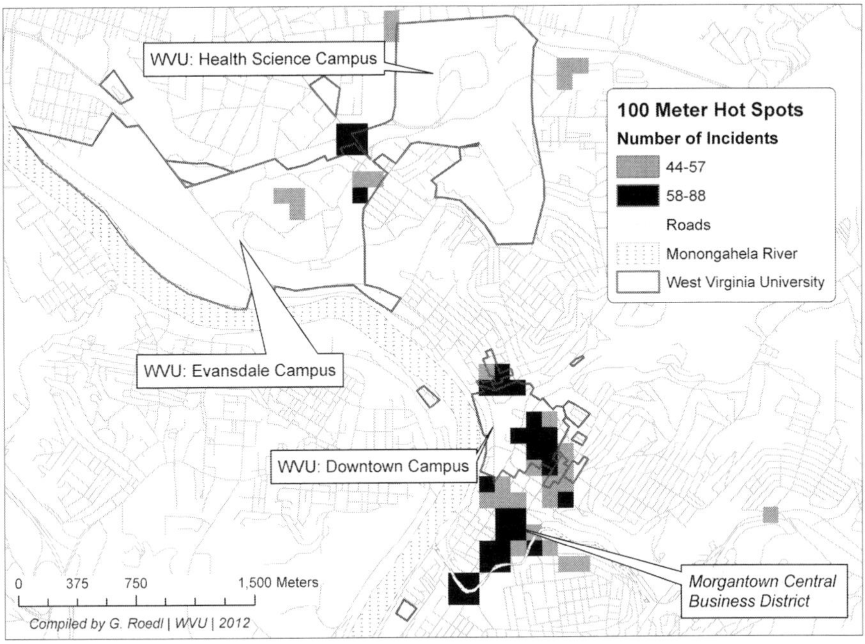

Fig. 17.6 Cross-jurisdictional density map depicting areas with large clusters of crime incidents

Ratcliffe 2006). Researchers (Johnson and Bowers 2004; Johnson et al. 2007, 2009a, b; Townsely et al. 2003) conducting space-time crime analyses concluded crime incidents are likely to happen within defined spatial and temporal proximity of previous incidents within jurisdictions of study. We quantified the spatio-temporal dimensions of clusters across jurisdictions to advance research and reaffirm the importance of cross-jurisdictional analyses. A complementary and alternative method to hot spot cluster analysis, well suited to quantifying both space and time interactions of crime data, is the Knox spatio-temporal interaction test (Grubesic and Mack 2008; Johnson and Bowers 2004; Townsley et al. 2003). The Knox index measures space-time interaction between discrete data points in terms of specified time and distance (Levine 2004; Knox 2002; Knox and Bartlett 1964; Kuldorf and Hjalmars 1999). To assess the statistical significance of Knox space-time interactions, a Monte Carlo simulation with 1,000 permutations was performed. The Knox index calculates observed clusters while the Monte Carlo simulation calculates expected clusters based on probabilities derived from the simulation distributions (See also Johnson et al. 2007 for a full description of the Monte Carlo simulation and permutation method). For this example analysis, observed and expected spatio-temporal clusters were calculated for a threshold of 2 days and a distance of 100 m for MPD and WVUPD crime incident data as separate jurisdictions and then merged together. As large buildings and parking lots are geo-coded by their geometric center, a distance of threshold of 100 m was selected based on a similar interval used by other researchers (Johnson and Bowers 2004; Johnson et al. 2007,

2009a; Townsley et al. 2003). The choice of 100 m ensured that large features located next to each other were analyzed as being spatially nearby. Ideally, spatial and temporal threshold distance values would be selected empirically from prior research; however, issues such as the effects of the spatial or temporal scale of features used in space-time interaction analyses have, as yet, received little recognition in the literature. Analyses often use a range of space-time bandwidths to identify thresholds (e.g., Johnson and Bowers 2004; Johnson et al. 2007, 2009a; Townsley et al. 2003). Next, the separate MPD and WVUPD observed clusters were summed and compared against the expected number of clusters (Knox 2002). The difference between the summed values and the expected values is accounted for by the space-time clusters that are not captured when doing spatio-temporal analyses of each LEA jurisdiction separately.

Results of the Knox space-time interaction test for LEA jurisdictions separately and combined are presented in Table 17.1; as well as the summed values and difference between the summed observed values and the combined expected values obtained from the Knox test. The difference column is the difference between the observed summed and estimated values and provides a count of the additional number of space-time clusters detected when performing a cross-jurisdictional analysis. Plainly, analysis of both jurisdictions simultaneously is able to capture more spatio-temporal clustering. Although in many respects such a statement appears to be one of common sense, it is an empirical example that geographic boundary effects often are not taken into consideration when analyzing a single jurisdiction. Practitioners, policy-makers, and criminologists performing analysis on spatial data should be aware that there are numerous statistical issues arising from geographic boundaries. The implication is that when jurisdictions share crime patterns on common borders, separate LEA jurisdiction analyses is an inaccurate assessment of crime risk; potentially resulting in misguided resource allocation and intervention strategies.

Comparison between the additional observed clusters for all crime-type incidents analyzed together, relative to incidents analyzed for crime types separately suggests that the type-of-crime clusters may be more random than the spatio-temporal case and this observation merits further examination. Clearly crimes of different types are clustering more often than crimes of similar types and contradict findings by Johnson et al. (2009b) who looked for spatio-temporal linkages across crime types. Although an increased volume of crime incidents analyzed resulted in more spatio-temporal clustering, the significance is that more prior incidents used in an analysis may lead to a greater confidence in identifying when and where a future incident may occur based on prior incidents. However, the additional incidents also offer less predictability in determining what type of crime may occur within the defined spatio-temporal threshold since the spatial and temporal aspects remain constant in the analysis while the range of incident types is increased. A better understanding of spatio-temporal clustering across various crime types can have important implications for policing methods and criminological theory.

Table 17.1 Results of Knox Interaction test for MPD, WVUPD, and MPD and WVUPD

	MPD (Knox)		WVUPD (Knox)		MPD & WVUPD (Knox)		MPD & WVUPD (Summed)		MPD & WVUPD (Difference)
Incident type	n=	Observed/ Expected	n=	Observed/ Expected	n=	Observed/ Expected	n=	MPD + WVUPD Observed	Knox minus Summed
All types	1874	305/177.55*	602	100/81.49 *	2476	435/271.77*	2476	405	30
Arson	5	0/0.00	3	0/0.00	8	0/0.00	8	0	0
Assault	360	22/12.75*	116	8/5.98	476	31/19.49*	476	25	1
Burglary	287	20/3.29*	12	0/0.03	299	20/3.31*	299	20	0
Destruction	537	39/21.40*	173	11/7.50*	710	55/31.56*	710	50	2
Robbery	35	0/0.06	7	0/0.00	42	0/0.07	42	0	0
Sexual	26	0/0.02	10	0/0.13	36	0/0.08	36	0	0
Theft	756	58/30.44*	283	24/18.75	1039	87/50.58*	1039	82	5
Veh. Theft	45	0/0.05	1	N/A	46	0/0.05	46	0	0

*p<0.0001

17.3.4 Spatial Regression Data Analysis

Having compared crime patterns for the Morgantown and WVU jurisdictions and illustrating space-time interaction, we conclude with an example of a cross-jurisdictional analysis incorporating ancillary demographic data to identify potentially explanatory relationships with crime. In a GIS, location is the common denominator between separate datasets, enabling them to be merged and managed to explore relationships between the data. GIS can be applied to crime data to map and analyze spatial patterns and trends (Boba 2005; Bruce 2001). This in turn can lead to intelligence-led and proactive policing leading to crime reduction and efficiency (Getis et al. 2000; Hirschfield and Bowers 2001; Markovic et al. 2006).

Combined incident reports collected during 2010 from MPD and WVUPD were joined spatially to demographic data to provide a robust dataset for spatial and statistical analysis across jurisdictions. US census blocks (2000) encompassing the Morgantown metropolitan statistical area (as outlined in Fig. 17.1) were used to provide the spatially contiguous demographic data and boundary delineations. As previously observed, crime was distributed throughout the study area in a number of clusters (Fig. 17.6). Local indicators of spatial autocorrelation (LISA) were generated for each of the 925 census blocks based on the recorded number of crime incidents occurring within each block. LISA is an ESDA method used to identify clusters of census blocks with incidents either higher or lower than that expected by random chance (see also Eck et al. 2005). A threshold was established to identify clusters with less than a 5% chance of occurring randomly ($p=0.05$). This yielded a total of 98 census blocks located adjacent to other census blocks which contained high numbers of incidents. Twenty census blocks were identified as being spatially-correlated with low number of incidents, while a further twenty blocks were identified as clusters of a high number of incidents adjacent to a low number of incidents. As locating and visually representing these scattered census blocks within the overall study area is difficult, at a small scale, a figure has not been included.

A spatial regression analysis was then used to identify possible associations between spatially-clustered census blocks and demographic predictor variables and to predict total crime counts for each census block. Prior to regression analysis, 74 demographic predictor variables were normalized and examined for correlation. Fourteen non-correlated demographic variables were then included in the analysis utilizing a spatial error regression model with maximum likelihood estimations. The regression equation yielded only two significant predictor variables. Clusters of high crime census blocks had a positive association with renter-occupied housing and a negative association with the total number of black residents (adjusted R-square = 29.2). Clusters of low crime census blocks indicated no association with demographic variables. Several significant predictor variables (adjusted R-square value = 60.1) were identified as contributing to the total number of incidents for each census block. Total number of white residents, number of families, total area, population in age bracket 30–64, owner-occupied housing, and renter occupied housing all had positive associations with the total number of reported incidents. Total number of black residents, residents of other race, population density, household size, and population in age bracket 65+ had negative associations with the total number of crime incidents.

Identifying empirically-demonstrated relationships with typical student demographic profiles has the potential to influence decision makers at colleges and within college towns. Inference from the demographic variables common to college students (e.g., age 18–24, renters, no children) suggests that census blocks with larger populations of college students have little statistical association with an increased number of crime incidents. Although additional analysis is required, these preliminary results suggest college students living on- and off-campus are unlikely to experience higher crime rates than the general population. For LEA practitioners, this type of analysis allows intelligent-led decisions to be made based on empirically-founded results. Intervention strategies can be targeted toward census blocks with high crime clusters. Knowing which demographic predictor variables are associated with increased crime provides intervention strategies with a concrete foundation. Likewise, we have also established a preliminary finding that some demographic characteristics apparently have little or no effect on crime.

Recognizing that crime is inherently geographic and analyzing where crime takes place is fundamental to tackling crime problems (Chainey et al. 2008). As demonstrated, a variety of different analyses may be performed to advance research knowledge, while improving the ability of LEAs to reduce crime and increase public safety. Analytical approaches should vary for each type of analysis based on place, purpose, spatial dependencies, type of crime, time, barriers, and the visual display of results (Eck et al. 2005). GIS and geospatial data used to map crime made these analyses possible and further support the three functions of crime mapping outlined by Boba (2005) It facilitates visual and statistical analyses of the spatial nature of crime and other types of events; 2) It allows analysts to link unlike data sources together based on common geographic variables; and 3) It provides maps that help to communicate analysis results.

17.4 Implications

Weisburd et al. (2002) described crime mapping across borders as a major issue for LEA problem-solving due to technological, organizational, political, and social barriers. The ISaCS partnership between LEA practitioners and researchers has demonstrated cross-jurisdictional collaboration can be achieved, resulting in an applied example of crime mapping and analysis across borders, which can be scaled-up to surrounding jurisdictions. Furthermore, we highlight a successful relationship between researchers and practitioners which has led to a mutually improved understanding of the role of research in LEA, bridging the gap between theory and practice. The culmination of a collaboration between a municipal LEA, campus LEA, and academic researchers identifying and solving problems together has led to an exploratory analysis of crime incidents recorded during the first year and has strengthened the overall understanding of crime patterns within the cross-jurisdictional study area. Although weaknesses in data, statistical assumptions, and the execution of the collaboration exist, the gaps between an effective and productive researcher-practitioner partnership and cross-jurisdictional collaborations are being narrowed while contributions to theory and practice have begun to emerge.

Spatio-temporal crime pattern is an under-researched area (Ratcliffe 2010) with some notable exceptions (e.g., Johnson and Bowers 2004; Johnson et al. 2007, 2009a, b; Townsely et al. 2003). Efforts to reduce victimization are dependent upon the empirically-validated existence of spatio-temporal clusters which can be used to anticipate increased risk (Johnson et al. 2007). A greater understanding of the factors contributing to when and where crime will occur has profound implications for criminological theories and policing efforts directed toward crime reduction. This research has demonstrated the utilization of a typical crime dataset (Townsley et al. 2003) in a progressive series of analyses that examined temporal, spatial, spatio-temporal, and contributing factors of crime for a community with a large student population to gain further insight into local crime patterns. Unlike much of the published research, we used a cross-jurisdictional approach that examined campus and municipal crime with the expressed objective of exploring risk across the community instead of within artificially created areal units.

Recapping and commenting additionally on some previously discussed exploratory findings, it was shown that campus crime rates were lower than the surrounding municipal crime rates (Bromley 1995; Henson and Stone 1999; NCVC 2009; Volkwein et al. 1995). A temporal analysis of crime events using ESDA graphing revealed campus and municipal crime patterns were similar in seasonal, daily, and hourly variations. Understanding when crime events are likely to occur across three temporal scales is relevant for LEA tactical and strategic planning in this partnership. From a criminological perspective, temporal insight into preferred activity times of offenders offers insight into understanding the underlying factors leading to crime. In this study, aggregated monthly crime statistics displayed decreases which coincided with periods of university recesses, suggesting the presence of college students could be an underlying contributor of crime either directly or indirectly. Additional research into the relationship between low or high periods of crime and the presence of college students is needed and could contribute to theories on offender motivation. An examination of spatial clusters using a simple point density map confirmed visually that WVUPD and MPD have a shared boundary problem. Researchers and practitioners need to be keenly aware of and understand the effects of common geographic problems, such as boundary effects, on analyses and interpretation. And finally, this research has explored some potential demographic associations contributing to higher crime areas. Although there was a significant relationship between high crime census blocks with renter-occupied housing, the relationship was weak. However, a regression analysis linking demographic variables to numbers of crime incidents occurring in census blocks across the study area produced both strong and significant relationships. The observed relationships indicated that typical student demographics had little effect on crime, which in turn suggests that college students are unlikely to experience an increased risk of victimization either on- or off-campus relative to that of the general population.

17.5 Current and Future Directions

As LeBeau and Leitner (2011) ask geographers to take up the quest for addressing the applied and basic research questions that need to be examined in the study of the geography of crime, we have taken our first step down that road; beginning with a small dataset and a familiar community. With crime rates well below national averages, an analysis of Morgantown and WVU crime is presented as a case study which may or may not be applicable to other geographic locations. Many more case studies need to be conducted for similar crime rate communities to be compared. Johnson et al. (2007), for example, examined burglaries in ten different cities from five countries, revealing both consistencies and inconsistencies, and suggested local differences have an important effect on spatio-temporal patterns. Their findings and ours emphasize the need for additional work to identify and validate factors contributing to spatio-temporal clustering across different types of crime as important contributions to criminological theory.

17.6 Conclusions

One of the primary purposes of the ISaCS partnership has been to use GIS as an analytical tool for identifying and setting up responses to crime trends; and as a research tool for identifying the potential underlying causes of crime. As such, campus and municipal LEAs have benefited from greater data sharing between agencies, increased information for crime reduction collaborations, and a greater appreciation of spatial technologies and mapping. Additionally, the researchers have gained a better understanding of both campus and municipal LEA practices, from daily operations to tactical and strategic planning. Conversely, LEA partners have demonstrated an increased trust and willingness to maintain an ongoing partnership with researchers, once it became evident the crime mapping and analysis research was not only useful for supporting and enhancing routine knowledge-led policing activities but was a beneficial contribution to problem-oriented decision making.

This applied research project has established geospatial technology in an LEA environment unlikely to have been able to do so without the support of the research-practitioner partnership. The physical infrastructure, data acquisition, mapping and analysis protocols, and establishment of researcher-practitioner trust are important steps in facilitating further integration of the derived spatial information into the daily practice of law enforcement. The ISaCS partnership has provided researchers with valuable insight into the analytical needs and daily operations of the practitioners while also giving the practitioners access to researchers predisposed to addressing the practitioners' overall desire for crime reduction. Thus far, knowledge-led decision making has been supplemented with intelligence-led decision making through the addition of cross-jurisdictional geospatial data in the operational routines of campus and municipal LEA jurisdictions. This crime mapping and analysis project has provided a valuable tool to LEA practitioners previously well-informed about the utility of crime maps for problem-oriented policing efforts, but without the capacity to initiate their own crime mapping and analysis program. Furthermore, community members and public officials are able to maintain a sense of security and confidence in the LEAs that embrace new technologies to reduce crime and victimization as well as providing the additional transparency in reporting, evident on online maps.

Acknowledgement This project was supported by Award No. 2009-IJ-CX-0205 awarded by the National Institute of Justice, Office of Justice Programs, U.S. Department of Justice. The opinions, findings, and conclusions or recommendations expressed in this publication are those of the author(s) and do not necessarily reflect those of the Department of Justice.

The authors wish to acknowledge the contributions of Chief Bob Roberts and Kate VanSant of WVU PD and Chiefs Phillip Scott and Ed Preston, and Jamie Michael of Morgantown PD. Furthermore, we would like to thank the anonymous reviewers for their valuable comments and recommendations; however, opinions, omissions and errors remain firmly the fault of the authors.

References

Anselin L, Cohen J, Cook D, Gorr W, Tita G (2000) Spatial analyses of crime. Crim Justice 4:213–262

Assunção R, Tavares A, Correa T, Kulldorff M (2007) Space-time cluster identification in point processes. Can J Stat 35(1):9–25

Backes B (2009) NIJ seeks to strengthen the practitioner-researcher bond. Correct Today 71(4):78–80

Baller RD, Anselin L, Messner SF, Deane G, Hawkins DF (2001) Structural covariates of US county homiciderates:incorporatingspatialeffects.Criminology39(3):561–590.doi:10.1111/j.1745-9125.2001.tb00933.x

Bartholomew B, Gibbs J, Mazeika D, Ahlin E, Joseph P, Miller N (2009) Trends in police research: a cross-sectional analysis of the 2006 literature. Police Pract Res 10(4):383–407. doi:10.1080/15614260903255038

Baum K, Klaus P (2005) Violent victimization of college students, 1995–2002. Bureau of Justice Statistics. U.S. Department of Justice, Office of Justice Programs, Washington, DC

Beal P, Kerlikowske RG (2010) Action research in Buffalo and Seattle. Police Pract Res 11(2):117–121. doi:10.1080/15614261003590852

Bernasco W, Block R (2009) Where offenders choose to attack: a discrete choice model of robberies in Chicago. Criminology 47(1):93–130

Bernasco W, Elffers H (2010) Statistical analysis of spatial crime data. In: Piquero AR, Weisburd D (eds) Handbook of quantitative criminology. Springer, New York, pp 699–724

Boba R (2005) Crime analysis and crime mapping, 1st edn. Sage Publications, Thousand Oaks

Boba R (2010) A practice-based evidence approach in Florida. Police Pract Res 11(2):122–128. doi:10.1080/1561426100359302

Bradley D, Nixon C (2009) Ending the 'dialogue of the deaf': evidence and policing policies and practices. An Australian case study. Police Pract Res 10(5/6):423–435

Braga AA (2001) The effects of hot spots policing on crime. Ann Am Acad Polit Soc Sci 578(1):104–125. doi:10.1177/000271620157800107

Braga AA (2005) Hot spots policing and crime prevention: a systematic review of randomized controlled trials. J Exp Criminol 1(3):317–342. doi:10.1007/s11292-005-8133-z

Bromley ML (1995) Comparing campus and city crime rates: a descriptive study. Am J Police 14(1):131–148

Bruce CW (2001) A thousand words for a picture: is the overvaluation of GIS disrupting a critical balance in crime analysis? Massachusetts Association of Crime Analysts Round Table 1 (Spring). http://www.macrimeanalysts.com/rt/f2cb1p.htm. Accessed 15 Feb 2012

Bruno L (2009) Campus crime stats don't tell whole story. DailyRecord.com. http://www.dailyrecord.com/article/20090622/COMMUNITIES/906220307/1005/NEWS01. Accessed 22 June 2009

Buerger ME (2010) Policing and research: two cultures separated by an almost-common language. Police Pract Res 11(2):135–143. doi:10.1080/15614261003593187

Chaiken MR (2001) COPS: Innovations in policing in American heartlands. LINC, Alexandria, VA. http://www.ncjrs.gov/pdffiles1/nij/grants/194604.pdf. Accessed 19 Jan 2010

Chainey S, Tompson L, Uhlig S (2008) The utility of hotspot mapping for predicting spatial patterns of crime. Sec J 21:4–28

Clarke RV, Eck J (2003) Become a problem solving crime analyst In 55 small steps. University College London, Jill Dando Institute of Crime Science, London

Coldren JRJ, Forde DR (2010) The memphis strategic approaches to community safety initiatives (SACSI) project: a case study. http://www.ncjrs.gov/pdffiles1/nij/grants/232198.pdf. Accessed 22 Oct 2010

Cordner G, White S (2010) The evolving relationship between police research and police practice. Police Pract Res 11(2):90–94. doi:10.1080/15614261003590753

Das DK (2010) From the Editor-in-Chief. Police Pract Res 11(2):87–89. doi:10.1080/15614261003665431

Davis RC (2010) A new approach in Dallas. Police Pract Res 11(2):129–131. doi:10.1080/15614261003593104

Eck JE (2002) Overcoming the barriers: crime mapping in the 21st century: crossing the borders of crime: factors influencing the utility and practicality of interjurisdictional crime mapping. vol 1. Police Foundation, Washington, DC. http://www.policefoundation.org/pdf/barriers-eck-2002.pdf. Accessed 16 Oct 2008

Eck JE, Chainey S, Cameron JG, Leitner M, Wilson RE (2005) Mapping crime: understanding hot spots. National Institute of Justice, Washington, DC

Engel RS, Whalen JL (2010) Police-academic partnerships: ending the dialogue of the deaf, the Cincinnati experience. Police Pract Res 11(2):105–116. doi:10.1080/15614261003590803

Fisher BS (1995) Crime and fear on campus. Ann Am Acad Polit Soc Sci 539:85–101

Fox JA, Hellman DA (1985) Location and other correlates of campus crime. J Crim Justice 13(5):429–444. doi:10.1016/0047-2352(85)90043-1

Getis A, Drummy P, Gartin J, Gorr W, Harries K, Rogerson P, Stoe D, Wright R (2000) Geographic information science and crime analysis. URISA J 12(2):7–14

Gilling D (2005) Partnership and crime prevention. In: Tilley N (ed) Handbook of crime prevention and community safety. Willan Publishing, Portland, pp 734–756

Goldstein H (1979) Improving policing: a problem-oriented approach. Crime Delinq 25(2):236–258. doi:10.1177/001112877902500207

Grubesic T, Mack E (2008) Spatio-temporal interaction of urban crime. J Quant Criminol 24(3):285–306. doi:10.1007/s10940-008-9047-5

Henson VA, Stone WE (1999) Campus crime: a victimization study. J Crim Justice 27(4):295–307. doi:10.1016/S0047-2352(99)00003-3

Hirschfield A, Bowers K (2001) Mapping and analysing crime data: lessons from research and practice. Taylor and Francis Inc, New York

Ikerd TE (2010) Putting POP to the pavement: captains in the Charlotte-Mecklenburg Police Department share their experiences. Police Pract Res 11(6):491–504. doi:10.1080/15614263.2010.497354

Janosik SM (2001) The impact of the Campus Crime Awareness Act of 1998 on student decision-making. NASPA J 38(3):348–360

Johnson SD, Bowers KJ (2004) The burglary as clue to the future. Eur J Criminol 1(2):237–255. doi:10.1177/1477370804041252

Johnson S, Bernasco W, Bowers K, Elffers H, Ratcliffe J, Rengert G, Townsley M (2007) Space–time patterns of risk: a cross national assessment of residential burglary victimization. J Quant Criminol 23(3):201–219. doi:10.1007/s10940-007-9025-3

Johnson S, Summers L, Pease K (2009a) Offender as forager? A direct test of the boost account of victimization. J Quant Criminol 25(2):181–200. doi:10.1007/s10940-008-9060-8

Johnson SD, Bowers KJ, Birks DJ, Pease K (2009b) Predictive mapping of crime by ProMap: accuracy, units of analysis, and the environmental backcloth. In: Weisburd D, Bernasco W, Bruinsma GJN (eds) Putting crime in its place. Springer, New York, pp 171–198. doi:10.1007/978-0-387-09688-9_8

Knox EG (2002) An epidemic pattern of murder. J Public Health 24(1):34–37. doi:10.1093/pubmed/24.1.34

Knox EG, Bartlett MS (1964) The detection of space-time interactions. J R Stat Soc C Appl Stat 13(1):25–30

Kulldorff M, Hjalmars U (1999) The Knox Method and other tests for space-time interaction. Biometrics 55(2):544–552. doi:10.1111/j.0006-341X.1999.00544.x

LeBeau JL, Leitner M (2011) Introduction: progress in research on the geography of crime. Prof Geogr 63(2):161–173. doi:10.1080/00330124.2010.547147

Levine N (2004) CrimeStat III: a spatial program for the analysis of crime incident locations. Ned Levine & Associates, Houston

Major Cities Chiefs Association (2009) Campus security guidelines: recommended policies for local and campus law enforcement agencies. Major Cities Chiefs Association, Columbia, MD. http://www.majorcitieschiefs.org/pdfpublic/MCC_Campus%20Security%20Guidelines.pdf.. Accessed 21 June 2010

Markovic J, Bueermann J, Smith K (2006) Coming to terms with geographical information systems. Police Chief 73(6)

McCarthy T, Ratcliffe J (2005) Garbage in, garbage out: geocoding accuracy and spatial analysis of crime. In: Wang F (ed) Geographic information systems and crime analysis. Idea Group Inc, Hershey, pp 45–59

McEwen T (1999) NIJ's locally initiated research partnerships in policing–factors that add up to success. Natl Inst Justice 238:2–10

McEwen T (2003) Evaluation of the locally initiated research partnership program. Institute for Law and Justice, Alexandria

McPheters LR (1978) Econometric analysis of factors influencing crime on the campus. J Crim Justice 6(1):47–52

Moore MH (1992) Problem-solving and community policing. Crime Justice 15:99–158

National Center for Victims of Crime (2009) National crime victim's rights week resource guide (NCVRW). U.S. Department of Justice, Office for Crime Victims, Washington, DC. http://ovc.ncjrs.gov/ncvrw2009/pdf/2009ResourceGuide.pdf. Accessed 23 June 2009

National Institute of Justice (2009) Solicitation: building and enhancing criminal justice researcher-practitioner partnerships. CFDA No. 16.560. US Department of Justice, National Institute of Justice, Washington, DC. https://www.ncjrs.gov/pdffiles1/nij/sl000884.pdf. Accessed 23 Sep 2011

Ratcliffe JH (2006) A temporal constraint theory to explain opportunity-based spatial offending patterns. J Res Crime Delinq 43(3):261–291. doi:10.1177/0022427806286566

Ratcliffe J (2010) Crime mapping: spatial and temporal challenges. In: Piquero AR, Weisburd D (eds) Handbook of quantitative criminology. Springer, New York, pp 5–24. doi:10.1007/978-0-387-77650-7_2

Rengert GF, Mattson MT, Henderson KD (2001) Campus security: situational crime prevention in high density environments. Criminal Justice Press, Monsey

Rivero DA, Pepper JP (2010) Proactive patrolling through the use of patrol scripts. Police Chief 77(9):66–68

Roberg R, Novak K, Cordner G (2008) Police and society, 4th edn. Oxford University Press, Oxford

Roedl G, Elmes G (2010) Geography, crime, spatial technologies, and safe campuses. Annual Meeting of the Association of American Geographers. AAG, Washington, DC

Rosenbaum DP (2010) Police research: merging the policy and action research traditions. Police Pract Res 11(2):144–149. doi:10.1080/15614261003593203

Scott MS (2010) Policing and police research: learning to listen, with a Wisconsin case study. Police Pract Res 11(2):95–104. doi:10.1080/15614261003590779

Sherman LW (1992) Attacking crime: police and crime control. Crime Justice 15:159–230

Sherman LW, Gartin PR, Buerger ME (1989) Hot spots of predatory crime: routine activities and the criminology of place. Criminology 27(1):27–56. doi:10.1111/j.1745-9125.1989.tb00862.x

Sloan JJ (1992a) Campus crime and campus communities: an analysis of crimes known to campus police and security. J Secur Admin 15(2):31–47

Sloan JJ (1992b) The modern campus police: an analysis of their evolution, structure, and function. Am J Police 11(2):85–104

Sloan JJ (1994) The correlates of campus crime: an analysis of reported crimes on college and university campuses. J Crim Justice 22(1):51–61

Stafford D, Rittereiser S (2007) Community policing, crime prevention and comprehensive training programs: key ingredients to reducing risk in campus public safety. Paper presented at the Stetson College of Law 28th annual national conference on law and higher education, February 17–21, 2007. http://justice.law.stetson.edu/excellence/Highered/archives/2007/CommunityPolicing.pdf. Accessed 04 Mar 2010

Tennessee Bureau of Investigation (2007) Crime on Campus 2006 Tennessee Bureau of Investigation, Nashville Crime Laboratory, Nashville, TN. http://www.tbi.state.tn.us/Info%20 Systems%20Div/TIBRS_unit/Publications/Crime%20On%20Campus%202006.pdf.. Accessed 06 July 2009

Townsley M, Homel R, Chaseling J (2003) Infectious burglaries. A test of the near repeat hypothesis. Br J Criminol 43(3):615–633

Uchida CD (1997) The development of the American police: an overview. In: Dunham RG, Alper GP (eds) Critical issues in policing: contemporary readings, 3rd edn. Waveland Press, Prospect Heights, pp 18–35

Virginia Tech Review Panel (2007) Mass shootings at Virginia Tech: report of the Review Panel. Commonwealth of Virginia. http://www.governor.virginia.gov/TempContent/techPanelReport-docs/FullReport.pdf. Accessed 19 Sep 2001

Volkwein JF, Szelest BP, Lizotte AJ (1995) The relationship of campus crime to campus and student characteristics. Res High Educ 36(6):647–670. doi:10.1007/BF02208249

Weisburd D, Neyroud P (2011) Police science: toward a new paradigm. Executive session on policing and public safety. Harvard Kennedy School's Program in Criminal Justice Policy and Management and U.S. Department of Justice, Office of Justice Programs, National Institute of Justice, Washington, DC. http://ncjrs.gov/pdffiles1/nij/228922.pdf. Accessed 1 Feb 2010

Weisburd D, Boba R, Groff E (2002) Introduction. In: Eck J (ed) Overcoming the barriers: crime mapping in the 21st century: crossing the borders of crime: factors influencing the utility and practicality of interjurisdictional crime mapping, vol 1. Police Foundation, Washington, DC, pp 1–4. http://www.policefoundation.org/pdf/barriers-eck-2002.pdf. Accessed 16 Oct 2008

Wood J, Bradley D (2009) Embedding partnership policing: what we've learned from the Nexus policing project. Police Pract Res 10(2):133–144. doi:10.1080/15614260802264586

Chapter 18
Construction of a Web-Based Crime Geointelligence Platform for Mexico City's Public Safety

Elvia Martínez-Viveros, José I. Chapela, Amílcar Morales-Gamas, Camilo Caudillo-Cos, Rodrigo Tapia-McClung, Mario Ledesma, and Fidel Serrano

Abstract This chapter describes an academia-government collaboration project aimed to develop a geointelligence capacity in order to insert the geospatial dimension in the information systems and decision making processes of Mexico City's Public Safety Ministry –a preventive police agency that deals with local jurisdiction crimes. The project was formalized by means of a 3 year contract comprising consultancy, training and technological developement. Framed in the project's purposes and goals, and as part of its products, processes and protocols identified for the implementation of a Geointelligence Laboratory in the Ministry, a geospatial data infrastructure (GDI) to enable the seamless integration of data from different sources, platforms and systems was implemented. An open-source interactive solution that retrieves and displays geospatial patterns and trends was developed, in order to feed decision making with results from analyses derived from the use of the GDI. Also, models for space and space-time analysis of crime incidence were implemented as part of the analytic processes routinely performed by the Ministry's internal users of the GDI.

Keywords Geointelligence • Geospatial Data Infrastructure • Public safety • Open-source spatial querying

---

E. Martínez-Viveros (✉) • A. Morales-Gamas • C. Caudillo-Cos • R. Tapia-McClung • M. Ledesma • F. Serrano
Centro de Investigación en Geografía y Geomática 'Ing. Jorge L. Tamayo' A.C., Contoy 137, Lomas de Padierna, Tlalpan, México D.F. 14249, México
e-mail: emartinez@centrogeo.org.mx; amilcar@centrogeo.org.mx; ccaudillo@centrogeo.org.mx; rtapia@centrogeo.org.mx; mledesma@centrogeo.org.mx; fidel@centrogeo.org.mx

J.I. Chapela
Centro de Inteligencia Competitiva, Citlaltépetl 24-101, Hipódromo, Cuauhtémoc, México D.F. 06100, México
e-mail: jichapelac@gmail.com

M. Leitner (ed.), *Crime Modeling and Mapping Using Geospatial Technologies*, Geotechnologies and the Environment 8, DOI 10.1007/978-94-007-4997-9_18,

18.1 Introduction

Control and regulation of public safety have become a primary concern for the governance of cities throughout the world and there is widespread agreement that geospatial information and knowledge play a key role in trying to achieve this. Most urban problems of crime and disorder are likely to fit into a spatial pattern and time trend. Space dimension is relevant when examining crime incidence geographies, the spatial behavior of criminals or the territories controlled by criminal organizations; it is also relevant to organize police forces in preventing and disrupting crime, or to locate and define alternative routes to divert people or vehicles from places affected by various events that disrupt order in the city. In fact, the spatial dimension of crime is an integral part of sociological and criminological studies (Carrabine et al. 2009; Bottoms 2007; Guerrien 2004; Prévôt-Schapira 2000; Rock 2007; Savelsberg 1984). Accordingly, police agencies should consider crime and disorder geographies and the characteristics of places in order to territorially organize protection and diminish the probability of crime occurrence. This chapter describes an academia-government collaboration project aimed at generating the capacity to manage the geospatial dimension in the policing processes of the institution responsible for public safety in Mexico City.

Mexico City is part of a megalopolis integrated by 76 counties populated, in 2010, by 20.1 million people; in itself this city hosted, that same year, 8.66 million inhabitants (INEGI 2011). This complex urban space is produced and reproduced by the entailment and crossing of many socio-spatial processes; crime and disorder being one of them. Mexico City's Public Safety Ministry is a preventive police agency dealing with local jurisdiction crimes: the types of crime related to multiple forms of robbery (larceny, burglary or vehicle theft, with or without violence) affecting transients as well as the safety of vehicles, public transport, households, and businesses, and in relation to assaults, homicides, kidnapping or sex offenses and numerous events leading up to injuries, death or street disorder. According to official population projections (Consejo Nacional de Población 2006) and data on delinquency (Secretariado Ejecutivo del Sistema Nacional de Seguridad Pública 2007, 2008, 2009), the annual incidence rate of these crimes per 1,000 inhabitants was 18.41 in 2007, 19.55 in 2008 and 21.30 in 2009. It is worth noting that federal crimes, such as trafficking of drugs, arms or persons or many types of organized crime, are under the jurisdiction of federal authorities. Also, this agency's main function is prevention; investigative functions are under the authority of another institution – Mexico City's Justice Attorney.

At Mexico City's Public Safety Ministry, interest in using the geospatial dimension in policing emerged as a result of a consultancy process initially advised by Rudolph Giuliani in 2002 (Davis 2007). Changes in the city's government authorities interrupted the process, but in 2008, the new authorities resumed this interest in an attempt to change a policing model that was mainly based upon reaction to incidents and to bureaucratic routines. An organizational restructuring took place in the Ministry, a criminal intelligence unit was created, a Compstat-style policing model was launched with the participation of top level commanders, and academic advising was sought in order to introduce the spatial dimension in the Ministry's intelligence processes. It was in this framework that a research group from CentroGeo -a Mexican public research institution specialized in Geomatics- undertook a 3 year joint project with Mexico City's Public Safety Ministry. The project was formalized by means of a collaboration contract comprising consultancy, training and technological development.

The purpose of the project was to develop an institutional geointelligence capacity in order to insert the geospatial dimension in the Ministry's information systems and decision making processes. Four main goals were recognized for the research team: (1) to identify processes and products related to the acquisition, management and analysis of geospatial data that support geo-intelligent policing decisions and actions at different hierarchical levels and territorial scales (Sect. 18.2); (2) to integrate a geospatial data infrastructure in a service-oriented architecture in order to enable the seamless integration of data from different sources, platforms and systems. (Sect. 18.3); (3) to develop interactive solutions in order to retrieve data and display patterns of crime incidence and statistics regarding the performance of the territorial police (Sect. 18.4); and (4) to integrate spatial and space-time analysis into decision making processes by approaching different scales of space-time geographies (Sect. 18.5). These goals, rather than representing an *a priori* plan of action, were proposed by the research team during the development of the initial stages of the project.

Lines of action were structured in order to attain such goals. Also, Ministry personnel were trained by CentroGeo's team in the use and management of technological developments and in space and space-time modeling. The establishment of a Geointelligence Laboratory was identified as the main tangible outcome of the collaboration. This laboratory was thought of as a physical and organizational space from which geospatial information and knowledge could be generated and managed in order to build and implement solutions to crime and disorder in the city from a territorial perspective. The main objective of the Ministry's authorities was to implement geointelligence processes in policing regulation and control at different territorial scales in order to reduce crime incidence rates, although a specific target was not made explicit. By 2010, when the project ended, it was still too early to thoroughly evaluate its direct impact upon such rates. At the entire city scale, rates for some crimes decreased while for others rates increased or remained constant. At a larger scale, some precincts showed effective results in decreasing crime rates, whereas some others could not do so. Fluctuations in crime incidence cannot be attributed as an effect of the project as a rigorous experimental design has not been done. In the remainder of this chapter a section will be devoted to briefly discuss the project's main lines of action.

18.2 Building Geointelligence Capacity

Before this project began Mexico City's Public Safety Ministry had several tasks that incorporated the geospatial perspective. Following the visit of Giuliani's consultancy team in 2002, some processes were launched. For instance, a geographical database of crime incidents was created in 2003, collecting data such as crime type, date, time, and location. However, this database lacked uniform geocoding criteria and metadata; it was kept in shape-file format, and used to generate crime pin maps that could only be consulted in the Ministry's headquarters on a desktop computer using proprietary desktop software. Other examples of spatial data generation were the geographical location of emergency calls and the position and movement of patrol cars through the use of real time monitoring, both hosted on closed proprietary systems. These examples point to a relevant production of spatial data; however the missing element was an integral framework for spatial information management that allowed the organization of information transactions among the Ministry's functional divisions, precincts and information systems, and between this institution and other agencies. There was a distinct lack of client service chains articulated with organizational levels and jurisdictions priorities.

Managing safety in Mexico City rests upon a territorial organization. The city is divided into 15 regions, in turn subdivided into 73 precincts. With the creation of the criminal intelligence unit in 2008, weekly meetings started to take place at central headquarters where statistics, graphs, and maps were discussed and territorial commanders were instructed by top authorities to reorganize operations in order to increase action efficiency rates. But the courses of action taken to try to improve this relied upon past experiences, instead of on innovative actions connecting police activities and patterns, and resulted in demands for more resources both human and material. Neither regions nor precincts had direct access to manipulate spatial data. Also, they did not have staff for analyses regarding the problems present in their jurisdictions. Recommendations about ways to improve action were not supported on spatial analysis results. Then, a Compstat process such as the one described by Willis et al. (2003) started to take place at this Ministry. A new impetus was given to the implementation of a policing process that was strategically structured, focused on patterns and trends, based upon data analysis, and oriented towards efficient and accountable operations in territorial police precincts, giving way to the demand of detecting management and analytical processes and systems needed to generate institutional geointelligence capacity and to start the implementation of the Laboratory.

The concept of geointelligence adopted in this project is drawn from the National Geospatial-Intelligence Agency definition and Ratcliffe's concept of criminal intelligence (Committee on Basic and Applied Research Priorities in Geospatial Science for the National Geospatial-Intelligence Agency 2006:9; Ratcliffe 2008:98–99,154). It encompasses technical and technological issues related to geospatial information acquisition and management, analysis of such information in knowledge generation processes, and the use of such knowledge to guide and influence decision making and action. It also acknowledges the complexities of translating analysis into decisions and actions. Most decision making or management models propose levels in order to differentiate purpose and action in time horizons, geographic scales, or the prioritization of problems. In order to deal with public safety threats and risks, levels are also proposed in terms of strategy, tactics, operations, and administration

(Bruce 2004:17; Ratcliffe 2008:99–100). Geointelligence processes at these levels were proposed as guidelines to steer the use of products emerging during the project development. These processes are summarized in Table 18.1.

A set of products, processes and protocols for the Geointelligence Laboratory implementation were identified. They impact the geointelligence levels transversally. Some of them were implemented throughout the project, while others remained pending for future action. They were organized into five themes, the relationships of which are summarized in Fig. 18.1. These themes are: informational base, interoperability, visualization, pattern identification and communications. Table 18.2 presents a summary of the products and systems, and progress of its implementation is reported on in the remaining sections.

18.3 Geospatial Data Infrastructure (GDI)

A GDI was designed and implemented in order to solve needs related to geospatial information management at organizational and technological levels. Not only did the Ministry have to deal with an important amount of information from different departments, formats, projections, and other institutions, but also with the need to export data and share information with other institutions. The GDI's conceptual model provides the basis and skills to promote the interaction among processes, systems and applications, thus enabling tactical operations and spatial analysis from the combination of vector and raster data in maps, emphasizing interactions between terrain features and the place context, and creating promising opportunities for crime abatement.

The purpose of the GDI is to share data across functional units, systems, and applications via Web services, interoperability, adoption of standards, and organizational agreements. The GDI enables the access, availability, analysis, documentation, management, and administration of the geospatial information in order to make geospatial Web information services fit into the different levels of geointelligence.

Web services can be oriented towards geointelligence levels and can be offered in isolation or integrated in packages. Among them one could list: geocodification, visualizations of statistics, maps, indicators, etc., spatial representations for communicating data or diagnostics, interactive maps, inventory management, spatial analysis results, and spatial scenarios.

A key element is the adoption of standards, which includes geospatial Web services based on worldwide standards, those sponsored by the Open Geospatial Consortium (OGC) and the World Wide Web Consortium (W3C) being the most important. They place emphasis on establishing the basis of interoperability, information sharing, knowledge, and processes, without forgetting aspects like security, transparency, stability, and redundancy of information flow. They provide guidance and infrastructure to enable timely, accurate, precise, and relevant geospatial information in order to support operations, analysis, and decision making.

Table 18.1 Processes at geointelligence levels and spatial knowledge needed to support them

Processes	Geospatial knowledge
Strategic level	
Public Policy design Long and medium term planning Risk factors identification Regional police organization Prevention programs Media relationships Informing society	Mexico City's socio-spatial structure Geospatial distribution of factors that contribute to crime and disorder (unemployment, precarious employment, informal economy, among others) Spatial distribution of risks in critical areas and locations with persistent problems Space-time analyses of vulnerability and exposure of victims Spatial prospective scenarios showing trends extrapolations or changes Design of periodic and specific reports to inform media and society
Tactical level	
Crime disruption Place-based prevention Allocation of surveillance resources (security cameras, alarms, lighting or patrol routes) Alternate surveillance and traffic route design, responding to events that affect daily life in the city (public demonstrations, major sporting events or traffic congestion, among others)	Crime patterns and their changes Space-time behavior of crime incidents Attributes of areas repeatedly victimized Prolific criminal activity inside precincts, at precinct boundaries or spilling over into neighboring precincts Spatial objects correlated with the incidence of specific crime types Spatial distribution of emergency calls Traffic volume along main roads, problematic crossroads and real time traffic monitoring
Operations level	
Priorities setting in terms of problems and places Design of precincts' regionalization Resource allocation Police operations analyses Programming, monitoring and accountability of daily territorial operations and resource use Design of tailor-made operations supporting tactical action Design of preventive actions with the media and citizenship	Statistics, graphs, maps and models identifying hot areas and hot spots by crime type and time dimension Networks and routes identification Geographical and functional allocation models Information and networking communications Indicators for evaluating achievements and lags and monitoring police operations
Administrative level	
Data, information and systems management processes Support information flow management inside and outside the agency	Data and information acquisition, organization and management; development, mastering and management of the technological base Data exchange protocols between precincts and police personnel

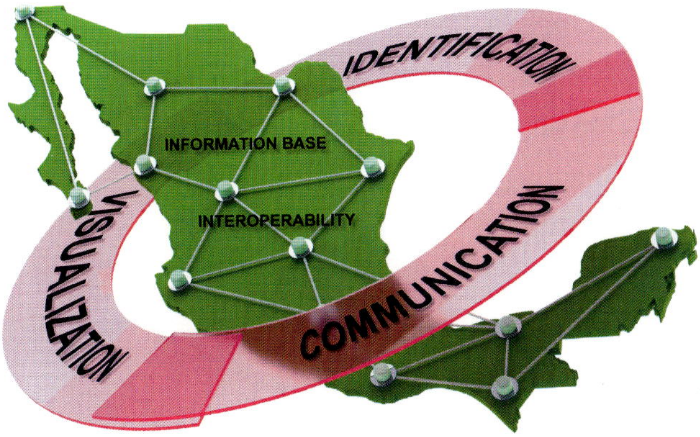

Fig. 18.1 Themes of the Geointelligence Laboratory implementation guide

The GDI needs to be accessed by many concurrent users for different reasons in compliance with the organizational agreements; therefore security was implemented to provide multilevel access, control, and management throughout the database. This mechanism relies on enabling security and privileges on both sides:

- Local network: responsible for identifying and allowing valid requests to reach specific terminals, in line with the firewall policy.
- Database: where the user registers, connects and uses permissions.

New users are added using the local network triad: username – password – domain name. Thus, only domain registered users can have access to the GDI. The next step assigns the user to the geospatial database users group via the database administration tool, and finally, specific permissions, such as select or edit, are granted for specific layers or groups of layers using the geospatial database administration tool.

This combination of security measures provides a way of guaranteeing who can view, change, update, and delete data (Poulsen 2003). Thus, the system controls access and identifies users and workstations that access information, according to the Ministry's security schema. GDI users are split into three levels –basic, intermediate, and advanced– based on their functionality requirements, operational needs, and permissions. Figure 18.2 shows these topics by user level.

An important topic associated with GDI management is the timing of data updates. On the one hand, GDI has base map layers, which provide the location reference or context for any application. Updating the layers depends on the release of new layers by the National Institute of Statistics, Geography, and Informatics (INEGI). On the other hand, it has operational layers that reflect the Ministry's areas of concern action, which support decision makers. These layers need to be updated monthly, weekly, and even daily to prepare reports, illustrate trends, and generate crime maps in order to enable crime prevention and management processes.

In order to manage the large quantity of data the GDI holds and to ensure its performance and scalability, the GDI was tuned and configured. This was done by

Table 18.2 Geointelligence Laboratory implementation guide

Themes	Products, processes and protocols
Informational Base	
Mexico City's basic geospatial representation Crime mapping and mapping capabilities of urban places and objects	Geo-coding processes (attributes registration) Address search algorithm Data homologation (geometric correction, analytic restitution, etc.) Integration of imagery information
Interoperability	
Same base map for all information systems and databases Information exchange with other information systems and databases Participation in intra-institutional and inter-institutional information networks Building a culture of collaboration	Implementation of a Geospatial data infrastructure Implementation of geospatial information Web services Adoption of Geospatial dimension by the agency's information systems and databases Adoption of standards and protocols for data exchange between information systems and databases
Visualization	
Web Geospatial visualization: easy navigation across the city scales, retrieval of attributes, analysis capabilities On-line construction of statistics, graphs, diagrams, spatial models and time dynamics Real-time visualizations	Map exploration and editing Information retrieval Combined spatial and graphic visualizations Spatial representation of model and network results Combined visualizations of different data types (texts, vectors, imagery, photograph, audio, etc.) Place visualization in space-time formats Time-series visualization
Pattern identification	
Spatial analysis Space-time analysis Expert spatial scenario building Spatial network analysis Socio-spatial structural context (socio-spatial causal factors of crime)	Spatial, space-time, serial, risk, and vulnerability analyses Detection of spatial variables associated to risks via image interpretation Spatial forecasting, prediction, and projection (extrapolation, qualitative, and participative techniques) Crime events and crime patterns simulation Mapping causation of crime
Communications	
Spatial monitoring of communication networks Spatial monitoring of surveillance networks Communications with civil society and citizenship	Data acquisition methods using GPS, satellite images, or other remote sensors Spatial representation of data flow in communications networks (voice, image, and data) Web publication of spatial information about delinquency and prevention Feedback from population on the Web

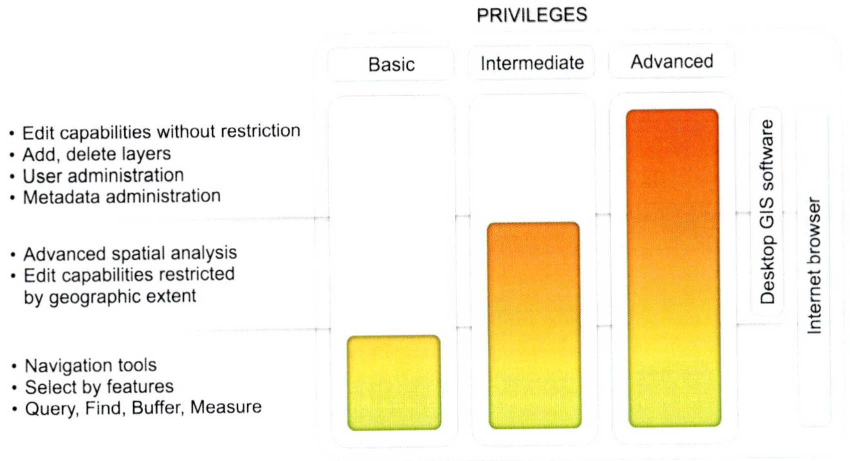

Fig. 18.2 Levels of user privileges and forms of access

making changes at both physical and logical levels and by an optimization process, which followed the steps listed below:

- Analysis: this process was based upon the analysis of each level's characteristics, taking into consideration the spatial and feature query needs for different user groups, the most common scales used for these queries, available disk space, the number of rows in the layer, and the growing storage forecast for each layer.
- Assessment: a value was assigned to each one of the layers' characteristics and a complete list of layers with assessment numbers in consecutive order from high to low was obtained.
- Indexing: a set of spatial indexes was created for each layer in order to assure the optimization of the response times per layer.
- Distribution: finally, each layer was divided in different files thus allowing the distribution of the database in a set of drives where the load balancing took place.

A catalog service was organized by the construction of metadata (based on the FGDC standard for geographic data) and a search service publication. Metadata provides detailed information on every layer stored in the geospatial database, which is grouped into four categories: identification information, data quality information, distribution information, and metadata reference information. The metadata catalog is managed by advanced users that create, publish, and manage metadata security. The GDI metadata publishes updates automatically every time a catalog update takes place.

The GDI enables the publishing of map services, which can be used by different devices and platforms in accordance with the OGC standards. The GDI architecture displays geospatial data and maps as Web services linked to demands related to building solutions. GDI architecture also plays an important role in the construction of interoperability and has a built-in "capability to communicate, execute programs, or transfer data among various functional units in a manner that requires the user to have little or no knowledge of the unique characteristics of those units" (OGC 2007).

This Web-based-interoperability service relies on OGC standards to provide a solid framework to create, share, and publish maps and data for GDI users. Its services are not tied to one specific vendor and it even employs open-source software suitable for use with mobile devices, internet browsers, and desktop GIS software. Moreover, the application is independent of hardware or software platforms. Any combination of commercial and free software supporting OGC and W3C specifications can be used.

Work in progress includes the daily operations of the territorial police and applications to spread relevant information. Also, it is planned the use of mobile devices to communicate location based data and mobile trajectories, and citizen participation via the Internet.

The growing number of social network users has created the opportunity for two-way communication between citizens and the Ministry. It can be embraced in order to create a more transparent and responsive culture and to enable citizens' cooperation to assist in solving many of the problems related to public safety in Mexico City. The GDI is ready to respond to these challenges by incorporating a management system for dynamic layers, created by internet users, which works in a real time environment, stores the information for future use and enables the process to send a fast response to the user.

Civilian generated information can support efforts in crime prevention by illustrating the impact of crime on the community and the impact of community efforts on crime. The GDI can facilitate informed decision-making on criminal activity and prevention by identifying the links between crime and other surrounding phenomena distributed in the geographic space (Getis et al. 2000).

The implementation of GDI enables information and knowledge to be shared across the organization, thus allowing for the development of applications like spatial querying (see Sect. 18.4), spatial analysis, and space-time analysis (see Sect. 18.5).

18.4 Spatial Querying for Automatic Comparative Graphic Information Retrieval: An Open-Source Solution

To prepare for the Compstat meetings, data had to be gathered from different systems and databases with non-compatible formats which, in some cases, were exchanged in non-electronic media. Data integration was carried out on desktop computers using Excel tables and graphs, maps were generated with MapInfo and presented as images without any interactivity, and presentations were integrated in PowerPoint. These weekly reports formed a very useful series of booklets, which evolved over time to include answers in the form of new data units, indicators, or maps, and that responded to many questions emerging at Compstat meetings. However, the integration of these weekly booklets became a heavy burden for the analysts and managers involved. The implementation of Compstat at headquarters was only the first step in the process, the next was to cascade the outcome of these meetings to 73 territorial police precincts, which required a way of accessing the database and obtaining graphs and maps automatically.

The utilization of a Geospatial Database Infrastructure is a great improvement for the collection, organization, and distribution of data. There is still, however, the need to process that data and to automate certain parts of the analyses and of the reporting. In fact, it was a customer requirement to provide an interactive and user-friendly way to access and retrieve crime data. A solution was developed to generate comparative crime rate reports, graphs, and maps, and to link these rates with operational police performance in territorial jurisdictions. This solution uses the Geospatial Database Infrastructure's Web information services and was developed as open-source. The design combines an interactive querying process with geographical space to give the user the possibility to view and compare statistics from different precincts and time trends.

The idea for this solution is for it to be used primarily at headquarters with further deployment at different individual police precincts. In order to ensure its functionality an open-source approach was taken. As previously mentioned, the database and map servers rely on proprietary software hosted at headquarters. As it is financially unviable to provide it to all police jurisdictions, the solution was developed using open-source software in order to benefit from the interoperability at which both proprietary and open-source software operate. Also, based on the future need to scale down the use of this solution outside headquarters, a Web-based solution was deemed appropriate.

On the client side, different pieces of open-source software were stitched together in a seamless Web application (Fig. 18.3): ExtJS (Sencha 2010) was chosen as the front-end framework, communication with the database server and dynamically generated content was done through PHP (Achour et al. 2010), maps were handled with OpenLayers (The Open Source Geospatial Foundation 2010) and GeoExt (GeoExt Community 2010), and graphics were generated using AmCharts (AmCharts 2010). On the server side, proprietary software was used: Microsoft's IIS Web server hosted the application, maps were served with ArcGIS Server and other database tables were stored in Microsoft SQL Server.

The aim of this application is to aid criminal analysts in accessing, visualizing, and analyzing trends and patterns of criminal spatial data. One of the goals of this application is to help the operator transition into the realm of spatial thinking, leaving behind the current state of affairs in which data are only regarded as a massive collection of names and numbers in a table. On the contrary, crime data are much more than just this as they must always occur somewhere and can be located on a map along with many more attributes that describe the type of event. It is when this special behavior is recognized that patterns on criminal records are observed and it becomes possible to either infer something about the process that gives place to what is presented on the map, or to try and develop a strategy that will help eradicate undesirable situations.

Fig. 18.3 Interoperability diagram

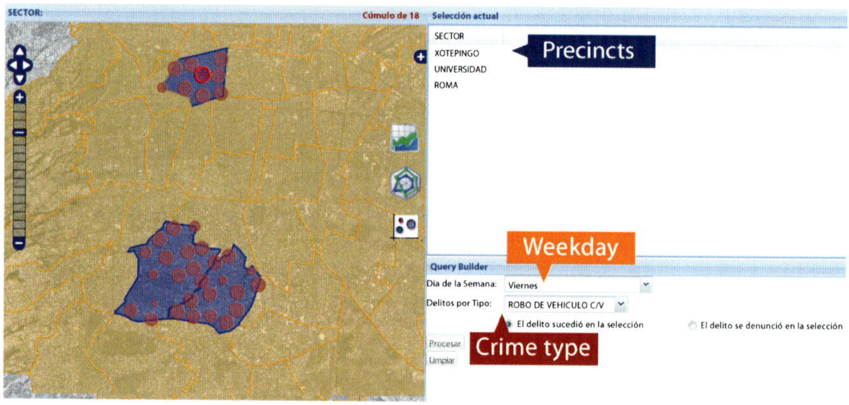

Fig. 18.4 Interface and user selection

It is therefore beneficial to move away from traditional thinking and towards a more territorial view, where the geographic space in question plays a central role. In order to achieve this, rather than letting the user select one or more jurisdictions from a list, the user must select them by clicking on a map, as operatives will be expected to be familiar with the spatial location of police precincts. If this is not the case, the user can hover over each polygon and the name will show up on the screen for identification. Selected polygons are highlighted and also displayed as a table on the right-hand side of the screen. Once the user has selected a region of interest, the database can be queried by criminal offense and by weekday (Fig. 18.4).

Depending on the scale of the map, in order to avoid cluttering, the results may or may not be shown as clusters. If they are clustered, hovering over each dot will tell the user how many hits were returned from querying the database. Zooming in or out will increase or decrease the number of dots shown on the map. This serves

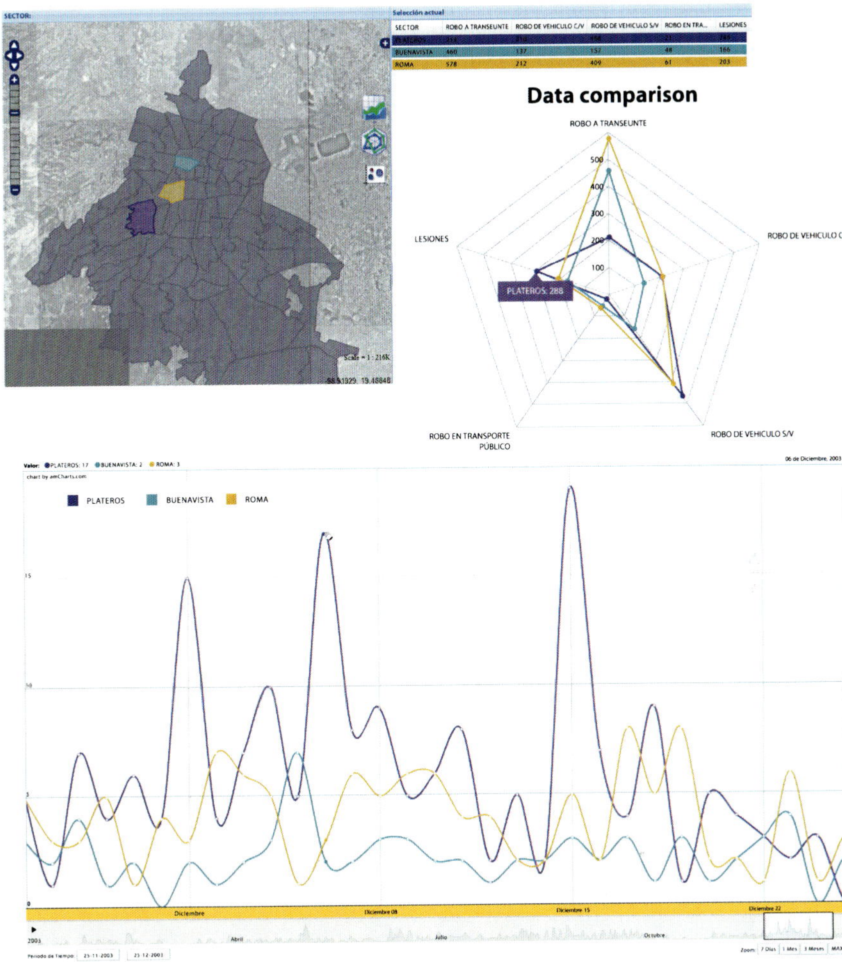

Fig. 18.5 Historical and radar plot

as a preliminary exploratory data analysis as it helps identify areas more likely to experience certain types of crime on a particular day of the week.

Once a selection has been made, the user can choose one of the three graphic types available: historical plot, radar plot, or efficiency plot. The first two plots let the user choose one or more offenses to be displayed and then ask for a range of dates in order to query the data. The historical plot shows data in a timeline and the user can scroll or zoom in and out of different time periods including days, weeks, months, and years. The radar plot shows data in a polygonal array depending on how many crime types were selected. If three types were selected, there will be a triangle; if four were selected, there will be a square, and so on. These two graphics are linked together in such a way that if a precinct is added or removed from the selection, they are updated to reflect the changes. Also, in each graphic, each precinct is represented by a color matching the one on the selection table to aid with its identification (Fig. 18.5).

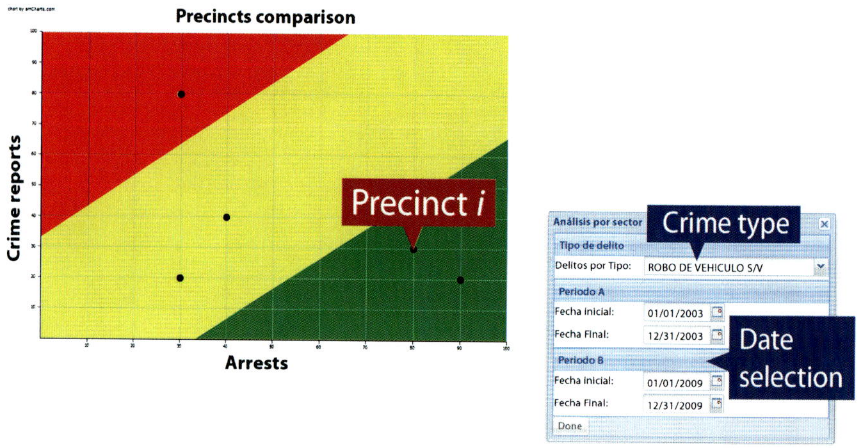

Fig. 18.6 Efficiency plot

The third graphic is a special type of scatter plot, which shows the evolution of different precincts throughout the selected period of time in terms of their efficiency. This is measured by the number of reports against the number of remissions and precincts are represented by a dot. This is a custom made graphic that helps authorities track the development of certain jurisdictions in a given range of dates. At first, it shows a static image of the precinct's efficiency at the beginning of the period and then, by clicking on each one, an animation is triggered showing the evolution of that particular precinct throughout the selected dates; it is also possible to animate all precincts at the same time. The plot background shows three different colors: red, yellow, and green. Depending on the reports to remissions ratio, each precinct is colored as follows: red corresponds to a poor ratio, yellow corresponds to a medium ratio, and green corresponds to a good ratio. When animating, the color of each precinct changes when crossing the threshold for each colored section on the background (Fig. 18.6).

It is worth remarking that displayed data are retrieved from the server at runtime. That is, the live application will be automatically updated whenever new information is added to the database and this will be immediately reflected without further intervention from the operatives. This is helpful in dividing labor between different members of the police department, as it simplifies the work-flow process: a team of individuals can be in charge of maintaining and updating the database while others are mining data from it without conflicts.

The spatial nature of crime incidence has been recognized for a long time in the literature of environmental criminology (Brantingham and Brantingham 1981). Spatial behavior varies over time because time and space are inseparable, as Harvey put it: "There are multiple spaces and times (and space-times) implicated in different physical, biological and social processes" (Harvey 1996:53 cited in Davoudi and Strange 2009:13). Criminal incidents in Mexico City occur in different areas at different times: they may form corridors at daytime, peak in some locations at rush hours, and attenuate at other times, perhaps during the night. There are places in the city where some types of crime happen more frequently at certain times of the day or on particular weekdays. We now have a situation where criminal records show both spatial and temporal distribution, which is paramount to properly analyzing and understanding the impact of crime containment programs and exercises.

18.5 Spatial and Space-Time Analysis: Hot Areas and Acute Hot Spots

Mexico's insecurity problems are not homogeneously distributed throughout its territory. Mexico City is the country's political power center and most populated city, and its insecurity problems are complex. The contribution of local jurisdiction crime types to the city's insecurity complexities is relevant, as crime affects the daily activities of the citizenship, undermining its economy and its sense of security. Besides, local crime feeds back different forms of organized crime, is a symptom of social decomposition, and a manifestation of structural economic problems. Among local jurisdiction crimes, the city's highest rates are those related to robberies: 11.06 incidents per 1,000 inhabitants in 2009 (Secretariado Ejecutivo del Sistema Nacional de Seguridad Pública 2009:11). The way in which the citizenship is affected by these types of crimes can be illustrated with examples of areas densely traveled by people with a high probability of becoming thieves' victims (Fig. 18.7a), busy roads where people traveling in public transportation are constantly assaulted (Fig. 18.7b), or places where violent disputes between gangs are constant (Fig. 18.7c).

Car robberies are the most frequent type of robbery with 26.4%, 44.1% of which involve some kind of violence (Secretariado Ejecutivo del Sistema Nacional de Seguridad Pública 2009). Figure 18.8 provides a timely overview of car robberies. Finally, in 2009 deaths related to rivalry between criminal groups -which are under federal jurisdiction- amounted to only 7.8% of the homicides considered under local jurisdiction (Presidencia de la República 2011; Secretariado Ejecutivo del Sistema Nacional de Seguridad 2009). Both kinds of events add to an 18.9 homicide rate per 100,000 inhabitants, which compares favorably with rates reported for New Orleans (51.7), Detroit (40.0), or Baltimore (37.3); but is a large lag in relation to the same rates for New York (5.6) or Los Angeles (8.1) (U.S. Census Bureau 2009). Figure 18.9 shows the city's areas where local jurisdiction homicides concentrated in 2009.

It is beyond the purpose of this section to analyze trends, spatial patterns, variations and refinements of crime incidence in Mexico City, as it was beyond the purpose of the project to give a detailed quantitative assessment of the spatial problems of Mexico City's crime incidence. The section's focus is rather on the way spatial and space-time modeling were inserted into the geointelligence process of the Ministry, in the hope that analysis would be performed within its organizational boundaries and spatial analysis and knowledge-supported decision-making might become part of its organizational culture.

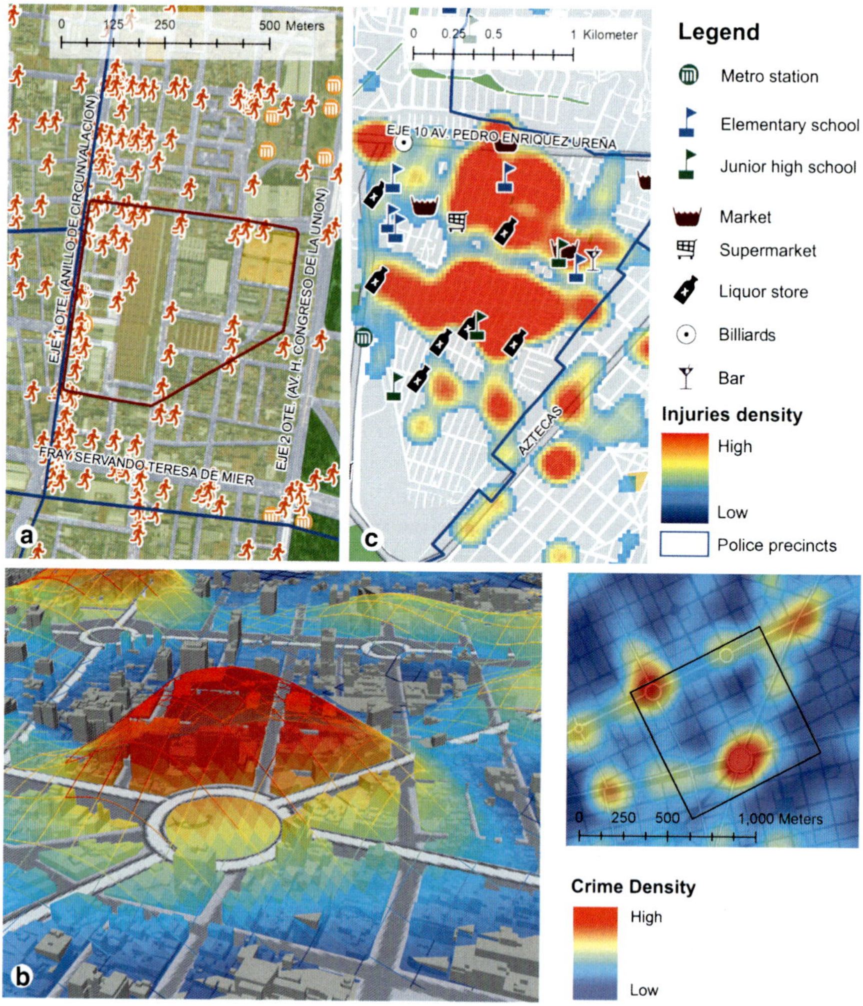

Fig. 18.7 Examples of places in Mexico City affected by different crime types. (**a**) Larceny theft incidence in *Congreso-Mixcalco-Heraldo* area with dense pedestrian flows due to the location of important markets (red outlined polygon). (**b**) Robberies in public transportation; the density pattern shows a hot spot in the *Glorieta Insurgentes*, a busy city landmark where three modes of public transportation converge (subway, bus and taxis). (**c**) Injuries derived from street fights in *Santo Domingo Coyoacan*, tend to concentrate near high schools and liquor stores (Source: Authors' elaboration based on data from Mexico City's Justice Attorney, contained in the Mexico City's Public Safety Ministry Geospatial Database)

The detection of crime hot areas and hot spots in urban spaces is not a new topic. The concentration of police resources in hot spots has proven its effectiveness by reducing crime incidence rates and emergency calls (Braga and Weisburd 2007:11–23; Sherman et al. 1997). Accordingly, there is a growing interest in hot spot policing

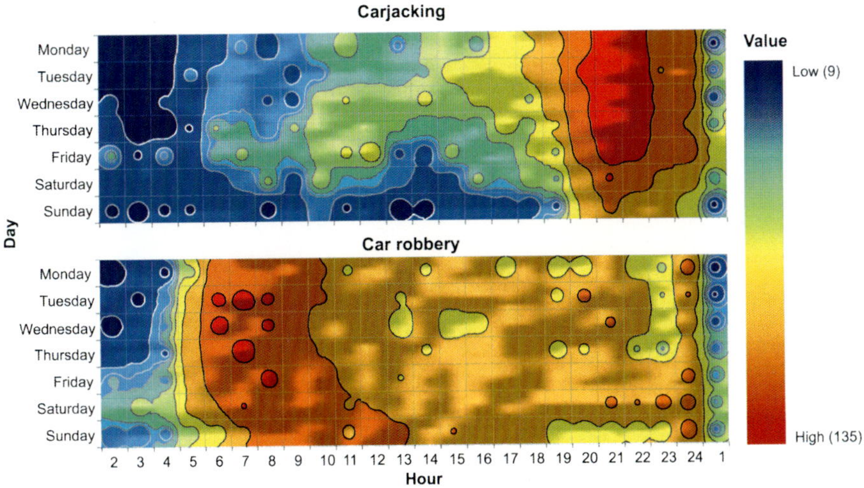

Fig. 18.8 Temporal pattern of carjacking and car robbery in 2009. The graph shows that carjacking involving some kind of violence, tend to concentrate during the evening reaching the critical time at late evening hours, while car robberies without violence show a more uniform distribution with clear concentrations in specific weekdays and daytimes (Source: Authors' elaboration based on data from Mexico City's Justice Attorney, contained in the Mexico City's Public Safety Ministry Geospatial Database)

and in research about processes and factors explaining the emergence and dynamics of these patterns, the detection of which is used in resource allocation and design of police operations in order to prevent further events from happening and to solve specific public safety problems (Weisburd and Braga 2006:225–244). In fact, hot spot detection has become part of most policing models and is an essential part of Compstat's processes that deal with monitoring and evaluating the effectiveness of police actions (Henry 2006:121–122; Willis et al. 2003:50–53).

However, hot spots policing in Mexico City emerged only recently, as part of the Compstat process launched by the intelligence unit of the Mexico City's Public Safety Ministry. An initial interest was to use the crime incidence data in order to detect the more conflictive areas. Data had been collected for 6 years and included at least the place, date and time of crime incidents disaggregated by type of crime. Spatial modeling was done by the research group applying a fuzzy logic algorithm which, according to a validity index, finds the best data fit to a given cluster partition. The algorithm performs a parametric optimization, starting with a large number of clusters and going on iteratively until the optimum fit is reached. The algorithm differentiates between central and peripheral concentric rings, the latter possibly being located in other clusters' peripheral areas. Each cluster center and rings have the same number of points. Fuzzy borders hence imply that an incident in space may belong simultaneously to several clusters but with a different degree of membership. Constructed hot spots follow a uniform distribution and have similar forms (López-Caloca et al. 2009:2–4). The algorithm was programmed in

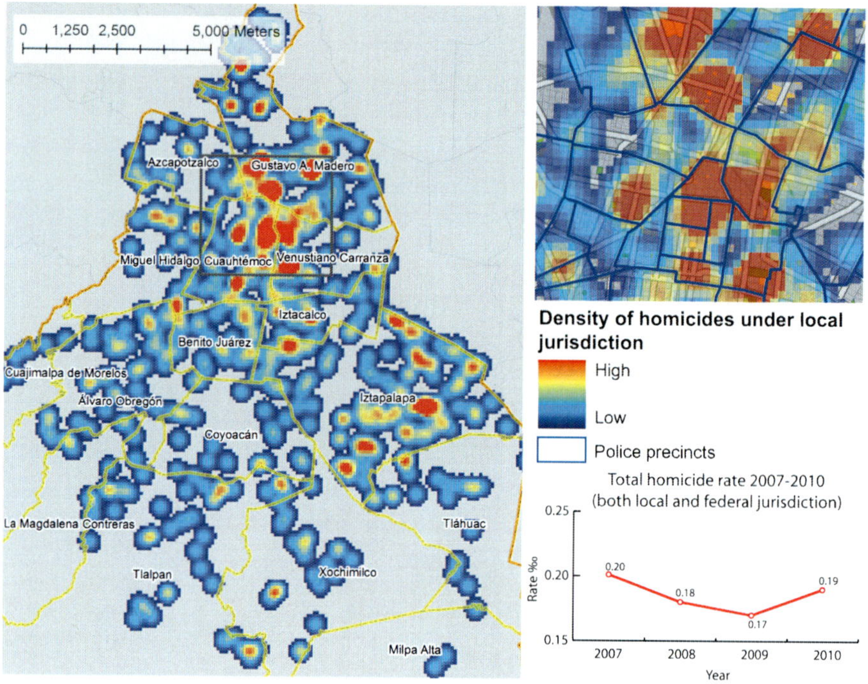

Fig. 18.9 Density pattern for homicides in 2009 (Source: Authors' elaboration based on data from Mexico City's Justice Attorney, contained in the Mexico City's Public Safety Ministry Geospatial Database)

MatLab, its results were transformed to vectors, and the model was applied to the incidence of several crime types at different territorial scales in Mexico City.

Figure 18.10 shows results of this algorithm for vehicle robbery at the city scale. The need to adapt the Ministry's administrative zoning became clear once it was detected that many of the most conflictive areas spilled over into adjacent police administrative territories or precincts, each of them with their own chain of command and all of them tending to neglect 'border incidents'. However the process did not lead to police administrative zoning being adapted to the clustering dynamics of crime. Instead, the Ministry became interested in working patterns inside each region's borders, in order to discuss them at weekly Compstat meetings taking place at central headquarters. Soon, the number of patterns demanded increased geometrically. The possible combinations of 15 regions and 9 crime types summarized from a wider classification, plus the requirement to differentiate patterns by year, month, daytime or last month's or week's incidents posed a problem for the implementation of the fuzzy logic algorithm, as it runs in an iterative process which was not programmed as a web service and took a lot of processing resources and time.

The number of patterns demanded by the Compstat process required the establishment of a pattern identification production line, operated by the Ministry's analysts, autonomous from CentroGeo's research group. The kernel density technique, for its ease of implementation and interpretation, was adopted. According to

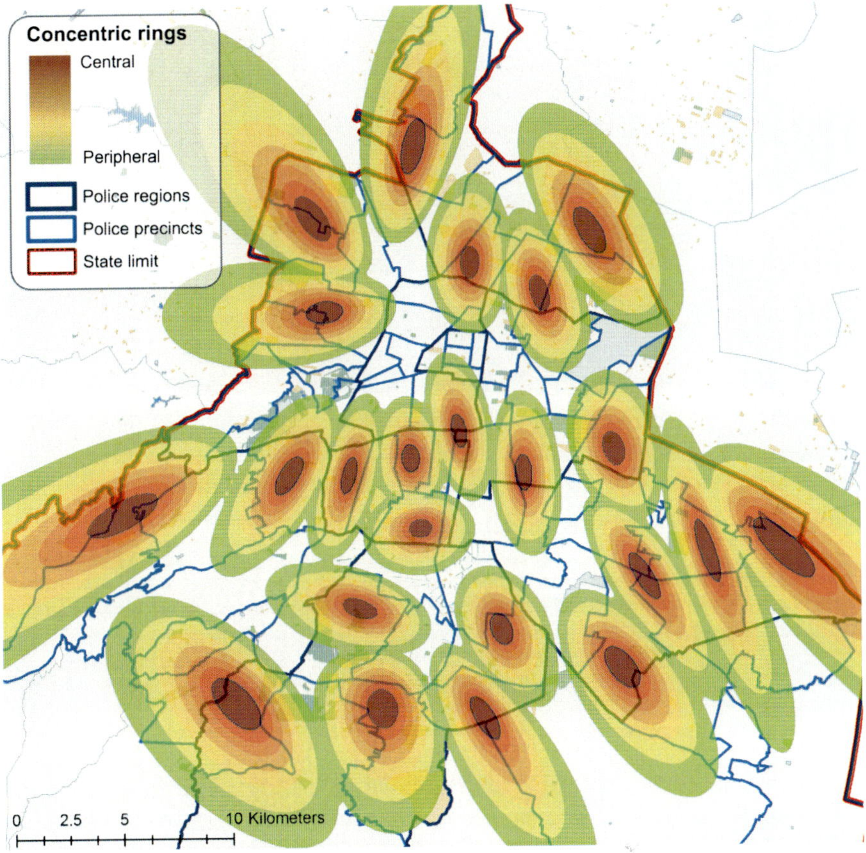

Fig. 18.10 Results of fuzzy logic spatial modeling for vehicle robbery from January to September, 2007. 23 city zones were identified as the most problematic, 13 of them cross the borders of police territorial precincts, posing a management problem in view of the scarce exchange of information among precincts (Source: Authors' elaboration based on data from Mexico City's Justice Attorney, contained in the Mexico City's Public Safety Ministry Geospatial Database)

Eck et al. (2005:26): "The most suitable method for visualizing crime data as a continuous surface is kernel density estimation".

The benefits of kernel density estimation are primarily practical because it can be used to study crime events where data on at-risk populations are missing (Assunção et al. 2007:10) and because it "allows analysts to visually simplify and examine complex point patterns of criminal incidents... [providing] greater flexibility in defining the borders for hot spots and analyzing hot spot areas" (Anselin et al. 2000:229). This last issue is of most importance because hot areas are not necessarily contained within administrative zones and landscape barriers can be reflected in the estimation results.

A production line of patterns based upon the kernel density could be implemented. Criminal incidence data were extracted from the GDI for analysis and models were run on desktop computers. Once the models' results were expressed in maps,

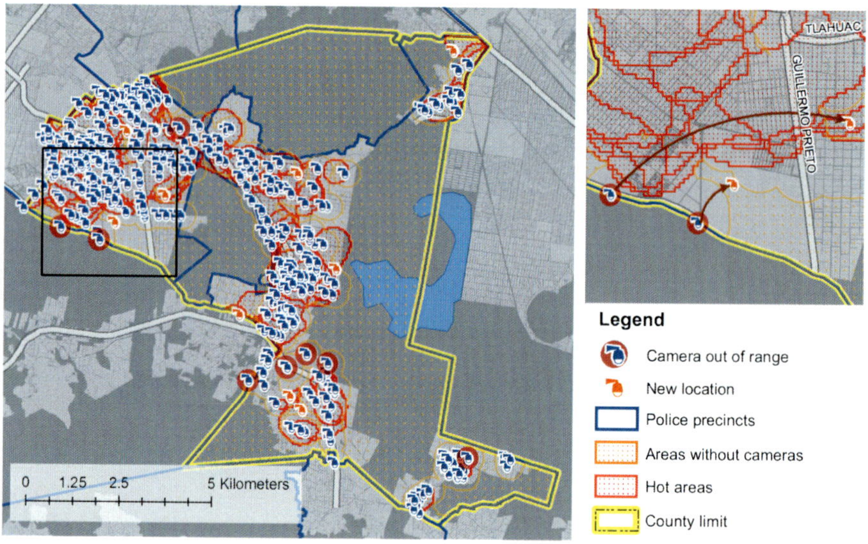

Fig. 18.11 Security cameras in Tláhuac's robbery hot areas. This is a place in which numerous cameras were planned to be located outside hot areas for robbery. They were relocated, either in a hot area or near relevant facilities (Source: Authors' elaboration based on data from Mexico City's Justice Attorney, contained in the Mexico City's Public Safety Ministry Geospatial Database)

they were incorporated into this infrastructure and distributed through map services. At Compstat meetings, the spatial trends monitoring of crime incidence went hand in hand with the spatial trends monitoring of precinct resources and of the action efficiency rates of police commanders, therefore supporting accountability processes. These analyses also supported some decisions such as installing security cameras in the main hot areas, connected to monitoring centers (Fig. 18.11).

Kernel density's main disadvantage, concurring with other interpolation techniques, is that it estimates wide high risk areas when in many instances the risk is highly localized in specific points such as crossroads, or in segments such as obscure sidewalks (Smith and Bruce 2008:70). At the Ministry's Compstat meetings the distribution of risk inside hot areas identified by kernel density was considered to be homogeneous. In order to further the geointelligence process, the need to build a new spatial consciousness among the Ministry's analysts was identified. This required awareness of the need to analyze different space-time geographies and to look for relationships between specific crime patterns and trends and specific place characteristics. Personnel from the Ministry enrolled in a Geomatics course at CentroGeo, the purpose of which was to develop the participants' ability to make recommendations to deal with spatially identified problems, therefore integrating spatial analyses into decision-making processes.

A key element in the training process was space-time analysis, aimed at the representation of street crime at the place scale, through the application of the Knox model (Knox 1964). This model was originally applied in research on epidemic

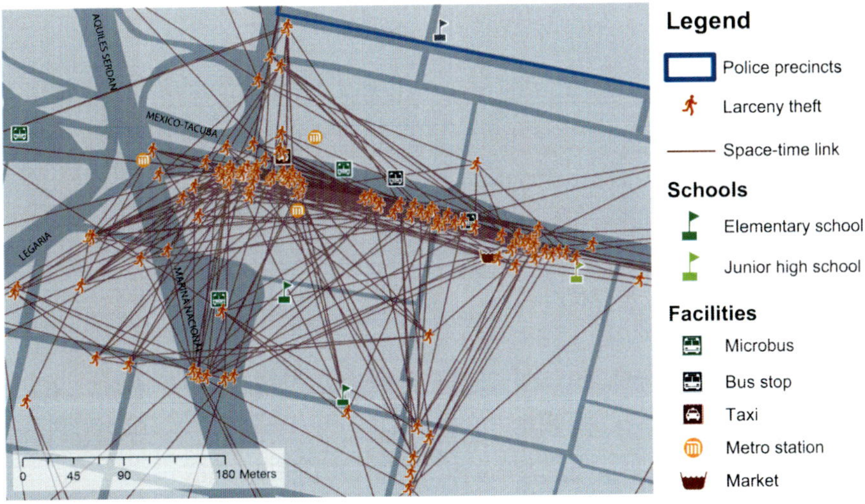

Fig. 18.12 Visualization of Knox's model results for larceny theft in *Tacuba*. The model's representation allowed police commanders to detect larceny 'footprints' left in places by space time related crimes. In such places daily larceny activity was analyzed and the movements of security cameras were programmed accordingly. Operators at the monitoring centers were given working orders specifying supervision timetables of the relevant cameras, and new patrol surveillance schemes with radio communications were programmed. This map shows an example of the model's visualization that supported such police operations (Source: Authors' elaboration based on data from Mexico City's Justice Attorney, contained in the Mexico City's Public Safety Ministry Geospatial Database)

emergence and it can be thought of as a statistical independence test. It is based on the number of events happening simultaneously inside a time and space threshold. The main hypothesis is that in a large number of events, the ones near in space tend to also be near in time. "The Knox test is an elegant and in many ways attractive method. For example, it is simple and straightforward to calculate the test statistic, and it requires knowledge only of cases with no need for controls" (Kulldorff and Hjalmars 1999:544). Disadvantages relate to the arbitrary selection of time and space thresholds (Assunção et al. 2007:10). The model was implemented using ClusterSeer software, the output of which is a space-interactions graph. This output was transformed into a more understandable visualization using simple metrics from graph theory (Jiang and Claramunt 2002:298). Figure18.12 presents the model results for larceny theft in Tacuba. The results of Knox's model were grasped by police commanders as criminal dynamic activities that left a 'footprint' or mark. These tracks underpinned the design of specific police surveillance actions which were closely monitored by commanding officers.

Considering space and time simultaneously requires seeking categories or typologies to frame and interpret clusters found. Grubezic and Mack (2008) conceptualize space-time patterns of interaction associated to criminal activity, maintaining one of both dimensions constant. If space is conceptually kept constant, they propose three temporal types: criminal series (a group of crimes perpetrated by the same individuals against one or several victims), criminal excursion (the same offenders committing various crimes along a relatively short period of time) and criminal trend (increasing or decreasing numbers of criminal events along a period of time). Knox's model output cannot be interpreted as criminal series or criminal excursion as it is a global statistical pattern, given specific thresholds, which does not guarantee that one is dealing with the same single offender; neither can it be seen as a criminal trend as it involves the spatial dimension. Ratcliffe (2004) proposes the 'acute' hot spot, which is a temporal pattern associated to particular activities (for example, larceny theft to people exiting a bar near closing time). This pattern does not correspond to any of the three aforementioned types, but results of Knox's model could fit better into this category.

18.6 Conclusions

Implementing a geointelligence process at Mexico City's Public Safety Ministry was influenced both by the concept of geointelligence and by the institutional will to introduce a more fitting policing model for public safety in Mexico City. However, this has not been a linear process; instead, it has proven to be a complex, changing process entangling research and technical development results with daily demands emerging from the dynamics of the police institution.

In Mexico, spatial perspective in criminal analysis is only recently emerging. Timely and high quality relevant spatial data are scarce and data generation and management processes face serious deficiencies. Also, the weakness of the institutional framework supporting spatial knowledge generation and analysis for public safety can be perceived in the lack of educational programs or professional and research associations that include a spatial perspective on crime or criminal analysis. As a paradox, technological firms offer last generation products and public safety institutions invest in expensive hardware and software, the full exploitation of which becomes limited by their lack of an appropriate organizational and knowledge base.

However, the idea of a Geointelligence Laboratory at Mexico City's Public Safety Ministry did become a reality; the space it occupied and the technologies acquired could be physically appreciated. Organizing both people and technology led to the implementation of some interesting geointelligence working processes and those that represent major efforts for CentroGeo's research team are: the integration of Geospatial Data Infrastructure, an interactive solution that combines graphics, maps and statistics for the automatic retrieval of information needed to respond to users' spatial queries; a spatial analysis production line in order to visualize hot spots and critical hot spots across the city, steering policing processes towards priority zones and prolific activity of place-based crime.

Among other products, processes or protocols that remain to be addressed, one could include: integrating applications which allow interactions between the Geospatial Data Infrastructure and devices equipped with global positioning system capabilities, such as patrol cars or mobile devices (phones, netbooks, tablets);

enabling real-time updating of spatial databases and harvesting of new incidents; simulating crime events and crime patterns in order to explore new criminal hypotheses and theories; analytic use of surveillance cameras, aerial photographs and satellite imagery in order to detect new spatial variables associated with places related to crime.

However, the most significant themes for furthering geointelligence advancement at the Ministry are the interoperability of all of its information systems, the cascading of data acquisition and analysis into police precincts, and the linking of spatial patterns with preventive measures.

The GDI has the capability to support crime mapping geocoding and visualization in a Web distributed environment, and to function as the Ministry's formal geographic data repository. It allows its database to be updated, integrating new spatial data and supporting online data sharing processes. It may also serve as a guide to the design of other information systems using the same database. However, to achieve full interoperability, the agency needs to reach a higher level of maturity regarding the incorporation of quality control processes on information flows among functional divisions, and of well-defined, documented and even measurable working processes. An interoperability plan framed in an information management model, with standards, organizational agreements, and maturity level milestones was proposed; however, the viability of its implementation is still uncertain.

The cascading issue is even more complex: human resources able to handle spatial data and technology are very scarce, human resources with analysis capabilities are even more so, and there is a high turnover of personnel already trained in Geomatics. Hence, training efforts need to be paramount and planned on a long-term permanent basis. Also, resources for the precinct's technological equipment will be needed and, in particular, a new spatial consciousness will be required as a component of the Ministry's organizational culture in order for space to be perceived as a key element in the design of police strategic and tactical operations.

Patterns and rhythms of criminal activity and disorder could describe a hot spot; to explain their emergence and evolution, further analysis is needed. Urban structure and processes need to be disclosed inside and between hot spots, as they relate to interactions, flows of people, social groups, and lead to spatial distributions of opportunities to criminal activities. The value of analysis results will lie in the possibility to motivate the involvement of institutions from other governmental sectors and organizations of the civil society into the design and implementation of place-based prevention programs.

References

Achour M, Betz F, Dovgal A, Lopes N, Magnusson M, Richter G, Seguy D, Vrana J (2010) PHP manual. The PHP Documentation Group. http://www.php.net/manual/en/. Accessed 3 Feb 2012

AmCharts (2010) AmCharts. http://www.amcharts.com/docs. Accessed 3 Feb 2012

Anselin L, Cohen J, Cook D, Gorr W, Tita G (2000) Spatial analyses of crime. Measurement and analysis of crime and justice. Crim Justice 4:213–262

Assunção R, Tavares A, Correa T, Kulldorff M (2007) Space-time cluster identification in point processes. Can J Stat 35(1):9–25

Bottoms A (2007) Place, space, crime and disorder. In: Maguire M, Morgan R, Reiner R (eds) The Oxford handbook of criminology. Oxford University Press, Oxford

Braga A, Weisburd D (2007) Police innovation and crime prevention: lessons learned from police research over the past 20 years.NCJRS https://www.ncjrs.gov/pdffiles1/nij/grants/218585.pdf. Accessed 25Aug 2011

Brantingham P, Brantingham B (1981) Environmental criminology. Sage Publications, Inc, Thousands Oaks

Bruce CW (2004) Fundamentals of crime analysis. In: Bruce C, Gwinn S, Cooper J, Hick S (eds) Exploring crime analysis. Readings on essential skills. The International Association of Crime Analysts, North Charleston, South Carolina

Carrabine E, Cox P, Lee M, Plummer K, South N (2009) Criminology. A sociological introduction. Routledge, Oxford

Committee on Basic and Applied Research Priorities in Geospatial Science for the National Geospatial-Intelligence Agency, Mapping Science Committee, Board on Earth Sciences and Resources Division on Earth and Life Studies, National Research Council of the National Academies (2006) Priorities for GEOINT research at the National Geospatial-Intelligence Agency. The National Academies Press, Washington, DC

Consejo Nacional de Población (2006) Proyecciones de población de México 2005–2050. http://www.conapo.gob.mx/index.php?option=com_content&view=article&id=36&Itemid=234. Accessed 1 Feb 2012

Davis D (2007) El Factor Giuliani: Delincuencia, la 'Cero Tolerancia' en el Trabajo Policiaco y la Transformación de la Esfera Pública en el Centro de la Ciudad de México. Estudios Sociológicos 25(75):639–681

Davoudi S, Strange I (2009) Space and place in twentieth-century planning: an analytical framework and an historical review. In: Davoudi S, Strange I (eds) Conceptions of space and place in strategic spatial planning, The RTPI library series. Routledge, London/New York

Eck J, Chainey S, Cameron J, Leitner M, Wilson R (2005) Mapping crime: understanding hot spots. NIJ special report. U.S. Department of Justice, Office of Justice Programs, National Institute of Justice

GeoExt Community (2010) GeoExt. http://www.geoext.org/. Accessed 23 July 2011

Getis A, Drummy P, Gartin J, Gorr W, Harries K, Rogerson P, Stoe D, Wright R (2000) Geographic information science and crime analysis. URISA J 12(2):7–14. http://cs.iupui.edu/~tuceryan/pdf-repository/Getis2000.pdf

Grubesic T, Mack E (2008) Spatio-temporal interaction of urban crime. J Quant Criminol 24(3):285–306

Guerrien M (2004) Transformation et Fragmentation des Espaces Urbains. Le Cas de la Zone Métropolitaine du Bassin de Mexico. L'Espace Géographique. http://www.cairn.info/revue-espace-geographique-2004-4.htm. Accessed 25 July 2011

Harvey D (1996) Justice, nature and geography of difference. Blackwell, Oxford.

Henry V (2006) Managing crime and quality of life using Compstat: Specific issues in implementation and practice. Visiting expert paper. 129th international senior seminar. Resource material series No. 68. UNAFEI Fuchu: Tokyo, Japan. http://www.unafei.or.jp/english/pdf/RS_No68/No68_12VE_Henry2.pdf. Accessed 25 July 2011

INEGI (2011) Censo de Población y Vivienda 2010: Consulta Interactiva de Datos. México. http://www.inegi.org.mx/lib/olap/consulta/general_ver4/MDXQueryDatos.asp?#Regreso&c=27770. Accessed 1 Feb 2012

Jiang B, Claramunt C (2002) Integration of space syntax into GIS: new perspectives for urban morphology. Trans GIS 6(3):295–309

Knox G (1964) The detection of space-time interactions. J R Stat Soc Series C (Appl Stat) 13(1):25–29

Kulldorff M, Hjalmars U (1999) The Knox method and other tests for space-time interaction. Biometrics 55(2):544–552

López-Caloca A, Martínez-Viveros E, Chapela-Castañares J (2009) Application of a clustering-remote sensing method in analyzing security patterns. Proc SPIE Int Soc Opt Eng 7344 734407:1–9

OGC (2007) OGC glossary of terms. http://www.opengeospatial.org/ogc/glossary/i. Accessed 17 Jul 2011

Poulsen E (2003) Designing a database for law enforcement agencies. In: Leipnik M, Albert D (eds) GIS in law enforcement: implementation issues and case studies. Taylor & Francis Group, New York

Presidencia de la República (2011) Sistema de Consulta de la Base de datos de Fallecimientos. http://www.presidencia.gob.mx/base-de-datos-de-fallecimientos/. Accessed 1 Feb 2012

Prévôt-Schapira MF (2000) Segregación, Fragmentación, Secesión. Hacia una nueva Geografía Social en la Aglomeración de Buenos Aires. Economía, Sociedad y Territorio 2(7):405–431

Ratcliffe JH (2004) The hot spot matrix: a framework for the spatio-temporal targeting of crime reduction. Police Pract Res 5(1):23

Ratcliffe JH (2008) Intelligence-led policing. Willan Publishing, Devon

Rock P (2007) Sociological theories of crime. In: Maguire M, Morgan R, Reiner R (eds) The Oxford handbook of criminology. Oxford University Press, Oxford

Savelsberg JJ (1984) Socio-spatial attributes of social problems: the case of crime and delinquency. Popul Environ 7(3):163–181

Secretariado Ejecutivo del Sistema Nacional de Seguridad Pública (2007, 2008 and 2009) Incidencia delictiva del fuero común. Secretaría de Gobernación. http://www.secretariadoejecutivosnsp.gob.mx/work/models/SecretariadoEjecutivo/Resource/131/1/images/CIEISP_2007.pdf. http://www.secretariadoejecutivosnsp.gob.mx/work/models/SecretariadoEjecutivo/Resource/131/1/images/CIEISP_2008.pdf. http://www.secretariadoejecutivosnsp.gob.mx/work/models/SecretariadoEjecutivo/Resource/131/1/images/CIEISP_2009.pdf. Accessed 1 Feb 2012

Sencha (2010) Ext JS 3 API documentation. http://www.sencha.com/products/extjs3 and http://dev.sencha.com/deploy/ext-3.4.0/docs/. Accessed 25 July 2011

Sherman LW, Gottfredson D, MacKenzie D, Eck J, Reuter P, Bushway S (1997) Preventing crime: what works, what doesn't, what's promising. A report to the United States Congress Prepared for the National Institute of Justice. NCJRS. http://www.ncjrs.gov/works/. Accessed 14 July 2011

Smith S, Bruce C (2008) CrimeStat III workbook. The National Institute of Justice (NIJ), Washington, D.C

The Open Source Geospatial Foundation (2010) OpenLayers. http://dev.openlayers.org/releases/OpenLayers-2.10/doc/apidocs/files/OpenLayers-js.html. Accessed 23 July 2011

U.S. Census Bureau (2009) Statistical abstract of the United States. Available at http://www.census.gov/compendia/statab/2012/tables/12s0309.pdf. Accessed 1 Feb 2012

Weisburd D, Braga A (2006) Police innovation. contrasting perspectives. Cambridge University Press, Cambridge

Willis J, Mastrofski S, Weisburd D (2003) Compstat in practice: an in-depth analysis of three cities. Police Foundation. http://www.policefoundation.org. Accessed 26 July 2011.

Index

M. Leitner (ed.), *Crime Modeling and Mapping Using Geospatial Technologies*, Geotechnologies and the Environment 8, DOI 10.1007/978-94-007-4997-9,